Meteorology

THE ATMOSPHERE AND THE
SCIENCE OF WEATHER

JOSEPH M. MORAN
MICHAEL D. MORGAN

University of Wisconsin–Green Bay

Meteorology

THE ATMOSPHERE AND THE SCIENCE OF WEATHER

In memory of

Joseph P. Moran

Lewis W. Morgan

Development Editor: Janilyn M. Richardson
Copy Editor: Betsey Rhame
Production Editor: Melinda Berndt
Cover photograph: Arjen and Jerrine Verkaik, Skyart Productions,
Islington, Ontario

Library of Congress Cataloging-in-Publication Data

Moran, Joseph M.
Meteorology : the atmosphere and the science of
weather.

Includes bibliographies and index.
1. Meteorology. 2. Weather. 3. Atmospheric
physics. I. Morgan, Michael D. II. Title.
QC861.2.M625 1986 551.5 85-29944
ISBN 0-8087-3241-2

Burgess Publishing
7110 Ohms Lane
Edina, MN 55435

J I H G F E D C B

Contents

Preface

This book is intended to introduce the college nonscience major to the fundamentals of atmospheric science. It is appropriate for an introductory course on the atmosphere, weather, or weather and climate. Our primary goal is to demonstrate how scientific principles govern the circulation of the atmosphere and the day-to-day sequence of weather events. In so doing, we also introduce the nonscience student to the nature of scientific inquiry and the methodology of science. Atmospheric science is especially well suited to achieving these goals because it is an applied science and lends itself readily to familiar illustrations.

Our approach is based on our combined 31 years of teaching science primarily to nonscience students with little or no background in science or math. Our aim in writing *Meteorology* has been to present the principles of meteorology as simply as possible in the context of everyday examples without sacrificing scientific integrity. By providing clear and logical explanations of the principles underlying meteorological phenomena and observations, it is not necessary to write around or avoid even the more sophisticated ideas. The basic principles are broken down into elementary components and arranged so that one concept builds on another. The geostrophic wind, for example, is introduced only after a detailed examination of the separate forces contributing to that wind. The atmosphere gradually emerges as a highly complex and interactive system governed by physical laws.

Our emphasis on scientific methodology allows the student a perspective on the accomplishments of atmospheric scientists and the challenges still facing them. The reader soon understands that weather is not an arbitrary act of nature, and yet weather forecasting has limits and the climatic future is uncertain. We have integrated topics of special contemporary interest including acid rains, potential climatic effects of rising levels of carbon dioxide, and threats to the ozone shield.

The chapters are arranged in a traditional sequence. Chapter 1 introduces the basic properties and structure of the atmosphere. Chapters 2, 3, and 4 cover energy in the atmosphere and focus on the global radiation balance. The main theme in these chapters is that weather is a response to heat imbalances within the earth-atmosphere system. Air pressure is defined in Chapter 5 and is related to other atmospheric properties. Water in the atmosphere is the subject of Chapters 6, 7, and 8, with special emphasis on saturation and precipitation processes. Having established the reasons for atmospheric circulation, we then proceed to examine the forces governing weather systems (Chapter 9). There follows a general description of global-scale circulation (Chapter 10), synoptic-scale weather (Chapter 11), and local and regional weather systems (Chapter 12). Chapters 13 and 14 deal with the genesis and characteristics of thunderstorms and severe weather phenomena such as tornadoes and hurricanes. The treatment of weather closes with a chapter on weather forecasting structured to integrate and apply the key concepts developed earlier. A special feature of the book is the inclusion of three final chapters on topics that are of particular contemporary interest: Chapter 16 on air pollution meteorology, Chapter 17 on world climates, and Chapter 18 on the climatic record and the variability of climate.

While it is desirable to cover Chapters 1 through 15 in sequence, certain sections of most chapters may be dropped without loss of continuity. For example, the sections on weather instruments and atmospheric optics may be deleted. Chapters 16, 17, and 18 may be covered in any order and any one or all may be omitted to fit the available time.

Special sections, set apart from the main flow of the chapters, have been included to provide considerable flexibility with regard to depth and breadth of topic coverage. Each chapter contains one or two supplementary *Special Topics*, such as "Why Is the Sky Blue?" and "Solar Power," that are related to a major theme of the chapter. *Mathematical Notes* at the ends of six chapters provide quantitative discussions of topics in the chapters, with the basic meteorological equations for those who are interested.

In addition, each chapter features the following elements, designed to guide student understanding:

- outline of the chapter
- summary statements
- key word list
- review questions
- points to ponder
- projects
- selected readings with annotations

Key words are boldfaced *and defined* at their first use in the chapter. These words (with the exception of those in the Special Topics and Mathematical Notes) are listed at the end of the chapter for review. The Glossary at the back of the book contains all terms on the key word lists.

Metric units are used throughout the book, with the English equivalents given in parentheses. Unit conversions, as well as weather map symbols, psychrometric tables, weather extremes, and climatic data are contained in the appendices.

Supplements

Instructor's Manual. This manual, prepared by Joseph M. Moran, provides assistance in using the book. Each chapter in the manual includes the following:

- a summary of the text chapter
- chapter objectives
- multiple-choice and completion questions, with answers, and essay questions

A list of audiovisual aids and information sources is also included.

A set of *Transparency Masters* of line drawings from the text is available to adopters.

A *Computerized Test Bank* for Apple IIe and IBM PC includes a program and test questions to generate customized tests. The test package is available to adopters.

Meteorology Exercise Manual and Study Guide, by Robert A. Paul, contains exercises and study questions that can be done in or out of a formal laboratory session.

Acknowledgments

In preparing the numerous drafts of this book, we have profited greatly from the creative ideas and constructive criticisms of many reviewers. We acknowledge the very valuable contributions of the following and give a special thanks to Professor Reid A. Bryson of the University of Wisconsin for his insight and example.

L. Dean Bark, Kansas State University; Arnold Court, California State University, Northridge; William A. Dando, University of North Dakota; Russell L. DeSouza, Millersville State College; Lee Guernsey, Indiana State University–Terre Haute; B. Ross Guest, Northern Illinois University; William C. Kaufman, University of Wisconsin–Green Bay; Bruce E. Kopplin, University of Nebraska–Lincoln; Garrick B. Lee, Butte College; William H. Long, Florida State University, Tallahassee; David W. Marczely, Southern Connecticut State University, New Haven; Shamin Naim, Illinois State University–Normal; T. R. Oke, The University of British Columbia; John E. Oliver, Indiana State University–Terre Haute; Robert A. Paul, Northern Essex Community College; Clayton Reitan, Northern Illinois University; Peter J. Robinson, University of North Carolina at Chapel Hill; Charles C. Ryerson, University of Vermont; Gregory E. Taylor, Creighton University; Charles L. Wax, Mississippi State University; Wayne Wendland, Illinois State Water Survey; George Wooten, Hillsborough Community College

We also have been fortunate in working with some of the most talented and enthusiastic professionals in college publishing. We thank John H. Staples, Nancy Flight, and Margaret Mason for their encouragement in the early days. With the guidance of Gary Brahms and Richard Abel at Burgess Publishing, the idea for this book took shape and matured. We owe much to the tireless efforts of our development editor, Janilyn M. Richardson: her dedication, fine attention to detail, and objectivity were essential to the successful evolution of this book. Betsey Rhame, our copy editor, transformed our writing into a style more palatable to nonscience majors.

We thank our skillful typist, Jeanne Broeren, for her patience, and we thank our wives, families, and friends for their understanding and encouragement. Finally, we acknowledge with gratitude the contributions of our students and colleagues at the University of Wisconsin–Green Bay.

A generation goes,
and a generation comes,
But the earth remains for ever.

The sun rises and
the sun goes down,
and hastens to the
place where it rises.

The wind blows to the south,
and goes round to the north;

round and round goes the wind,
and on its circuits the wind returns.

ECCLESIASTES 1:4–6

The atmosphere is a thin envelope of gases, clouds, and particles that surrounds the globe. Weather in its myriad forms occurs primarily in the lowest 10 km (6.2 mi) of the atmosphere. (NASA photograph)

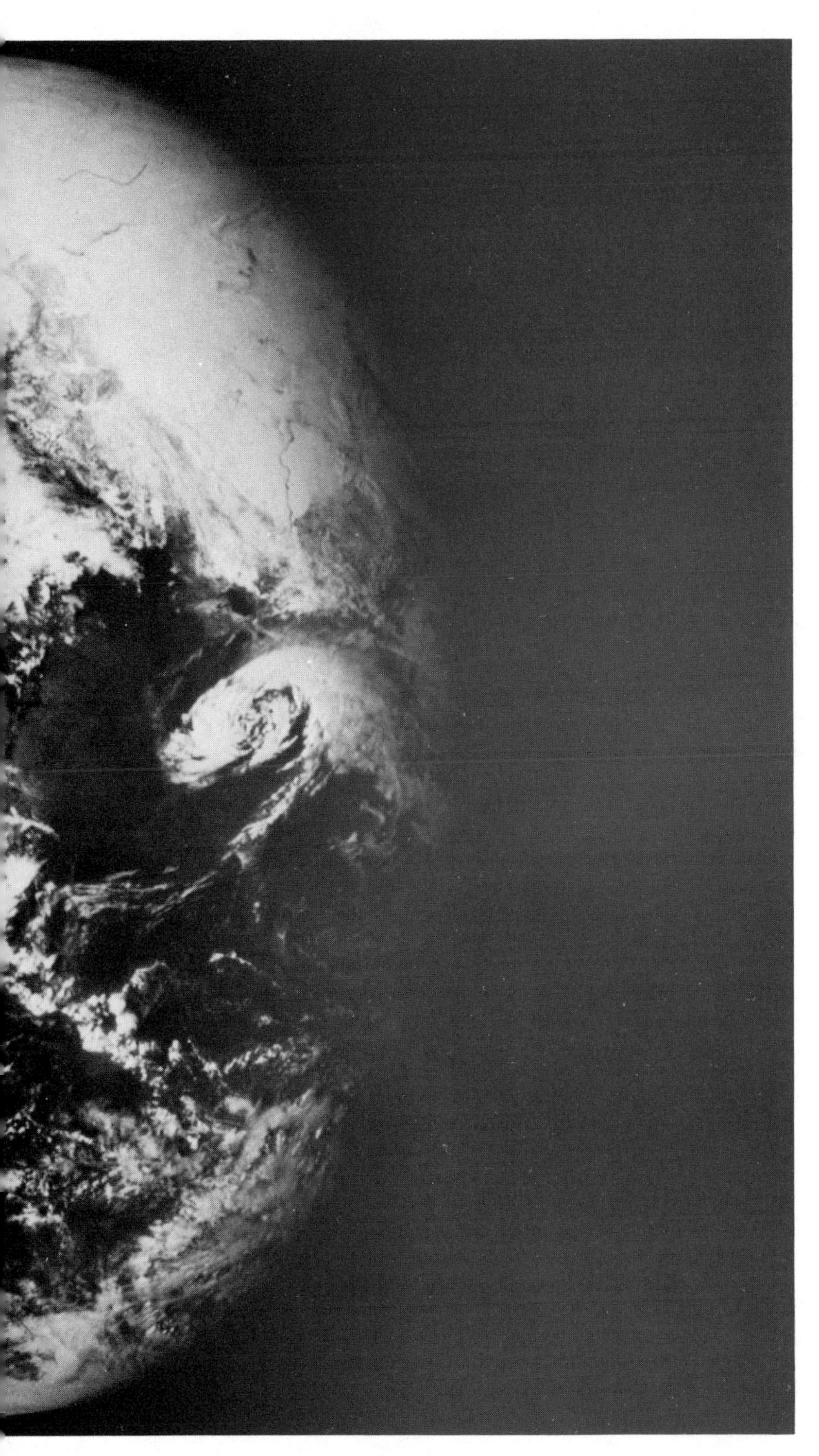

ONE

Atmosphere: Composition and Structure

EVERYONE seems to be interested in the weather. It affects virtually every facet of our daily lives—what we wear, whether we can go swimming or skiing, the price of oranges and coffee in the grocery store. Tranquil, pleasant weather allows us to enjoy a variety of outdoor recreational activities. Sometimes, however, the weather turns stormy, and although the rains or snows are beneficial, our plans may change abruptly. We know from personal experience that variability is one characteristic of weather.

Weather is the state of the atmosphere at some place and time, described in terms of such variables as temperature, cloudiness, precipitation, wind, pressure, and radiation. **Meteorology** is the study of the atmosphere and the processes that cause weather. **Climate** is traditionally defined as the weather conditions at some locality averaged over a given time period, but climate is more than this. Departures from the average, variability, and extremes in weather are also important aspects of climate. For example, it may be useful to know not only the long-term average temperature for January but also the lowest and highest temperatures ever recorded in any January. Climate can be considered the ultimate environmental control. Climate determines, for example, what crops can be cultivated, the long-term water supply, and the average heating and cooling requirements for homes.

The **atmosphere**, where weather takes place, is composed of a mixture of gases in which tiny solid and liquid particles are suspended. Some of these particles—water droplets and ice crystals—are visible as clouds. The atmosphere surrounds the earth as a relatively thin envelope. The portion of the atmosphere that contains 99 percent of the atmospheric mass has a thickness of only about 0.25 percent of the earth's diameter. In thickness, the earth's atmosphere is like the skin on an apple!

The atmosphere is essential to the functioning of our environment. It shields living things from exposure to hazardous ultraviolet radiation; it contains the gases required for the life-sustaining processes of cellular respiration and photosynthesis; and it supplies the water needed by all life.

What we understand today about the nature of the atmosphere, weather, and climate is the culmination of centuries of painstaking inquiry by scientists from many disciplines. Physicists, chemists, astronomers, and others discovered and applied basic principles to unlock the mysteries of the atmosphere. Although much progress has been made, many questions remain. Modern atmospheric scientists (meteorologists and climatologists) continue the efforts of their predecessors and, although armed with more sophisticated tools like satellites and computers, they still rely on the scientific method as a means of investigation. The **scientific method** is a systematic form of inquiry involving observation, speculation, and reasoning. (Readers unfamiliar with the scientific method are referred to the Special Topic in this chapter for a description.)

We begin our study of the atmosphere by surveying its basic characteristics, origin, composition, and structure.

Atmosphere: Origin and Evolution

The atmosphere that currently envelops our planet is the product of a lengthy evolutionary process that began at the earth's birth approximately 4.8 billion years ago. Astronomers scanning the solar system and geologists analyzing meteorites and fossilized rock fragments have given us a partial scenario of the origins of the atmosphere.

Primeval phase

The earth and the entire solar system are believed to have developed out of an immense cloud of dust and gases within the Milky Way galaxy. In the beginning, the earth was an aggregate of dust and meteorites surrounded by a gaseous envelope of hydrogen and helium. There was no gaseous oxygen. For millions of years, the earth's mass grew by accretion as the planet swept up cosmic dust. Bombardment by meteorites warmed the earth and its atmosphere, driving off most of the original atmospheric gases.

The earth then became geologically active as volcanoes spewed forth huge quantities of lava, ash, and a variety of gases, much as Mount Saint Helens did in the spring of 1980. Evidence suggests that then, as now, water vapor (H_2O), carbon dioxide (CO_2), and nitrogen (N_2) were important constituents of volcanic emissions. Gaseous, or free, oxygen was notably absent, although oxygen was bound to other elements in various chemical compounds as, for example, in carbon dioxide. Millions of years of volcanic activity eventually produced an atmosphere rich in nitrogen and carbon dioxide. The breakdown of water vapor, a chemical compound of hydrogen and oxygen, contributed only minor amounts of oxygen to the early atmosphere. Radioactive decay of an isotope of potassium in the earth's crustal rock added the inert gas argon to the evolving atmosphere. Then, with the formation of seas and the coming of life, important changes took place in the atmospheric composition.

More than 3 billion years ago, volcanic activity subsided, and the earth and its atmosphere gradually cooled. Cooling caused some of the water vapor to condense into clouds, yielding torrential rains that gave rise to the first rivers, lakes, and seas. In these seas, primitive forms of life appeared, perhaps 2 to 3 billion years ago. Photosynthesis began with the coming of the first marine plants. **Photosynthesis** is the process whereby plants use sunlight, water,

---

SPECIAL TOPIC

The scientific method

The scientific method is illustrated by an investigation of an acid rain problem. In the 1970s, ponds and lakes in the Adirondack Mountains of New York State were observed to be losing their fish populations. In the past, these same lakes had supported abundant aquatic life. Some local businesses were concerned, because barren lakes meant no fishing, and no fishing meant fewer tourist dollars. Environmentalists speculated that the lakes were yet another casualty of pollution.

Biologists proposed initially that toxic materials (poisons) were entering the lakes and were creating conditions intolerable to aquatic life. If this were true, what were the toxins? Where were they coming from? Testing of lake water samples revealed that the waters did not contain hazardous concentrations of toxins, but the waters were abnormally acidic. (Laboratory studies have shown that excessively acidic waters are lethal to young fish.) On the basis of this observation, a new hypothesis was formulated: that acidic rainwater was turning the lake waters more acidic.

Normal rainwater is known to be slightly acidic, because rainwater dissolves some of the carbon dioxide in the air to form a weak and harmless carbonic acid. Rainwater samples collected in the vicinity of the Adirondack lakes and tested in the laboratory, however, were over 100 times more acidic than normal. Why was the rainwater so acidic? Further laboratory study identified sulfuric and nitric acids in the rainwater. Such acids form when rainwater dissolves oxides of sulfur and nitrogen—common industrial air pollutants. The Adirondacks are downwind of some major industrial sources of these air

and carbon dioxide to manufacture their food, composed chiefly of carbohydrates. A byproduct of this process is oxygen, which is released to the atmosphere. Through subsequent millions of years, photosynthesis added oxygen to the atmosphere until, eventually, oxygen became the second most abundant atmospheric gas after nitrogen.

While all of this was taking place, the concentration of atmospheric carbon dioxide fell significantly. Photosynthesis removed some, but most of the carbon dioxide was dissolved in ocean waters. Eventually carbon dioxide made up only a small fraction of the atmospheric gases.

---

pollutants, for example, coal-fired power plants.

By this reasoning, the loss of fish in Adirondack Mountain lakes was at least circumstantially linked to industrial air pollution. The hypothesis that excessively acidic rainwater killed the fish is consistent with what we know about the tolerance of fish for acidic waters, the chemical reactions involving rainwater and air pollutants, and the type of air pollutants that are transported to the Adirondack region.

The *scientific method* can be viewed as a sequence of steps in which scientists (1) identify a specific question, (2) propose an answer to the question in the form of an "educated guess," (3) state the educated guess in such a way that it can be tested, that is, formulate a hypothesis, (4) predict what the consequences would be if the hypothesis were correct, (5) test the hypothesis by checking to see if the prediction is correct, and (6) revise or restate the hypothesis if the prediction is wrong.

Note that these steps are not always in this order, and some steps, such as 3 and 4, may be combined. Indeed, in actual practice a scientist does not follow this scheme cookbook style. The discrete steps are usually thoroughly integrated as a single avenue of inquiry. The use of the scientific method is not in itself a recipe for creativity. The method does not provide the key idea, the "hunch," the educated guess that becomes the hypothesis. The method is, rather, a technique to test the validity, or worth, of that creative key idea—however or wherever the idea originates.

As in the acid rain example, a hypothesis serves merely as a working assumption that may eventually be accepted, rejected, or modified. Indeed, inquiry, creative thinking, and imagination are stifled when hypotheses are considered immutable. A new hypothesis (or an old, resurrected one) may be subjected to considerable discussion, debate, and disagreement within the scientific community. Disagreement among scientists on a particularly controversial issue is often widely publicized and may confuse the general public. The prevailing reaction may be: "Well, if the so-called experts can't agree among themselves, who am I to believe? Is there really a problem after all?" Debate and disagreement are actually a usual step in arriving at scientific understanding.

Modern phase

These long, evolutionary processes yielded the modern atmosphere, which is a mixture of many different gases. Because the lower atmosphere undergoes continual mixing, atmospheric gases occur almost everywhere, in about the same relative proportions, up to an altitude of approximately 80 km (50 mi).* This portion of the atmosphere is sometimes called the **homosphere**. We can thus travel anywhere on the earth's surface and confidently breathe essentially the same type of air. Above 80 km, gases are stratified. This means that concentrations of the heavier gases decrease more rapidly with altitude than do concentrations of the lighter gases. The atmospheric region above 80 km is known as the **heterosphere**.

Not counting water vapor (which has a highly variable concentration), nitrogen normally constitutes 78.08 percent of the volume of the lower atmosphere (below 80 km), and oxygen constitutes 20.95 percent of the volume. The next most abundant gases are argon (0.93 percent) and carbon dioxide (0.03 percent). The atmosphere also contains small quantities of neon, helium, methane, krypton, hydrogen, xenon, ozone, and many other gases. The volume of some of these gases (for example, carbon dioxide and water vapor) varies from place to place with time. Table 1.1 shows

*Metric units, with the English-unit equivalents in parentheses, are used throughout this book. For unit conversions refer to Appendix I.

TABLE 1.1

Relative Proportions of Gases Composing Dry Air in Lower Atmosphere*

GAS	PERCENT BY VOLUME	PARTS PER MILLION
Nitrogen	**78.08**	**780,840**
Oxygen	**20.95**	**209,460**
Argon	**0.93**	**9,340**
Carbon dioxide	**0.03**	**330**
Neon	**0.0018**	**18**
Helium	**0.00052**	**5.2**
Methane	**0.00014**	**1.4**
Krypton	**0.00010**	**1.0**
Nitrous oxide	**0.00005**	**0.5**
Hydrogen	**0.00005**	**0.5**
Ozone	**0.000007**	**0.07**
Xenon	**0.000009**	**0.09**

*Below 80 km.

the relative proportions of the principal gases in the lower atmosphere, excluding water vapor.

In addition to gases, the atmosphere contains minute liquid and solid particles, collectively termed **aerosols**. Most aerosols are found in the lower atmosphere near their source, the earth's surface. They originate through forest fires, from wind erosion of soil, as tiny sea-salt crystals from ocean spray, in volcanic eruptions, and from human industrial and agricultural activities. Some particles, such as meteoric dust, also enter the atmosphere from above.

It may be tempting to dismiss as unimportant the substances that make up only a small fraction of the atmosphere, but the significance of an atmospheric gas or aerosol is not necessarily related to its relative abundance. For example, water vapor, carbon dioxide, and ozone occur in minute concentrations, yet they are essential for life. By volume, no more than about 4 percent of the lowest kilometer of the atmosphere is water vapor—even in the warm, humid air over tropical oceans and rain forests. Without water vapor, however, there would be no rain or snow to replenish soil moisture, rivers, lakes, and seas. Carbon dioxide (only 0.03 percent of the atmosphere by volume) is necessary for photosynthesis. Together, water vapor and carbon dioxide act as a blanket over the earth's surface—causing the lower atmosphere to retain warmth and making the planet more amenable to life. Ozone (O_3) is formed in the atmosphere by the action of radiant energy from the sun on free oxygen (O_2) molecules. Although the volume percentage of ozone is minute, this vital gas shields living things from exposure to potentially lethal intensities of ultraviolet radiation.

The aerosol concentration in the atmosphere is also relatively small, yet these suspended particles participate in important processes. Some aerosols act as nuclei for the development of clouds and precipitation, and some influence air temperatures by interacting with sunlight.

Air pollutants

Although concentrations of the major atmospheric components essentially stabilized more than 600 million years ago, the atmosphere continues to evolve slowly. With the coming of human beings, and especially since the industrial revolution, the composition of the atmosphere has been modified by air pollution.

Air pollutants are gases and aerosols in the air that threaten the well-being of living organisms (especially human beings) or that disrupt the orderly functioning of the environment. Many of these substances occur naturally in the atmosphere. Sulfur dioxide (SO_2) and carbon monoxide (CO), for example, are normal minor components of the atmosphere that become pollutants only when their concentration approaches or exceeds the tolerance limits of organisms. Certain air pollutants, however, do not occur naturally in the atmosphere, and some of these are hazardous in any concentration. An example is asbestos fibers, which are known to cause cancer.

Air pollutants are products of both natural events and human activities. Natural sources of air pollutants include forest fires, dispersal of pollen, wind erosion of soil, organic decay, and volcanic eruptions. The single most important human-related source of atmospheric pollutants is the internal combustion engine. According to the United States Environmental Protection Agency (EPA), transportation vehicles yearly emit more than 100 million metric tons (110 million tons) of the major air pollutants. Many industrial sources also contribute to pollution of the atmosphere. Pulp and paper mills, iron and steel mills, oil refineries, smelters, and chemical plants are prodigious producers of air pollutants. Additional pollutants come from fuel combustion by industrial and domestic furnaces, from refuse burning, and from various agricultural activities such as crop dusting. In the United States, almost 160 million metric tons (175 million tons) of the chief air contaminants are emitted into the atmosphere each year as the result of human activity—almost 0.7 metric ton per person (Table 1.2).

Some substances are pollutants immediately upon emission into the atmosphere. These are designated **primary air pollutants**. In addition, within the atmosphere, chemical reactions involving these primary pollutants—both gases and aerosols—produce **secondary air pollutants**. Examples are acid mists and smog. In

TABLE 1.2

Estimated Emissions of Principal Air Pollutants in the United States During 1980

AIR POLLUTANT	MILLIONS OF METRIC TONS PER YEAR
Carbon monoxide	**85.4**
Volatile organic compounds (hydrocarbons)	**21.8**
Oxides of nitrogen	**20.7**
Sulfur oxides	**23.7**
Total suspended particulates	**7.8**
Total	**159.4**

Source: *1982–83 Statistical Abstract of the U.S.*, U.S. Department of Commerce, Bureau of the Census

some cases, the environmental impact of the individual primary pollutants is less severe than the effects of the secondary pollutants that they form. We have much more to say about air pollution in Chapter 16.

Probing the Atmosphere

Much of what is known about the composition and properties of the atmosphere is derived from direct sampling and measurement. Through the years, exploration of the atmosphere has progressed from tentative probing with primitive instruments to sophisticated remote sensing that employs space-age technology. At first, scientists investigated the atmosphere from the ground, climbing rugged mountain peaks to sample the rarified air. In 1743, the intrepid Benjamin Franklin conducted his famous kite-flying experiment to investigate the electrical properties of the atmosphere. Franklin attached a brass key to his kite string in order to attract an electrical current during a rainstorm. (Fortunately, a lightning bolt was not attracted to the key, for it surely would have been lethal.) From then until the 1930s, kites equipped with thermometers were used to compile temperature profiles of the lower atmosphere.

In 1804, the French scientist J. L. Gay-Lussac ushered in the age of manned balloon exploration of the atmosphere. He took air samples and measured temperature and humidity. On one ascent he reached an altitude of 7 km (4.3 mi). By the close of the nineteenth century, invention of recording instruments made possible unmanned balloon monitoring of the atmosphere. As a balloon ascended, its instrument package profiled temperature, pressure, and humidity. Eventually, the balloon burst, and the instrument package parachuted back to earth where the recordings were read. Through the early part of the twentieth century, weather instruments borne by balloons, kites, and aircraft provided information on the lowest 5 km (3 mi) of the atmosphere.

A leap forward in atmospheric monitoring came in the 1930s when the first **radiosonde**, a small instrument package equipped with a radio transmitter, was carried aloft by a helium-filled balloon. These devices transmit to a ground station continuous altitude measurements, called **soundings**, of temperature, pressure, and relative humidity. With radiosondes, the data are received immediately; no recovery of a recording instrument is needed. Today, radiosondes are sent up simultaneously at 12-hour intervals from hundreds of ground stations around the world (Figure 1.1). Information is monitored up to the altitude at which the balloon bursts (typically about 30 km, or 19 mi), and then the instrument package descends to the surface on a parachute. Some radiosondes are recovered, refurbished, and reused. Radiosonde movements usually are tracked from the ground by radar, thus giving an indication of variations in wind direction and speed with altitude. A radiosonde observation of this kind is called a **rawinsonde**.

A

B

FIGURE 1.1

(A) Launch of a radiosonde, a balloon-borne instrument package (B) that provides vertical profiles of air temperature, pressure, and relative humidity. (Photographs by Mike Brisson)

Direct exploration of the middle and upper atmosphere was greatly aided after World War II by the development of rocketry and, later, by the introduction of satellites. Until the first rockets penetrated altitudes above 30 km (19 mi), our understanding of the upper reaches of the atmosphere was based on speculation and indirect—albeit ingenious—ground-based studies. The age of satellites and remote sensing began in 1957 when the Soviet Union launched Sputnik I. These "eyes in the sky" are invaluable tools for weather observation. They provide data on cloud patterns, winds, temperature, humidity, and radiation. The United States orbited its first experimental weather satellite, TIROS (Television and Infrared Observational Satellite), in 1960. Since then, the United States has launched a series of increasingly sophisticated weather satellites. An example of one such satellite is shown in Figure 1.2.

FIGURE 1.2

Artist's view of a synchronous meteorological satellite (SMS) positioned over the equator at about 37,000 km (23,000 mi) altitude. This satellite orbits at the same rate as the earth rotates, so the satellite always surveys the same portion of the planet. From its relatively high orbit, it "sees" about one quarter of the earth's surface. (NASA photograph)

Probing the earth's atmospheric envelope has provided a detailed picture of the characteristics of the atmosphere. The remainder of this chapter describes the vertical profile of atmospheric temperature and the electrical properties of the upper atmosphere.

Thermal Structure of the Atmosphere

For convenience of description, the atmosphere is usually subdivided into concentric layers, shown in Figure 1.3, according to the vertical profile of the average air temperature. Most weather occurs in the bottom layer, the **troposphere**, which extends from the earth's surface to an altitude ranging between 20 km (13 mi) at the equator to 8 km (5 mi) at the poles. Normally, but not always, the temperature within this region decreases steadily with altitude. Air temperatures on mountain tops are therefore usually lower than in surrounding valleys. The upper boundary of the troposphere is called the **tropopause**, and is a transition zone between the troposphere and the next higher layer, the stratosphere.

The **stratosphere** extends from the tropopause up to about 50 km (31 mi). On average, in the lower portion of the stratosphere, the temperature does not change with altitude. When temperature is constant, the condition is described as **isothermal**. From the top of the troposphere up to about 20 km (13 mi), isothermal conditions prevail in the stratosphere. Above about 20 km (13 mi), the tem-

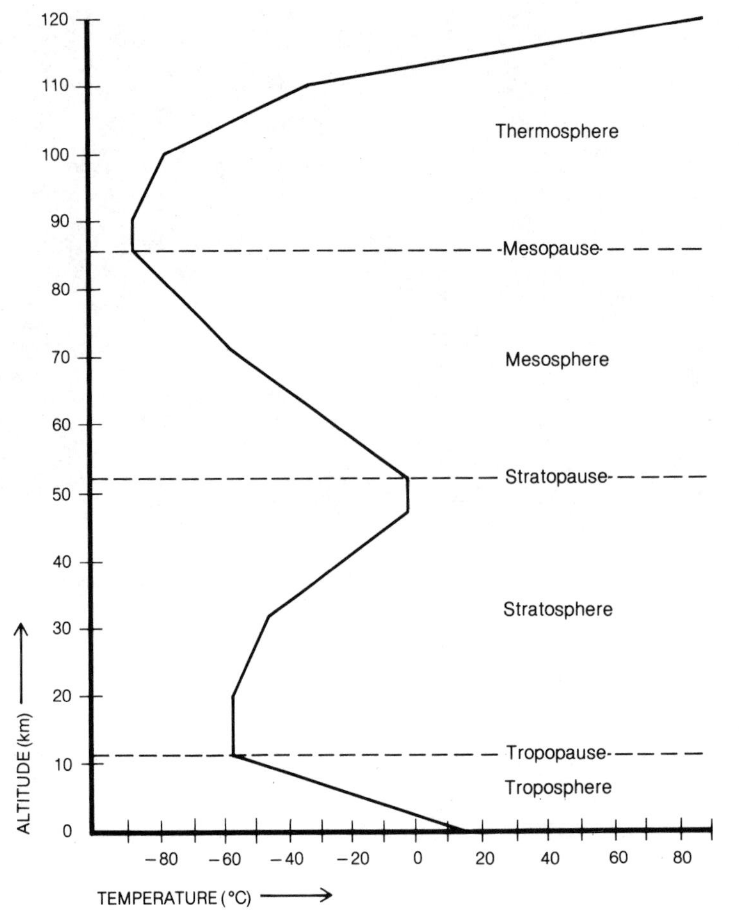

FIGURE 1.3

The average temperature of the atmosphere varies with altitude. Based on this variation, the atmosphere is subdivided into four zones.

perature increases with altitude up to the top of the stratosphere, the stratopause.

The stratosphere is ideal for jet aircraft travel because it is above the weather, and so has excellent visibility and generally smooth flying conditions. Since the early 1970s, scientists have been concerned about possible detrimental effects of pollutants entering the stratosphere. Because little exchange of air takes place between the troposphere and stratosphere, pollutants that enter the lower stratosphere tend to stay there for long periods. Gases thrown into the stratosphere during violent volcanic eruptions, for example, may remain there for several years. Pollutants that threaten to erode the protective ozone layer within the stratosphere have become a major environmental issue.

The **stratopause** is the transition zone between the stratosphere and the next higher layer, the **mesosphere**. In this layer, the temperatures once again decrease with increasing altitude. The mesosphere extends up to the **mesopause**, which is about 80 km (50 mi) above the earth's surface, and features the lowest mean temperatures in the atmosphere (−90 °C, or −130 °F). Above this is the **thermosphere**, where temperatures at first are isothermal and then increase rapidly with altitude.

The Ionosphere

The **ionosphere** is located primarily within the thermosphere, from altitudes of about 80 to 900 km (50 to 600 mi). The region is named for its relatively high concentration of ions. An **ion** is an atomic-scale particle possessing an electrical charge. High-energy solar radiation entering the upper atmosphere strips electrons (negatively charged subatomic particles) from oxygen and nitrogen atoms and molecules, leaving them as positively charged ions. Actually, the ions tend to concentrate within the lower thermosphere.

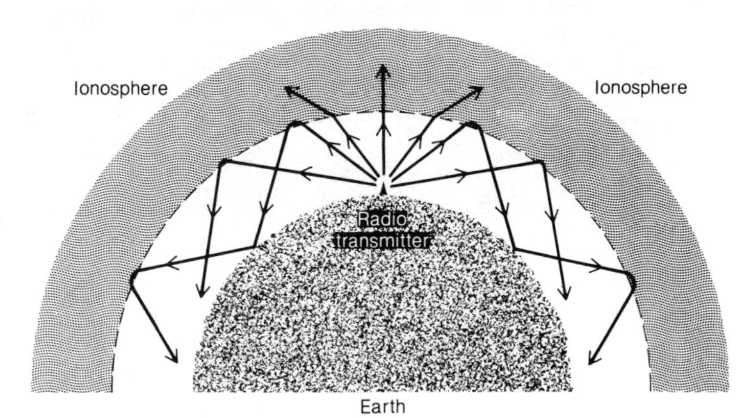

FIGURE 1.4
Within the ionosphere, layers of charged subatomic particles reflect outgoing radio waves. Multiple reflections involving both the ionosphere and the earth's surface greatly extend the range of radio transmissions.

Although the upper atmosphere does not greatly influence day-to-day weather, the ionosphere is important for long-distance radio transmissions. Ions reflect radio waves and thereby extend the range of radio transmissions (Figure 1.4). Radio signals travel in straight lines, and bounce back and forth between the earth's surface and the ionosphere. By repeated reflections, a radio signal may travel completely around the globe.

FIGURE 1.5

The aurora borealis viewed in an Alaskan night sky. (Photograph courtesy of S. I. Akasofu, Geophysical Institute, University of Alaska)

The ionosphere is also the site of the spectacular **aurora borealis** (northern lights) in the Northern Hemisphere and of the **aurora australis** (southern lights) in the Southern Hemisphere. Auroras appear in the polar night sky as overlapping ribbons or curtains of blue-green light, occasionally fringed with red and pink (see Figure 1.5). These awesome displays are caused by gigantic disturbances on the sun called solar flares (Figure 1.6), which send forth high-velocity streams of electrically charged subatomic particles, protons and electrons. These streams of particles are known as the **solar wind**. When these solar particles reach the ionosphere, they collide with oxygen and nitrogen molecules, energizing them and causing them to emit light.

The earth's magnetic field channels the solar particles toward the earth's geomagnetic poles, so the auroral display is visible only in high latitudes. In North America, the **auroral zone**, where the aurora is visible, is centered on the northwest tip of Greenland (at latitude 78.5 degrees N and longitude 69 degrees W). Auroral activity varies with the sun's activity. When the sun is quiet, the aurora is confined to a relatively small area at very high latitudes. When the sun is active, the auroral zone expands toward the equator, and the aurora may be seen across southern Canada and the northern tier states of the United States.

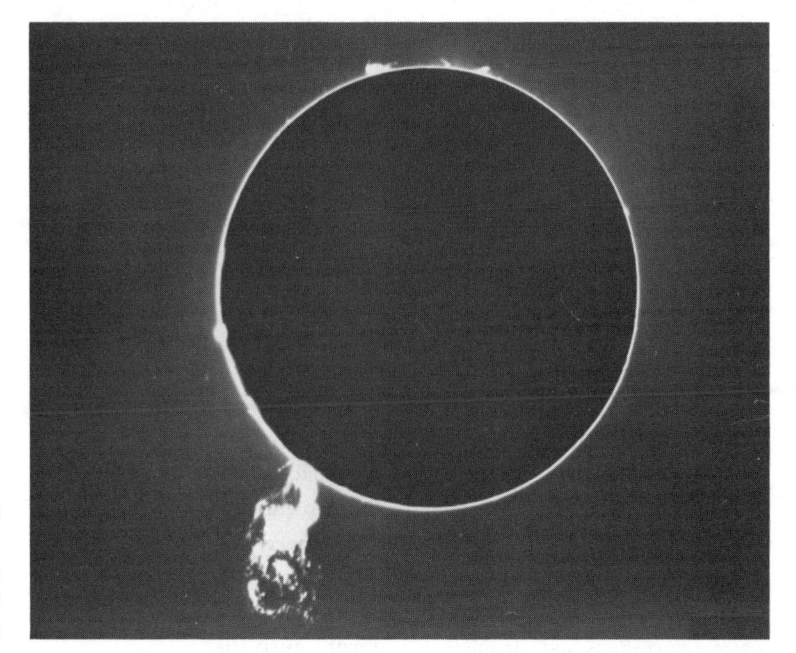

FIGURE 1.6
Solar flares emit high-velocity streams of electrically charged subatomic particles. (NCAR/NSF photograph)

Conclusions

This chapter considered the basic compositional and structural characteristics of the atmosphere. The focus of this book is the troposphere, the principal domain of weather. Our next major objective is to examine the reasons for weather. To do this, an understanding of heat and the transfer of heat is necessary, because imbalances in heating within the troposphere are what trigger weather phenomena.

SUMMARY STATEMENTS

- The atmosphere is a mixture of gases in which solid and liquid particles are suspended.
- The earth's present atmosphere is the product of a long evolutionary process that began billions of years ago and involved the development of continents, seas, and life.
- The significance of an atmospheric gas or of an aerosol is not necessarily related to its relative concentration in the atmosphere.
- Human beings pollute the atmosphere by increasing the concentrations of gases and aerosols to levels that adversely influence the well-being of living things, or to levels that disrupt the orderly functioning of the environment, or both.
- Over the years, our tools for investigating the reaches of the atmosphere have progressed from instrumented kites and manned balloons to rockets, rawinsondes, and satellites.
- The atmosphere may be subdivided into concentric layers on the basis of the vertical temperature profile.
- The ionosphere is located within the thermosphere. It contains a relatively high concentration of charged particles, and is the site of aurora displays.

KEY WORDS

weather
meteorology
climate
atmosphere
scientific method
photosynthesis
homosphere
heterosphere
aerosols
air pollutants
primary air pollutants
secondary air pollutants
radiosonde
soundings
rawinsonde
troposphere
tropopause
stratosphere
isothermal
stratopause
mesosphere
mesopause
thermosphere
ionosphere
ion
aurora borealis
aurora australis
solar wind
auroral zone

REVIEW QUESTIONS

1 What is the difference between weather and climate?

2 Explain why a description of climate, given only in terms of average values of various weather elements, is incomplete and perhaps misleading.

3 Provide several illustrations of how the earth's atmosphere is essential to the functioning of living organisms.

4 What role have volcanic eruptions played in the evolution of the earth's atmosphere?

5 Explain how the evolution of the earth's atmosphere is related to the development of oceans and the coming of life on earth.

6 Oxygen is the second most abundant gas composing the atmosphere. Explain its origins.

7 Distinguish between the homosphere and the heterosphere.

8 Most atmospheric aerosols are products of processes occurring at the earth's surface. Identify several of these processes.

9 Present several examples of how "minor" constituents (gases and aerosols) of the atmosphere play important roles in the functioning of the environment.

10 Most "air pollutants" are actually normal components of the atmosphere. Explain this apparent inconsistency.

11 Distinguish between natural sources of air pollutants and air pollutant sources related to human activities.

12 Describe the various methods whereby scientists have investigated the properties of the middle and upper atmosphere.

13 List some of the advantages of satellite observations of the atmosphere compared with other techniques of atmospheric monitoring.

14 What is the basis for subdividing the atmosphere into the troposphere, stratosphere, mesosphere, and thermosphere?

15 What advantages does the stratosphere offer over the troposphere for jet aircraft travel?

16 Why do pollutants that enter the lower portion of the stratosphere tend to remain there for long periods?

17 What is the source of ions in the ionosphere?

18 Why is the aurora visible only at high latitudes?

19 How and why does auroral activity vary with solar activity?

20 Why is the auroral zone not centered at the geographical pole?

POINTS TO PONDER

1 Photosynthesis occurs chiefly during the growing season. Speculate on how variations in the rate of photosynthesis through the course of a year influence the concentration of carbon dioxide in the atmosphere.

2 Why does a radiosonde balloon eventually burst as it ascends to altitudes above about 30 km (19 mi)?

3 The sun is the ultimate source of heat for the earth. Mountain tops are closer to the sun than lowlands and yet mountain tops are colder than lowlands. Why?

PROJECTS

1 Keep a daily log of weather conditions in your area. Use your own (or the school's) weather instruments or rely on television or radio weather reports.

2 Keep track of the changing patterns of weather on a national scale by watching televised weather summaries or by listening to the NOAA (National Oceanic and Atmospheric Administration) weather radio reports on a daily basis. Is the extent of stormy weather greater or less than the extent of fair weather across the nation?

SELECTED READINGS

Conover, J. H. "The Blue Hill Observatory." *Weatherwise* 37 (1984):296–303. *A historical sketch of a weather observatory founded near Boston in 1885; site of the world's first continuous recording of air-temperature variation with altitude.*

Frisinger, H. H. *The History of Meteorology: To 1800.* New York: Science History Publications, 1977. 148 pp. *Chronicles the people and events that are the roots of modern meteorology.*

Hughes, P. "Weather Satellites Come of Age." *Weatherwise* 37 (1984):68–75. *A historical review of the advances in global weather monitoring provided by satellites.*

Huschke, R. E., ed. *Glossary of Meteorology.* Boston: American Meteorological Society, 1959. 638 pp. *Definitions of more than 7000 terms used in weather and climate studies.*

Oliver, V. J. "Using Satellites to Study the Weather." *Weatherwise* 34 (1981):164–170. *A well-illustrated discussion of the perspective on weather systems provided by weather satellites.*

Walker, J. *Light From the Sky.* San Francisco: W. H. Freeman, 1980. 78 pp. *Reprints of Scientific American articles on various atmospheric optical phenomena, including the aurora.*

Maycomb was an old town, but it was a tired old town when I first knew it. . . . Somehow, it was hotter then: a black dog suffered on a summer's day; bony mules hitched to Hoover carts flicked flies in the sweltering shade of the live oaks on the square. Men's stiff collars wilted by nine in the morning. Ladies bathed before noon, after their three-o'clock naps, and by nightfall were like soft teacakes with frostings of sweat and sweet talcum.

HARPER LEE
To Kill a Mockingbird

Heat and temperature are two different concepts. The temperature of different substances responds differently to the addition of the same quantity of heat. This is one reason why, at the beach in summer, the sand feels warmer than the water. (Photograph by R. Hamilton Smith)

TWO

Heat and Temperature

Temperature Scales

Temperature Measurement

Heat Units

Transport of Heat

Conduction

Convection

Radiation

Specific Heat

Heating and Cooling Degree-Days

"TEMPERATURE" is one of the most important and common terms used to describe the state of the atmosphere. Air temperature is an essential element in any weather observation or weather forecast. From our everyday experience, we know that air temperature is variable, changing with time and from one place to another. We also know that heat and temperature are related concepts. When we heat a pan of soup on the stove, the soup's temperature rises. When we drop an ice cube into a drink, the temperature of the drink drops. Granted that the two concepts are closely linked, what is the precise distinction between heat and temperature?

All substances are composed of a multitude of molecules or atoms, tiny microscopic and submicroscopic particles, that are in constant motion. **Temperature** is a measure of the average energy of this motion, called **kinetic energy**. The faster the motion, the higher the temperature becomes. **Heat** is the total molecular energy of a given amount of a substance, a gram, for example. Temperature is thus directly proportional to heat:* when heat is added to a substance, the temperature of the substance rises.

Temperature Scales

For most scientific purposes, temperature is described in terms of the Celsius scale. Established by the Swedish astronomer Anders Celsius in 1742, the Celsius temperature scale has the numerical convenience of a 100-degree interval between the freezing and boiling points of pure water. The United States is virtually the only nation that still uses the Fahrenheit temperature scale for everyday measurements, including weather reports. Introduced in 1714 by a German scientist, Gabriel Fahrenheit, the Fahrenheit scale is numerically more cumbersome than the Celsius scale. If a thermometer graduated in both scales is immersed in a glass containing a mixture of ice and water, the Fahrenheit scale will read 32 °F, whereas the Celsius scale will read 0 °C. In boiling water at sea level, the readings will be 212 °F and 100 °C.

*Note that heat is a characteristic of a substance and is not a separate entity. Strictly speaking, therefore, it is imprecise to refer to heat apart from the particular substance that possesses the heat. For example, heat does not rise, but heated air does rise. In our discussions of heat, we refer either explicitly or implicitly to the substance possessing the heat.

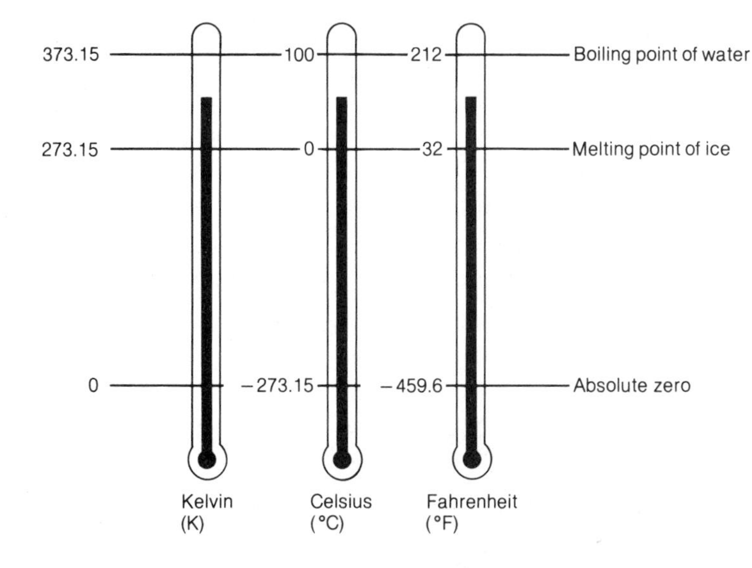

FIGURE 2.1

A comparison of the three temperature scales: Kelvin, Celsius, and Fahrenheit.

Molecular activity is less in cold substances than in hot substances. There is, theoretically, a temperature at which all molecular activity ceases. It is called **absolute zero**, or 0 kelvins (K), and corresponds to −273.15 °C (−459.67 °F). Actually, there is some atomic-level activity at 0 K, but an object at 0 K would not emit electromagnetic radiation (Chapter 3 discusses this further). On the Kelvin scale, temperature is the number of degrees above absolute zero.* The Kelvin scale is therefore a more direct measure of molecular activity, or heat energy, than the Fahrenheit and Celsius scales. Since nothing can be colder than absolute zero, there are no negative temperatures on the Kelvin scale. A 1-degree interval on the Kelvin scale corresponds precisely to a 1-degree increment on the Celsius scale. The three scales are compared in Figure 2.1.

Temperature Measurement

A **thermometer** is the usual instrument for monitoring variations in air temperature. Perhaps the most common type of thermometer consists of a liquid-in-glass tube (Figure 2.2) that was invented in the midseventeenth century. Typically, the liquid is either mercury (which freezes at −39 °C or −38 °F) or alcohol (which freezes at

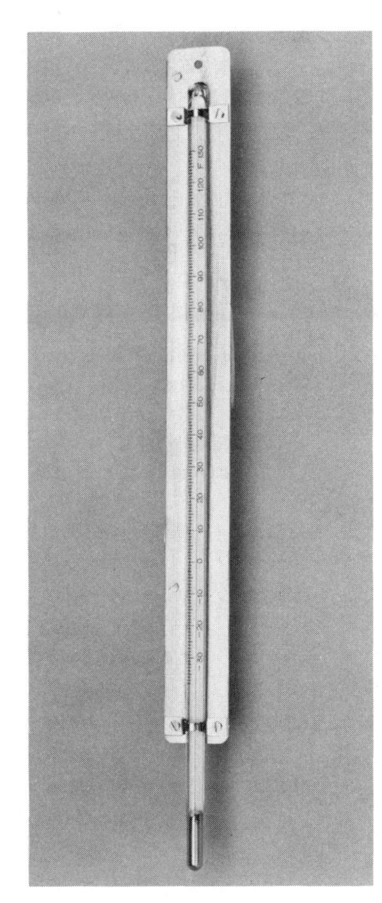

FIGURE 2.2

A liquid-in-glass thermometer. (Courtesy of Qualimetrics)

*Whereas units of temperature are expressed as "degrees Fahrenheit" on the Fahrenheit scale and "degrees Celsius" on the Celsius scale, on the Kelvin scale they are expressed simply as "kelvins."

−117 °C or −179 °F). As the air warms, the liquid expands and rises in the glass tube; as the air cools, the liquid contracts and drops in the tube. Some liquid-in-glass thermometers are designed to indicate the maximum and minimum temperatures over a specified period (Figure 2.3).

A second type of thermometer uses a metal coil to take advantage of the expansion and contraction that accompany the warming and cooling of a metal. A modification of this approach employs a bimetallic strip, consisting of two different metals welded side by side. The two metal strips have different rates of thermal expansion, so one expands more than the other in response to the same amount of heating. Because the two metals are bonded together, the bimetallic strip bends. The greater the heating is, the greater the bending is also. A series of gears, or levers, or both, translates the response of the coil or bimetallic strip to a pointer and a dial calibrated to read in degrees. Alternatively, these devices may be rigged to a pen and a clock-driven drum to give a continuous tracing of temperature with time (Figure 2.4). This device is called a **thermograph**.

FIGURE 2.3
A thermometer designed to record the maximum and minimum temperature, usually for a 24-hour period. (Courtesy of Belfort Instrument Company)

Another type of thermometer depends on certain materials that show changes in electrical resistance as the temperature fluctuates. Variations in electrical resistance are calibrated in terms of temperature. Such an instrument may be designed to give remote temperature readings by mounting the sensor at the end of a long wire joined to the instrument.

Regardless of the type of thermometer used, the two essential considerations in selecting an instrument are accuracy and response time. For most meteorological purposes, a thermometer that can

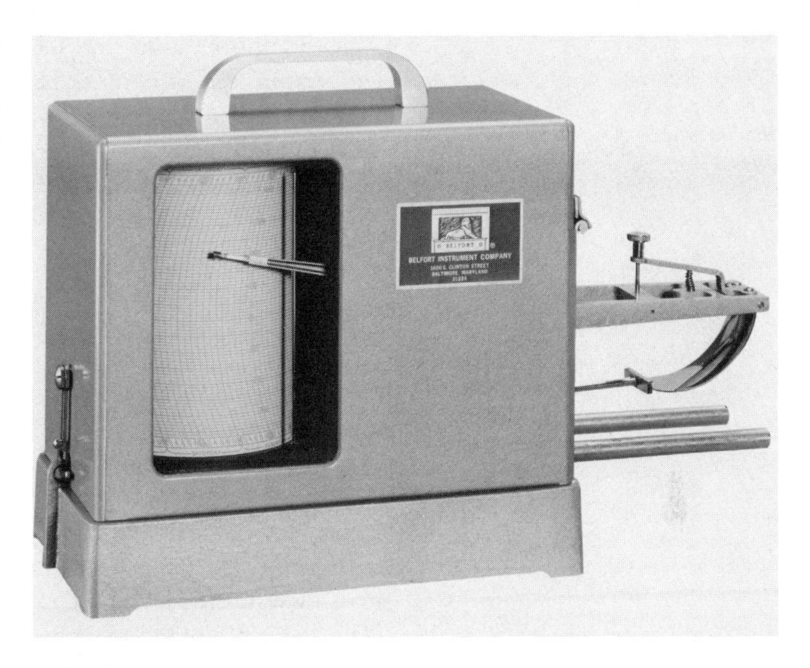

FIGURE 2.4

A thermograph provides a continuous tracing of fluctuations in air temperature. (Courtesy of Belfort Instrument Company)

detect temperature changes of 0.5 °C is sufficient. Response time refers to the instrument's capability in registering rapid oscillations in temperature. Most liquid-in-glass and electrical resistance thermometers have rapid response times, whereas metallic thermometers tend to be more sluggish. For reliable performance, thermometers should be mounted in a place shielded from direct sunlight and precipitation, and should be adequately ventilated (Figure 2.5).

FIGURE 2.5

A National Weather Service standard weather instrument shelter houses instruments in a white louvered box that provides ventilation and protection from sun and precipitation. Cylindrical instruments to the right of the shelter are rain and snow gauges. (Photograph by J. M. Moran)

In addition to using the standard thermometers described, air temperature can sometimes be deduced in unconventional ways. One interesting and surprisingly accurate approach is to count cricket chirps. For temperatures above about 12 °C (54 °F), the number of cricket chirps heard in an 8-second period plus 4 approximates the air temperature in degrees Celsius.

Heat Units

Although temperature is a convenient way to describe relative heating, we can quantify heat energy directly. Meteorologists traditionally measure heat energy in units called calories. A **calorie** is defined as the amount of heat needed to raise the temperature of 1 gram (g) of water 1 Celsius degree (from 14.5 °C to 15.5 °C, for example). Note that this definition specifies a substance (water), a mass of water (1 g), and a temperature change (1 Celsius degree). (The "calorie" used to measure the energy content of food is actually 1000 heat calories, or 1 Kcal.) The standard international unit for energy of any form, including heat, is the joule, abbreviated J. One calorie equals 4.1868 J.

In the English system, heat is quantified as British thermal units (Btus). A **Btu** is defined as the amount of heat required to raise the temperature of 1 lb of water 1 °F (from 62 °F to 63 °F, for example). One Btu is equivalent to 252 cal and to 1055 J.

Transport of Heat

Within the atmosphere, one place is usually warmer or colder than another, and within our environment, the ground is usually warmer or colder than the overlying air. These temperature differences result from imbalances in the heating and cooling processes that are the subject of Chapter 3. A change in temperature over a certain distance is known as a **temperature gradient**. A familiar temperature gradient is the difference between the temperature at the hot equator and the temperatures at the cold poles (a horizontal temperature gradient). Another well-known gradient is the temperature difference between the earth's surface and the troposphere (a vertical temperature gradient).

In response to temperature gradients, heat is transported from relatively warm places to relatively cold places. This redistribution of heat tends to reduce temperature gradients to zero. Within the atmosphere, heat is transported by conduction, convection, and radiation.

Conduction

Conduction occurs within a substance or between substances that are in direct physical contact. In **conduction**, the kinetic energy of atoms or molecules (that is, heat) is transferred by collisions with neighboring atoms or molecules. This is how a spoon placed in a steaming cup of coffee is heated. As the faster atoms of the hot coffee collide with the slower atoms of the initially cool spoon, some kinetic energy is transferred to the atoms of the spoon. These atoms then transmit some of their energy, via collisions, to their neighbors, so heat is eventually conducted up the handle of the spoon and the handle becomes hot to the touch.

Some substances conduct heat much more readily than others. As a rule, solids are better conductors than liquids, and liquids are better conductors than gases. At one extreme, metals are excellent conductors of heat, and at the other extreme, air is a very poor conductor of heat. When fiberglass fibers are used to insulate an attic, it is primarily the air trapped between the fibers that inhibits heat loss. In time, as the fiberglass fibers settle and become matted down, the air is excluded, and the material loses much of its insulating value. A fresh snow cover has an extremely low thermal conductivity because of the air trapped between the individual snow flakes. A thick snow cover (20–30 cm, or 8–12 in.) can thus inhibit or prevent the freezing of the underlying soil, even though air temperatures may drop well below freezing. In time, however, like the fiberglass, the snow cover will lose some of its insulating property as the snow settles and the air escapes.

Heat is conducted from warm ground to cooler overlying air, but because the air has a low heat conductivity, conduction is significant only in a very thin layer of air that is in immediate contact with the earth's surface. In heating the atmosphere, convection is much more important than conduction.

Convection

While conduction takes place in solids, liquids, or gases, convection occurs in liquids or gases only. **Convection** is the transport of heat within a fluid medium through movement of the medium. In the atmosphere, convection often develops because of density* differences that are triggered by the conduction of heat from the earth's surface. As the warm ground heats the cooler overlying air, the air becomes warmer and less dense than the surrounding air. In equal volumes, warm air is lighter than cold air. The warmer, lighter air then rises in the atmosphere until it reaches an altitude where its temperature and density match that of the surrounding air. Meanwhile, the colder, denser air sweeps in to take the place of the rising air, and is warmed by contact with the earth's surface. The process is then repeated. In this way, as shown in Figure 2.6, a convective circulation of air develops and transports heat into the atmosphere.

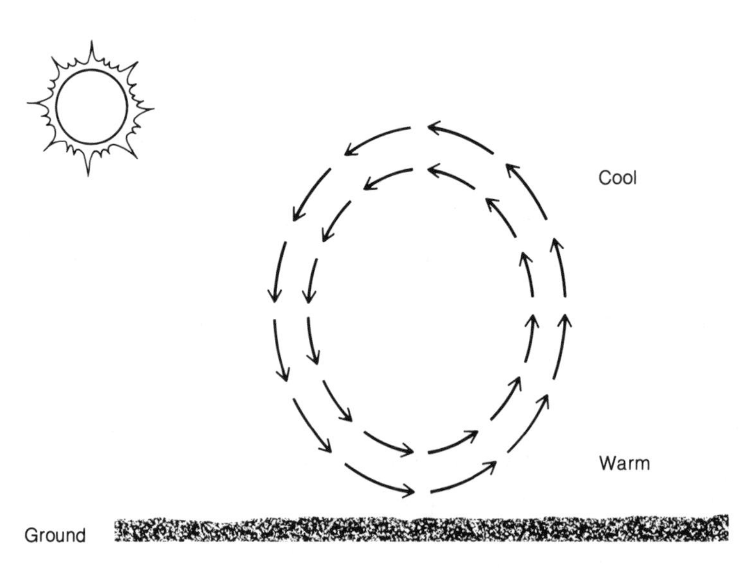

FIGURE 2.6

Convection currents transport heat from the earth's surface into the atmosphere.

This same process is visible in a pan of boiling water on a stove. We can actually see the circulating water that is redistributing the heat from the bottom of the pan to the water.

Radiation

Radiation, the third mechanism of heat transport, is unlike the other mechanisms in that it does not require an intervening physical medium. It can take place in a vacuum. **Radiation** consists of electromagnetic waves traveling at the speed of light (300,000 km or 186,000 mi per second). Radiation transports energy in wavelike

*Density is mass per unit volume.

motions analogous to the way sea waves transport energy. Because radiation is the principal means by which the earth-atmosphere system* receives and loses heat, we elaborate on the properties of radiation in Chapter 3.

Specific Heat

Whether by conduction, convection, or radiation, the transport of heat from one object to another is accompanied by changes in temperature. Unless heat is used to bring about a change in phase (the evaporation of water, for example), an object that loses heat exhibits a drop in temperature, and an object that gains heat exhibits a temperature rise. The temperature response to an input (or output) of a specified quantity of heat does, however, differ from one substance to another.

The amount of heat required to change the temperature of 1 g of a substance by 1 Celsius degree is the **specific heat** of that substance. Two different materials registering the same temperature do not necessarily possess the same amount of heat energy. Different quantities of heat may also be required to raise or lower the temperature, by 1 degree, of equal amounts of two different substances. The specific heat of all substances is measured relative to that of water, which is 1 cal per gram per degree Celsius (at 15 °C). The specific heat of some familiar materials are listed in Table 2.1.

*The phrase "earth-atmosphere system" refers to the earth's surface and atmosphere considered together.

TABLE 2.1

Specific Heat of Some Familiar Substances

SUBSTANCE	SPECIFIC HEAT*
Water	**1.000**
Ice (at 0 °C)	**0.478**
Wood	**0.420**
Aluminum	**0.214**
Brick	**0.200**
Granite	**0.192**
Sand	**0.188**
Dry air†	**0.171**
Gold	**0.031**

*In calories per gram per °C.
†At constant volume.

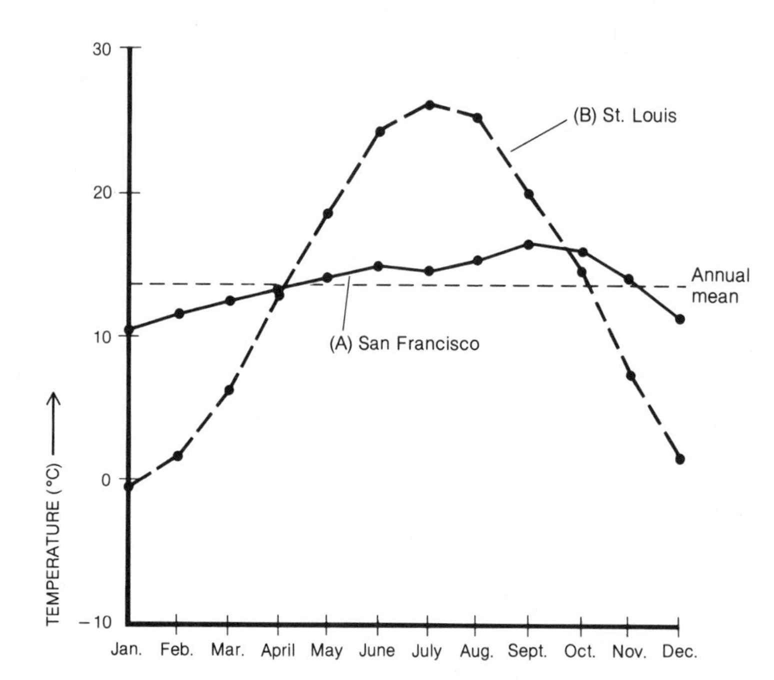

FIGURE 2.7

March of mean monthly temperature for (A) maritime San Francisco and (B) continental St. Louis. (Data from NOAA)

Note that water has the greatest specific heat of any naturally occurring substance. For example, the specific heat of water is about five times that of dry sand. One calorie will raise the temperature of 1 g of water by 1 °C, while 1 cal will raise the temperature of 1 g of sand about 5 °C. This is why at the beach in the summer the sand feels hot relative to the water. On exposure to the same heat source, substances with low specific heats warm up more than those substances with high specific heats.

The surface temperature of the land exhibits a greater variability with time than does the surface temperature of a body of water such as a lake or sea. For example, the land heats up more during the day and during the summer and it cools down more at night and during the winter than an adjacent body of water. The primary reason for this contrast between land and water is the water's greater specific heat, but differences in heat transport are also important. While heat is circulated readily through large volumes of water, heat is conducted very slowly into the soil. Solar radiation penetrates water to significant depths, but strikes only the land surface. For the same surface area exposed, heat is thus transported to greater volumes of water than land. In other words, an identical input of heat would cause a land surface to warm up more than an equivalent area of a water body.

The more stable sea surface temperatures have important implications for climate, because air temperatures are regulated to a large extent by the temperature of the surface over which the air resides or travels. Localities that are immediately downwind of an ocean (maritime localities) exhibit smaller seasonal temperature variations than do localities that are situated well inland (continental localities). An illustration of this effect is the comparison of the average monthly temperatures of San Francisco and St. Louis (Figure 2.7). Although the cities are at the same latitude, summers are cooler and winters are milder in maritime San Francisco than in continental St. Louis.

Climatologists use an **index of continentality** to describe the degree of maritime influence on continental air temperature. Several different indexes are available, but indexes are typically based on the difference between average winter and summer temperatures. A generalized index for North America is shown in Figure 2.8. Note that because winds blow primarily from west to east, western North America is more maritime (less continental) than eastern North America.

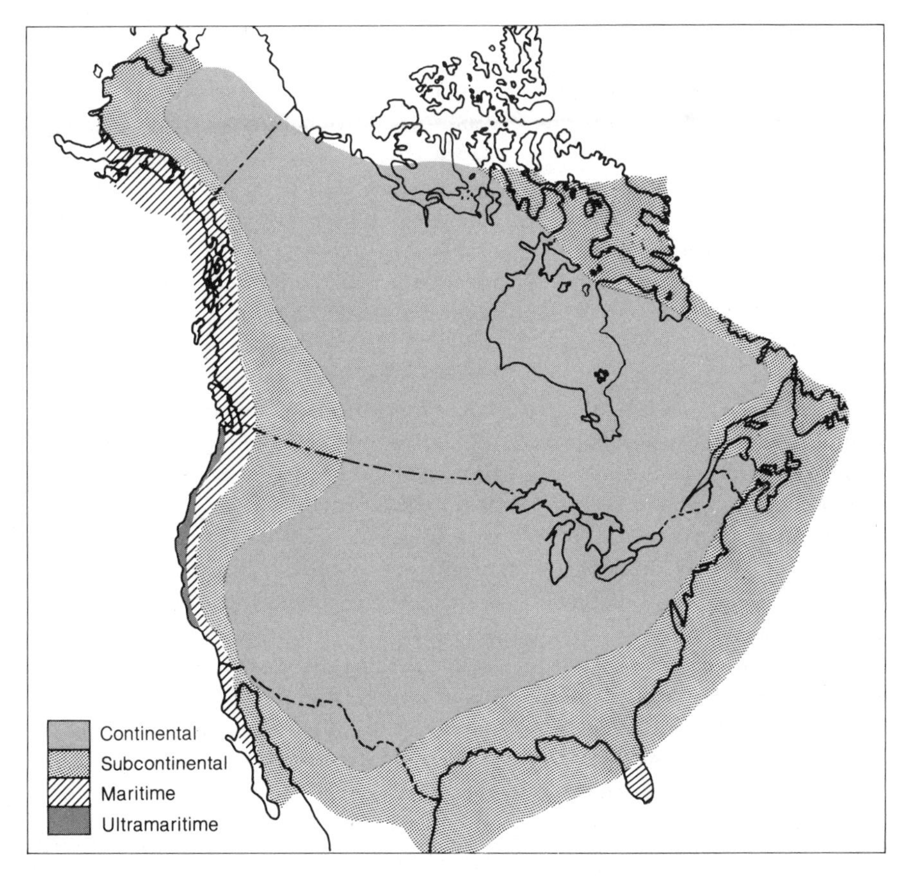

FIGURE 2.8

Indexes of continentality gauge the influence of oceans on air temperature over continents. In this scheme, North America is divided into zones designated with increasing maritime influence: continental, subcontinental, maritime, and ultramaritime. The greater the maritime influence, the less the seasonal temperature variation. (Modified after Currey, D. R. "Continentality of Extratropical Climates." *Annals of the Association of American Geographers* 64, no. 2 (1974):274.)

SPECIAL TOPIC

Heat and human comfort

To understand how air temperature influences human comfort, we must first understand that humans are *homeothermic*. This means that we regulate our internal or core temperature within ± 2 °C (± 3.6 °F) of 37 °C (98.6 °F) despite much larger variations in the temperature of the surrounding air (i.e., the *ambient temperature*). The *core* refers to those regions of the body where vital organs, such as the brain, heart, lungs, and digestive tract, are located. If the vital organs are not maintained at a nearly constant temperature, they do not function properly. Other parts of the body, such as the legs and arms, however, may experience considerable temperature changes.

At ambient air temperatures of 20 °C to 25 °C (68 °F to 77 °F), someone who is fully clothed and indoors will feel comfortable at rest. In this temperature range, the body can maintain a core temperature of 37 °C without resorting to special temperature-regulating mechanisms.

When we are exposed to ambient temperatures outside the 20 °C to 25 °C range, the body must initiate processes that maintain the core temperature at 37 °C. For example, if we stand in the sun on a day when air temperatures hit 30 °C (86 °F), our core temperature will begin to rise. In response, we begin to sweat. Heat at the skin surface evaporates the sweat, and as a consequence, the skin cools. We have all experienced the cooling effects of evaporation as we step out of a shower or leap out of a swimming pool. Evaporative cooling reduces skin temperature, which, in turn, normally leads to a reduction in core temperature.

If we are exposed to air temperatures below 20 °C (68 °F), the core temperature begins to decline, and we start to shiver. This increased muscle activity produces additional heat, which helps to raise the core temperature back to 37 °C. Sweating and shivering are examples of *thermoregulation*—mechanisms that assist in maintaining a nearly constant core temperature regardless of environmental temperatures.

Another means of thermoregulation involves the regulation of blood flow. The major heat transfer from the body core to the skin occurs via the circulatory system. When we are exposed to air temperatures below 20 °C, our bodies can limit heat loss by restricting blood flow to the skin. Under the direction of the nervous system, many of the tiny blood vessels in the skin constrict. As ambient air temperatures decline, additional blood vessels constrict, further reducing blood flow to the body surface. This phenomenon can be observed by putting your hand in a container of ice water; the skin becomes paler. The reduced blood flow places a thicker insulative layer between the heat-producing core and the skin surface.

In contrast, ambient air temperatures above 25 °C (77 °F) trigger increased blood flow to the body surface. Blood vessels near the skin surface dilate, giving the skin a reddish or flushed appearance. The greater flow of blood to the body surface increases skin temperature. As a consequence, the body-to-air temperature gradient increases and enhances cooling by radiation, conduction, and convection. Greater blood flow to the skin also increases evaporative cooling by supplying more water for sweating.

In addition to these physiological processes, our behavioral responses assist in thermoregulation. For example, if we feel hot, we shed clothing, seek shelter from the sun, or turn on the fan or air conditioner.

Under some conditions, thermoregulatory processes are insufficient to maintain a 37 °C core temperature. For example, if a person hiking in the woods is drenched by a cold rain, and if the soaked hiker then overexerts himself physically and becomes exhausted, thermoregulatory mechanisms may not be able to compensate for heat loss. The core temperature consequently drops, and hypothermia may ensue.

Hypothermia refers to those responses that occur when the human core temperature drops below 35 °C. Initially, shivering becomes more violent and uncontrollable. In addition, the victim begins to have difficulty speaking and becomes apathetic and lethargic. As the core temperature falls below 32 °C (90 °F), shivering is replaced by muscular rigidity, and muscle coordination becomes difficult. Mental abilities are highly impaired, and the victim is generally unable to help himself. At a core temperature of 30° C (86 °F), the person drifts into unconsciousness. Death usually occurs at core temperatures between 24 °C and 26 °C (75 °F and 79 °F), because the heart rhythm becomes uncontrollably irregular (ventricular fibrillation) or uncontrollably halted (cardiac arrest).

Once hypothermia is established, the victim is in serious trouble. With only a 3 °C (5.7 °F) drop in core temperature, the ability of the body to regulate its core temperature is already greatly impaired. If the core temperature drops to 29 °C (85 °F), thermoregulation is essentially ineffective. The first signs of hypothermia should never be ignored, and action should be taken immediately. Treatment takes two forms—prevention of further heat loss and the addition of heat. The former includes replacing wet clothing with dry clothing, finding shelter, and insulating the person from the ground so that body heat is not conducted to the colder ground surface. The body can be heated by an external source, such as a space heater or other human bodies. Administering hot drinks, if the victim is conscious, will help warm the core from the inside.

In some situations, thermoregulatory processes may be unable to prevent a rise in core temperature. For instance, a person exposed to hot desert conditions without an adequate supply of water will eventually experience an increase in core temperature. If core temperature continues to rise, hyperthermia may ensue. *Hyperthermia* refers to those responses that take place when human core temperature reaches 39 °C (102 °F). As core temperature climbs to 41 °C (105 °F), thermoregulatory controls break down, and a person may suddenly and quite unexpectedly collapse. The victim also experiences muscle cramps or spasms, and slips into unconsciousness. Sweating ceases, although it is not known if this is a cause or a result of hyperthermia. With serious heat stress, the individual usually dies within a few hours unless the core temperature can be lowered artificially. These responses are collectively identified by various names, including heatstroke, sunstroke, and heat apoplexy.

The victim of hyperthermia must be treated promptly, because once thermoregulatory controls fail, the core temperature rises rapidly. To save the victim, core temperature must be regulated from outside the body. If possible, the victim should be moved to a cooler environment. In addition, the body should be placed in cold water, preferably ice water. Alternatively, sponging the body with alcohol enhances cooling as the alcohol evaporates.

The human body possesses amazing capabilities to adjust to changing air temperatures and to provide a sense of comfort. These capabilities are, however, limited. If the core temperature begins to deviate from normal, the individual or his or her companions must take corrective action quickly. Failure to do so may be fatal.

Heating and Cooling Degree-Days

With the contemporary concern for energy conservation, television and newspaper weather summaries routinely report heating or cooling degree-day totals in addition to temperature. Heating and cooling degree-days are measures of household energy consumption for space heating and cooling, respectively.

Heating degree-day units are computed only for days when the mean outdoor air temperature is lower than 65 °F (18 °C). Heating engineers who formulated this index early in this century found that when the mean outdoor temperature fell below 65 °F, space heating was required in most buildings to maintain an indoor air temperature of 70 °F (21 °C). The mean daily temperature is the simple arithmetic average of the 24-hour maximum and minimum air temperatures. Subtracting the daily mean temperature from 65 °F yields the number of heating degree-day units for that day. For example, suppose that this morning's low temperature was 36 °F (2 °C), and this afternoon's high temperature was 52 °F (11 °C). Today's mean temperature would then be 44 °F (7 °C), giving a total of (65 °F minus 44 °F) 21 heating degree-day units. It is usual to keep a running total of heating degree-day units, that is, adding degree-day units for successive days through the heating season (actually from July of one year through June of the next).

Fuel distributors and power companies monitor cumulative degree-day unit totals closely. Fuel oil dealers base fuel use rates on degree-day units, and schedule home deliveries accordingly. Natural gas and electrical utilities anticipate power demands on the basis of degree-day unit totals, and implement priority use policies on the same basis when capacity fails to keep pace with demand.

Figure 2.9 shows the average annual heating degree-day totals over the United States. Outside of mountainous areas, regions of equal heating degree-day totals tend to be aligned from east to west. The degree-day totals increase poleward. As an example, the annual space heating requirement in Chicago (6100 heating degree-day units) is about four times that of New Orleans (1500 heating degree-day units).

Cooling degree-day units are computed only for days when the mean outdoor air temperature is higher than 65 °F.* Supplemental air conditioning may be needed on such days. Again, a cumulative total is maintained through the cooling (summer) season. As a general rule, however, air conditioning is needed only in

*Higher base temperatures are sometimes used.

FIGURE 2.9

Average annual heating degree-day unit totals over the United States. (EIDS data)

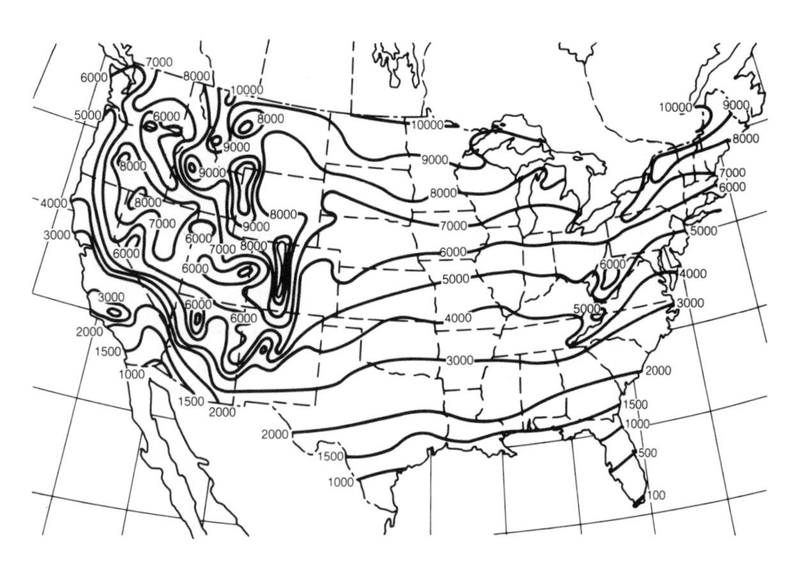

those localities where cumulative cooling degree-day unit totals exceed 700 (Figure 2.10).

Note that indexes of heating and cooling requirements are based on outside air temperatures and do not take into account other weather elements, such as wind speed and humidity, which influence human comfort and demands for space heating and cooling. Heating and cooling degree-day units are therefore only approximations of our residential fuel demands for heating and cooling. (For information on human responses to heat stress, see the Special Topic, Heat and Human Comfort.)

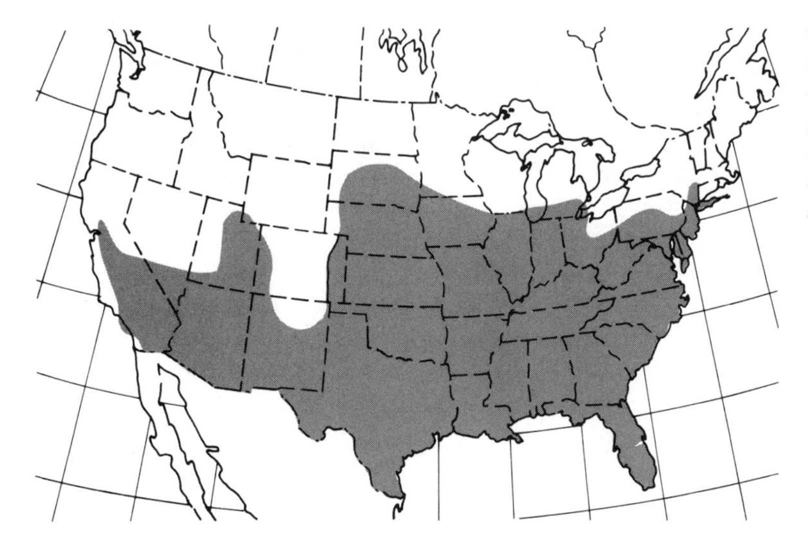

FIGURE 2.10

Shaded area of the United States is where average annual cooling degree-day units total more than 700 (with a base of 65 °F, or 18 °C). Air conditioning is desirable in this area. (NOAA data)

SPECIAL TOPIC

Heat and crop yields

There is an old midwestern saying that in spring corn should not be planted before the leaves of an oak tree reach the size of a squirrel's ear. While this is a colorful guide, farmers today use more sophisticated and reliable indexes of crop-climate relationships to help optimize crop yields. The local growing season length and the average annual growing-degree units are two important factors.

The length of time during the year when air temperatures remain sufficiently mild to permit plant growth is a critical determinant of agricultural productivity. Usually the *growing season* is described as the number of days between the dates of the last killing frost in spring and the first killing frost in fall. However, *killing frost* is an ambiguous term. Whether or not a frost kills a plant depends on the plant species and its life stage, the duration of freezing temperatures, and the rate of freezing. Thus, no single temperature reading can fully convey the actual impact on agriculture in a region, particularly where a variety of crops are grown. Nonetheless, for ease of comparison, the period most commonly used to delineate the growing season is the freeze-free period, that is, the time between the last day of a recorded instrument shelter temperature of 0 °C (32 °F) in the spring and the first date of 0 °C (32 °F) in the autumn.

The average length of the growing season shortens with increasing latitude so that the types of crops that can be successfully grown also change. For example, the shortest maturing field corn requires a growing season of approximately 85 days. Hence, corn cannot be grown in those areas of the northern Midwest and southern Canada where the growing season is less than 85 days. Even then, reported growing season lengths are average values, and often the growing season is shorter than the average. Thus, even if the maturity time of corn matches the average growing season length, growing the corn is a risky venture.

The growing season is lengthened locally by the moderating influence of nearby large bodies of water. A good example is in the state of Wisconsin, where localities bordering Lake Michigan can have a growing season that is 20 to 40 days longer than sites at the same latitude, but 100 km (62 mi) inland. Topography can also influence growing season length. Because cold air is relatively dense, it tends to flow downhill so that the growing season typically is several weeks shorter in valleys than in the surrounding hillsides. For that reason, vineyards and orchards are usually sited on hillsides rather than in valley bottoms.

One problem with relying on growing season length as a potential measure of agricultural success is that temperatures above 0 °C (32 °F) are not equally effective in promoting crop growth. For example, corn grows very little at temperatures below 10 °C (50 °F) or above 30 °C (86 °F). Hence, the actual air temperature during the growing season can either accelerate or retard plant growth and often is as significant in affecting yields as the actual length of the growing season.

To better gauge the effect of temperature on plant growth during the growing season, an index called *growing-degree units* has been developed. Because one of its most successful applications is in growing corn, we will use corn as an example of how the index works. To determine the growing-degree units (GDUs) for a particular day, the average daily temperature is computed by adding the highest and lowest temperature in °F for the day (24 hours) and dividing by two. Because corn essentially grows only when air temperatures are between 50 °F and 86 °F (10 °C and 30 °C), any average daily temperature above 86 °F is counted as 86 °F and any average daily temperature below 50 °F is counted as 50 °F. The number 50 (the lower threshold for growth of corn plants) is then subtracted from the daily average temperature to obtain the number of GDUs for that day. GDUs are then summed from one day to the next to give a cumulative total.

There are several important applications of GDUs. One is to match the GDU requirement of a hybrid with the long-term average number of GDUs for a region. For example, a corn farmer living in central Illinois or Indiana typically selects a hybrid that requires 3000 to 3400 GDUs. But if the planting season is delayed (perhaps by early spring rains) or if fields have to be replanted (perhaps because of poor germination due to low soil moisture after planting), the farmer may have to switch to a shorter season hybrid (that is, a hybrid that requires fewer GDUs). By computing the number of GDUs that have already accumulated during the growing season and subtracting this total from the number of GDUs for an average growing season, the farmer obtains a good estimate of what hybrid to use so that the corn reaches maturity before the first killing freeze.

The GDU system has been criticized for a number of reasons: the threshold temperature may vary with the maturity of the crop; not all temperatures within the growth range affect growth equally; and day-night temperature shifts may be more important than average daily temperatures. For example, day-night temperature shifts are critical for growing wine grapes in the Napa Valley in California. Because of these limitations, GDUs do not always work. But for some grains, such as corn and oats, and for some fruit crops, such as apples, pears, and peaches, GDUs provide an effective, if not precise, guide to choosing the best varieties and planning planting and harvest times.

Heat is also a critical factor in agricultural success. Of all the climatic elements, the availability of heat is the single most important consideration in determining where a crop can be grown. For information on this, see the Special Topic, Heat and Crop Yields.

Conclusions

Conduction, convection, and radiation redistribute heat within the earth-atmosphere system. Depending primarily on the specific heat of the medium, the temperature of air, water, and land changes in response to this heat redistribution. What is the origin of heat energy? Ultimately, almost all of the heat on earth comes from the sun via electromagnetic radiation. This process is described in the next chapter.

SUMMARY STATEMENTS

- Heat is the total molecular energy contained in a specified amount of some substance.
- Temperature is a commonly used but inexact way of describing the relative heat energy of substances. We use the Fahrenheit, Celsius, and Kelvin temperature scales.
- Heat energy is quantified directly as calories, joules, or British thermal units.
- In response to gradients in temperature, heat is transported by conduction, convection, and radiation.
- In heating the atmosphere, convection is much more important than conduction. Recall, however, that the heat transported by convection was originally conducted from the earth's surface to the atmosphere.
- The temperature response to an input (or output) of heat differs from one substance to another, depending on the specific heat of the substance.
- The warming and cooling rates of large bodies of water and land have important implications for climate. Water bodies, such as seas and lakes, exhibit less variability in temperature than do continents.
- An index of continentality is used to represent the degree of maritime influence on continental temperatures. The greater the maritime influence is, the less the seasonal temperature contrast will be.
- Heating and cooling degree-days are measures of household energy consumption for space heating and cooling, respectively.

KEY WORDS

temperature
kinetic energy
heat
absolute zero
thermometer
thermograph
calorie
Btu
temperature gradient
conduction
convection
radiation
specific heat
index of continentality
heating degree-day units
cooling degree-day units

REVIEW QUESTIONS

1 Distinguish between heat and temperature.

2 Why is temperature an inexact way of describing the heat energy of different substances?

3 Why is the Celsius temperature scale considered to be more convenient than the Fahrenheit temperature scale for most scientific purposes?

4 What is the significance of "absolute zero," that is, zero kelvins?

5 In some localities, winter temperatures dip below −40 °F. In those areas, what type(s) of thermometer is used to measure temperatures?

6 What is meant by a thermometer's response time? Why is it an important consideration in selecting a thermometer?

7 Present some examples of temperature gradients (a) within the same substance and (b) between different substances.

8 Describe three processes by which heat can be transferred from one place to another. Provide a common example of each.

9 Air has a low thermal conductivity. How is this property of air used in insulating a home?

10 Why is convection much more important than conduction in the transport of heat from the earth's surface into the atmosphere?

11 Compare the heat conductivity of solids, liquids, and gases.

12 Why can a thick snow cover inhibit the freezing of the underlying soil even though air temperatures drop well below the freezing point?

13 Define specific heat.

14 Referring to Table 2.1, determine the temperature change of 1 g of each of the following substances upon an addition of 1.0 cal of heat: (a) ice, (b) dry air, (c) aluminum.

15 How and why do maritime climates differ from continental climates?

16 Heating degree-day units are computed using a base of 65 °F. Why?

17 Why do large bodies of water, such as lakes or seas, reduce the temperature fluctuations of the overlying air?

18 Convert the following:

0 °C = ______ °F = ______K
20 °C = ______ °F = ______K
65 °F = ______ °C = ______K
300 K = ______ °C = ______°F

19 Today's high temperature was 20 °F and this morning's low temperature was minus 5 °F. Compute today's heating degree-day units.

20 Describe the natural responses of the human body to very low air temperatures and to very high air temperatures.

POINTS TO PONDER

1 Compute the average temperature gradient between the earth's surface and the tropopause. How does the magnitude of this gradient compare with that of the average horizontal temperature gradient between the equator and poles?

2 In winter, thermalpane windows or storm windows are used to cut heat loss from buildings. Speculate on how these windows reduce conductive and convective heat flow.

3 Why do you suppose the specific heat differs from one substance to another?

4 How do the thermal conductivity and specific heat of dry air compare in magnitude with the same properties of materials composing the earth's surface?

PROJECTS

1 Determine whether your location has a continental or maritime climate. You may wish to plot mean monthly temperatures in order to determine the winter-to-summer temperature contrast. These data are available from the nearest National Weather Service office, but first check the government documents section of your library.

2 Maintain a running total of heating degree-day units and cooling degree-day units for your area. Is air conditioning really necessary in your area?

3 Design an experiment to determine the temperature response of samples of dry and wet sand to the same heat input. Explain any difference in thermal response.

SELECTED READINGS

Asimov, I. *Understanding Physics: Motion, Sound, and Heat.* New York: New American Library Mentor Book, 1969. 248 pp. *A well-written and lucid explanation of basic physical principles.*

Griffiths, J. F. "A Chronology of Items of Meteorological Interest." *Bulletin of the American Meteorological Society* 58 (1977): 1058–1067. *Major landmarks in the science of meteorology from 1066 B.C. to 1974 A.D.*

Quayle, R., and F. Doehring. "Heat Stress: A Comparison of Indices." *Weatherwise* 34, no. 3 (1981):120–124. *A comparison of four different indexes.*

Splendid with splendor hid
you come, from your Arab
abode, a fiery topaz
smothered in the hand of a
great prince who rode
before you, Sun—whom
you outran,
piercing his caravan.

MARIANNE MOORE
Sun

The sun supplies the energy that drives the atmosphere's circulation. The most intense portion of the solar energy that reaches earth is visible as sunlight. (Photograph by Arjen Verkaik)

THREE

Radiation

THE SUN supplies heat to the earth-atmosphere system by electromagnetic radiation. Heat also leaves the earth-atmosphere system and travels off into space by means of electromagnetic radiation. Ultimately, this form of energy drives the circulation of the atmosphere and sustains weather. This chapter examines the properties of electromagnetic radiation: how radiation interacts with the components of the earth-atmosphere system, and its conversion to heat. We first consider the nature of electromagnetic radiation in general, and we then describe some of the specific properties of its various forms.

Electromagnetic Radiation

The world is bathed continually in **electromagnetic radiation**, so named because it exhibits both electrical and magnetic properties. All objects with a temperature above absolute zero (0 K) emit electromagnetic radiation. Most of the electromagnetic energy that reaches the earth originates in the sun. Only a small and insignificant amount comes from more distant stars and other celestial objects. The many types of electromagnetic energy together make up the **electromagnetic spectrum** illustrated in Figure 3.1. Light, which is visible radiation, represents only a very small portion of this spectrum. Other types of electromagnetic energy include radio waves, microwaves, infrared radiation, ultraviolet radiation, X rays, and gamma radiation.

Electromagnetic radiation travels in the form of waves, which are usually described in terms of wavelength or frequency. The length of a wave, or one **wavelength**, is the distance from wave crest to wave crest or from wave trough to wave trough, as shown in Figure 3.2. A wave's **frequency** is defined as the number of crests (or troughs) that pass a given point in a given period of time, usually 1 second. Passage of one complete wave is called a cycle, and a frequency of one cycle per second is termed one **hertz** (Hz). A wave's frequency is inversely proportional to its wavelength, that is, the higher the frequency, the shorter the wavelength. Some radio waves have frequencies of just a few hundred hertz and wavelengths hundreds of kilometers long. Gamma rays, in contrast, have frequencies as high as 10^{24} (a trillion trillion) Hz and wavelengths as short as 10^{-14} (a hundred trillionth) m.

Electromagnetic waves can travel through space as well as through gases, liquids, and solids. In a vacuum, all electromagnetic

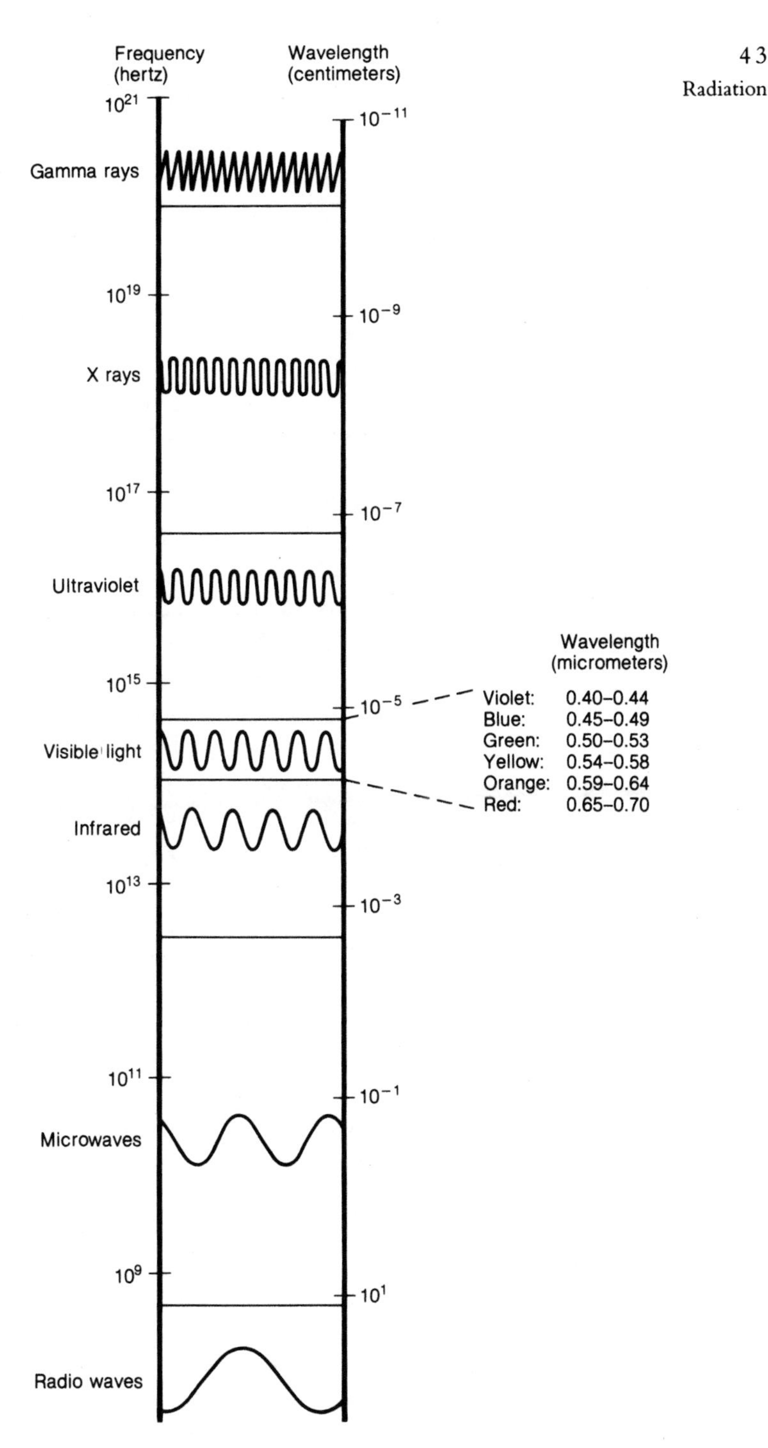

FIGURE 3.1

The electromagnetic spectrum consists of many types of radiation that are distinguished on the basis of wavelength, frequency, and energy level.

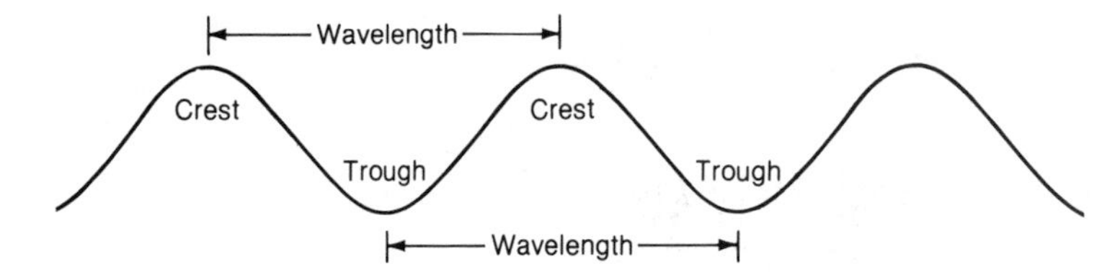

FIGURE 3.2

The wavelength of an electromagnetic wave is the distance between successive crests, or the distance between successive troughs.

waves travel at their maximum speed, 300,000 km (186,000 mi) per second. All forms of electromagnetic radiation slow down when passing through materials, their speed depending on the wavelength and the type of material. As electromagnetic radiation passes from one medium to another, it may be reflected or refracted (i.e., bent) at the interface. This happens, for example, when solar radiation strikes the ocean surface: some is reflected and some is bent upon penetrating the water. Electromagnetic radiation can also be absorbed, as when solar radiation is absorbed at the earth's surface and converted to heat.

Although the electromagnetic spectrum is continuous, we assign different names to different segments of this energy spectrum because we detect, measure, generate, and use different segments in different ways. All types of electromagnetic radiation are alike in every respect except wavelength, frequency, and energy. Wavelength and energy are dependent on frequency—the higher the frequency is, the higher the energy level. The different types of electromagnetic radiation do not begin or end at precise points along the spectrum. For example, red light shades into invisible infrared radiation (infrared, *below* red) on the frequency scale. At the other end of the visible portion of the electromagnetic spectrum, violet light shades into invisible ultraviolet radiation (ultraviolet, *beyond* violet).

At the low-energy (low-frequency, long-wavelength) end of the electromagnetic spectrum are **radio waves**. Their wavelength ranges from hundreds of kilometers to a small fraction of a centimeter, and their frequency ranges up to a billion hertz. FM (frequency modulation) radio waves, for example, range from 88 million to 108 million Hz, hence the familiar 88 and 108 at opposite ends of the FM radio dial. In Chapter 1, we saw how radio waves are reflected by the ionosphere. For more information on long distance transmission of radio signals, refer to the Special Topic, Radio Waves and the Ionosphere.

Next comes the **microwave** portion of the electromagnetic spectrum, which has wavelengths ranging from 300 mm to 0.1 mm. Some microwave frequencies are used for radio communication, in microwave ovens, and for tracking weather systems (radar).

Infrared radiation is between microwaves and visible light. We cannot see infrared radiation, but we can feel the heat generated if the infrared radiation is intense enough, as it is, for example, when emitted by a hot stove. Small amounts of infrared radiation are emitted by every known object or material, no matter how cold. Some naturally occurring infrared radiation comes directly from the sun, but most infrared radiation is converted by the earth and atmosphere from solar radiation.

At its uppermost frequencies, infrared radiation shades into the lowest frequency of visible radiation, red light. Wavelengths of **visible light** range from about 0.70 micrometers (μm) at the red end to about 0.40 μm at the violet end of the spectrum. (A **micrometer** is a millionth of a meter.) Visible light is essential for photosynthesis and for many activities of plants and animals. In plants, light coordinates the opening of buds and flowers and the dropping of leaves. In some animals, light regulates reproduction, hibernation, and migration.

Beyond visible light on the electromagnetic spectrum and in order of increasing frequency, increasing energy level, and decreasing wavelength, are **ultraviolet radiation**, **X rays**, and **gamma radiation**. All three of these types of radiant energy occur naturally, and all can be produced artificially. All have medical uses: ultraviolet radiation is a potent germicide; X rays are a powerful diagnostic tool; and both X rays and gamma radiation are used to treat cancer patients.

These three highly energetic types of radiation can be dangerous as well as useful. Ultraviolet rays can cause irreparable damage to the retinal cells of the eye. A person can be permanently blinded by looking at the sun during a partial solar eclipse without using a filter to block ultraviolet rays. Overexposure to ultraviolet rays, X rays, or gamma rays can also cause sterilization, cancer, genetic mutations, and tissue damage to a fetus.

Fortunately for us, the atmosphere blocks out most incoming ultraviolet radiation and virtually all X radiation and gamma radiation. Without this protective atmospheric shield, all life on earth would be quickly destroyed.

SPECIAL TOPIC

Radio waves and the ionosphere

While driving at night and listening to your car's AM radio, you may have been surprised to find a radio station from a city more than 1000 km (620 mi) away—one that during the day would produce only static.

Reception of distant radio signals at night is not at all unusual. Late night radio listeners in Illinois and Wisconsin routinely pick up WBZ in Boston (1030 on the AM dial) even though the station's transmitter is more than 1500 km (930 mi) away.

Appearance of distant radio signals at night and their subsequent disappearance during sunlit hours is due to interaction of radio waves with the ionosphere, the portion of atmosphere above the stratosphere. Radio signals are long electromagnetic waves that travel in straight paths in all directions away from their source transmitters and are received directly by a radio on the earth's surface. The curvature of the earth, however, limits the distance from the transmitter that direct radio waves can be picked up. Radio waves travel in straight paths and hence with distance the earth's surface gradually curves under and away from those direct signals. About 100 km (62 mi) from the transmitter is a quiet zone where radio waves are not received. At night, however, beyond the quiet zone, reception resumes, because radio waves are reflected back to the earth's surface from the upper region of the ionosphere.

Recall from Chapter 1 that the ionosphere is an area of ions and free electrons in the upper atmosphere. Highly energetic ultraviolet radiation and X rays emanating from the sun cause the atmosphere's molecular nitrogen (N_2) and molecular oxygen (O_2) to split into atoms, positively charged ions, and free electrons. The production rate of ions and electrons depends on two factors that vary with altitude: (1) the density of atoms and molecules available for ionization, which decreases rapidly with altitude, and (2) the intensity of solar radiation, which increases with altitude. Combined, these two factors maximize the concentration of ions and free electrons in the ionosphere.

Traditionally, the ionosphere is subdivided vertically into several layers. From lowest to highest, those are designated as D (60 to 85 km), E (85 to 140 km), and F (above 140 km). The original basis for this subdivision was the belief that each layer is a distinct zone of maximum electron density. Measurements by rockets and satellites, however, show that the ionosphere is not made up of separate layers; rather, electron density increases nearly continuously with altitude to a maximum at a level close to 300 km (186 mi). Hence, the D, E, and F labels used in this discussion merely refer to specific altitude regions within the ionosphere.

Radio waves that enter the ionosphere interact with the free electrons of the D, E, and F regions and are absorbed or reflected toward the earth's surface depending on the amount of solar radiation.

At night, in the absence of ionizing radiation, the D region virtually disappears as ions and electrons recombine into neutral particles. The recombination rate depends on air density, that is, the denser the air, the greater the likelihood of collision of particles and the capture of electrons by positive ions. In the E and F regions, the air is so rarefied that collisions are infrequent and, although the E region weakens, the two regions persist through the night. Radio waves that enter the F region are reflected toward the earth's surface. (Exceptions are radio waves that reach the F region at nearly a right angle; these waves pass through the F region and into space.) At night, because of F-region reflection, radio signals are propagated many hundreds of kilometers from their point of origin.

During daylight, with the return of ionizing radiation, the D region redevelops. Most radio waves that reach the D region are absorbed rather than reflected. Waves that do penetrate the D region are reflected by the overlying F region to the D region where they are absorbed. Consequently, during sunlit hours, radio-wave propagation is not aided by F region reflection.

In summary, then, radio signals travel greater distances at night because direct radio waves are reflected in the upper ionosphere.

Because the ionosphere is generated by ionizing radiation from the sun, any solar activity that disturbs the flow of this radiation may affect the ionosphere and, subsequently, radio communication on earth. Sudden ionospheric disturbances (SIDs), typically lasting 15 to 30 minutes, are the consequence of bursts of ultraviolet radiation from the sun. Ionization temporarily increases, causing radio transmissions to fade. The same solar activity responsible for auroral displays (Chapter 1) increases ionization and causes radio fade-out.

Not all radio signals are affected by the ionosphere. In fact, waves in the 8-mm to 20-m wavelength range pass unimpeded through the ionosphere. Television waves are in this category, and long distance transmission of television signals relies on artificial satellites orbiting synchronously with the rotating earth to reflect the signal from one place to another over great distances.

Solar Energy Input

The sun, our closest star, is a huge gaseous body composed almost entirely of hydrogen and helium and featuring internal temperatures that may exceed 15 million K. The ultimate source of solar energy is a continuous nuclear fusion reaction in the sun's interior. Simply put, in this reaction four hydrogen nuclei fuse to produce one helium nucleus. However, the mass of the four hydrogen nuclei is about 1 percent greater than the mass of one helium nucleus. This excess mass is converted into energy according to Einstein's equation, $E = MC^2$, whereby mass, M, is related to energy, E, and C is the speed of light (300,000 km per second). Because C^2 is such an enormous number, a very small mass is converted into a tremendous amount of energy. Some of the energy produced by nuclear fusion is used to bind the helium nucleus together. The rest of the energy is radiated and convected to the sun's surface and from there, energy is radiated off to space.

The visible surface of the sun, known as the **photosphere**, is much cooler than the sun's interior with temperatures near 6000 K. Relatively hot spots, called **faculae**, and relatively cool spots, called **sunspots**, dot the photosphere and, as we will see in Chapter 18, may affect solar energy output and the earth's climate. Above the photosphere is the **chromosphere** consisting of ionized hydrogen and helium at 4000 to 40,000 K. Beyond this is the outermost portion of the sun's atmosphere, the **corona**, a region of hot (1 to 2 million K) and high rarefied gases that extends millions of kilometers into space. Solar flares erupt from the photosphere into the corona and trigger auroral displays in the earth's atmosphere.

The solar radiation that reaches the earth is called **insolation**.* About 41 percent is visible as sunlight. The remainder consists primarily of near-infrared (46 percent) and ultraviolet (9 percent) radiation. Of the enormous energy radiated by the sun—about 5.6×10^{27} cal per minute—only about one half of one billionth of the total amount is intercepted by the earth.

We know from experience that the intensity of solar radiation varies significantly over the course of a year. In winter, the sun is lower in the sky, and the sun's rays are weaker than in summer. In winter, the days are also shorter than they are in summer. Even in the course of a single day, noticeable changes in insolation occur: The noon sun is brighter than the rising or setting sun.

It is evident, then, that the angle of the sun above the horizon

*Insolation refers to *in*coming *sol*ar radi*ation*.

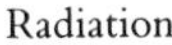

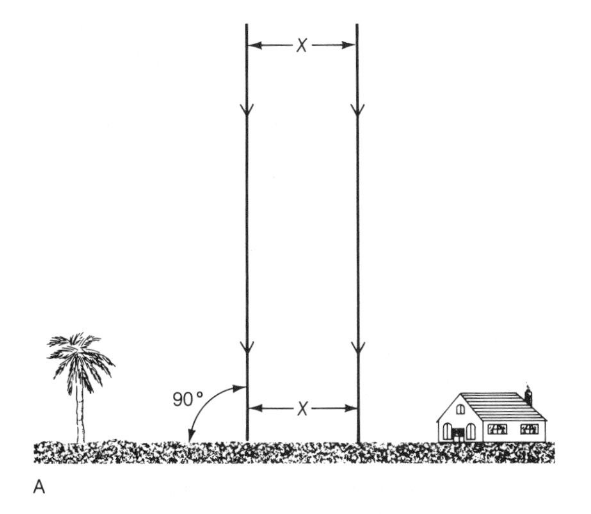

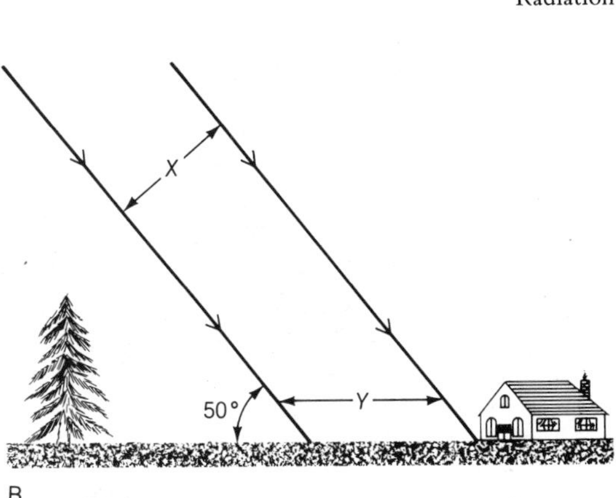

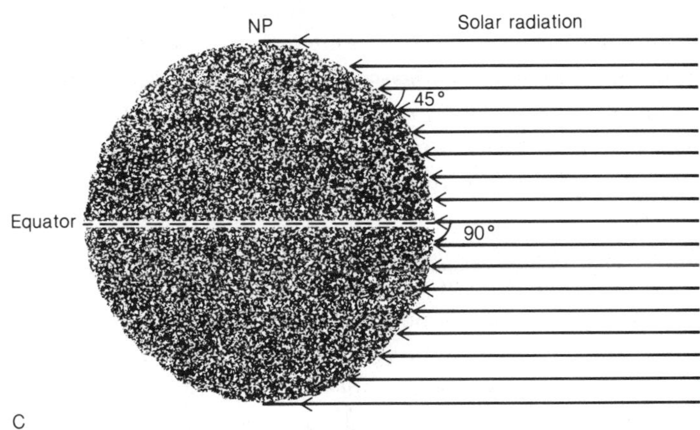

FIGURE 3.3

The intensity of solar radiation that strikes the earth's surface varies with solar altitude (the sun's angle above the horizon). (A) Solar radiation is most intense when the sun is directly overhead (solar altitude of 90 degrees). (B) With lower solar altitude, solar radiation received at the earth's surface is spread over an increased area (*Y* is greater than *X*) so that radiation intensity decreases. (C) On any given day, solar altitude will vary with latitude, because the curved surface of the earth intercepts parallel beams of solar radiation having uniform intensity. In this case, the most concentrated solar radiation is at the equator. Solar altitude and solar radiation intensity decrease with latitude.

influences the intensity of solar radiation received at the earth's surface. This angle is known as the **solar altitude**. The sun is, on average, 150 million km (93 million mi) away from the earth. The distance is so great that solar radiation reaches the earth as parallel beams of energy that are uniform in intensity. (The intensity of solar radiation might be expressed, for example, in units of calories striking an area of 1 cm^2 per minute.) As shown in Figure 3.3A, when the noon sun is directly overhead (solar altitude of 90 degrees), the sun's rays reaching the earth's surface are most concentrated and therefore most intense. As the sun moves lower in the sky and as the solar altitude decreases (Figure 3.3B), solar radiation is spread over a larger area and thus becomes less intense at the earth's surface. Because the earth presents a curved surface to incoming solar radiation (Figure 3.3C), the solar altitude and radiation intensity always vary with latitude.

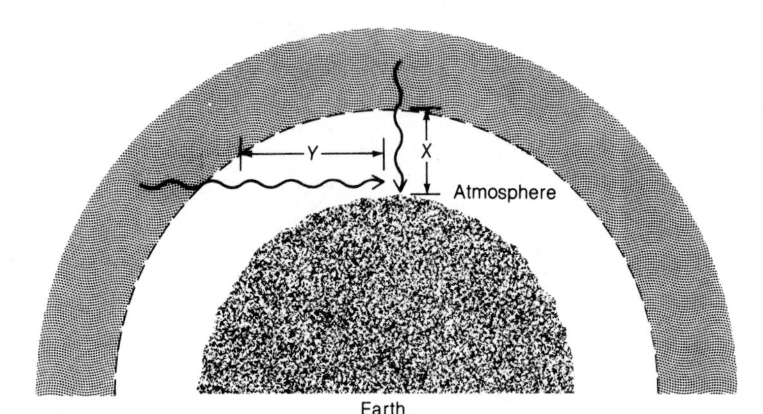

FIGURE 3.4
The path of solar radiation through the atmosphere lengthens as the solar altitude decreases, that is, as the sun moves lower in the sky. *X* is the solar radiation path length with high solar altitude, and *Y* is the path length with low solar altitude.

Solar altitude also influences the interaction between insolation and the atmosphere. With decreasing solar altitude, the path of the sun's rays through the atmosphere lengthens (Figure 3.4). As path length increases, there is greater interaction* between the solar radiation and the component gases and aerosols of the atmosphere. The result of this interaction is a weakening of the solar ray intensity. Thus, even with clear skies, the longer the path of the solar radiation, the less intense the radiation is when it reaches the earth's surface.

While solar altitude has an important influence on the intensity of solar radiation striking the earth's surface, the length of day also influences the total amount of radiant energy that is received. For some average intensity of solar radiation, those areas of the earth with longer daylight hours will receive more total calories of energy. Variations in both solar altitude and in length of day accompany the annual march of the seasons. Before examining these relationships, however, let us first consider the fundamental motions of the earth in space: rotation of the planet on its axis and the planet's revolution about the sun.

Earth's motions in space

The earth's axis is an imaginary line drawn through the north and south poles. Rotation of the earth on its axis accounts for day and night. Approximately once every 24 hours, the earth makes one complete rotation. At any given point in time, half of the planet is in darkness (night) and the other half is illuminated by the sun (day).

*The nature of this interaction (absorption, reflection, and scattering) is discussed later in this chapter.

In one year, which is actually 365.25 days, the earth makes one revolution about the sun in a slightly elliptical orbit (Figure 3.5). The earth's orbital eccentricity, that is, its departure from a circular orbit, is so slight that the earth-sun distance varies little throughout the year. The earth is closest to the sun (147 million km, or 91 million mi) on about January 3, and farthest from the sun (152 million km, or 94 million mi) on about July 4. These are the dates of **perihelion** and **aphelion**, respectively. In the Northern Hemisphere, the earth is therefore closest to the sun in winter and farthest from the sun in summer. The eccentricity of the earth's orbit does not, therefore, explain the seasons. What, then, does account for them?

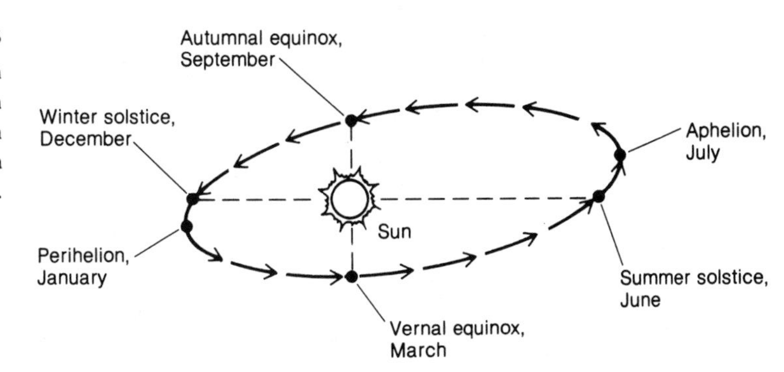

FIGURE 3.5

The earth's orbit is an ellipse, with the sun located at a focus. The earth is closest to the sun at perihelion (about January 3) and farthest from the sun at aphelion (about July 4).

The seasons

The seasons are attributed to the 23 degree 27 minute tilt of the earth's rotational axis to the plane defined by the earth's orbit (Figure 3.6). This tilt causes the earth's orientation to the sun to change continually as the planet revolves about the sun. The Northern Hemisphere thus leans away from the sun during winter and toward the sun in summer. At the same time that the Northern Hemisphere leans away from the sun, the Southern Hemisphere is leaning toward the sun. When it is winter in the Northern Hemisphere, it is therefore summer in the Southern Hemisphere, and vice versa. As the earth's orientation to the sun changes, so, too, do the solar altitude and length of day and hence the amount of solar radiation received. In the winter hemisphere, solar altitudes are lower, days are shorter, and there is less solar radiation. In the summer hemisphere, solar altitudes are higher, days are longer, and

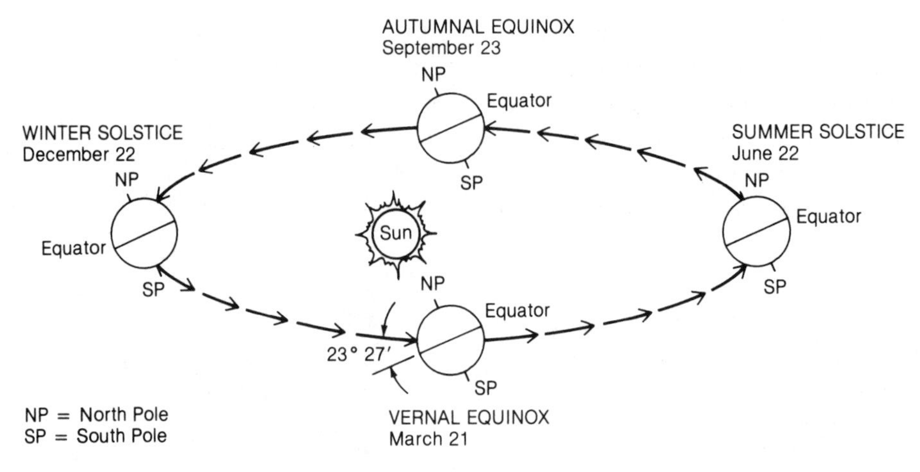

FIGURE 3.6

The seasons change because the earth's equatorial plane is inclined (at 23 degrees 27 minutes) to its orbital plane. The seasons given are for the Northern Hemisphere.

there is more solar radiation. Less solar radiation in winter than in summer means that winters are colder than summers.

If there were no axial tilt, the earth's rotational axis would be perpendicular to its orbital plane, and the earth would always have the same orientation to the sun. Without the earth's axial tilt, there would be no seasons.

How does the earth's orientation with respect to the sun change over the course of a year? Viewed from earth, the sun's most intense radiation (solar altitude of 90 degrees) moves from 23 degrees 27 minutes south* of the equator to 23 degrees 27 minutes north of the equator, and then back to 23 degrees 27 minutes south. On March 21 or 22,† and again on September 22 or 23,† the sun's noon position is directly over the equator. Day and night are of equal length (12 hours) everywhere (Figure 3.7). For this reason, these dates are labeled **equinoxes**, from the Latin for *equal nights*.

Following the equinoxes, the sun continues its journey toward maximum poleward locations, its **solstice** latitudes. On June 21 or 22,† the sun's noon rays are vertical at 23 degrees 27 minutes N, the **Tropic of Cancer**. As shown in Figure 3.8, daylight is continuous north of the **Arctic Circle** (66 degrees 33 minutes N), and absent south of the **Antarctic Circle** (66 degrees 33 minutes S). Elsewhere, days are longer than nights in the Northern Hemisphere

*Latitude is distance on the earth's surface measured in degrees and minutes north or south of the equator. The latitude of the equator is 0 degrees, and that of the poles is 90 degrees N and 90 degrees S; 1 degree equals 60 minutes.

†Because of leap years, these dates vary.

FIGURE 3.7

At autumnal and vernal equinoxes, insolation is maximum at the equator, and day and night are of equal length everywhere.

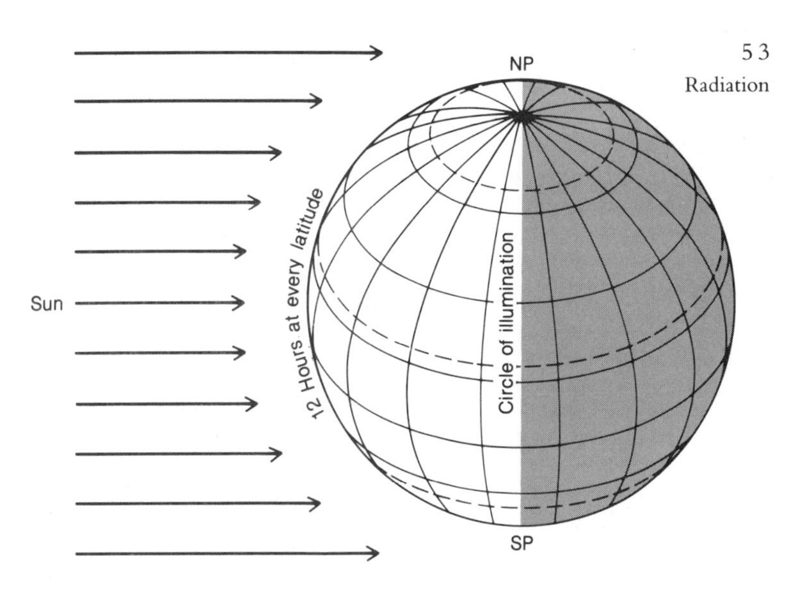

where it is summer, and days are shorter than nights in the Southern Hemisphere where it is winter.

On December 21 or 22, the noon sun is directly over 23 degrees 27 minutes S latitude, the **Tropic of Capricorn**, and the situation is reversed (Figure 3.8). Daylight is continuous south of the Antarctic Circle and absent north of the Arctic Circle. Elsewhere, nights

FIGURE 3.8

At the Northern Hemisphere winter solstice, maximum insolation is at 23 degrees 27 minutes S, and days are shorter than nights north of the equator. At the Northern Hemisphere summer solstice, maximum insolation is at 23 degrees 27 minutes N, and days are longer than nights north of the equator.

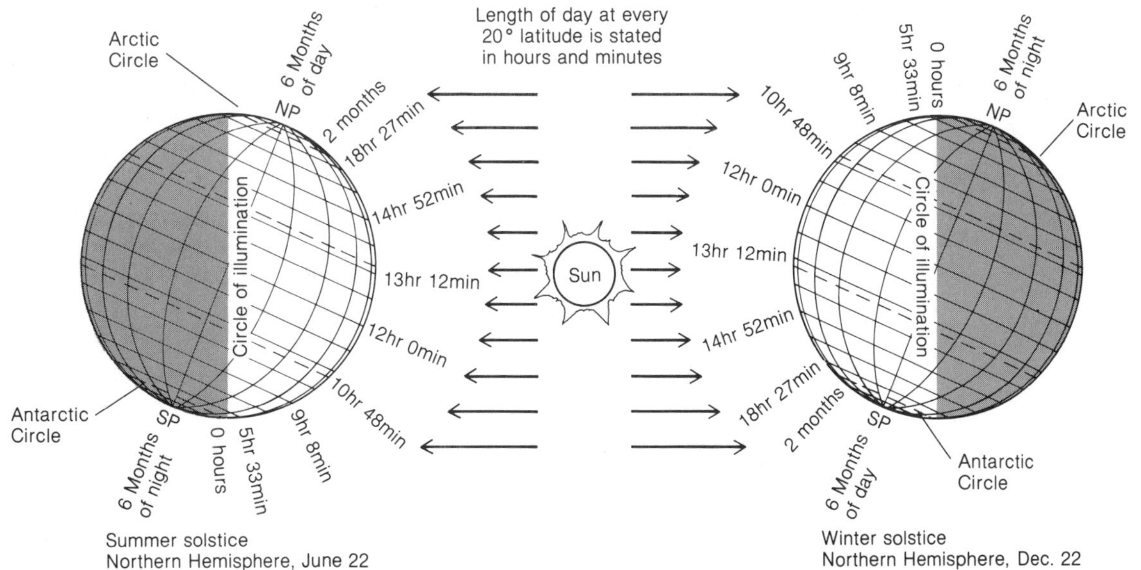

are longer than days in the Northern Hemisphere where it is winter, and days are longer than nights in the Southern Hemisphere where it is summer.

Solar radiation is of maximum intensity where the noon sun is directly overhead. North and south of this latitude, the intensity of solar radiation diminishes as the solar altitude decreases. For example, at the equinoxes, insolation is most intense at the equator at noon and decreases with latitude to zero at the poles. At the Northern Hemisphere summer solstice, solar radiation decreases to zero at the Antarctic Circle, and at the Northern Hemisphere winter solstice, solar radiation decreases to zero at the Arctic Circle.

The solar constant

For convenience of study, the rate of solar energy input into the earth-atmosphere system is expressed as the solar constant. The **solar constant** is defined as the rate at which solar radiation falls on

---

SPECIAL TOPIC

Why is the sky blue?

Visible light is composed of the entire spectrum of colors—red, orange, yellow, green, blue, and violet. If a beam of sunlight is passed through a glass prism and onto a screen, the light beam is refracted (bent) by the glass, and we see the color spectrum. The more energetic violet end of the spectrum is bent the most, and the less energetic red end is bent the least.

Refraction is not the only way that visible light is dispersed into its component colors. The scattering of sunlight produces the same effect and is responsible for the color of the sky. Scattering occurs in the atmosphere when particles in the atmosphere interact with light waves and send those light waves in random directions. If the radii of the scattering particles are much smaller than the wavelength of the scattered light, then the amount of scattering varies with wavelength. This is the case, for example, when visible light is scattered by the gas molecules composing the atmosphere. In a now classic experiment performed in 1881, Lord Rayleigh demonstrated that this scattering of light is inversely proportional to the fourth power of the wavelength. This means that violet light, at the short-wavelength end of the visible spectrum, is scattered much more than red light, which is at the

a surface located at the top of the atmosphere and positioned perpendicular to the sun's rays when the earth is at its mean distance from the sun. The "constant" designation is actually misleading, because solar energy input fluctuates by a few tenths of a percent over a year and exhibits some longer-term variations (discussed in Chapter 18). Nevertheless, we can approximate the solar constant as 2.00 cal per cm^2 per minute, or 1367 watts per m^2.*

Atmosphere-radiation interactions

As solar radiation travels through the atmosphere, it interacts with the atmospheric gases and aerosols. Some radiation—about 25 percent—is absorbed by oxygen, ozone, water vapor, ice crystals, water droplets, and dust particles. Through absorption, this radiant energy is converted to heat energy, and the air is warmed to some

*For the distinction between energy units and power units, see Appendix I.

---

long-wavelength end. As sunlight travels through the atmosphere, the various colors are therefore scattered selectively out of the solar beam: violet is scattered more than blue, blue more than green, green more than yellow, and so forth.

The dependence of scattering on wavelength implies that scattered sunlight is mostly violet in color. Why, then, does the sky appear blue rather than violet? The principal reason is that the human eye is more sensitive to blue light than to violet light, so the sky appears bluer to the human eye than it really is. Another factor contributing to the sky's blueness may be the dilution of violet light by all of the other scattered colors. Although the other colors are scattered less than violet, they tend to wash violet into blue.

All of this explains why the sky appears black at night, for without sunshine, no scattering occurs. If we were on the moon, which has a highly rarefied atmosphere, the sky would appear black and stars would be visible even in bright sunshine: With a rarefied atmosphere, scattering is inconsequential.

A different effect occurs when radiation is scattered by particles that have radii approaching or exceeding the wavelength of the radiation being scattered. In these instances, the scattering is not wavelength dependent. Instead, radiation of all wavelengths is scattered equally. The particles composing clouds (tiny ice crystals or water droplets or both) and most atmospheric aerosols are sufficiently large to scatter sunlight in this way. For this reason, clouds appear white, and when the atmosphere contains a considerable accumulation of aerosols, the entire sky is a hazy white.

extent. For example, warming in the upper stratosphere is due to absorption of solar ultraviolet radiation by ozone and oxygen. Some solar energy is reflected—primarily by clouds—back into space, and some is scattered, or dispersed, in all directions. In fact, as described in the Special Topic, Why Is the Sky Blue?, the scattering by nitrogen and oxygen molecules of the blue-violet portion of visible sunlight is what gives the sky its blue color. Solar radiation that is not reflected, or scattered back into space, or absorbed by gases and aerosols, then reaches the earth's surface. Solar radiation that is transmitted directly through the atmosphere to the earth's surface constitutes **direct insolation**. This is augmented by **diffuse insolation**: solar radiation that is scattered or reflected to the surface or both.

The solar radiation, both direct and diffuse, that reaches the earth's surface is either reflected or absorbed by it. The amount reflected varies from one place to another, depending on the nature of the surface. Light surfaces have higher reflectivity for solar radiation than do dark surfaces. For example, skiers who have been sunburned on the slopes know that snow is very reflective. The measure of surface reflectivity is the surface's **albedo**. Fresh-fallen snow typically has an albedo of between 75 and 95 percent, that is, 75 to 95 percent of the solar radiation that strikes a snow cover is reflected. At the other extreme, the albedo of a dark surface, such as a blacktopped road or a green forest, may be as low as 5 percent. The portion of insolation that is not reflected is absorbed by the earth's surface and converted to heat. The earth's surface is thus warmed. The albedo of some common surfaces are listed in Table 3.1.

The solar radiation budget

Recent satellite measurements indicate that about 30 percent of the incoming solar radiation is reflected, or scattered, or both, by the **earth-atmosphere system** and is thus lost to space. This reflected light, which has been seen as "earthshine" by U.S. astronauts on the moon, is known as the earth's **planetary albedo**. The remaining 70 percent of the solar radiation intercepted by the earth-atmosphere system is absorbed, converted to heat, and ultimately is involved in the functioning of the environment.

Of the total solar radiation intercepted by the earth-atmosphere system, only 19 percent is absorbed directly by the atmosphere. In other words, the atmosphere is relatively transparent to solar radia-

FIGURE 3.9

In a satellite view of North America, note how the highly absorptive oceans appear somewhat darker than the more reflective continents. Note also how highly reflective the clouds are. The large white swirl on the east coast is the cloud pattern associated with an intense storm. (NOAA photograph)

tion. The remaining 51 percent of solar radiation (30 percent was reflected) is absorbed by the earth's surface, primarily because of the low albedo of the ocean waters covering nearly three quarters of the globe. The low reflectivity of the oceans (and hence their high absorption of solar radiation) is evident in the satellite image shown in Figure 3.9. Because of their low reflectivity, the oceans appear

TABLE 3.1

Reflectivity (or Albedo) of Some Common Surface Types for Visible Solar Radiation

SURFACE	ALBEDO (PERCENT REFLECTED)
Grass	**16–26**
Deciduous forest	**15–20**
Coniferous forest	**5–15**
Crops	**15–25**
Tundra	**15–20**
Desert	**25–30**
Blacktopped road	**5–10**
Sea ice	**30–40**
Fresh snow	**75–95**
Old snow	**40–70**
Glacier ice	**20–40**
Water (high sun)	**3–10**
Water (low sun)	**10–100**

somewhat darker than the adjacent continental landmass. The distribution of solar radiation in the earth-atmosphere system is summarized in Figure 3.10.

The earth's surface is the principal recipient of solar heating, and the earth's surface, in turn, continuously radiates heat back to the atmosphere, which eventually radiates it back into space. The earth's surface is thus the main source of heat for the atmosphere. This is evident in the vertical temperature profile of the troposphere. Air is normally warmest close to the earth's surface, and air temperature drops with increasing altitude, that is, away from the main source of heat.

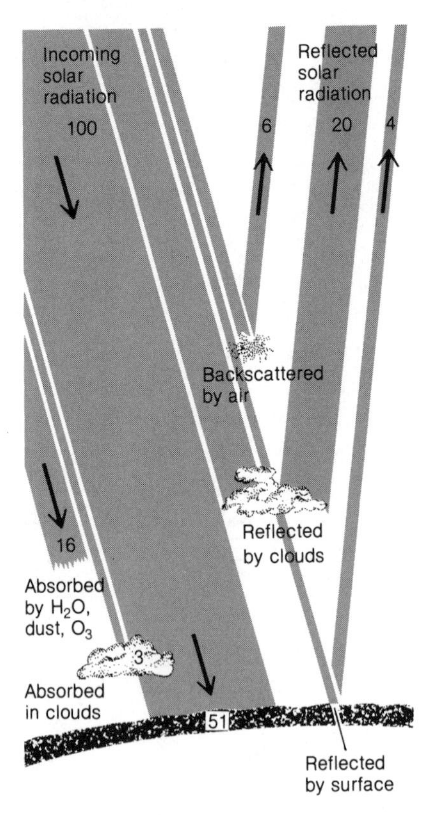

FIGURE 3.10

The disposition of solar radiation is shown as it interacts with the atmosphere and the earth's surface. Note that of the 100 units of solar energy entering the atmosphere, 16 units are absorbed by water vapor, dust, and ozone, 3 units are absorbed by clouds, 51 units are absorbed at the earth's surface, 6 units are scattered back into space by the atmosphere, 20 units are reflected back into space by clouds, and 4 units are reflected into space by the earth's surface. Hence, 16 + 3 + 51 + 6+ 20 + 4 = 100. (From Ingersoll, A. P. "The Atmosphere." ***Scientific American*** **249, no. 3 (1983):164. Copyright © 1983 by Scientific American, Inc. All rights reserved.)**

The Infrared Response

If solar radiation were continually absorbed by the earth-atmosphere system without any compensating flow of heat out of the system, the air temperature would rise steadily. In reality, the average global air temperature changes little from year to year, because an equal amount of heat leaves the earth-atmosphere system, mainly in the form of infrared radiation. While solar radiation is being supplied only to the illuminated portion of the globe, infrared radiation is emitted ceaselessly, both day and night, by the entire earth-atmosphere system.

Why does the earth-atmosphere system emit radiation in the infrared portion of the electromagnetic spectrum? The explanation is found in Wien's displacement law, one of the physical laws that describe the behavior of electromagnetic radiation. (These physical laws are summarized in the Mathematical Note at the end of the chapter.)

Wien's displacement law holds that the higher the surface temperature of a radiating object (such as a hot stove, the earth-atmosphere system, or the sun), the shorter the wavelengths of its radiated energy. Hot objects thus emit radiation of relatively short wavelength, while cold objects emit radiation of relatively long wavelength. The effective radiating temperature of the sun is about 6000 K (11,000 °F). The electromagnetic waves emitted by the sun are therefore relatively short. The most intense solar radiation, which is within the visible portion of the electromagnetic spectrum, is emitted by waves of about 0.50 μm in length (green). Figure 3.11 shows how the intensity of solar radiation varies by wavelength. The much cooler surface of the earth, radiating at an

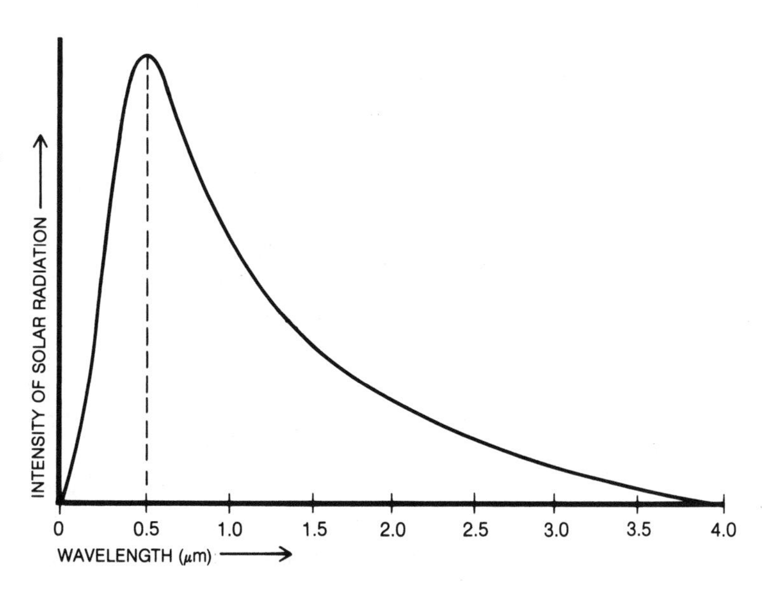

FIGURE 3.11

The intensity of solar radiation is a function of wavelength. Peak energy intensity is emitted by waves 0.50 μm long, in the green region of the visible spectrum.

average temperature of about 288 K (59 °F), emits longer waves, as shown in Figure 3.12. In fact, the earth-atmosphere system emits infrared radiation of maximum intensity at a wavelength of about 10 μm. The earth-atmosphere system thus responds to solar radiant heating by emitting comparatively long-wave infrared radiation.

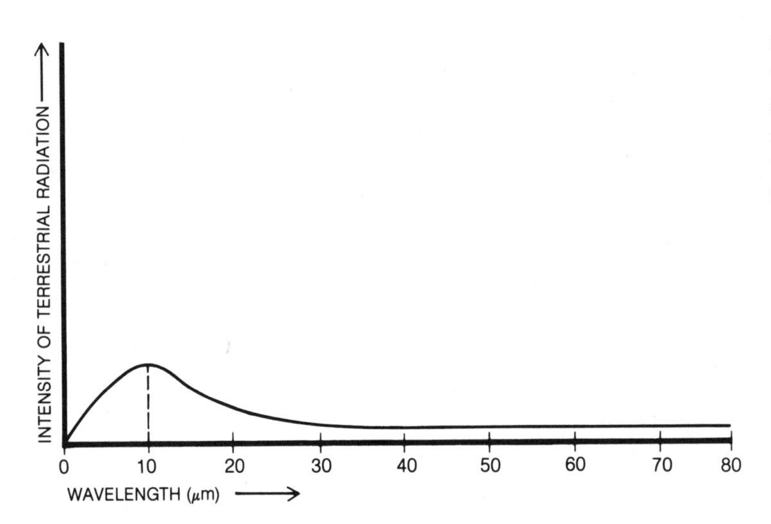

FIGURE 3.12

The intensity of terrestrial radiation is a function of wavelength. Peak energy intensity is emitted by waves about 10 μm long, in the infrared region of the spectrum.

The area under the curve in Figure 3.11 represents the total radiational energy emitted at all wavelengths by the sun. The area under the curve in Figure 3.12 represents the total radiational energy emitted by the earth-atmosphere system at all wavelengths. Note, however, that the vertical scales in these two figures are not comparable because the sun emits considerably more radiational energy than does the earth-atmosphere system. This contrast is described by the **Stefan-Boltzmann law:** the total energy radiated by an object is proportional to the fourth power of its absolute temperature (T^4). The sun radiates at a much higher temperature than does the earth-atmosphere system. The Stefan-Boltzmann law predicts that the sun's energy output per square meter is more than 10^5 times that of the earth-atmosphere system. As solar radiation travels through space, its intensity diminishes rapidly so that by the time solar radiation reaches the earth, it has the same intensity as the radiation emitted to space by the earth-atmosphere system.

Because solar radiation and terrestrial radiation peak in different portions of the electromagnetic spectrum, they have different properties, and their interactions with the atmosphere also differ. As noted earlier, atmospheric constituents absorb only about 19 percent of the incoming solar radiation. In contrast, the atmosphere absorbs a relatively large percentage of the infrared radiation emitted by the earth's surface. The atmosphere, in turn, reradiates some infrared radiation back to the earth's surface (see Figure 3.13). This reradiation by the atmosphere slows the escape of heat into space, and causes the lower atmosphere to have a higher average temperature than the upper atmosphere. The lower atmosphere is thus more hospitable to life. In fact, thanks to this reradiation, the earth's average surface temperature is more than 30 °C (54 °F) higher than it would otherwise be.

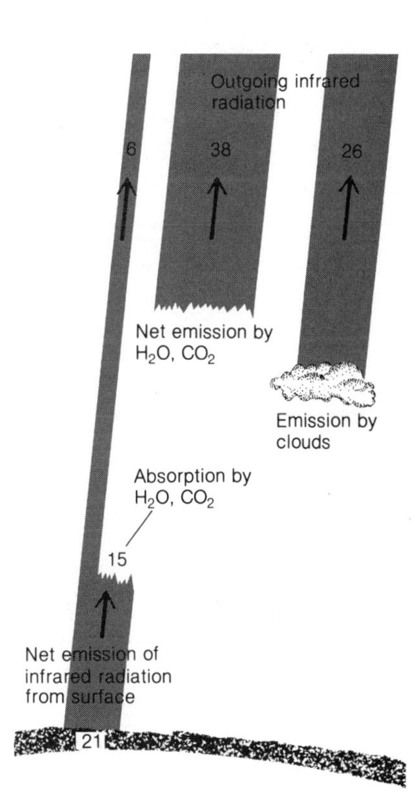

FIGURE 3.13

Infrared radiation interacts with components of the atmosphere. The numbers are based on 100 units of incident solar radiation. Note absorption by CO_2 and H_2O. The 70 units of solar energy that were absorbed by the earth-atmosphere system are emitted to space as infrared radiation. (From Ingersoll, A. P. "The Atmosphere." ***Scientific American*** **249, no. 3 (1983):164. Copyright © 1983 by Scientific American, Inc. All rights reserved.)**

Atmospheric gases, primarily water vapor and carbon dioxide and to a lesser extent ozone, methane, and nitrous oxide, absorb infrared radiation and thus impede its loss to space. The percentage of infrared radiation absorbed varies with wavelength, as shown in Figure 3.14. Interestingly, the percentage absorbed is very low in the wavelength bands around 8 and 11 μm, which includes the wavelength of peak infrared intensity. It is through these

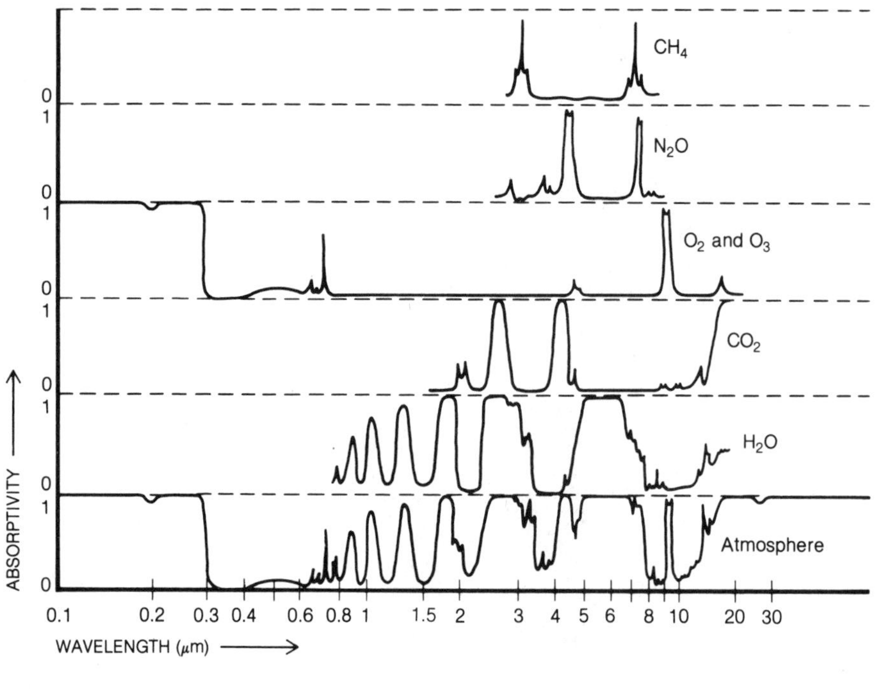

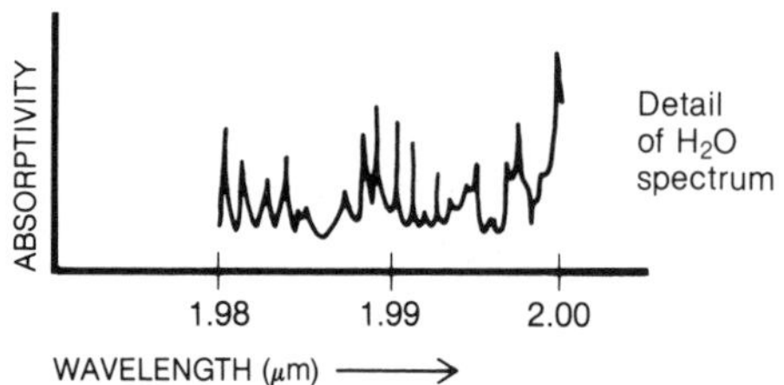

FIGURE 3.14

Absorption of radiation by certain components of the atmosphere and by the atmosphere as a whole is shown as a function of wavelength. Absorptivity is the fraction of the radiation absorbed and ranges from 0 to 1 (100 percent absorption). Absorptivity is very low or near zero in atmospheric windows. Note the infrared "windows" near 8 and 10 μm. (From Fleagle, R. G., and J. A. Businger. *An Introduction to Atmospheric Physics*. New York: Academic Press, 1963, p. 153.)

atmospheric windows, wavelength bands in which there is little or no absorption of radiation, that most heat eventually escapes into space as infrared radiation.

Like the earth's atmosphere, glass transmits solar radiation readily but slows the transmission of infrared radiation. Greenhouses, where plants are stored or grown, are designed to take advantage of this property of glass by being constructed almost entirely of glass panes. Because of the similarity in absorption and reradiation properties, the behavior of atmospheric gases is often referred to as the "**greenhouse effect**." The analogy is not, however, strictly correct. The reduced transmission of infrared radiation by glass is only part of the reason why greenhouses retain internal heat. The principal reason is that the greenhouse cuts the conductive and convective loss of heat by acting as a shelter from the wind. Because reference to the "greenhouse effect" is so common in discussions of atmospheric heat balance, we refer to the term, but always in quotation marks. Some meteorologists argue that the term **atmospheric effect** would be more appropriate in describing this behavior of the atmosphere.

To illustrate the "greenhouse effect" in the atmosphere, compare the typical summer weather of the southwestern United States with that of the Gulf of Mexico coast. Both areas are at the same latitude and therefore receive about the same intensity of sunlight. Both areas consequently have typical afternoon temperatures above 30 °C (86 °F). At night, however, air temperatures in the two areas may differ markedly. In the Southwest there is relatively little water vapor in the air to impede the escape of infrared radiation. Heat is readily lost to space, and the earth-atmosphere system cools rapidly. Surface air temperatures may fall below 15 °C (59 °F) by dawn. Along the Gulf Coast, however, the air is more humid, and thus absorbs more infrared radiation. Because a portion of this heat is reradiated back toward the earth's surface, the temperature may only fall into the 20s on the Celsius scale (the 70s, Fahrenheit).

Clouds, which are composed largely of water droplets, also produce a "greenhouse effect." Nights are therefore usually colder when the sky is clear than when the sky is cloud covered.

Radiation Measurement

The **Eppley pyranometer** is the standard instrument for measuring the intensity of solar radiation striking a horizontal surface. The instrument consists of a sensor enclosed in a transparent hemisphere (Figure 3.15) that transmits total (direct plus diffuse) short-wave

FIGURE 3.15

The Eppley pyranometer is the standard instrument used for solar radiation measurements. (Science Associates photograph)

(less than 3.5 μm wavelength) insolation. The sensor is a disc consisting of alternating black and white wedge-shaped segments that form a starlike pattern. The black wedges are highly absorptive and the white wedges are highly reflective of solar radiation. Differences in absorptivity and reflectivity mean that the black and white portions of the sensor respond differently to the same inten-

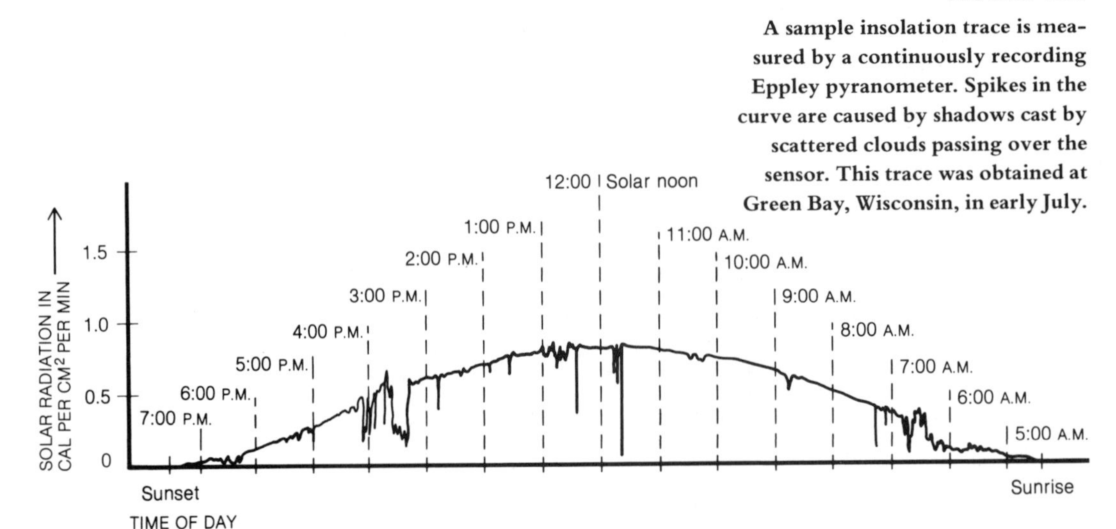

FIGURE 3.16

A sample insolation trace is measured by a continuously recording Eppley pyranometer. Spikes in the curve are caused by shadows cast by scattered clouds passing over the sensor. This trace was obtained at Green Bay, Wisconsin, in early July.

MATHEMATICAL NOTE

Radiation laws

Several laws describe properties of the electromagnetic radiation emitted by a *perfect radiator*. A perfect radiator, usually called a *black body*, is a hypothetical object that absorbs all of the radiation that strikes it, that is, a perfect radiator neither reflects nor transmits any radiation that strikes it. In reality, no perfect radiators exist, but the sun and the earth-atmosphere system radiate *approximately* as black bodies would radiate. We can therefore apply black body radiation laws to solar and terrestrial radiation, with some qualifications. Here, in brief, are the basic black body radiation laws.

***Kirchhoff's law* holds that a perfect absorber of a given wavelength is also a perfect emitter of radiation at that same wavelength. A black body thus emits, as well as absorbs, all incident radiation at all**

sity of solar radiation. The temperature difference between the black and white segments is calibrated in terms of radiation flux (calories per square centimeter per minute, for example). An Eppley pyranometer may be linked electronically to a pen recorder, which traces a continuous record of insolation (Figure 3.16), or the instrument's output signal may be recorded on a magnetic tape cassette for processing and storage.

Special care must be taken in mounting and maintaining an Eppley pyranometer. The instrument should be situated where it will not be affected by shadows, by any highly reflective surfaces nearby, or by other sources of radiation. The glass bulb must also be kept clean and dry.

Conclusions

The amount of incoming solar radiation is balanced by the amount of infrared radiation emitted to space by the earth-atmosphere system. Emission of infrared radiation causes cooling, and absorption of solar radiation causes warming. Within the earth-atmosphere system, however, the rates of radiational heating and cooling are not the same everywhere. The reasons for these energy imbalances and their weather implications are covered in the next chapter.

---

wavelengths. In general, for all objects the efficiency of radiation absorption, called *absorptivity*, equals the efficiency of radiation emission, called *emissivity*. A good absorber is therefore a good emitter, and a poor absorber is a poor emitter. The emissivity and absorptivity of a black body are both 100 percent.

The *Stefan-Boltzmann law* states that the rate at which a black body radiates energy across all wavelengths (called *emittance*, E) is directly proportional to the fourth power of the *absolute temperature*, T (measured in kelvins), of the radiating body. The mathematical statement of this law is

$$E = \sigma T^4$$

where σ is the Stefan-Boltzmann constant equal to 5.67×10^{-8} watts/m^2 K^4.

***Planck's law* states that the rate at which radiation is emitted by a black body depends on the absolute temperature (K) of the body and the specific wavelength of the radiation.**

***Wien's displacement law* holds that the wavelength at which a black body emits the maximum intensity of radiation, λmax, is inversely proportional to the absolute temperature, T, of the black body, that is,**

$$\lambda max = C/T$$

where C is the constant of proportionality equal to 2898 if λmax is expressed in micrometers (μm).

SUMMARY STATEMENTS

- Visible and infrared radiation represent small segments of a continuous spectrum of electromagnetic radiation.
- Forms of electromagnetic radiation are distinguished on the basis of wavelength, frequency, and energy level.
- As a result of the earth's elliptical orbit, spherical shape, and tilted rotational axis, insolation is unevenly distributed over the earth's surface and changes in the course of a year.
- Solar radiation not absorbed by the atmosphere or returned to space via reflection and scattering reaches the earth's surface, where it is either reflected or absorbed, depending on the surface albedo.
- The earth's surface and atmosphere emit infrared radiation; the surface is the primary heat source for the lower atmosphere.
- Water vapor, carbon dioxide, and other gases absorb and reradiate infrared energy back toward the earth's surface, thereby moderating the temperature of the lower atmosphere. This so-called "greenhouse effect" is also produced by clouds.

KEY WORDS

electromagnetic radiation
electromagnetic spectrum
wavelength
frequency
hertz
radio waves
microwave
infrared radiation
visible light
micrometer
ultraviolet radiation
X rays
gamma radiation
photosphere
faculae
sunspots
chromosphere
corona
insolation
solar altitude
perihelion
aphelion
equinoxes
solstice
Tropic of Cancer
Arctic Circle
Antarctic Circle
Tropic of Capricorn
solar constant
direct insolation
diffuse insolation
albedo
earth-atmosphere system
planetary albedo
Wien's displacement law
Stefan-Boltzmann law
atmospheric windows
"greenhouse effect"
atmospheric effect
Eppley pyranometer

REVIEW QUESTIONS

1 What is the relationship between the wavelength and frequency of electromagnetic radiation?

2 Describe how energy level varies within the electromagnetic spectrum.

3 In the Northern Hemisphere, we are closer to the sun during the winter than during the summer. Why, then, is winter colder than summer?

4 In middle latitudes, days are longer than nights between the spring and fall equinoxes. Why?

5 What is the significance of the Tropic of Cancer and the Tropic of Capricorn?

6 What is the significance of the Arctic Circle and the Antarctic Circle?

7 How and why does the solar altitude affect the intensity of solar radiation received at the earth's surface?

8 Define in detail the solar constant.

9 Absorption of solar radiation by constituents of the atmosphere is an energy conversion process. Explain what is meant by this statement.

10 Why is the sky blue and why are clouds white?

11 In your locality, are there seasonal changes in surface albedos? What are the implications of these changes for air temperatures?

12 Although insolation reaches its maximum intensity at noon, surface air temperatures typically do not reach a maximum until several hours later. Please explain.

13 How does the relatively high albedo of a snow cover influence air temperature?

14 The atmosphere is relatively transparent to solar radiation. Elaborate on this statement.

15 For the entire globe, why must incoming solar radiation balance outgoing infrared radiation? What would be the implications for global climate if this energy balance did not prevail?

16 What is the relationship between the temperature of a radiating object and the wavelength of emitted radiation?

17 What is meant by an atmospheric *window* for infrared radiation?

18 What is the significance of the "greenhouse effect" for temperatures at the earth's surface?

19 Are air temperatures likely to be lower on a clear night or a cloudy night? Explain your choice.

20 Explain why the night-to-day temperature difference is typically much greater in a warm and dry locality (such as the Nevada desert) than in a warm and humid locality (such as New Orleans).

POINTS TO PONDER

1 Radiation intensity decreases as the inverse square of the distance traversed. If the mean earth-sun distance were *three times* what it is today, the solar constant would be reduced to what fraction of its present magnitude?

2 If there were no atmosphere, the total solar radiation received per day at the earth's surface would reach a maximum in summer at the South Pole. Explain why the maximum is at the pole and not at the solstice latitude.

3 Speculate on how the albedo of the moon compares with the planetary albedo of the earth.

4 What natural or human-related activities might cause a change in the earth's planetary albedo? What are the implications of such changes for global climate?

5 What basic assumptions are made when black body radiation laws are applied to the sun or the earth-atmosphere system?

PROJECTS

1 Determine whether solar radiation data are available for your locality. If available, find out how variations in cloud cover influence the radiation.

2 For your area, determine whether there is a relationship between cloud cover at night and the early morning low temperature. For reasons that are discussed in the next chapter, limit your observations to days when the wind is light or calm.

SELECTED READINGS

Byers, H. R. *General Meteorology.* New York: McGraw-Hill, 1974. 461 pp. *An advanced quantitative text featuring a section on radiation in the atmosphere.*

Walsh, J. E. "Snow Cover and Atmospheric Variability." *American Scientist* 72, no. 1 (1984):50–57. *The influence of regional snow cover on weather and climate.*

Then the sea
And heaven rolled as one
and from the two
Came fresh transfigurings
of freshest blue.

WALLACE STEVENS
Sea Surface Full of Clouds

Imbalances in heating and cooling trigger processes that redistribute heat within the earth-atmosphere system. Evaporation of water at the earth's surface and its subsequent condensation as clouds is a crucial heat transfer process. (Photograph by Mike Brisson)

FOUR

Heat Imbalances and Weather

WEATHER is not a capricious act of nature. It is a response to unequal radiant heating and to unequal radiant cooling within the earth-atmosphere system. Circulation of the atmosphere redistributes heat from warmer areas to cooler areas.

Heat Imbalance: The Atmosphere Versus the Earth's Surface

If we measured the annual rate of radiant heating and the annual rate of radiant warming for the earth's atmosphere, and if we then did the same for just the earth's surface, we would find an important heat imbalance. In the atmosphere, the rate of cooling due to infrared emission is greater than the rate of warming due to the absorption of solar radiation. At the earth's surface, however, the rate of warming due to the absorption of solar radiation exceeds the rate of cooling due to infrared emission. This heat imbalance is summarized in Figure 4.1, which combines the distribution of

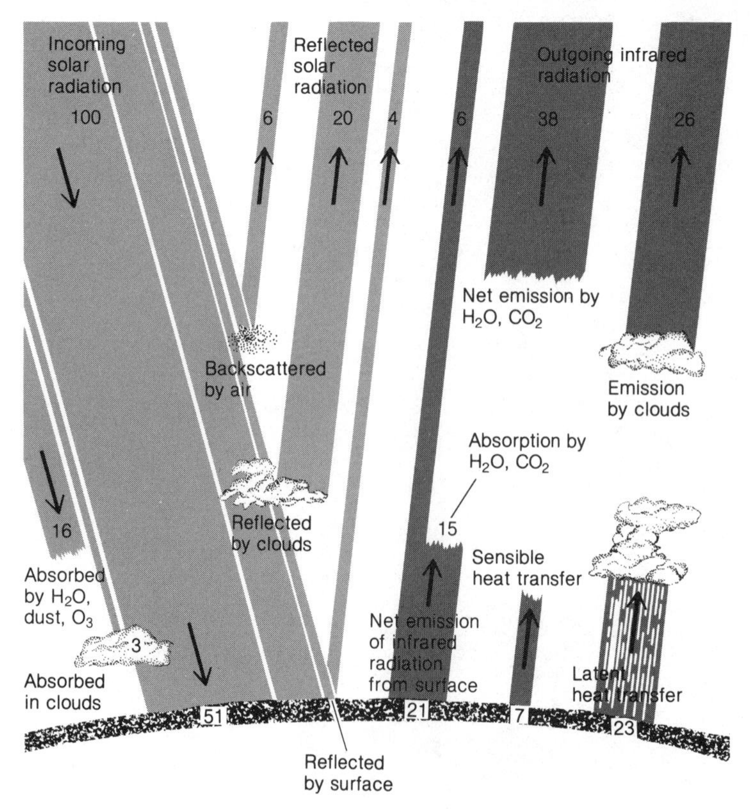

FIGURE 4.1

The distribution of 100 units of incoming solar radiation and of infrared radiation on a global scale indicates excess heating at the earth's surface. This excess heat is transferred to the atmosphere via sensible and latent heating. (From Ingersoll, A. P. "The Atmosphere." *Scientific American* 249, no. 3 (1983): 164. Copyright © 1983 by Scientific American, Inc. All rights reserved.)

the incoming solar radiation with that of the outgoing infrared radiation.

The radiant energy distribution implies a net cooling of the atmosphere and a net warming of the earth's surface, yet the atmosphere is in reality *not* cooling relative to the earth's surface. A transfer of heat must be occurring on a global scale from the earth's surface to the atmosphere. How is this heat transfer accomplished? The flow of heat energy from the earth's surface to the atmosphere is brought about by sensible heat transfer (about 23 percent) and by latent heat transfer (about 77 percent). In Figure 4.1, 30 units of heat energy are transferred from the earth's surface to the atmosphere: 7 by sensible heating (7/30 or 23 percent) and 23 by latent heating (23/30 or 77 percent).

Sensible heating

The term *sensible* is used to describe this heat-transfer mechanism because heat redistribution brought about by sensible heating can be monitored, or "sensed," as temperature changes. **Sensible heat transfer** involves two processes: conduction and convection. Heat is conducted from the earth to the atmosphere through direct contact between the cool air of the atmosphere and the earth's warm surface. As the lowest portion of the atmosphere is heated by the ground, the warm air becomes lighter than the cooler surrounding air and begins to rise. The surrounding air then sweeps in, replacing the warmer air and transferring heat by convective currents from the surface of the earth to the atmosphere.

Latent heating

Latent heat transfer is the movement of heat from one place to another as water changes phase. It arises from the differences in molecular activity represented by the three physical phases of water. In the solid phase (ice), the molecules are relatively inactive and vibrate about fixed locations. In the liquid phase, the molecules move with greater freedom, so that liquid water takes the shape of its container. In the vapor phase, the molecules exhibit maximum activity and diffuse readily throughout the entire volume of a container. A change of phase is thus a change in energy level, or level of molecular activity. This in turn means that during a phase change, heat must be either added to or released from the substance undergoing the change.

As an illustration, follow the fate of an ice cube as it is heated (Figure 4.2). The specific heat of ice is about 0.5 cal per gram per degree Celsius, which means that 0.5 cal of heat must be supplied for every degree of temperature rise. Once 0 °C (32 °F) is reached, an additional 80 cal of heat per gram (called the **heat of fusion**, or the **heat of melting**) must be supplied to break the forces that bind molecules in the ice phase. The temperature of the water and ice remains at 0 °C until all of the ice is melted. The specific heat of

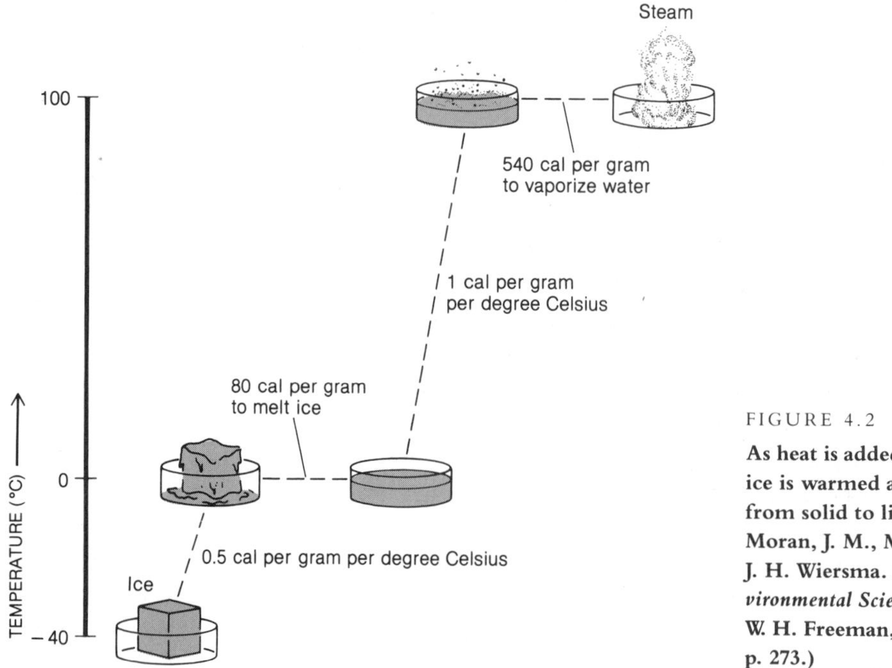

FIGURE 4.2

As heat is added to an ice cube, the ice is warmed and changes phase from solid to liquid to vapor. (From Moran, J. M., M. D. Morgan, and J. H. Wiersma. *Introduction to Environmental Science.* San Francisco: W. H. Freeman, Copyright © 1980, p. 273.)

liquid water is 1.0 cal per gram per Celsius degree, so from that point on, only 1.0 cal is needed to raise the temperature of 1 g of liquid water 1 °C. A phase change from liquid water to water vapor requires the addition of much more heat than the phase change from ice to liquid water. The **heat of vaporization** is temperature dependent. It varies from about 600 cal per gram at 0 °C to 540 cal per gram at 100 °C (212 °F). If the sequence is reversed—that is, if the water vapor is cooled until it becomes liquid and then ice—the temperature drops, or phase changes take place, or both occur as equivalent amounts of heat are lost.

Applying this phase-change concept to the earth-atmosphere system illustrates latent heat transfer. As the earth's surface absorbs solar radiation, some of the heat is used to vaporize water from oceans, lakes, rivers, soil, and vegetation. Conversely, when water changes from the vapor phase back to the liquid phase of water droplets or to the solid phase of ice crystals, and forms clouds in the atmosphere, heat is released, thus warming the atmosphere. The heat required for vaporization is supplied at the earth's surface, and that same heat is subsequently released to the atmosphere during cloud development.

The formation of familiar convective clouds combines sensible heat transfer with latent heat transfer to channel heat from the earth's surface into the atmosphere. Rising columns of relatively warm air in convective currents often produce **cumulus clouds**, which resemble puffs of cotton floating in the sky (Figure 4.3). These clouds are sometimes referred to as "fair weather" cumulus, because they are seldom accompanied by rain or snow. On the

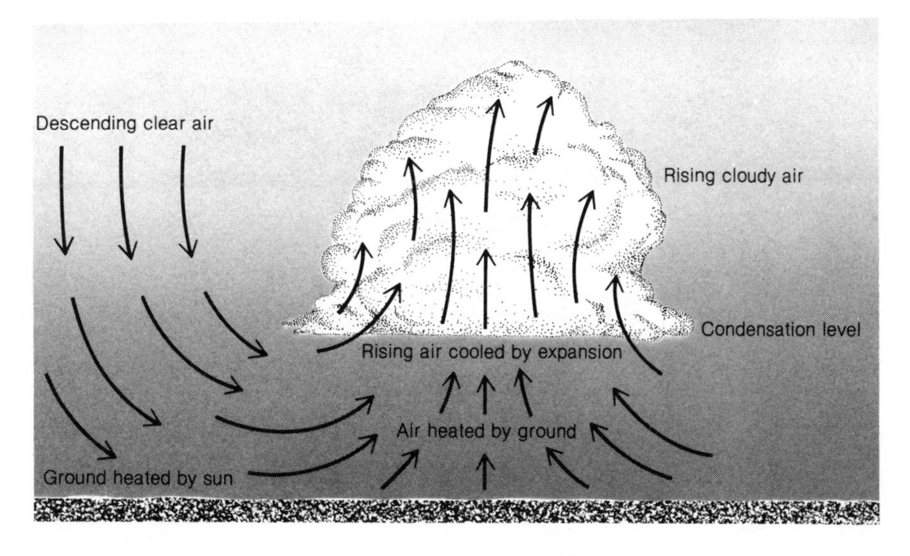

FIGURE 4.3

Cloud development transports excess heat at the earth's surface into the atmosphere via conduction, convection currents, and latent heat transfer. (After Neiburger, M., J. G. Edinger, and W. D. Bonner. *Understanding Our Atmospheric Environment.* San Francisco: W. H. Freeman, 1973, p. 6. Copyright © 1973. All rights reserved.)

other hand, when certain atmospheric conditions develop (see Chapter 13), convective currents surge to great altitudes, and cumulus clouds billow upward to form towering **cumulonimbus clouds**, known also as thunderstorm clouds (Figure 4.4). In retrospect, then, an important heat transfer was occurring when that thunderstorm washed out your ball game, or sent you scurrying home from the beach last summer.

Since ocean waters cover 70.8 percent of the earth's surface, it is not surprising that latent heat transfer is more significant on a

FIGURE 4.4

When convection currents extend deep into the atmosphere, cumulus clouds may billow upward to form thunderclouds. (Photograph by Mike Brisson)

global scale than sensible heat transfer. As shown in Table 4.1, the ratio of sensible heating to latent heating, called the **Bowen ratio**, varies from one locality to another and depends on the amount of surface moisture. The ratio of sensible heating to latent heating is about 1:10 for oceans, and about 2:1 for a relatively dry area like the desert interior of Australia. The drier the land surface, the less important latent heating becomes and the more important sensible heating becomes.

We noted earlier that nearly all weather is confined to the troposphere, the lowest subdivision of the atmosphere. This implies that sensible and latent heat transfer operate primarily within the troposphere. Heat and temperature distributions above the troposphere are determined by radiative processes alone. Also important is the fact that in some places heat transport is from the atmosphere to the earth's surface, which is the reverse of the average global situation. This occurs, for example, when warm air blows over cold, snow-covered ground or when warm air blows over a relatively cool ocean or lake surface.

Heat Imbalance: Latitudinal Variation

On a global scale, imbalances in radiant heating and radiant cooling occur not only vertically, between the earth's surface and the atmosphere, but also horizontally, with latitude. Because the earth is nearly spherical, parallel beams of incoming solar radiation strike lower latitudes more directly than higher latitudes (Figure 3.3). In higher latitudes, solar radiation spreads over a greater area and is less intense than in lower latitudes. The output of infrared radiation varies less with latitude. Consequently, over the course of a year, the rate of infrared cooling in higher latitudes exceeds the rate of warming caused by the absorption of solar radiation. In lower latitudes, however, the reverse is true: the solar radiation warming rate is greater than the infrared cooling rate.

For many years, scientists assumed that the 38-degree latitude circles marked the division between net radiative cooling and net

TABLE 4.1

Bowen Ratio* for Various Geographical Areas

GEOGRAPHICAL AREA	BOWEN RATIO
Europe	**0.62**
Asia	**1.14**
North America	**0.74**
South America	**0.56**
Africa	**1.61**
Australia	**2.18**
Atlantic Ocean	**0.11**
Indian Ocean	**0.09**
Pacific Ocean	**0.10**
All land	**0.96**
All oceans	**0.11**

Source: Sellers, W. D. *Physical Climatology*. Chicago: The University of Chicago Press, 1965, p. 105

*The Bowen ratio is the ratio of heat used for conduction and convection (sensible heating) to heat used for vaporization of water (latent heating).

radiative warming. Recent satellite measurements, shown in Figure 4.5, suggest that the division lies closer to the 30-degree latitude circles.

The implication is that higher latitudes experience net cooling while tropical latitudes are sites of net warming, but in fact, lower latitudes do not become progressively warmer relative to higher latitudes. Heat must therefore be transported from the tropics to the

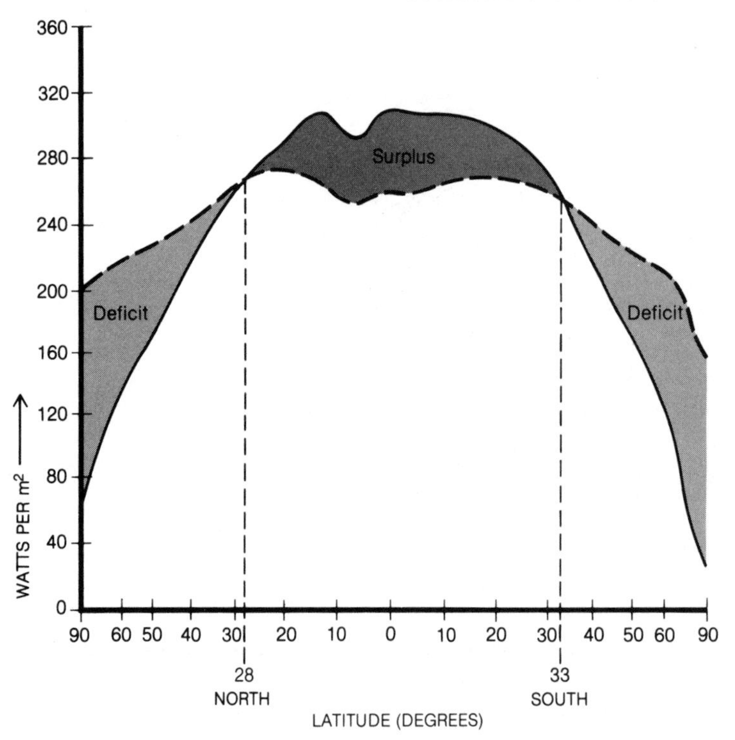

FIGURE 4.5

Variation by latitude of absorbed solar radiation (solid line) and outgoing long-wave radiation (dashed line) was obtained by satellite measurement from June 1974 to February 1978. Radiative cooling and warming rates are equal at about 28 degrees N and 33 degrees S. (From Winston, J. S., et al. *Earth-Atmosphere Radiation Budget Analyses Derived From NOAA Satellite Data June 1974–February 1978.* Washington, D.C.: NOAA Meteorological Satellite Laboratory, 1979.)

middle and high latitudes. This **poleward heat transport** is brought about primarily by the exchange of air masses*: warm air masses that form in lower latitudes flow poleward and are replaced by cold air masses that flow toward the equator from higher latitudes. In this way, sensible heat is transported poleward.

Air mass exchange accounts for about half of the total poleward heat transport. The remaining poleward heat transport is due to storms (about 30 percent) and to ocean currents (about 20 percent). In low latitudes, water that evaporates from the warm ocean surface is drawn into the circulation of a developing storm. As the storm travels poleward, some of that water vapor condenses as clouds,

*An air mass is a huge volume of air, covering hundreds of thousands of square kilometers, that is relatively uniform in temperature and water vapor content.

thereby releasing latent heat. The heat of vaporization acquired in lower latitudes is thus transported into higher latitudes. In addition, cold ocean currents drift toward the tropics, while warm ocean currents drift poleward. In the tropics, a relatively cool ocean current is a "heat sink," that is, heat is conducted from the air to the cold ocean water. In middle and high latitudes, on the other hand, a relatively warm ocean current is a heat source for the atmosphere, that is, heat is conducted from sea to air.

Weather: Response to Heat Imbalances

Weather redistributes heat within the earth-atmosphere system. The flow of cold and warm air masses from one place to another, cloud formation, storms, and convection currents all counter imbalances in radiant heating and cooling. A cause-effect chain thus operates in the earth-atmosphere system starting with the sun as the prime energy source and resulting in weather.

We have seen that within the earth-atmosphere system, some solar radiant energy is converted to heat through absorption, and some of this heat is emitted back to space as infrared radiation from the earth's surface and atmosphere. Some solar energy is also converted to **kinetic energy**, the energy of motion, in the circulation of the atmosphere. Kinetic energy is manifested in winds, in convection currents, and in the exchange of air masses. Circulation (weather) systems do not last indefinitely, however. The kinetic energy of atmospheric circulation ultimately is dissipated as frictional heat when the winds blow against the earth's surface. This heat is in turn emitted into space as infrared radiation. Figure 4.6 is a schematic diagram of the major energy transformations operating in the earth-atmosphere system. (Technology also taps solar energy and transforms it into heat and electricity for space heating and cooling. See the Special Topic, Solar Power.)

Conversion of solar radiation into heat, kinetic energy, or electricity follows the **law of energy conservation**. Simply stated, energy is neither created nor destroyed but can change from one form to another. The law of energy conservation also means that on a global scale, incoming solar radiation must balance outgoing infrared radiation.

In summary, the sun drives the atmosphere in two respects: (1) Imbalances in solar heating spur atmospheric circulation (weather), which redistributes heat. (2) Solar energy is the source of kinetic energy, which is manifest in the circulation of the atmosphere.

SPECIAL TOPIC

Solar power

The total amount of solar energy that falls yearly on U.S. lands is about 600 times greater than the amount of energy the United States consumes during the same period. Assuming that solar energy could be collected and sold at average electricity rates, more than $5000 worth of energy shines on the roof of a small house (93 m^2, or 1000 ft^2) in the course of a year.

The collection of solar radiation, however, presents some problems, because unlike fossil fuels (coal, natural gas, oil) or nuclear fuels, sunlight is diffuse and reduced by cloudiness. In fact, the solar power reaching U.S. soil averages annually only about 190 watts per m^2. (For reference, recall that the solar constant is about 1367 watts per square meter.) The area with the greatest potential for solar power is in the southwestern United States where cloudiness is minimal and insolation is intense.

To tap solar radiation, we use *solar collectors*, panels that collect and concentrate the sun's rays. At present, solar collectors are generally used for space or water heating in small buildings such as homes, apartments, schools, and small businesses. In the near future, solar-powered air conditioners may cool these same places. Solar panels do not produce extremely high temperatures, and these devices are therefore not adequate for most industrial purposes.

Solar collectors are framed panels of glass that trap solar energy. The sunlight is usually allowed to pass through two layers of glass before it is absorbed by a blackened (low albedo) metal plate. The absorbed heat energy is then transferred from the absorbing plate to either air or a liquid, which is conveyed by fans or pumps to wherever the heat is needed. Figure 1 is an example of the type of solar panels currently in use. Solar collectors typically capture 30 percent to 50 percent of the solar energy that reaches them.

All technologies based on the collection of solar radiation are limited by the fact that such radiation is not continuous. Not only does the sun not shine at night, but its intensity

FIGURE 1

Solar collectors on the roof of the visitors' center at Mt. Rushmore, South Dakota. (U.S. Department of Energy photograph from Honeywell)

varies seasonally, geographically, and with cloud cover. Seasonal changes are especially troublesome in middle and high latitudes. For example, in eastern Washington state, a relatively sunny rain-shadow locality, the average annual insolation is 194 watts per m^2, but the march of monthly mean values varies from a low of 50 to a high of 343 watts per m^2, a sevenfold difference. This seasonal variation underscores the desirability of developing long-term (summer to winter) storage systems for such localities. In tropical latitudes, seasonal differentials in insolation are considerably less.

To reduce the variability of insolation due to changes in solar altitude, solar collectors are tilted, and some are designed to track the sun so the collectors are always perpendicular to the solar beam. The advantage of tilted and tracking collectors over collectors that are fixed and horizontal depends on the average cloud cover and latitude of the site. An optimal situation occurs in winter at a midlatitude locality favored by clear skies. There, tilted and tracking solar collectors could double the amount of absorbed insolation.

Scientists and engineers are currently studying ways to convert solar energy to electricity on a large scale. In one conversion system—called a *power tower system*—computer-controlled mirrors, or *heliostats*, track the sun and focus its energy on a single heat-collection point on a tower. Concentrated sunlight in these systems can produce temperatures up to 480 °C (900 °F), temperatures high enough to convert water into high-pressure steam for driving turbine generators. A large solar-powered system of this sort requires a land area filled with tracking mirrors (Figure 2). In fact, the generation of 50 megawatts of electricity (enough for 15,000 homes) would require about 1.6 km^2 (1 mi^2) of land, and would render the land unusable for other purposes. An experimental power tower is now operating in California's Mojave Desert. Southern California

FIGURE 2

A power tower stands amid a field of mirrors. Computer-controlled mirrors (called heliostats) track the sun and focus solar radiation on a heat exchanger in the tower. This radiation is converted to heat which produces steam in a boiler. The steam is then used to drive a turbine and generate electricity. This facility in Barstow, California, is operated by Southern California Edison. (U.S. Department of Energy photograph.)

(continued)

Edison's Solar One facility consists of an array of 1818 heliostats and can generate 10 megawatts of electricity.

An alternative to solar-driven turbines is photovoltaic cells, that is, *solar cells*. These cells convert a small amount of visible sunlight directly into electricity. Two very thin layers of specially treated metal crystals (silicon metal is commonly used) are sandwiched together to form a solar cell. A schematic drawing of a silicon solar cell is presented in Figure 3. When sunlight strikes the junction between the two crystal layers, a direct electrical current flows through wires connecting the layers. About 40 solar cells must be linked to produce the 12-volt potential of a car battery.

Usable quantities of electrical energy are produced when many small solar cells are connected to form a solar electricity-generating panel. An example is shown in Figure 4.

FIGURE 4

Solar electricity-generating panel composed of solar cells. (U.S. Department of Energy photograph)

Despite the simplicity of solar cells, they still are extremely expensive because of costly manufacturing processes. New technology is rapidly lowering the cost, but the cells remain two to three times more expensive than conventional electricity-generating systems.

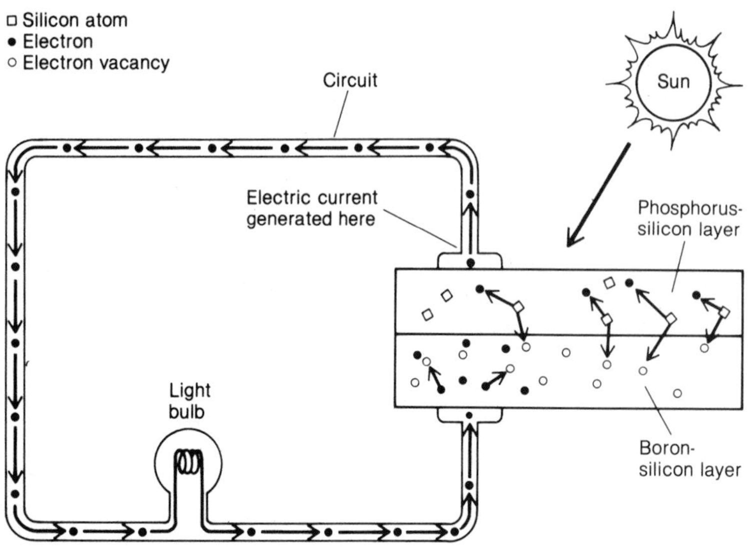

FIGURE 3

Schematic drawing of a silicon solar cell. Sunlight frees electrons from silicon atoms, and the electrons then flow through the circuit to generate an electrical current. (From Federal Energy Administration)

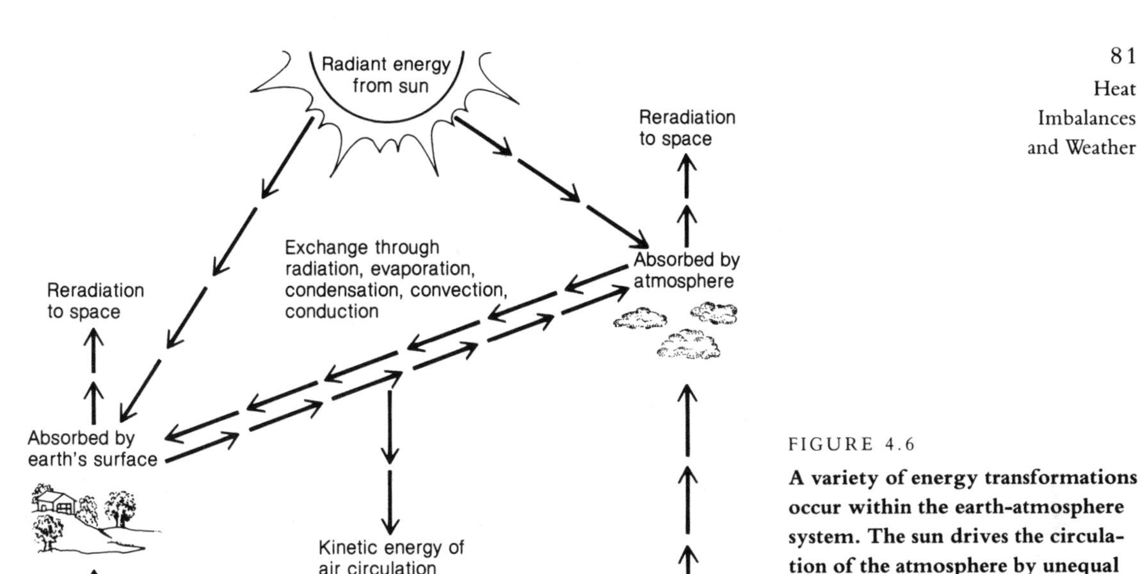

FIGURE 4.6

A variety of energy transformations occur within the earth-atmosphere system. The sun drives the circulation of the atmosphere by unequal heating. (From Miller, A., et al. *Elements of Meteorology*, 4th edition. Columbus, Oh.: C. E. Merrill, 1983, p. 48. Used with permission of the publisher.)

Seasonal contrasts

Because heat redistribution varies with the season, weather across North America changes throughout the year. When horizontal temperature gradients are great across the continent, the weather tends to be energetic: storm systems are large and intense, winds are strong, and the weather is changeable. Such weather is typical of winter, when it is not unusual for daily temperatures in the southern United States to be more than 30 °C (54 °F) warmer than temperatures across southern Canada.

In contrast, when minimal temperature variation occurs across the continent, as in summer, the weather tends to be more tranquil, and large-scale weather systems are generally weak and not well defined. Nevertheless, summer weather may sometimes be active, and intense heating of the ground by the hot summer sun often triggers strong convection and the development of thunderstorms. Some of these weather systems spawn destructive hail, forceful winds, and heavy rains. These systems are, however, usually shorter lived and more localized than winter storms.

Scales of weather systems

Although the atmosphere is a continuous fluid,* for the convenience of study we view atmospheric circulation as being composed of discrete weather systems, grouped according to the spatial scale over which they operate. The large-scale wind systems of the globe (polar easterlies, westerlies, and trade winds) are features of the **global-scale circulation. Synoptic-scale weather** is continental or oceanic. Migrating storms, air masses, and fronts, usually highlighted on television and on newspaper weather maps, are important components of this scale. **Mesoscale systems** include thunderstorms and sea and lake breezes—phenomena that may influence the weather in one section of a city and leave adjacent areas unaffected. The circulation of air within a small environment—for example, a tornado—represents the smallest spatial subdivision of atmospheric motion, or **microscale weather**. In later chapters, weather systems of all scales are examined in some detail.

Each small-scale weather system is part of and dependent on larger scale weather systems. For example, extreme nocturnal radiant cooling, the rapid nighttime loss of infrared radiation, requires a synoptic weather pattern that favors clear skies and light winds or calm. At the microscale level, such weather conditions may be accompanied by frost formation on your tomatoes.

Regardless of the scale of operation, all atmospheric circulation redistributes heat from one place to another. Atmospheric circulation, together with local conditions of radiant heating and radiant cooling, thus control the temperature of air.

Variation of Air Temperature

Air temperature is variable. Temperature fluctuates from night to day, from one day to the next, with the seasons, and from one place to another. Our discussion of the causes of weather provides some insight into why temperature varies. Air temperature is regulated by local radiative conditions and by the movement of air masses. Although in reality these two factors work in concert, we first consider them separately for the purposes of study.

*"Fluid" can refer to either liquids or gases. In this case, *fluid* refers to the mixture of gases composing the atmosphere.

Radiative controls

Local radiative conditions influencing air temperature include (1) the intensity of incoming solar radiation, as determined by solar altitude and thus by time of year and time of day, (2) the cloud cover, since cloudiness affects the local radiation balance both during the day and at night, and (3) the nature of the surface cover, because surface characteristics determine the albedo and the percentage of absorbed solar radiation used for sensible heating and latent heating. Air temperature is usually higher in June than in January, during the day than at night, under clear afternoon skies rather than under cloudy afternoon skies, when the ground is bare instead of snow covered, and when the ground is dry rather than wet.

As an illustration of local radiative controls, consider the influence of ground characteristics on air temperature. All other factors being equal, in response to the same insolation, the air over a dry surface warms up more than it would if that surface were moist. When the surface is dry, the absorbed solar radiation is used primarily for sensible heating of the air (conduction and convection of heat from the surface and into the air). Hence, the Bowen ratio is relatively high. On the other hand, when the surface is moist, much of the absorbed radiational energy is used to evaporate water and the Bowen ratio is lower. This suggests a simple means of reducing summer air conditioning needs: where water is in plentiful supply, shallow pools of water placed on rooftops reduce solar heating of the building.

Because of the heat required for vaporization of snow, a snow cover also reduces sensible heating of the overlying air. In addition, the relatively high albedo of snow substantially decreases the amount of available solar radiation that is absorbed at the surface of the snow cover and converted to heat. Consequently, a snow cover lowers the day's maximum temperature. Snow is also an excellent radiator of infrared and hence, nocturnal radiational cooling is extreme where the ground is snow covered—especially when skies are clear and winds are calm. On such nights, the air temperature near the surface may drop 10 °C (18 °F) or more lower than it would if the ground were bare of snow. The net effect of a snow cover, then, is to reduce the day's average temperature.

The temperature cycle, called the march of mean monthly temperatures, clearly reflects the systematic variation in incoming solar radiation in the course of a year. In the latitude belt between the tropics of Cancer and Capricorn, solar radiation varies little in the course of a year, and the variation in average monthly air temperature during the year exhibits no significant seasonal contrasts (Figure 4.7). In the middle latitudes, solar radiation has a pronounced annual maximum and minimum. In the high latitudes poleward of the Arctic and Antarctic circles, the seasonal difference

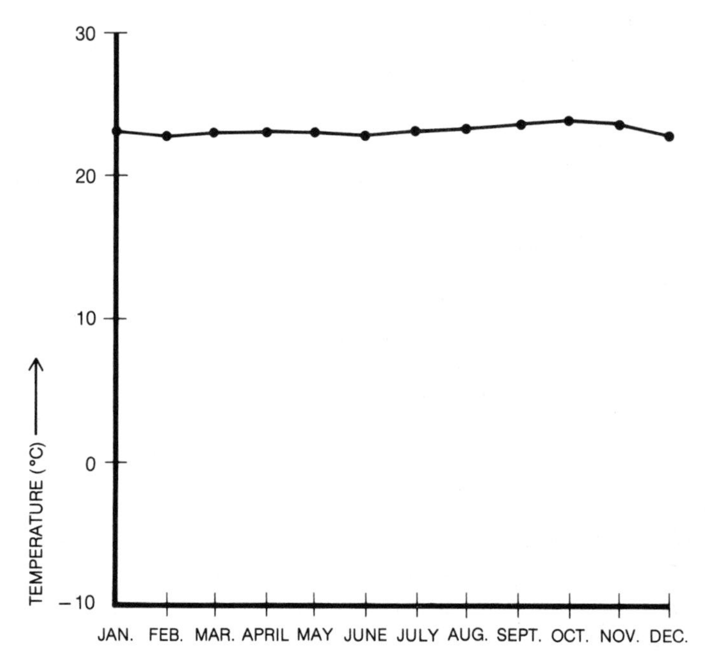

FIGURE 4.7
The march of monthly mean temperatures for Clevelandia, Amazon Basin (4 degrees N, 52 degrees W). At this near-equatorial location, little temperature change occurs during a year because of minimal variation in solar radiation and length of daylight. (World Meteorological Organization data)

in solar radiation is extreme, varying from zero in winter to a maximum in summer. This marked periodicity of incident solar radiation outside of the tropics accounts for the distinct winter-summer temperature contrasts observed in middle and high latitudes (Figure 4.8).

In middle and high latitudes, the march of mean monthly temperatures lags behind the monthly variation in insolation, so that times of maximum and minimum solar radiation typically do not coincide with the warmest and coldest months of the year, respectively. This is because the troposphere's thermal structure takes time to adjust to the changing solar energy input. The warmest portion of the year thus occurs typically about a month after the summer solstice, and the coldest part of the year usually occurs about a month after the winter solstice. In the United States, the temperature cycle lags the solar cycle by an average of 27 days. However, in coastal localities with a strong maritime influence (Florida,

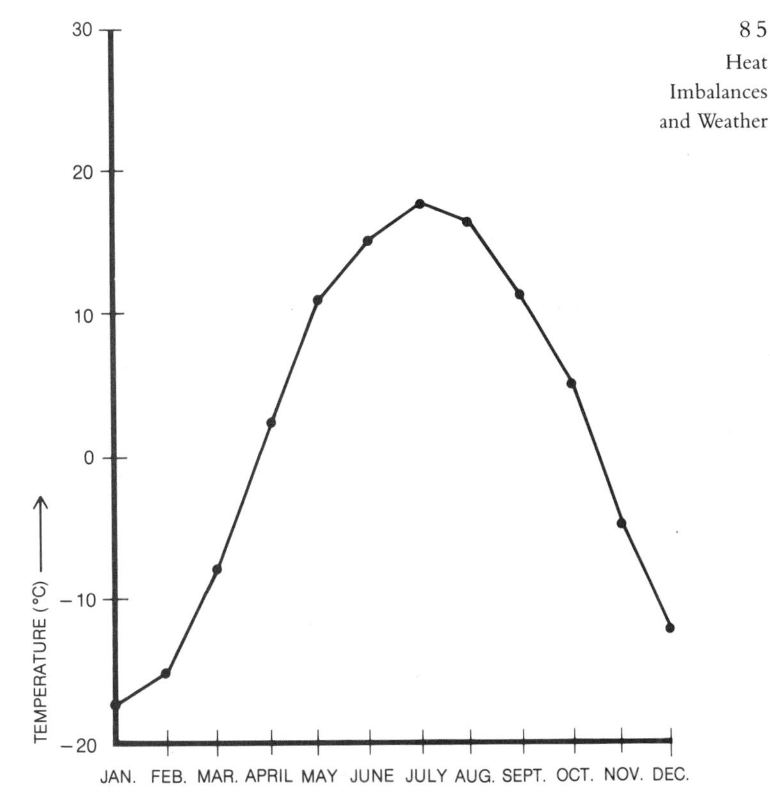

FIGURE 4.8
The march of mean monthly temperatures for Regina, Saskatchewan, Canada (50 degrees 26 minutes N, 104 degrees 40 minutes W). The temperature regime at this highly continental midlatitude locality is strongly influenced by the seasonal variation of incoming solar radiation. (Atmospheric Environment Service [of Canada] data)

the shoreline of New England, and California, for example), the average lag time is up to 36 days.

To some extent, the day-night (diurnal) variation in insolation is reflected in the variation of air temperature over the course of a 24-hour day. An example of a midlatitude diurnal temperature variation is presented in Figure 4.9 along with plots of incoming solar radiation and outgoing infrared radiation. In this example, the day's lowest temperature occurs near sunrise as the culmination of 12 hours of radiative cooling. The day's highest temperature is recorded in midafternoon, even though the peak insolation occurs at noon. The time required for the temperature of the lower troposphere to adjust to the day-night variation in radiation accounts for the lag of several hours between radiative input and the response of air temperature.

This time lag explains why in summer the greatest risk of sunburn is at noon and not during the warmest time of day. Solar

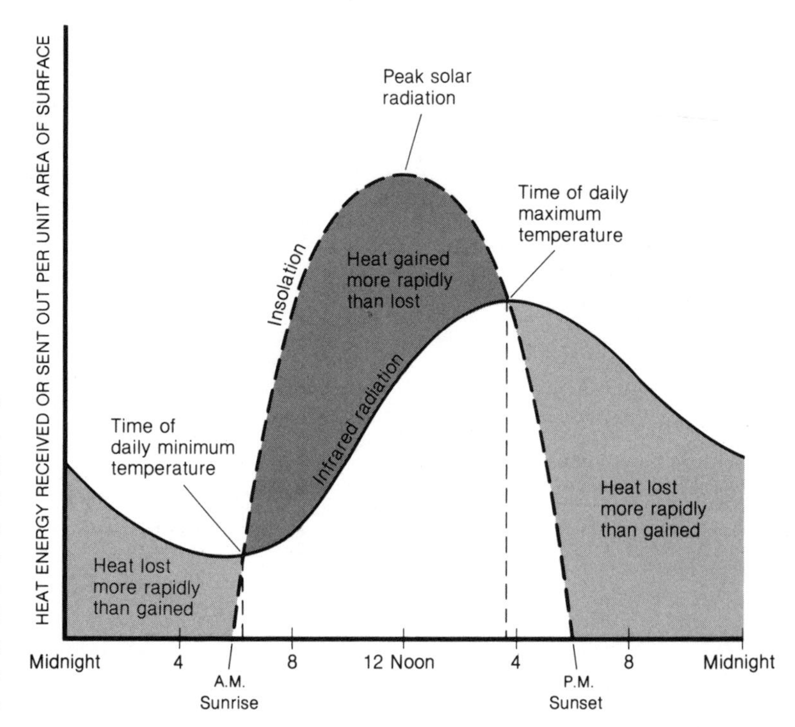

FIGURE 4.9

Variation during the course of a day in incoming solar radiation and outgoing infrared radiation. Note that the temperature is at minimum near sunrise and at maximum during midafternoon. This is a midlatitude locality with clear skies and no air mass advection. (From Ward's Natural Science Establishment, edited by A. N. Strahler)

radiation, the cause of sunburn, is most intense at noon, but the air temperature usually reaches a maximum several hours later.

Air mass controls

Air mass advection refers to the movement of an air mass from one locality to another. **Cold air advection** occurs when the wind blows across regional isotherms* from a cold area to a warmer area (arrow *A* in Figure 4.10), and **warm air advection** takes place when the wind blows across regional isotherms from a warm area to a colder area (arrow *B* in Figure 4.10). Air mass advection thus occurs when one air mass replaces another air mass having different thermal characteristics.

In terms of temperature variations at a given locality, the significance of air mass advection depends on the initial thermal characteristics of the new air mass, as well as on the degree of modification the air mass undergoes as it travels across the earth's surface. For example, a surge of bitterly cold arctic air loses much of its punch when it travels over ground that has no snow, because the

*Lines on a map drawn through localities having the same air temperature.

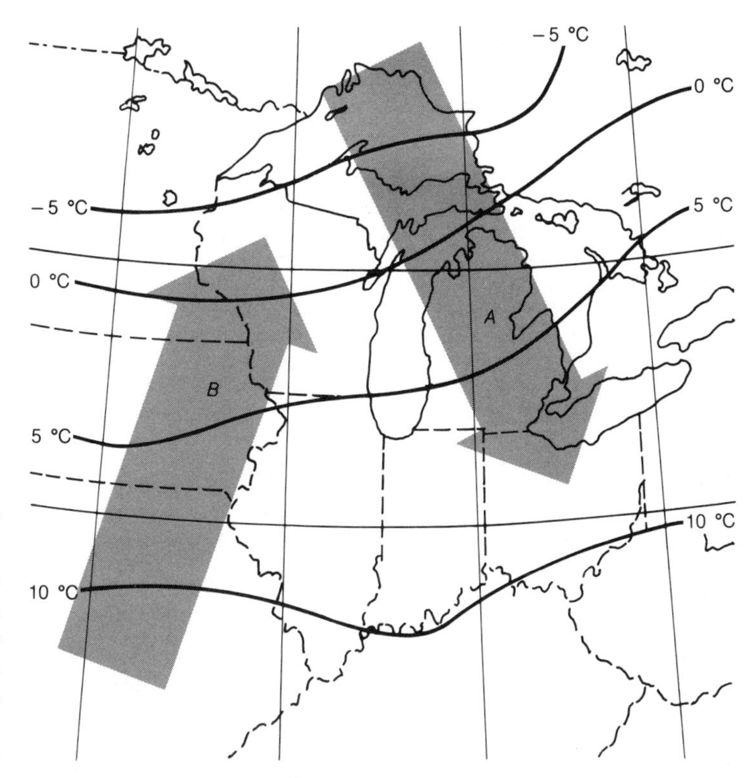

FIGURE 4.10

Cold air advection occurs when (A) the horizontal wind blows across isotherms from cold areas toward warmer areas, and warm air advection occurs when (B) the horizontal wind blows across isotherms from warm areas toward colder areas. Solid lines are isotherms.

arctic air mass is warmed from below by sensible heating (conduction and convection). In contrast, sensible heating and air mass modification are minimized when the arctic air moves over a cold, snow-covered surface.

Although we have discussed local radiative conditions and air mass advection separately, the two regulate air temperature together. For example, air mass advection may compensate for or even overwhelm local radiative influences on temperature. Local radiative conditions usually favor rising air temperatures from an early morning low to an early or midafternoon high. This typical pattern can change, however, if an influx of cold air occurs during the same period. Depending on how cold the incoming air is, air temperatures may climb more slowly than usual, may remain steady, or may even fall. If cold air advection is extreme, temperatures may drop precipitously throughout the day, in spite of bright, sunny skies. In another example, air temperatures may climb through the evening hours as a consequence of strong warm air advection, so the day's high temperature occurs just before midnight.

Conclusions

Imbalances in radiant heating and radiant cooling are ultimately responsible for the circulation of the atmosphere. One important consequence of atmospheric circulation is the formation of clouds that can produce rain and snow. Before we examine cloud and precipitation processes in detail, however, we must examine briefly another variable of the atmosphere: air pressure.

SUMMARY STATEMENTS

- Heat imbalances develop between the earth's surface and the atmosphere, and between tropical and high latitudes, because of differences in the rates of solar heating and infrared cooling.
- Excess heat is transported from the earth's surface to the atmosphere by conduction, convection, and vaporization of water.
- Excess heat is transported poleward from the tropics by air mass exchanges, storm systems, and ocean currents.
- Vertical and poleward heat transport in the earth-atmosphere system is brought about by various weather systems spanning a wide range of spatial scales.
- The sun drives the circulation of the atmosphere by producing heat differentials within the earth-atmosphere system and by being the ultimate source of kinetic energy.
- Air temperature is regulated by local radiative controls and by advection of air masses.

KEY WORDS

sensible heat transfer
latent heat transfer
heat of fusion
heat of melting
heat of vaporization
cumulus clouds
cumulonimbus clouds
Bowen ratio
poleward heat transport
kinetic energy
law of energy conservation
global-scale circulation
synoptic-scale weather
mesoscale systems
microscale weather
air mass advection
cold air advection
warm air advection

REVIEW QUESTIONS

1 What is meant by radiative heating and radiative cooling within the earth-atmosphere system?

2 On a global scale and over the course of a year, compare the radiative heating and cooling of the earth's surface with the radiative heating and cooling of the troposphere.

3 How is the excess heat energy at the earth's surface transported into the troposphere? Which of these processes is most important and why?

4 What processes are involved in sensible heat transfer?

5 How does molecular activity vary with the three physical phases of water?

6 Explain why heat is either added or released when water undergoes a phase change.

7 What happens to the temperature of water during a phase change?

8 When clouds form, heat is released to the atmosphere. Why?

9 Define the Bowen ratio.

10 The Bowen ratio for Asia is greater than that for North America. Why?

11 What is the global average Bowen ratio?

12 Under what conditions is the net flow of heat from the troposphere to the earth's surface?

13 Describe how radiative heating and radiative cooling vary with latitude.

14 What processes are involved in poleward heat transport? Which process is most important?

15 State the law of energy conservation.

16 Why is air temperature variable?

17 On a winter day, is the air temperature more likely to be higher if the ground is snow covered or if it is bare? Explain your answer.

18 Why is the winter solstice usually not the coldest day of the year?

19 Explain why in summer the greatest risk of sunburn is not during the warmest time of day.

20 Under what conditions might the day's high temperature be recorded at 11 P.M.?

POINTS TO PONDER

1 In your own words, explain the following statements: (a) The sun drives the atmosphere. (b) The atmosphere is heated from below.

2 List some of the advantages of solar power over other more conventional energy sources. What might be some disadvantages?

3 To evaluate the solar power potential of a given locality, what records of climatological elements should we consult?

4 In spite of clear skies, the air temperature can remain steady throughout the day. Explain this fact.

5 Speculate on whether there might be seasonal changes in the poleward transport of heat.

6 What is the basic reason for heat transport in the earth-atmosphere system? Why does the atmosphere circulate?

7 How does the law of energy conservation apply to the circulation of the atmosphere?

PROJECTS

1 Design a simple dwelling that uses a few passive solar features. The building must be warm in winter and cool in summer.

2 How extensively are active and passive solar heating systems utilized in your community? You may wish to arrange a tour of a public facility.

SELECTED READINGS

Ingersoll, A. P. "The Atmosphere." *Scientific American* 249, no. 3 (1983):162–174. *A concise summary of atmospheric characteristics and the global radiation balance.*

Lehr, P. E., R. W. Burnett, and H. S. Zim. *Weather.* New York: Golden Press, 1975. 160 pp. *An exceptionally well-illustrated and lucid description of weather phenomena.*

Trenberth, K. E. "What are the Seasons?" *Bulletin of the American Meteorological Society* 64 (1983): 1276–1282. *An analysis of the distinction between the meterological seasons and the astronomical seasons.*

"Yes, what a climb that was! I was scared to death, I can tell you.
Sixteen hundred metres—that is over five thousand feet, as I reckon it. . . ."
And Hans Castorp took in a deep, experimental breath of the strange air. It was fresh, and that was all. It had no perfume, no content, no humidity; it breathed in easily, and held for him no associations.

THOMAS MANN
The Magic Mountain

Air pressure drops rapidly with increasing altitude. With a drop in air pressure, the air also thins. At high elevations, the air is thin enough that mountain climbers may require a supplementary oxygen supply. (Union Pacific Railroad photograph)

FIVE

Air Pressure

TELEVISION and radio weather reports typically include the latest air pressure reading along with air temperature and relative humidity. Although we usually are physically aware of changes in temperature and humidity, we do not sense changes in air pressure as readily. If we follow air pressure reports over a period of time, however, we quickly learn that important shifts in weather accompany variations in air pressure.

What is air pressure? Air exerts a force on the surface of objects that it contacts. **Air pressure** is a measure of that force per unit of surface area. Air is made up of molecules in rapid, random motion, and each molecule exerts a force as it collides with the surface of a solid or liquid. In a millionth of a second, each square centimeter of the earth's surface, for example, is bombarded by billions upon billions of gas molecules. The total air pressure exerted, then, is the cumulative force of a multitude of molecules colliding with any object in contact with air.

The amount of pressure produced by the gas molecules composing air depends on (1) the mass of the molecules, (2) the pull of **gravity**,* and (3) the kinetic energy of the molecules. Often, air pressure is described simply as the weight of the atmosphere over some area—a square centimeter at the earth's surface, for example. Note, however, that since

$$\text{weight} = \text{mass} \times \text{gravity}$$

this description is incomplete, because it does not account for the influence of molecular kinetic activity on air pressure. In this chapter, the properties of air pressure and the reasons for the spatial and temporal variability of air pressure are examined.

Pressure Balance

The average air pressure at sea level is about 1 kg per cm^2 (or about 14.7 lb per $in.^2$). This means that the pressure of the atmosphere on the roof of a typical three-bedroom ranch-style house at sea level is about 2.1 million kg (4.6 million lb), a pressure equivalent to the combined weight of 1500 full-size autos. Why doesn't the roof collapse under this enormous pressure? The reason the roof does not collapse is that the air pressure is the same at any point in all directions—up, down, or sideways. The air pressure within the

*Gravity is the force that holds you and all other objects on the earth's surface.

house therefore exactly counterbalances the external air pressure. This pressure balance (or equilibrium) is the usual condition in the atmosphere except during very violent storms, such as tornadoes.

Variation with Altitude

We know from experience with a bicycle tire pump that air is compressible, that is, its volume and density are changeable. The pull of gravity compresses the atmosphere so that maximum **air density*** occurs at the earth's surface. In other words, the gas molecules are spaced most closely at the earth's surface, and the spacing between molecules increases with increasing altitude. The number of gas molecules per unit volume thus decreases with altitude above the earth's surface. This "thinning" of the air is so rapid that at an altitude of only 16 km (10 mi), air density is only about 10 percent of the average sea-level value. The declining air density with altitude means fewer molecular collisions and a declining air pressure (Figure 5.1). The rapid drop of air pressure with increasing altitude was first verified in 1646 when mountain climbers took

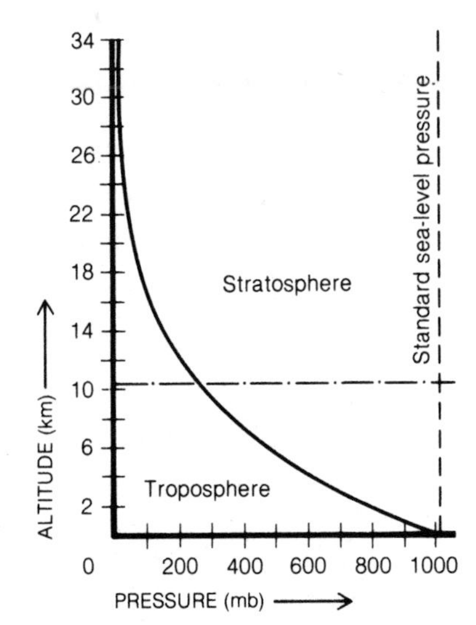

FIGURE 5.1

Air pressure drops rapidly with altitude. The unit of pressure is the millibar (mb) with the average air pressure at sea level equal to 1013.25 mb.

barometer† readings as they ascended a mountainside in France.

Although air pressure and air density decrease rapidly with altitude, specifying an altitude at which the earth's atmosphere defi-

*Density is mass per unit volume.

†Barometers, the instruments used to monitor air pressure, are described later in this chapter.

nitely ends is impossible. No point can be clearly identified as the beginning of "space." Rather, we describe the vertical extent of the atmosphere in terms of the relative distribution of its mass. Half of the atmospheric mass lies between the earth's surface and an altitude of about 5.5 km (3.4 mi). About 99 percent of the atmosphere lies below 32 km (20 mi). Above 80 km (50 mi), the relative proportion of atmospheric gases changes markedly, and beyond approximately 950 km (590 mi), the atmosphere merges with the highly rarefied interplanetary gases, hydrogen and helium.

From a slightly different perspective, the atmosphere thins so rapidly with altitude that at an altitude of only 32 km (20 mi), air pressure is less than 1 percent of the air pressure at sea level. This rapid pressure drop means that we experience appreciable changes in air pressure as we travel up mountains and down into valleys. For example, the average air pressure at Denver, a mile above sea level, is about 83 percent of the average air pressure at Boston, barely above sea level.

The expansion and thinning of air at higher altitudes that accompany the drop in air pressure (with altitude) trigger physiological adjustments in humans. Visitors to mountainous areas may complain of dizziness, headache, and shortness of breath because of the lower concentration of oxygen in the relatively thin air. After a week or two, however, these distressing symptoms usually disappear. The number of red blood cells, which carry oxygen, increases. This allows visitors to adjust gradually, or acclimatize, to low oxygen levels. Some people are better able to acclimatize than others because of differences in individual genetic makeup. An upper altitude limit does exist, however, and above that limit no human can adjust. At the summit of Mount Everest, the earth's highest peak at 8850 m (29,028 ft), the oxygen concentration is so low that only those who are both genetically inclined and carefully acclimatized can survive without an auxiliary oxygen supply.

Above about 5500 m (18,000 ft), aircraft cabins must be pressurized. For example, the air pressure outside a transcontinental jet aircraft cruising at 10,000 m (32,800 ft) is only about 25 percent of the mean sea-level pressure. The cabin is typically pressurized to about 75 percent of its sea-level value. Nonetheless, as the aircraft changes altitude, passengers notice changes in air pressure by the "popping" sensation in their ears. The reasons for this phenomenon are discussed in the Special Topic.

Very low air density at high altitudes also has interesting implications in terms of air temperature and heat transfer. In the ther-

SPECIAL TOPIC

Ear-popping

During rapid ascent or descent in an airplane or elevator, we commonly feel the effects of changing air pressure on our eardrums. As Figure 1 indicates, the eardrum separates the outer ear from the middle ear chamber. As air pressure in the outer ear changes, the eardrum becomes distorted if air pressure in the middle ear does not equalize with the air pressure in the outer ear. For example, going up in an elevator lowers air pressure in the outer ear. If the middle ear chamber does not experience an equal decline in air pressure during the ascent, its air pressure will be greater than the pressure in the outer ear. As a consequence, the eardrum bulges outward. Such deformation not only causes physical discomfort, but the bulging eardrum does not vibrate efficiently and sounds seem muffled. If the air pressure difference between the middle and outer ear continues to increase, the eardrum could rupture, and a permanent hearing loss might ensue.

Fortunately, a way exists to equalize air pressure in the middle ear. A tube (called the auditory or eustachian tube) connects the middle ear to the upper throat region, which leads to the outside via the oral and nasal cavities (Figure 1). The tube is normally closed where it enters the throat, but it will open eventually if a sufficient difference in air pressure builds up between the middle ear and the throat. To reduce discomfort, however, the opening of the auditory tube can be induced by yawning or swallowing. Air travelers often chew gum and subsequently swallow to more quickly equalize air pressure on both sides of the eardrum.

When the auditory tube opens, air pressure in the middle ear quickly equalizes with the external air pressure, and the eardrum pops back to its normal shape. The vibrations of the eardrum, which are associated with its rapid change in shape, are what we hear as "ear-popping"—our body's way of preventing a permanent hearing loss when we experience a major change in air pressure.

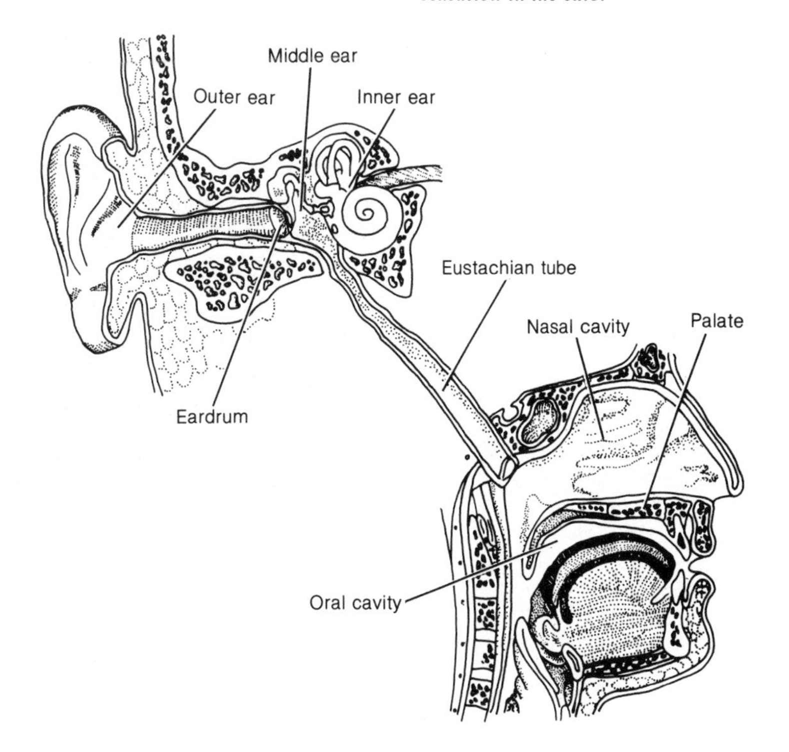

FIGURE 1

Pressure differences between the middle ear and the outer ear distort the eardrum. Opening of the eustachian tube equalizes pressure and causes a popping sensation in the ears.

mosphere, atoms and molecules move about with considerable activity that is indicative of very high temperatures (refer to the temperature profile of the atmosphere in Figure 1.3). There are so few atoms and molecules per unit volume, however, that the *total* molecular energy (that is, heat) is relatively low. In spite of thermosphere temperatures that approach 1200 °C (2200 °F), heat is not readily conducted to cooler bodies. Satellites orbiting at these altitudes, for example, do not acquire such temperatures.

Horizontal Variations

Air pressure differs from one place to another, and variations are not always due to differences in the elevation of the land. In fact, meteorologists are more interested in air pressure variations that arise from factors other than land elevation. Weather stations usually adjust local air pressure measurements to what the air pressure would be if the station were actually located at sea level. When this **reduction to sea level** is carried out everywhere, air pressure still varies from one place to another, and fluctuates from day to day and even from hour to hour (Figure 5.2).

Although spatial and temporal fluctuations in surface air pressure (reduced to sea level) are relatively slight, they can accompany some important changes in weather. In middle latitudes, weather is dominated by a continuous procession of different air masses that bring about changes in air pressure and changes in weather. Recall that an **air mass** is a huge volume of air that is relatively uniform in temperature and water vapor concentration. As air masses move from place to place, surface air pressures fall or rise, and the weather changes. As a general rule, weather deteriorates when air pressures fall and improves when air pressures rise.

Why do some air masses exert greater pressure than other air masses? One reason is differences in air density that arise from differences in air temperature, or from differences in water vapor content, or from both. Recall from Chapter 2 that temperature is a measure of the average kinetic energy of molecules. As air temperature increases, the constituent molecules move about with greater activity. If the heated air were inside a closed container, such as a rigid metal can, we would expect the air pressure on the internal walls of the container to rise as the increasingly energetic molecules bombarded the walls with greater force. The air density inside the container would not change, because no air is added to or removed from the container and the air volume is constant. In

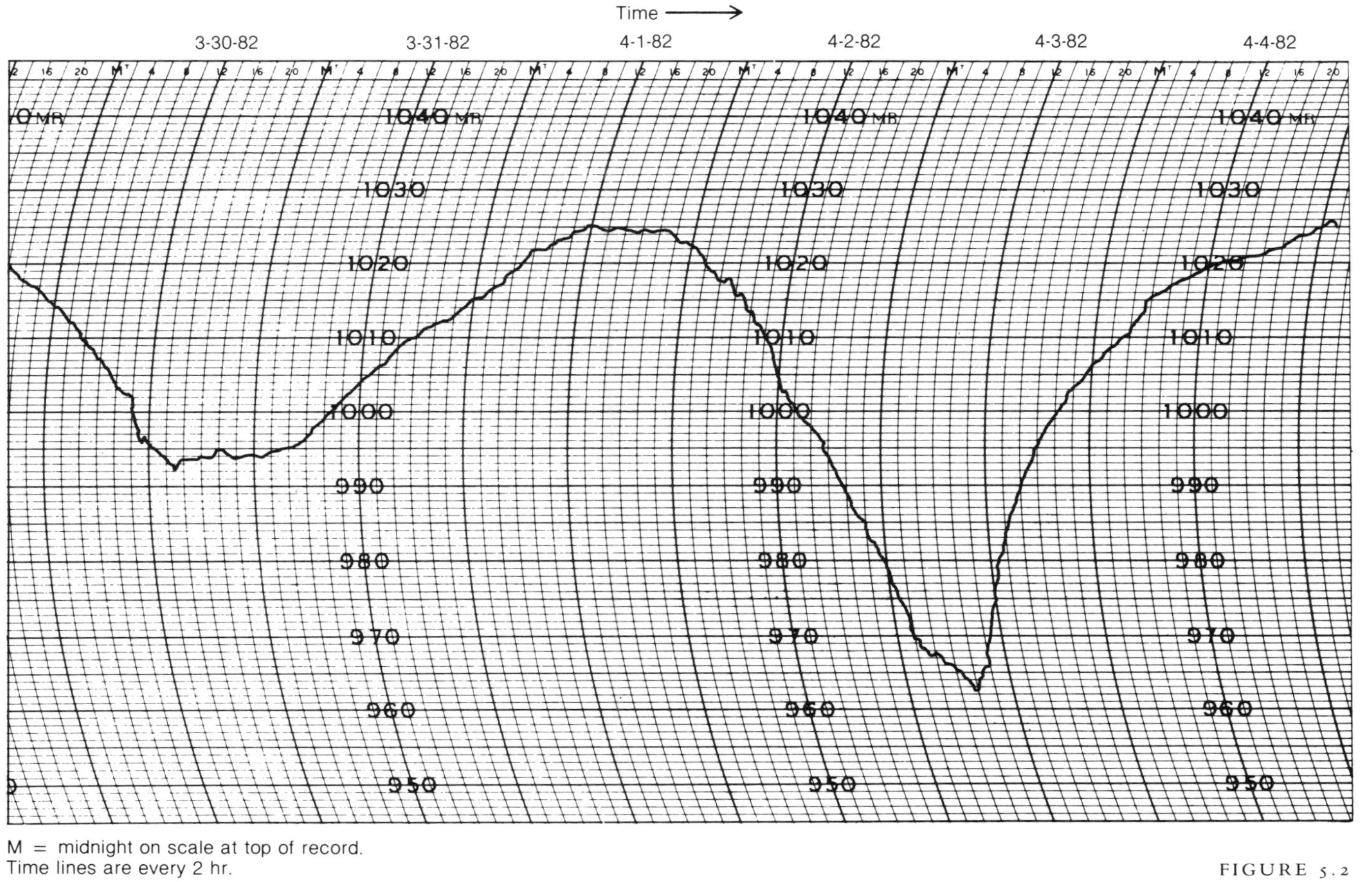

FIGURE 5.2

A tracing from a barograph showing variation in air pressure in millibars reduced to sea level at Green Bay, Wisconsin, from 30 March through 4 April 1982. Note that significant changes in air pressure occur from day to day and even from one hour to the next.

contrast, the atmosphere is not confined by walls, so the air is free to expand and contract, and the air density is changeable. When, therefore, the air is heated within the atmosphere by conduction, convection, or radiation, the net effect is for air pressure to decrease (*not* increase) with rising temperature. This is because the greater activity of the heated molecules increases the spacing between neighboring molecules and thus reduces air density. The decreasing air density then lowers the pressure exerted by the air. Warm air is thus lighter (less dense) than cold air and consequently exerts less pressure.

Water vapor molecules are lighter than the average mass of the other gaseous molecules composing air. When water molecules enter the air by vaporization, they displace heavier molecules and make the mixture lighter. The greater the water vapor content of air, therefore, the less dense the air is. At equal volumes and temperatures, then, a humid air mass exerts less pressure than a relatively dry air mass.

Cold, dry air masses are accompanied by higher surface pressures than warm, humid air masses. Warm, dry air in turn causes higher pressures than an equally warm but more humid air mass. The replacement of one air mass by another can mean changes in air pressure and weather, but surface air pressures can fluctuate even without an exchange of air masses, because air pressure can fall or rise as the air is locally heated or cooled.

In addition to air pressure changes caused by variations of temperature and water vapor content, air pressure can also be influenced by the circulation pattern of air. Consider some examples. Suppose that at the earth's surface, horizontal winds blow radially away from a central point, as in Figure 5.3A. This is an example of

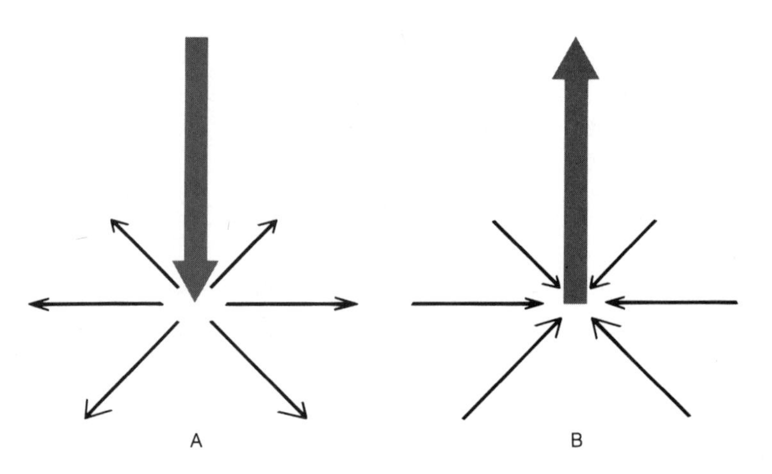

FIGURE 5.3

Divergence of winds at the earth's surface causes air to descend from aloft (A). Convergence of winds at the earth's surface causes air to ascend (B). Such patterns of airflow can cause changes in air density and air pressure.

divergence of air. At the center, air descends from above and takes the place of the diverging air. If more air diverges at the surface than descends from aloft, then the air density and air pressure decrease. On the other hand, suppose that at the earth's surface, horizontal winds blow radially inward toward a central point, as in Figure 5.3B. This is an example of **convergence** of air. If more air converges at the surface than ascends, then air density and air pressure increase. The air pressure changes caused by the divergence and convergence of air are the subject of later chapters.

We now have some insight into the meaning of those "H" and "L" symbols on newspaper and television weather maps (Figure 5.4). After air pressure readings are reduced to sea level, an "H" (or "high") is used to designate places where sea-level air pressure is relatively high compared with the pressure of surrounding areas, and an "L" (or "low") is used to indicate regions where air pressure

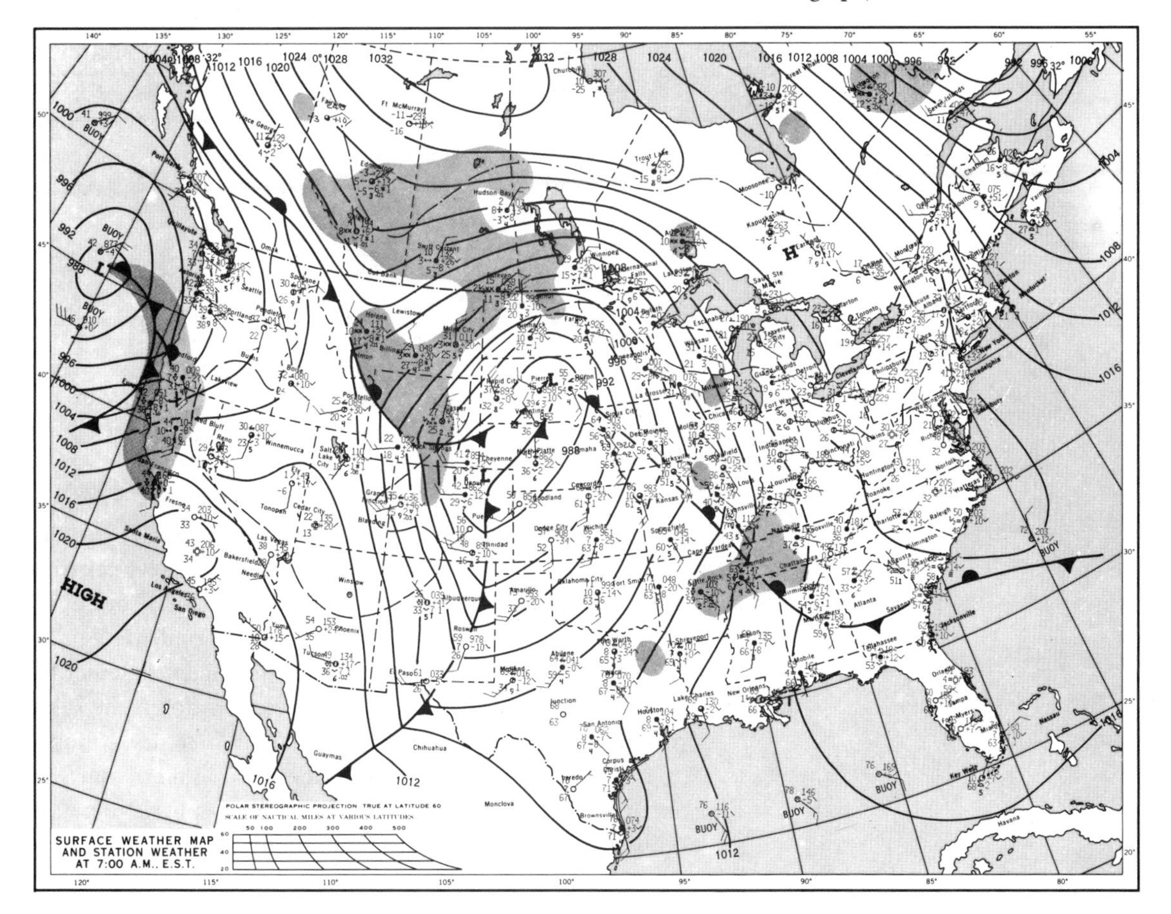

FIGURE 5.4
A typical surface weather map showing variations in air pressure from one place to another. Solid lines are isobars, lines joining locales of the same air pressure. Darkened areas are where rain or snow is falling. (NOAA photograph)

is relatively low by comparison. These spatial variations in surface air pressure can often be attributed to differences in temperature or to differences in the water vapor content of air or both.

Pressure Units

On television and radio weather reports, air pressure readings are usually given in units of length (millimeters or inches), according to the type of instrumentation used to measure the pressure. It is more appropriate, however, to express air pressure in units of pressure. Physicists use the pascal as the metric unit of pressure and have determined the average air pressure at sea level to be 101,325 pascals, or 101.325 kilopascals. Meteorologists, on the other hand, traditionally designate air pressure in **millibar** units, with 1 millibar (mb) equal to 100 pascals. The average sea-level air pressure reading is thus 1013.25 mb, and the usual worldwide range in sea-level pressure is about 980 to 1040 mb. The lowest air pressure ever recorded at sea level was 870 mb, which occurred during a Pacific typhoon on 12 October 1979. The highest pressure ever recorded at sea level was 1084 mb at Agata, USSR, on 31 December 1968, and was associated with an extremely cold air mass.

Air Pressure Measurement

A **barometer** is the instrument used to monitor changes in air pressure. There are two basic barometer types: mercurial and aneroid.

The more accurate but the more cumbersome of the two is the **mercurial barometer**, invented in 1643 by Evangelista Torricelli, an Italian physicist and student of Galileo. The instrument consists of a glass tube a little less than 1 m (39 in.) long, sealed at one end, open at the other end, and filled with mercury, a very dense liquid. The open end of the tube is inverted into a small open container of mercury, as shown in Figure 5.5. Mercury settles down the tube (and into the container) until the weight of the mercury column exactly balances the weight of the atmosphere acting on the surface of the mercury in the container. The average atmospheric pressure at sea level will support the mercury column in the tube to a height

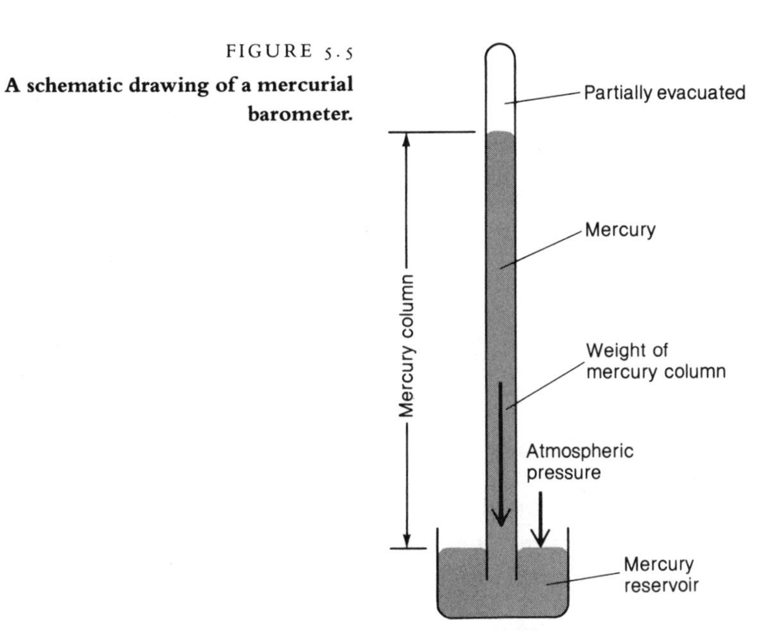

FIGURE 5.5
A schematic drawing of a mercurial barometer.

of 760 mm (29.92 in.). When air pressure changes, however, the height of the mercury column changes. Falling air pressure allows the mercury column to drop, and increasing air pressure forces the mercury column to rise.

The **aneroid** (nonliquid) **barometer**, pictured in Figure 5.6, is less precise but more portable than the mercurial barometer. It

FIGURE 5.6
An aneroid barometer, an instrument used to monitor variations in air pressure. (Courtesy of Belfort Instrument Company)

consists of a coil of tubing from which much of the air has been evacuated. A spring keeps the chamber from collapsing. As air pressure changes, the coil flexes, bending in when pressure rises and bowing out when pressure drops. A series of gears and levers transmits these movements to a pointer on a dial, which is calibrated to read in equivalent millimeters (or inches) of mercury or to read directly in units of air pressure.

Some aneroid barometers, especially those designed for home use, have dials with legends, such as "fair," "changeable," and "stormy," corresponding to certain ranges of air pressure. These designations should not be taken too literally, since a given air pressure reading does not always correspond to a specific type of weather.

Much more useful than these legends for local weather forecasting is **air pressure tendency**, that is, the change in air pressure with time. Rising air pressure usually means improving weather, whereas falling air pressure generally signals deteriorating weather. For determining pressure tendency, some aneroid barometers are equipped with a second pointer that serves as a reference marker. By turning a knob on the barometer face, the user sets the second pointer to correspond to the current air pressure reading. At a later time, the user can observe the new pressure reading and compare it with the set reading to determine the pressure tendency.

An aneroid barometer may also be linked to a pen that records on a clock-driven drum chart, as shown in Figure 5.7. This instrument, called a **barograph**, provides a continuous record of air pressure variations with time. Because air pressure drops with altitude, an aneroid barometer can be calibrated to read elevation. Such an instrument is called an **altimeter**. Some altimeters are so accurate that they can detect an elevation change of less than 1 m. Because the rate of pressure drop with altitude is also affected both by changes in air temperature and by changes in water vapor content of air, a correction must be applied to the altimeters used on aircraft or the altimeter reading could be inaccurate.

The Gas Law

To this point, we have described the atmosphere in terms of variations in temperature, pressure, and density. These three important properties, collectively known as **variables of state**, change in magnitude from one place to another across the earth's surface,

FIGURE 5.7

A barograph provides a continuous trace of air pressure. (Courtesy of Belfort Instrument Company)

change with altitude above the earth's surface, and change with time. The three variables of state are also interrelated through the **gas law**. Although the gas law applies to a single "ideal" gas, the law provides a reasonably accurate description of the behavior of the atmosphere, which, as noted earlier, is a mixture of many gases. Simply put, the gas law states that air pressure is proportional to the product of air density and temperature, that is,

air pressure = constant* × density × temperature

(The physical basis for this law is presented in the Mathematical Note at the end of this chapter.)

The dependence of air pressure on two interdependent variables, density and temperature, complicates matters. Since within the atmosphere the volume of air can change, then temperature variations can affect the air density. In terms of the gas law equation, this means that raising the air temperature does not always increase the pressure, nor does lowering the air temperature always decrease

*A constant is an experimentally derived number that changes a proportional relationship to an equation.

the pressure. For example, in the winter it is usual for temperatures to drop (not rise) as the surface air pressure rises. The gas law is still satisfied in this case, because air density increases at the same time as the temperature drops. In another example, as we learned earlier, both air pressure and air density decline rapidly with altitude, but the temperature drops with altitude through some layers of the atmosphere and rises with altitude through other layers. Air temperature and air density therefore usually vary in opposite directions.

MATHEMATICAL NOTE

The gas law

Although the atmosphere is a mixture of many gases, it behaves much as if it were a single *ideal gas*. By definition, an ideal gas follows the kinetic-molecular theory precisely, that is, an ideal gas is made up of a very large number of minute particles, called *molecules*, that are in rapid and random motion. As they move about, molecules experience perfectly elastic collisions, so they lose no momentum. They are so small that the attractive forces between them are negligible.

An ideal gas follows Charles's law and Boyle's law exactly, whereas a real gas behaves only approximately as these laws dictate. *Charles's law* holds that with constant pressure, *P*, the absolute temperature (K) of an ideal gas, *T*, is inversely proportional to the density, ρ, of the gas. That is,

$$T \propto 1/\rho$$

As a sample of gas is warmed, the gas expands, and its density decreases. According to *Boyle's law*, on the other hand, when the temperature is held constant, the pressure and density of an ideal gas are directly proportional. That is,

$$P \propto \rho$$

As the pressure on an ideal gas increases, its volume decreases and its density increases.

Conclusions

The range of air temperature and pressure variations in the earth-atmosphere system allows water to be present in all three phases. Earlier we saw how the phase changes of water help to redistribute heat within the atmosphere (latent heat transfer). Air temperature changes also trigger the phase changes of water that cause clouds to either form or dissipate. These processes are the subject of the next chapter.

Charles's and Boyle's laws are combined as the *ideal gas law*, that is,

$$P \propto \rho T$$

The constant of proportionality, R, varies depending on the specific gas. The ideal gas law is thus expressed as an equation relating variables of state:

$$P = R\rho T$$

This so-called *equation of state* describes approximately the behavior of dry air (air minus water vapor) when we assign a certain value for R. With a slight adjustment in density, the equation can also be used for humid air in which no condensation takes place. Condensation, the change in phase of water from a vapor to a liquid, is accompanied by a release of heat, which elevates air temperature and lowers density.

If we apply the equation of state to humid air, then the density must be adjusted downward depending on the concentration of water vapor. (Recall that the addition of water vapor lowers the density of air.) In the equation of state, this adjustment is not made by lowering the density directly. Instead, the temperature is adjusted upward by an amount that would bring about the appropriate density reduction. This adjusted temperature, called the *virtual temperature*, is never more than a few degrees higher than the actual air temperature. For humid air, the equation of state thus becomes:

$$P = R\rho T_v$$

where T_v is the virtual temperature.

SUMMARY STATEMENTS

- The pressure exerted by air depends on the pull of gravity and the mass and kinetic energy of the gas molecules that compose air.

- Air pressure and air density decrease very rapidly with altitude, but the atmosphere has no clearly defined upper boundary. The atmosphere gradually merges with the gases of interplanetary space.

- The thinning of air with altitude has adverse physiologic effects on humans.

- Differences in air mass characteristics cause the surface air pressure (reduced to sea level) to differ from one place to another. Changes in temperature or in water vapor concentration or in both affect air density and air pressure.

- Important weather changes can accompany relatively small changes in surface air pressure.

- The variables of state (temperature, pressure, and density) are interrelated through the gas law.

KEY WORDS

air pressure
gravity
air density
reduction to sea level
air mass
divergence
convergence
millibar
barometer
mercurial barometer
aneroid barometer
air pressure tendency
barograph
altimeter
variables of state
gas law

REVIEW QUESTIONS

1 Define air pressure.

2 Why are we not particularly aware of day-to-day variations in air pressure?

3 Why does the atmosphere's greatest density occur adjacent to the earth's surface?

4 Is there a clearly identifiable top to the earth's atmosphere? Explain your response.

5 Why must the cabins of commercial jet aircraft be pressurized at altitudes above about 15,000 ft?

6 Why do weather stations routinely adjust air pressure readings to sea-level values?

7 As a general rule of thumb, how does the weather change as air pressure rises and falls?

8 How does temperature influence the pressure exerted by a sample of air?

9 Explain why increasing the concentration of water vapor reduces the density of air?

10 In equal volumes, which air mass exerts a greater pressure: a warm and humid air mass or a cold and dry air mass?

11 On a particularly hot and muggy evening, a sportscaster comments that baseballs hit to the outfield will not carry far in the *heavy* air. How valid is this observation?

12 What is the meaning of the "H" and "L" symbols that are displayed on newspaper and television weather maps?

13 Air pressure readings are often reported in units of length (millimeters or inches) rather than units of pressure. Why?

14 Explain the principle of the mercurial barometer.

15 What are the advantages of an aneroid barometer over a mercurial barometer?

16 Why is a barometer also an altimeter?

17 Explain why air pressure tendency is useful for local weather forecasting.

18 As an auto is driven, its tires heat up because of friction. From the gas law, predict what (if anything) will happen to the tire pressure as the tires heat up.

19 In your own words, provide a statement of the gas law.

20 Cold air masses are accompanied by higher surface air pressures than are warm air masses. Explain how the gas law is still satisfied.

POINTS TO PONDER

1 The atmosphere thins rapidly with altitude, and yet the proportion of oxygen and nitrogen is constant throughout the troposphere. What does this statement imply about the ascending distribution of oxygen and nitrogen molecules within the atmosphere?

2 A jet aircraft is flying at the 400-mb level, that is, at the altitude where air pressure is 400 mb. What fraction of atmosphere mass is below the aircraft?

3 What is an ideal gas? What assumptions are made in applying the gas law to the atmosphere?

4 What is the purpose of the virtual temperature? The virtual temperature is never more than a few degrees higher than the actual air temperature. What does this imply about the relative concentration of water vapor in the atmosphere?

PROJECTS

1 If you have access to a barometer, or better yet a barograph, determine the relationship between air pressure tendency and major weather changes.

2 Design an experiment that demonstrates that in equal volumes relatively dry air exerts more pressure than relatively humid air.

SELECTED READINGS

Battan, L. J. *Weather in Your Life*. San Francisco: W. H. Freeman, 1983. 230 pp. *A well-written survey of applied aspects of weather and climate.*

Perlman, E. "For a Breath of Thin Air." *Science 83*, no. 4 (1983):52–58. *Description of a scientific expedition to the summit of Mount Everest to study human response to severe oxygen deprivation.*

All the rivers run into the sea, yet the sea is never full; unto the place from whence the rivers come thither they return again.

ECCLESIASTES 1:7

Water exists in three phases: ice, liquid, and vapor. Water changes phase as it circulates between the earth's surface and the atmosphere. This circulation is an important part of the global water cycle that supplies us with all of our fresh water. (Photograph by Mike Brisson)

SIX

Humidity and Stability

WITHIN the atmosphere, water occurs in all three phases—as invisible water vapor and as tiny ice crystals and water droplets visible as clouds. The total water content of the atmosphere is very small, and most of that content is in the lower portion of the troposphere. Indeed, if at any moment all water were removed from the atmosphere as rain, the global average total rainfall would be only about 2.5 cm (1 in.). Water continually cycles into the atmosphere as vapor from reservoirs of water at the earth's surface, and water continually leaves the atmosphere and returns to the earth's surface as rain, snow, and other forms of precipitation. On average, the residence time of a water molecule in the atmosphere is about 10 days. This cycling is an essential component of a much larger system known as the hydrologic cycle.

The Hydrologic Cycle

The water on earth is distributed among oceanic, terrestrial, and atmospheric reservoirs (Table 6.1). The ceaseless flow of water among the reservoirs, known as the **hydrologic cycle**, is illustrated in Figure 6.1. In brief, water vaporizes from sea and land to form clouds from which rain and snow fall back to the earth, thus supplying rivers, which flow back to the seas. The endlessness of the hydrologic cycle is expressed in the verse from Ecclesiastes on the preceding page: "All the rivers run into the sea, yet the sea is never full; unto the place from whence the rivers come thither they return again." Here we will focus on the critical link in the cycle, the link that joins the atmosphere to the oceanic and terrestrial reservoirs of water.

TABLE 6.1
Water Stored in Global Reservoirs

RESERVOIR	PERCENT OF EARTH'S TOTAL WATER
World oceans	**97.2**
Ice sheets and glaciers	**2.15**
Groundwater	**0.62**
Lakes (freshwater)	**0.009**
Inland seas, saline lakes	**0.008**
Soil water	**0.005**
Atmosphere	**0.001**
Streams	**0.0001**

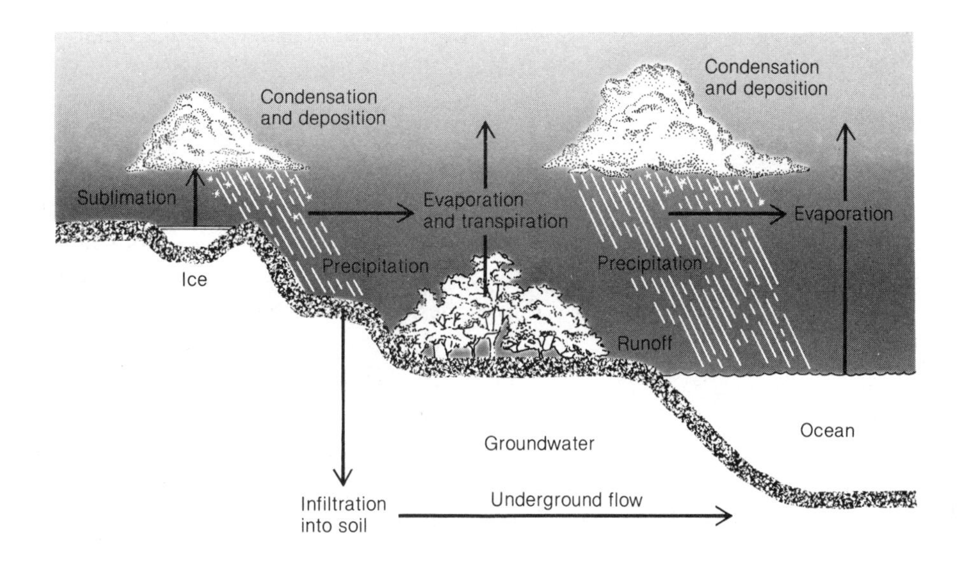

FIGURE 6.1
The hydrologic cycle involves a continuous transfer of water among terrestrial, oceanic, and atmospheric reservoirs. (From Moran, J. M., M. D. Morgan, and J. H. Wiersma. *Introduction to Environmental Science.* San Francisco: W. H. Freeman, copyright © 1980, p. 156.)

Water moves from the earth's surface into the atmosphere by evaporation, transpiration, and sublimation. **Evaporation** is the process by which water changes from a liquid to a vapor at a temperature below the boiling point of water. Evaporation occurs from all open bodies of water, as well as from the wetted surfaces of plant leaves and stems and from the soil. Direct evaporation of water from the oceans is the principal source of atmospheric water vapor.

Transpiration is the process by which water taken up through plant root systems is released as vapor through tiny pores located on the underside of green leaves. On the continents, transpiration is considerable and is often more important than direct evaporation from lakes, streams, and the soil surface. In fact, a single hectare (2.5 acres) of corn typically transpires 35,000 l (8750 gal) of water per day. Measurements of the direct evaporation and transpiration that occur on land are usually combined as **evapotranspiration**.

Sublimation is the process by which water changes directly from a solid to a vapor without passing through an intervening liquid phase. The gradual shrinkage of snowbanks, even though the air temperature remains well below the freezing point, is a consequence of sublimation. Evaporation and sublimation require an addition of heat, which, as we saw in Chapter 4, is supplied by solar radiation.

Water returns to the land and sea from the atmosphere by condensation, deposition, and precipitation. **Condensation** is the process by which water changes from a vapor to a liquid in the form of droplets. **Deposition** is the process by which water changes directly from vapor into solid ice crystals. In the atmosphere, water droplets and ice crystals produced by condensation and deposition, respectively, are visible as clouds. Condensation or deposition on exposed surfaces at ground level is visible as dew or frost respectively. With condensation and deposition, water shifts from a higher to a lower state of molecular activity, and heat is released during these phase changes. **Precipitation**—rain, drizzle, snow, ice pellets, and hail—returns a major portion of atmospheric water from clouds to the earth's surface, where most of it vaporizes back into the atmosphere.

The evaporation-condensation and sublimation-deposition sequences purify water. As water vaporizes, suspended and soluble substances like sea salts are left behind. Through these cleansing mechanisms, water from the sea eventually falls on land as freshwater precipitation, which replenishes terrestrial reservoirs.

When water moves from the atmosphere back to the land and sea as precipitation, an essential subcycle in the global water cycle is completed. Comparing the movement of water into and out of the terrestrial reservoirs with the movement of water into and out of the sea is instructive. The balance sheet for the inputs and outputs of water to and from the various global reservoirs is called the **water budget** (Table 6.2). Each year on the continents the total precipitation exceeds evapotranspiration by about one third. At sea, however, the annual evaporation exceeds precipitation. Over the course of a year, therefore, the water budget shows a net gain of water on land and a net loss of water from the oceans, with the excess on land approximately equal to the ocean's deficit. The land,

TABLE 6.2

Global Water Budget*

SOURCE	CUBIC METERS PER YEAR (GALLONS PER YEAR)
Precipitation on sea	**3.24×10^{14} (85.5×10^{15})**
Evaporation from sea	**3.60×10^{14} (95.2×10^{15})**
Net loss from sea	**-0.36×10^{14} (-9.7×10^{15})**
Precipitation on land	**0.98×10^{14} (26.1×10^{15})**
Evaporation from land	**0.62×10^{14} (16.4×10^{15})**
Net gain on land	**$+0.36 \times 10^{14}$ (9.7×10^{15})**

*Note that the excess of water on land equals the deficit of water at sea.

however, is not getting any soggier, nor are the world's oceans drying up. Why not? The explanation is that the excess precipitation on land drips, seeps, and flows from the land back to the seas.

Once precipitation reaches the earth's surface, it follows various routes (Figure 6.2). Depending on the intensity of the precipitation and on the vegetation, topography, and physical properties of the soil, a portion of the precipitated moisture vaporizes, while the remainder either seeps into the ground or runs off as streams or rivers to the sea. A river drains water from a fixed geographical region called a **drainage basin**, or **watershed**. Climate, vegetation, topography, and various activities, both natural and human, in the drainage basin will affect both the quantity and the quality of river water.

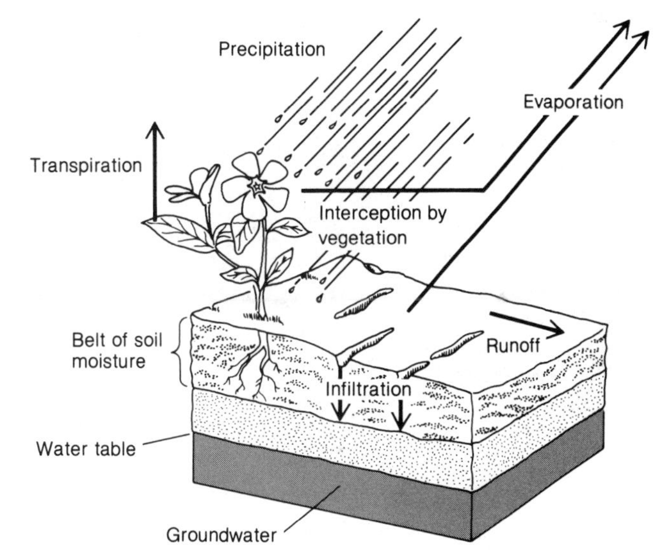

FIGURE 6.2

The various pathways taken by precipitation that falls on the land. (From Moran, J. M., M. D. Morgan, and J. H. Wiersma. *Introduction to Environmental Science*. San Francisco: W. H. Freeman, copyright © 1980, p. 163.)

How Humid Is It?

"It's not the heat, it's the humidity." This popular statement ascribes the discomfort we feel on a hot, muggy day to the water vapor content of the air. Water vapor concentration in air is an important determinant of our physical comfort, as discussed in this chapter's Special Topic. We know from experience that the water vapor content of the air is quite variable. In this section, let us examine some ways to quantify the water vapor content of air: vapor pressure, mixing ratio, and relative humidity.

Vapor pressure

Vapor pressure is simply the pressure exerted by the water vapor in a sample of air. It is dependent on the exchange of water molecules occurring at the interface between air and water or between air and ice. At the air-water interface, some water molecules escape from the water surface and enter the air as vapor, while other molecules leave the vapor phase and return to the water surface as liquid. Net evaporation occurs if more water molecules enter the air than return to the liquid water, and net condensation takes place if more water molecules return to the liquid water than enter the atmosphere as vapor. This interaction is illustrated in Figure 6.3.

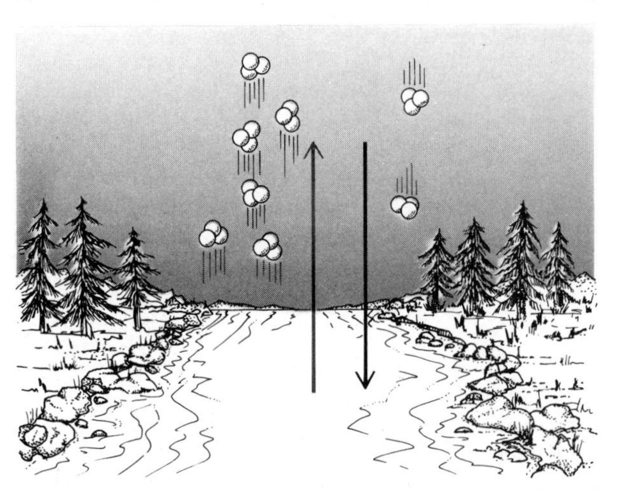

FIGURE 6.3

A continuous exchange of water molecules takes places at the air-water or air-ice interface. In this case, more molecules leave the water surface as vapor than enter the water as liquid. Net evaporation is the consequence. (From Moran, J. M., M. D. Morgan, and J. H. Wiersma. *Introduction to Environmental Science*. San Francisco: W. H. Freeman, copyright © 1980, p. 157.)

The same type of two-way exchange takes place at the interface between ice and air, except that water molecules cannot escape from an ice surface as readily as from a liquid water surface. In the solid phase, water molecules are less energetic than in the liquid phase. Net sublimation occurs when more water enters the vapor phase than returns to the ice, and net deposition takes place when more water returns to the ice than vaporizes.

When water molecules enter the air as vapor, the molecules disperse rapidly, spreading throughout the air. In the process, water vapor mixes with the other atmospheric gases and thus contributes to the total pressure exerted by air. Basic laws describing the behavior of a mixture of gases indicate that we can consider the contribution of water vapor to the total air pressure separately from the contributions of the other gases comprising air. The partial pressure exerted by water vapor alone, the vapor pressure, is only a small fraction of the total air pressure.

TABLE 6.3

Variation of Saturation Vapor Pressure With Temperature

TEMPERATURE		SATURATION VAPOR PRESSURE*	
°C	°F	OVER WATER	OVER ICE
50	122	123.40	
40	104	73.78	
30	86	42.43	
20	68	23.37	
10	50	12.27	
0	32	6.11	6.11
−10	14	2.86†	2.60†
−20	−4	1.25	1.03
−30	−22	0.51	0.38
−40	−40	0.19	0.13

*Unit is millibars (mb).
†Note that for temperatures below freezing, two values are given, one over supercooled water and the other over ice.

There is an upper limit to the concentration of water vapor in air. When air contains its maximum water vapor concentration, the air is said to be *saturated*. As shown in Table 6.3, the vapor pressure at saturation, called the **saturation vapor pressure**, increases as temperature rises. Simply put, warm air can accommodate more water vapor than cold air. The temperature dependence of the saturation vapor pressure is not surprising, because temperature regulates the escape of water molecules from a liquid water or an ice surface. The higher the temperature, the greater the molecular activity, and the more readily the water molecules escape the water surface by vaporizing.

Mixing ratio

The water vapor concentration of air can also be described by the mass* of water vapor per mass of dry air (usually expressed as grams of water vapor per kilogram of dry air). This measure of water vapor concentration is called the **mixing ratio**, because it specifies how much water vapor is mixed with the other gases in air. The **saturation mixing ratio** is the maximum concentration of water vapor in a given mass of dry air. As with the saturation vapor pressure, the saturation mixing ratio is directly proportional to the temperature (Table 6.4).

*Mass (M) is weight (W) divided by the acceleration of gravity (g), since $W = M \times g$.

TABLE 6.4

Variation of Saturation Mixing Ratio With Temperature

TEMPERATURE		SATURATION MIXING RATIO*	
°C	°F	OVER WATER	OVER ICE
50	122	88.12	
40	104	49.81	
30	86	27.69	
20	68	14.95	
10	50	7.76	
0	32	3.84	3.84
−10	14	1.79†	1.63†
−20	−4	0.78	0.65
−30	−22	0.32	0.24
−40	−40	0.12	0.08

*Unit is grams of water vapor per kilogram of dry air.

†Note that for temperatures below freezing, two values are given, one over supercooled water and the other over ice.

Relative humidity

Relative humidity is perhaps the most familiar way of describing the water vapor content of air. It is the water vapor measurement most often quoted by television and radio weathercasters. Relative humidity is determined at a particular temperature by comparing the actual concentration of water vapor in the air with the concentration of water vapor in saturated air at that same temperature. Relative humidity is usually expressed as a percentage, and can be calculated approximately from either the vapor pressure or the mixing ratio. That is,

$$\text{relative humidity} = \frac{\text{vapor pressure}}{\text{saturation vapor pressure}} \times 100\%$$

or

$$\text{relative humidity} = \frac{\text{mixing ratio}}{\text{saturation mixing ratio}} \times 100\%$$

Consider an example. On a summer afternoon, the air temperature is 30 °C (86 °F), and the vapor pressure is 10 mb. Table 6.3 indicates that the saturation vapor pressure of air at 30 °C is 42.43 mb. Using the formula, we compute the relative humidity to be about 25 percent. This means that the air is capable of holding about four times as much water vapor before becoming saturated.

The direct dependence of relative humidity on air temperature can sometimes be confusing and can lead to misconceptions. Even if the actual concentration of water vapor in the air does not change, the relative humidity will rise or fall depending on how the air temperature changes. Consider some illustrations.

On a clear, calm day we would expect the air temperature to rise from a minimum near sunrise to a maximum during early afternoon and to fall off thereafter. If no net change in vapor pressure (or mixing ratio) occurs, then the relative humidity changes inversely with temperature: the relative humidity is highest when the temperature is lowest and vice versa. Hence, during the late morning, the air warms and the relative humidity drops, but the air is really not drying out. The air still contains the same concentration of water vapor.

In winter, a furnace heats the air that is drawn in from outdoors, and in the process, the relative humidity indoors declines. In very cold localities, the drop in relative humidity may be extreme, perhaps necessitating artificial humidification. When outdoor air at −20 °C (−4 °F) and relative humidity of 50 percent is drawn indoors and heated to 20 °C (68 °F), the relative humidity drops to about 3 percent. A humidifier increases the relative humidity of indoor air to more comfortable levels by vaporizing water into the air and thereby raising the actual mixing ratio (or vapor pressure).

Perhaps surprisingly, the mean annual water vapor content of the troposphere is about the same over the desert area of the southwestern United States as it is over the Great Lakes region. There are, however, significant differences in relative humidity. Higher average temperatures in the Southwest mean lower relative humidities and the reduced likelihood of cloudiness and precipitation.

Humidity Measurement

Leonardo da Vinci, a creative genius of the fifteenth century, was probably the first to conceive of an instrument to gauge the water vapor content of air. His design was a simple balance. A small wad of dry cotton on one side of the scale was exactly balanced by a weight on the other side of the scale. As the cotton absorbed water vapor from the air, an imbalance would develop with the amount of imbalance being a measure of humidity.

Today, an instrument commonly used to measure relative humidity is the **psychrometer**. It usually consists of two identical mer-

SPECIAL TOPIC

Humidity and human comfort

Why do we feel uncomfortable on a hot, muggy day? Our physical comfort during such weather depends, in part, on our ability to lose heat via evaporative cooling. This ability depends, in turn, on the vapor pressure gradient between our skin (the evaporative surface) and the surrounding air. The greater the vapor pressure of the ambient air, the smaller the gradient is, and hence the smaller the loss of water vapor and the less the evaporative cooling.

In light of these facts, we can better understand why people often experience greater discomfort when the weather is *both* hot and humid. As air temperatures rise above 25 °C (77 °F), sweating and consequent evaporative cooling increase heat loss from the body. If the air is also quite humid, however, the rate of evaporation is reduced, which hampers the body's ability to maintain a constant core temperature (37 °C, or 98.6 °F). In contrast, in desert areas, the drier air facilitates evaporative cooling. At an air temperature of 32 °C (90 °F), we feel more comfortable under dry conditions than under humid conditions.

Scientists at the National Oceanic and Atmospheric Administration (NOAA) estimate that heat waves accompanied by high humidities kill approximately 150 people each year in the United States. Everyone is adversely affected to some extent by high temperatures and humidities. As the weather causes greater discomfort, people become more irritable and are less able to perform physical and mental tasks. In hot, humid weather, the efficiency of factory workers declines, and students do not concentrate as well on their studies.

In recent decades, a variety of indexes have been developed that combine temperature and humidity to indicate their im-

TABLE 1

Apparent Temperature Index

RELATIVE HUMIDITY (%)	AIR TEMPERATURE (°F) 70	75	80	85	90	95	100	105	110	115	120
	Apparent Temperature (°F)										
0	64	69	73	78	83	87	91	95	99	103	107
10	65	70	75	80	85	90	95	100	105	111	116
20	66	72	77	82	87	93	99	105	112	120	130
30	67	73	78	84	90	96	104	113	123	135	148
40	68	74	79	86	93	101	110	123	137	151	
50	69	75	81	88	96	107	120	135	150		
60	70	76	82	90	100	114	132	149			
70	70	77	85	93	106	124	144				
80	71	78	86	97	113	136					
90	71	79	88	102	122						
100	72	80	91	108							

Source: National Weather Service, NOAA

pact on human comfort and to advise people of the potential danger from heat stress. The *heat index*, or *apparent temperature index*, developed by R. G. Steadman in 1979, has been reported regularly by the National Weather Service since the summer of 1984. As Table 1 illustrates, high temperature and high humidity combine to produce an apparent temperature that is considerably greater than the actual temperature. For example, a person exposed to an air temperature of 38 °C (100 °F) and a relative humidity of 50 percent experiences the same discomfort and stress as if the air temperature were 49 °C (120 °F), because the high humidity retards evaporative cooling from the skin.

The heat index is divided into four categories based on the severity of impact on human well-being. These categories are listed in Table 2.

When apparent temperatures reach between 41 °C and 54 °C (105 °F and 130 °F), heat cramps and heat exhaustion are likely with prolonged exposure and physical activity. At apparent temperatures over 54 °C (130 °F), heatstroke (hyperthermia) is imminent. Keep in mind that the actual degree of heat stress experienced will vary with individual age, health, and body characteristics.

At lower humidities, we experience the apparent temperature as lower than the actual temperature, indicating again the role of evaporative cooling. At lower humidities, the temperature feels cooler than it actually is. It is therefore desirable to increase room humidity during the winter heating season. By increasing the humidity, we feel comfortable at a lower room temperature, and can turn down the thermostat and save on fuel bills.

TABLE 2

Hazards Posed by Heat Stress by Range of Apparent Temperature*

CATEGORY	APPARENT TEMPERATURE	HEAT SYNDROME
I	54 °C or higher 130 °F or higher	Heatstroke or sunstroke *imminent.*
II	41 °C to 54 °C 105 °F to 130 °F	Sunstroke, heat cramps, or heat exhaustion *likely*. Heatstroke *possible* with prolonged exposure and physical activity.
III	32 °C to 41 °C 90 °F to 105 °F	Sunstroke, heat cramps, and heat exhaustion *possible* with prolonged exposure and physical activity.
IV	27 °C to 32 °C 80 °F to 90 °F	Fatigue *possible* with prolonged exposure and physical activity.

Source: National Weather Service, NOAA

*Apparent temperature combines the effects of heat and humidity on human comfort.

cury-in-glass thermometers mounted side by side, as shown in Figure 6.4. The bulb of one thermometer is wrapped in a muslin wick. Readings are taken by first soaking the wick in distilled water, and then whirling the thermometers (or aerating them by a fan) until the wet-bulb thermometer reading steadies. The dry-bulb thermometer measures the actual air temperature. Air streaming past the wet-bulb thermometer vaporizes water from the muslin sleeve, causing evaporative cooling. The drier the air, the greater the evaporation, and the lower the reading will be on the wet-bulb thermometer compared with the dry-bulb reading. The temperature difference between the two thermometers, called the **wet-bulb depression**, is calibrated in terms of percent relative humidity on a psychrometric table (see Appendix II).

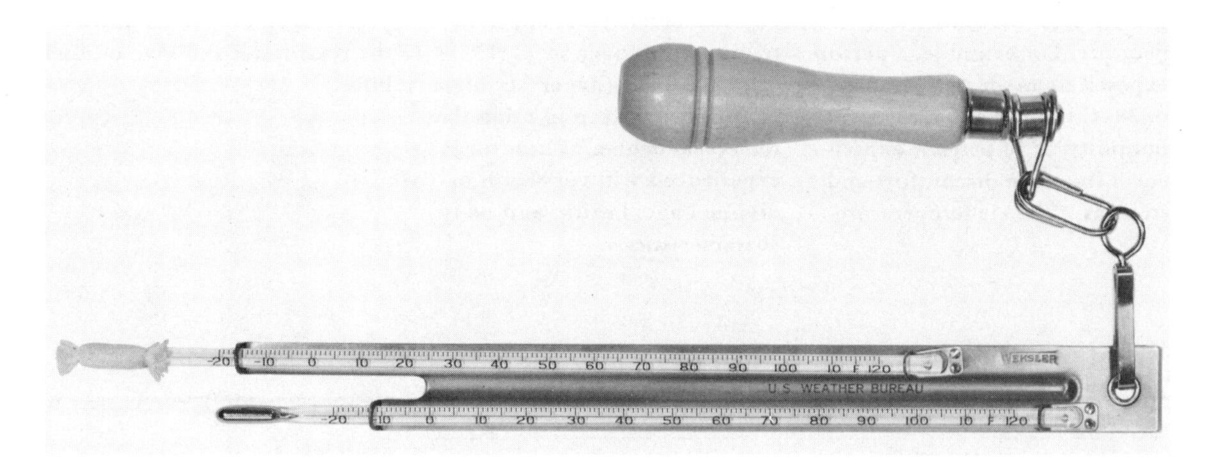

FIGURE 6.4
A sling psychrometer, an instrument used to measure relative humidity. (Courtesy of Belfort Instrument Company)

For example, if the air temperature (dry bulb) is 20 °C (68 °F), and the wet-bulb depression (dry-bulb reading minus wet-bulb reading) is 5 °C (9 °F), the relative humidity is 58 percent. If there is no wet-bulb depression, the air is saturated, that is, the relative humidity is 100 percent.

An instrument that is less accurate than a psychrometer but measures relative humidity more directly and conveniently is the **hair hygrometer**. As the name implies, this device uses hair, which lengthens slightly as the relative humidity increases and contracts as the relative humidity drops. Usually a sheaf of blond hairs is linked mechanically to a pointer on a dial that is calibrated to read in percent relative humidity. Hair typically changes length by about 2.5 percent over the full range of relative humidity from 0 to 100 percent.

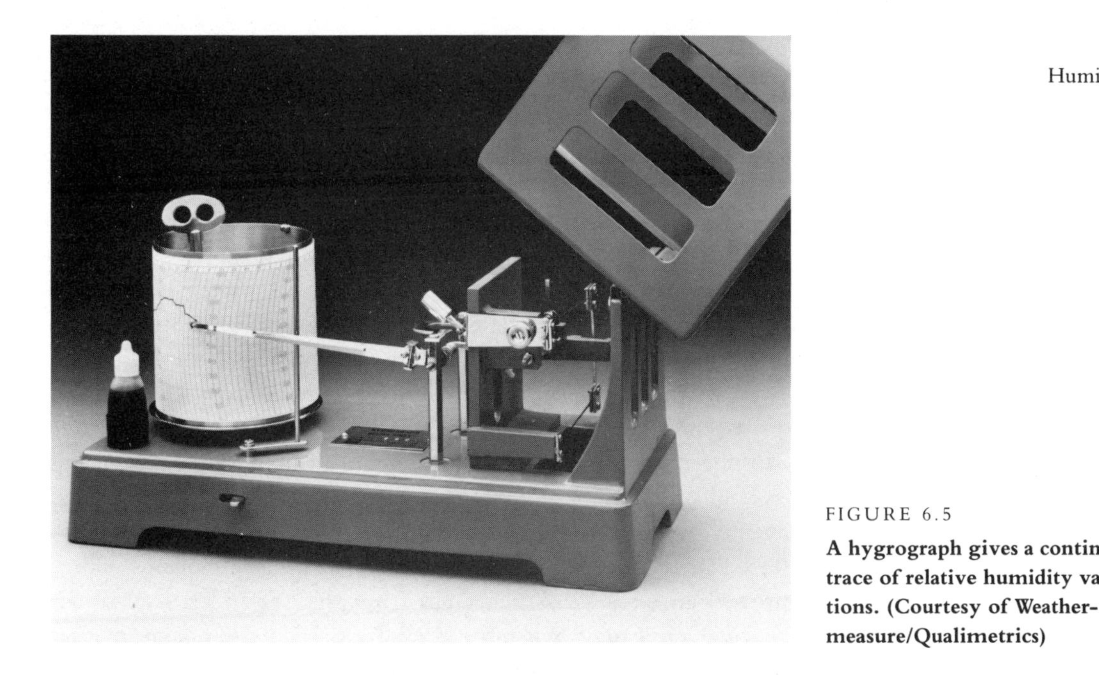

FIGURE 6.5

A hygrograph gives a continuous trace of relative humidity variations. (Courtesy of Weathermeasure/Qualimetrics)

A hair hygrometer may be designed to activate a pen on a clock-driven drum, as shown in Figure 6.5. This instrument, called a **hygrograph**, traces a continuous record of fluctuations in relative humidity during a given time period.

The type of hygrometer used in radiosondes (described in Chapter 1) is based on changes in the electrical resistance of certain chemicals as they adsorb* moisture from the air. The adsorbing chemical may be the lithium chloride coating on a strip of plastic. The more humid the air, the more moisture will be adsorbed, and the greater the change will be in the resistance of an electric current passing through the lithium chloride. Differences in resistance are calibrated in terms of percent relative humidity.

Expansional Cooling and Saturation

The relative humidity is 100 percent at saturation, but it is possible for air to achieve supersaturation, that is, for the relative humidity to be greater than 100 percent. As the relative humidity approaches or exceeds saturation, however, the condensation or deposition of

*Adsorption refers to the action of a gas or soluble substance in being held or condensing on the *surface* of some object.

water vapor becomes more and more likely. Since clouds are composed of condensation products or deposition products or both, this means that cloud development becomes more likely.

How, then, does the relative humidity increase? Relative humidity increases when (1) air is cooled, thus reducing its capacity to hold water vapor, or when (2) more water vapor is added to the air. In this chapter, we focus on the first process.

One way in which air cools and increases its relative humidity is by expansion. In fact, expansional cooling is the principal means of cloud formation in the atmosphere. **Expansional cooling** takes place when the pressure on a volume of air drops, as it does when air ascends in the atmosphere. (Recall from Chapter 5 that atmospheric pressure declines rapidly with altitude.) As air ascends, it expands in the same way that a helium-filled balloon expands (and eventually bursts) as it drifts skyward. Now, if you have ever released air from a tire, you know that air cools as it expands. Compressed air expands as it escapes from the tire and is cool to the touch. Conversely, when air is compressed, its temperature rises and its relative humidity drops. A familiar illustration of this phenomenon is the warming of the cylinder wall of a bicycle tire pump as air is pumped (compressed) into a tire. **Compressional warming** thus occurs when air sinks in the atmosphere.

Adiabatic processes

Expansional cooling and compressional warming within the atmosphere are **adiabatic processes**. To understand adiabatic processes, think of upward- and downward-moving currents of air as composed of streams of discrete volume units, called **air parcels**. If a rising air parcel is unsaturated and is not heated or cooled from outside, the expansional cooling is **dry adiabatic**. This means that radiation and conduction are not effective in adding heat to or removing heat from the air parcel. As the parcel rises and expands, it pushes against the surrounding air, and cools at a rate equivalent to the energy expended. This cooling rate, termed the **dry adiabatic lapse rate**, amounts to 10 °C for every 1000 m of ascent (or 5.5 °F for every 1000 ft) (Figure 6.6).

An air parcel descending in the atmosphere undergoes compressional warming. Whether the parcel is unsaturated or initially saturated, the compressional warming is adiabatic if warming occurs without any radiative or conductive heat exchange between the air parcel and the surrounding air. The atmosphere compresses the air

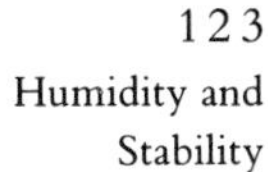

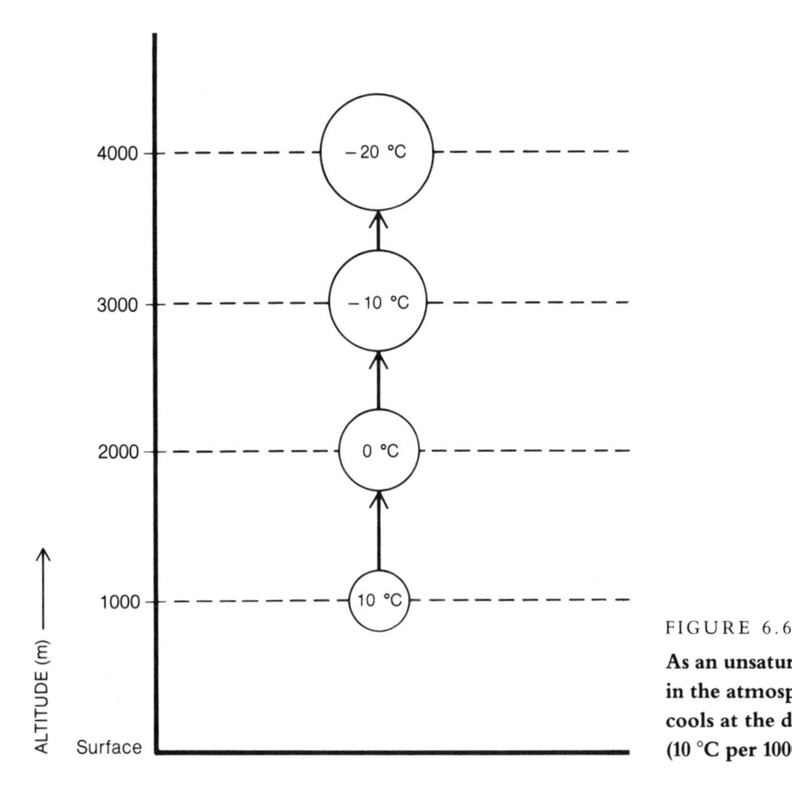

FIGURE 6.6
As an unsaturated parcel of air rises in the atmosphere, it expands and cools at the dry adiabatic lapse rate (10 °C per 1000 m of ascent).

parcel, and the parcel temperature rises 10 °C for every 1000 m of descent. (For more on energy conservation and the adiabatic process, refer to the Mathematical Note at the end of the chapter.)

Should rising air parcels cool to such a point that condensation or deposition takes place, the ascending air then no longer cools at the dry adiabatic rate (Figure 6.7). When condensation or deposition occurs, latent heat is released and partially counters adiabatic cooling. As a consequence of latent heat release, a saturated ascending air parcel cools more slowly than the dry adiabatic lapse rate, and the rate of cooling depends on the water vapor content of the air parcel. This lower cooling rate is called the **moist adiabatic lapse rate**, and varies from about 3 °C per 1000 m (2 °F per 1000 ft) for very humid air, to about 9 °C per 1000 m (5 °F per 1000 ft) for air having a very low mixing ratio. An average value for the moist adiabatic lapse rate may be chosen for convenience, that is, 7 °C per 1000 m (3.5 °F per 1000 ft).

As long as an unsaturated air parcel moves upward, its temperature will drop and its relative humidity will approach saturation. Forces arising from density differences within the atmosphere

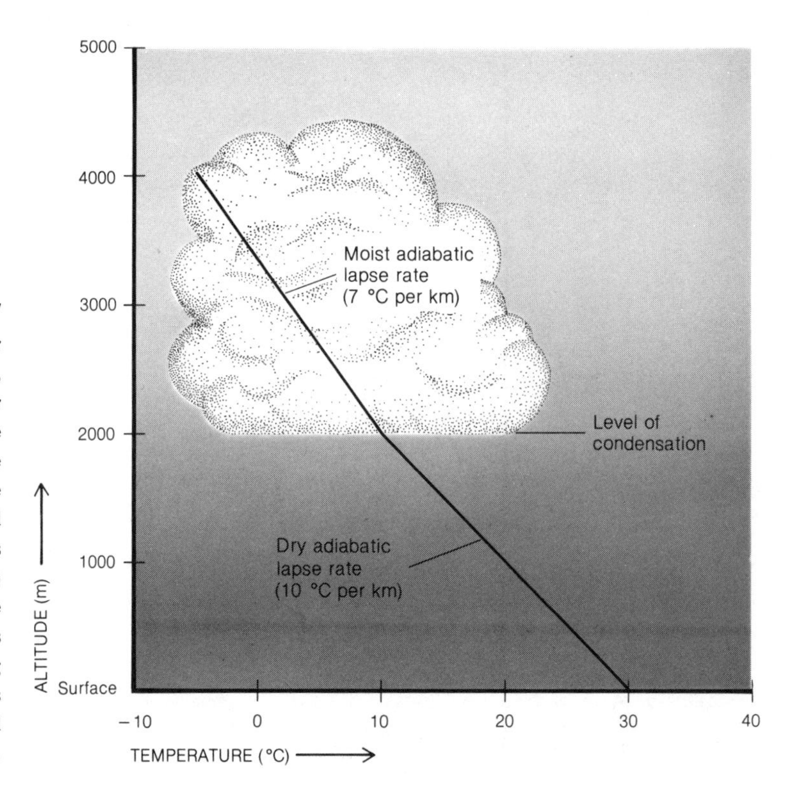

FIGURE 6.7

Rising parcels of unsaturated air cool at the dry adiabatic lapse rate, but as the rising parcels cool, their relative humidity increases. At the level of condensation, the relative humidity is 100 percent, and the parcels are saturated. Continued lifting of the saturated parcels causes condensation (or deposition) and the release of latent heat. The latent heat release partially offsets the adiabatic cooling, so that rising saturated air parcels do not cool as rapidly as rising unsaturated parcels.

may either enhance or suppress this vertical motion of air. The net effect depends on the stability of the atmosphere.

Atmospheric stability

An air parcel is subject to the buoyant forces that arise from density differences between the parcel and the surrounding air. Recall that at constant pressure, the warmer an air parcel is, the lower its density will be. Parcels that are warmer (lighter) then the surrounding **ambient air** therefore tend to rise, and parcels that are cooler (denser) than the ambient air tend to sink. An air parcel continues to rise or sink until it reaches air of the same temperature (or density). We determine the **stability** of an air layer by comparing the temperature change of an ascending or descending air parcel with the thermal profile, or sounding, of the air layer in which the parcel ascends or descends.

The cooling rate of a rising air parcel depends on whether the parcel is saturated or unsaturated. In either case, atmospheric stability is determined by comparing the air parcel temperature change with altitude to the vertical temperature profile of the ambient air. Recall from Chapter 1 that the vertical temperature profile (or sounding) of ambient air is measured by a balloon-borne radiosonde. An air layer is **stable** when an air parcel ascending within the layer becomes cooler than the ambient air, or when a descending air parcel becomes warmer than the ambient air. For either upward or downward displacements, an air parcel in stable air thus returns to its original altitude. An air layer is **unstable** when an ascending air parcel becomes warmer than the ambient air and continues to ascend, or when a descending air parcel becomes cooler than the ambient air and continues to descend.

Although both the adiabatic lapse rates for unsaturated and saturated air are fixed, the temperature profile (sounding) of the troposphere is not. The temperature profile—and hence the atmospheric stability—changes significantly from season to season, from day to day, and even from hour to hour. As an illustration, Figure 6.8 shows the change in thermal profile from 6 A.M. to noon on a summer day. By sunrise, nocturnal radiational cooling has stabilized the air layer closest to the ground, but by noon, the bright summer sunshine has warmed the ground and destabilized the air layer. A further complication is that stability can change with al-

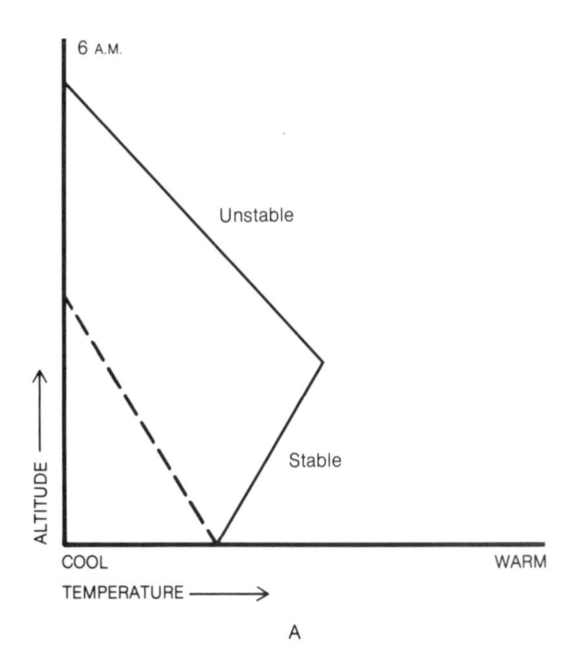

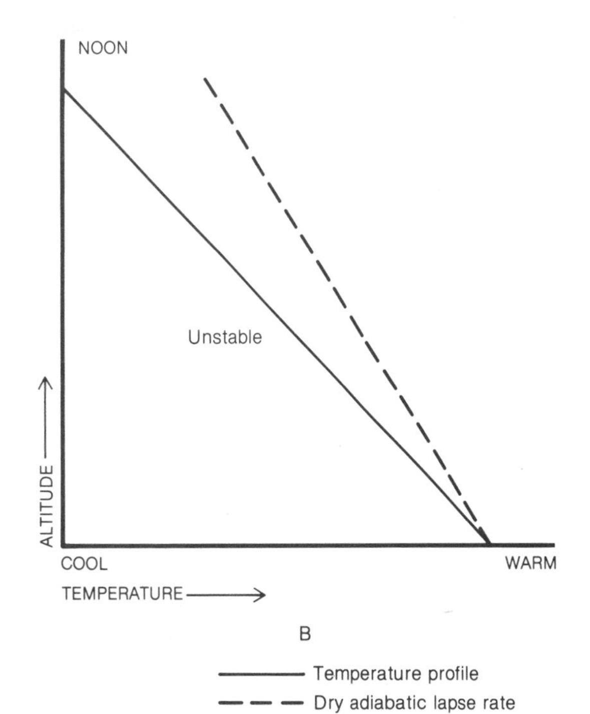

FIGURE 6.8

Nocturnal radiational cooling generates a surface temperature inversion by dawn (A). By noon, however, bright sunshine has warmed the ground and destabilized the overlying air layer (B).

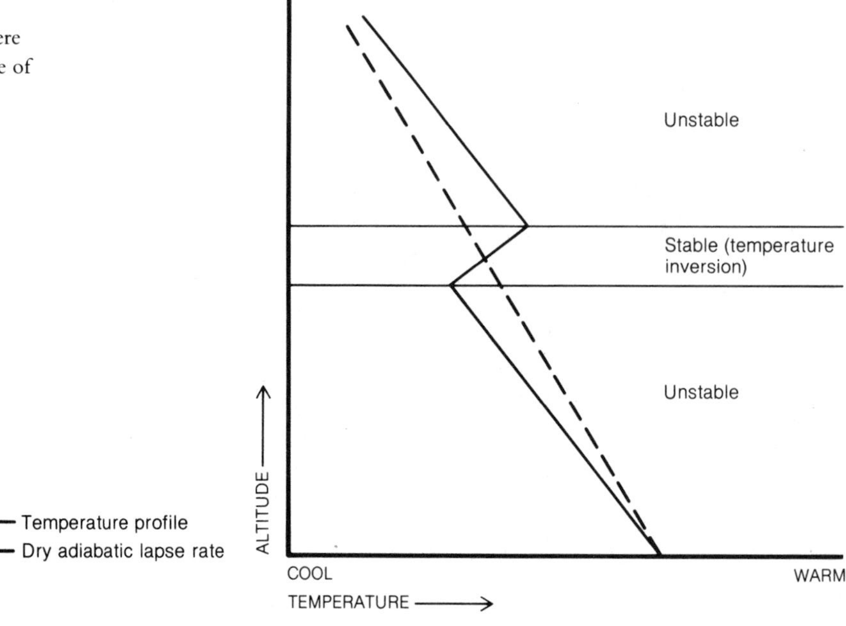

FIGURE 6.9

Atmospheric stability can change with altitude. In this case, a stable air layer (a temperature inversion) is sandwiched between lower and upper unstable air layers.

titude. As shown in Figure 6.9, a stable air layer may occur over an unstable air layer.

Figure 6.10 summarizes a number of possible atmospheric stability conditions, along with the dry adiabatic and average moist adiabatic lapse rates. If the sounding indicates that the temperature of the ambient air is dropping more rapidly with altitude then the dry adiabatic lapse rate (that is, more than 10 °C per 1000 m), then the ambient air is unstable for both saturated and unsaturated air parcels. This situation is called **absolute instability**. If the sounding lies between the dry adiabatic and moist adiabatic lapse rates, **conditional stability** prevails, that is, the air layer is stable for unsaturated air parcels and unstable for saturated air parcels. An air layer is stable for both saturated and unsaturated air parcels when the sounding indicates one of the following conditions:

1. The temperature of the ambient air drops more slowly with altitude than the moist adiabatic lapse rate.
2. The temperature is constant with altitude (called **isothermal**).
3. The temperature increases with altitude (a **temperature inversion**).

These three are conditions of **absolute stability**.

It is evident that atmospheric stability influences weather by affecting the vertical motion of air. Stable air suppresses vertical mo-

FIGURE 6.10

Atmospheric stability is determined by comparing vertical temperature profiles (solid lines) with the dry adiabatic lapse rate in the case of unsaturated air parcels and with the moist adiabatic lapse rate in the case of saturated air parcels.

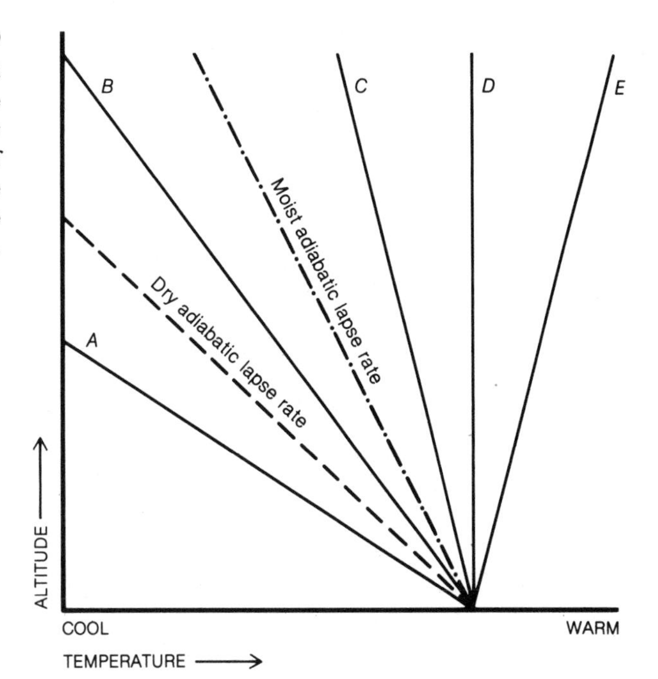

A Absolute instability
B Conditional stability
C Absolute stability (lapse)
D Absolute stability (isothermal)
E Absolute stability (inversion)

tion, and unstable air enhances vertical motion, convection, and cloud development. Because stability also affects the rate at which polluted air mixes with clean air, stability must be considered in assessing air pollution potential. This point is examined in Chapter 16.

Lifting Processes

Expansional cooling occurs through orographic lifting, frontal uplift, and convective currents. **Orographic lifting** occurs when air is forced upward by topography, the physical relief of the land. As winds sweep the land, topographic irregularities force the moving air to alternately rise and sink. If relief is sufficiently great, the resulting expansional cooling and compressional warming of the air will affect cloud and precipitation patterns. For example, a mountain range lying perpendicular to the direction of the prevailing wind forms a natural barrier that results in heavier precipitation on the side of the mountain range facing the wind. As air is forced to rise, it expands, and its temperature falls. The relative humidity also increases, condensation may occur, and eventually precipitation,

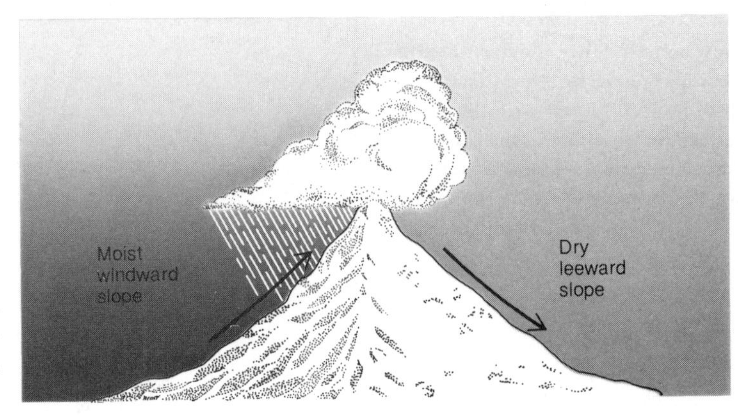

FIGURE 6.11

A mountain range that intercepts a flow of humid air may induce a cloudy, rainy climate on its windward flanks and a dry climate on the leeward flanks.

FIGURE 6.12

A rain forest in the Olympic National Park of northwestern Washington (A) is in marked contrast to the cold desert vegetation of eastern Oregon (B). A mountain range between the two areas accounts for the disparity in precipitation. (Photograph A from National Park Service; photograph B from Oregon Department of Transportation)

A

B

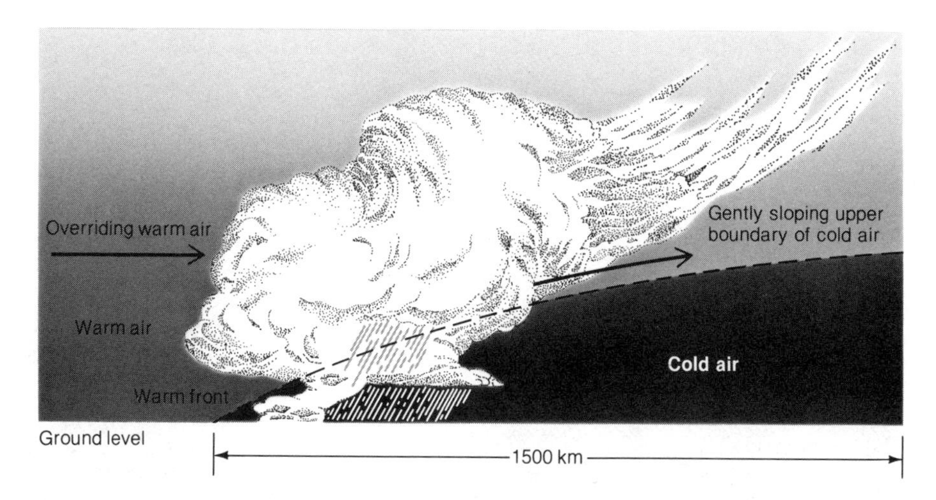

FIGURE 6.13

As warm, light air displaces cooler, heavier air, the warm air overrides the cool air along a gently sloping front. The warm air cools to the saturation point, and clouds develop. (Modified from Anthes, R. A., et al. *The Atmosphere*, 3rd edition. Columbus, Oh.: Charles E. Merrill, 1981, p. 207.)

called **orographic precipitation**, may develop. On the opposite, leeward side of the range (away from the wind), air descends and is warmed, which reduces the relative humidity and the likelihood of cloud formation and precipitation. The mountain range thus establishes two contrasting climatic zones: a moist climate on the windward side and a dry climate on the leeward side (Figure 6.11).

This orographically induced disparity in precipitation is especially apparent from west to east across Washington and Oregon, where the north-south Cascade Range intercepts the prevailing humid airflow off the Pacific. The result is exceptionally rainy conditions in the western portions of these states and arid conditions in the eastern portions. Figure 6.12 shows two contrasting Pacific Northwest landscapes that exhibit this disparity. As a consequence, markedly different plant and animal communities are found on opposite slopes of the mountain barrier. The disparity in precipitation also has a direct impact on domestic water supply, the type of crops grown, and the type of shelter that must be built. For example, Denver, located on the leeward side of the front range of the Rocky Mountains, must obtain some of its water from the wet western side of the range by means of tunnels.

Clouds and precipitation are often triggered by **frontal uplift**, which occurs when warm and cool air masses meet. Because an advancing, warm, humid air mass is lighter, it rides over a cold, dry air mass (Figure 6.13). The leading edge of the warm air at the earth's surface is called a **warm front**. The denser, cold, dry air displaces the warm, humid air by pushing under it (Figure 6.14). The leading edge of cold air at the earth's surface is called a **cold front**. The net effect of the replacement of one air mass by another

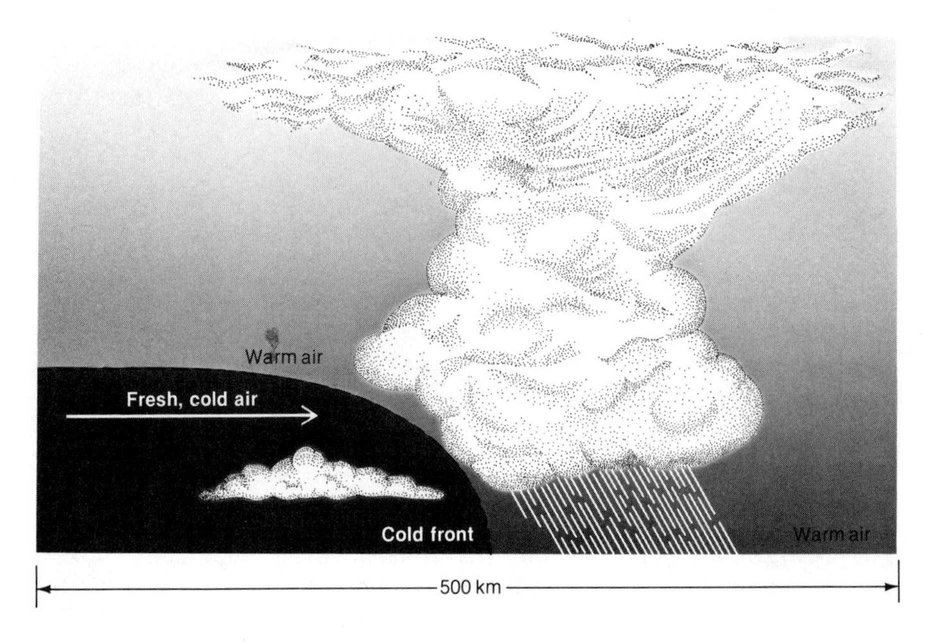

FIGURE 6.14
As cool air displaces lighter warm air, the cool air slides under the warm air, thus causing the warm air to rise along a steeply sloping front. The warm air eventually cools to saturation, and clouds develop. (Modified from Anthes, R. A., et al. *The Atmosphere*, 3rd edition. Columbus, Oh.: Charles E. Merrill, 1981, p. 210.)

air mass is the lifting of air, which, in turn, leads to expansional cooling, cloud development, and perhaps rainfall or snowfall.

Convection can also lead to the formation of clouds and precipitation. It is an important means of heat transfer (Chapter 4) that develops when the sun heats the earth's surface, and the warm surface in turn heats the air in contact with it. The warmed air rises, expands, and cools. Eventually, the rising air becomes so cool and heavy that it sinks back to the surface, thus completing the convective cell circulation. Clouds may form where convection currents are surging upward, but clouds dissipate where the currents are downward. In general, the higher the convection current reaches into the atmosphere, the greater the expansional cooling will be, and the more likely the occurrence of clouds and precipitation. Convection currents that reach to great altitudes spawn thunderstorms.

Conclusions

In this chapter, we explained how expansional cooling can lead to the saturation of air. Radiational cooling and increases in water vapor concentration can also raise the relative humidity to saturation. These processes are described in Chapter 7, along with the mechanics of cloud development and the classification of clouds.

MATHEMATICAL NOTE

Energy conservation and the dry adiabatic process

Energy cannot be created or destroyed, but it can change from one form to another. Another way to express this *law of energy conservation* is to observe that in any physical or biological process, regardless of the transformations that take place, all of the original energy can be accounted for. The application of this law to atmospheric processes provides us with valuable information about the workings of weather, particularly those processes that are adiabatic.

Heat added to an air parcel is either added to the parcel's store of internal energy or is used to do work on the parcel. Conversely, heat released by an air parcel is either subtracted from the internal energy of the parcel or is the consequence of the parcel's work on its surroundings. In order to understand this concept, let us first examine separately (1) internal energy, and (2) work done on or by an air parcel. We will then combine the two in one expression of energy conservation.

Internal energy refers to the molecular activity of the mixture of gases comprising an air parcel. In an air parcel, molecular activity is directly proportional to heat energy, so an addition of heat to the parcel accelerates molecular activity, and a subtraction of heat slows molecular activity. A change in the internal energy, or heat energy, of an air parcel (ΔQ) is therefore proportional to a change in the temperature (ΔT) of the parcel. That is,

$$\Delta Q \propto \Delta T$$

As we saw in Chapter 2, however, internal energy and temperature are related through the specific heat of the air (C). So,

$$\Delta Q = C \times \Delta T$$

Actually, the specific heat of dry air can be evaluated at constant pressure (C_P = 1005 J per km per K) or, as in this case, at constant volume (C_V = 718 J per kg per K). Hence,

$$\Delta Q = C_V \times \Delta T$$

Work done on or by an air parcel consists of compression or expansion of the air parcel. An illustration will demonstrate this concept. Suppose we have a sealed cylindrical container of air as in Figure 1. At one end, the cylinder is equipped with a piston that can compress the air sample or allow the air sample to expand. If we compress the air, we use energy to do work on the air sample. If we then release the piston, the air sample expands and works against the piston by

(continued)

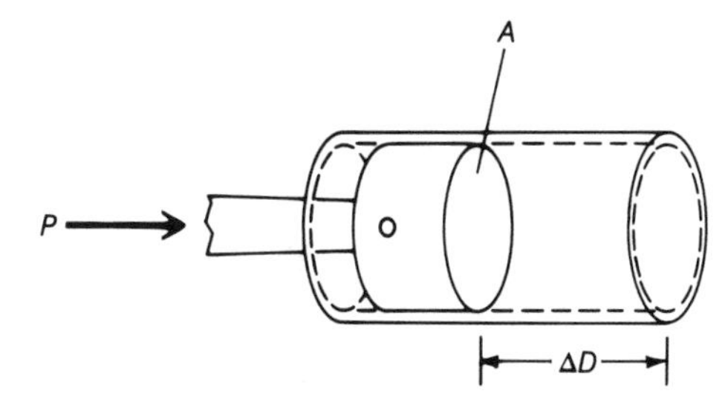

FIGURE 1

A piston is used to compress a sample of air or to allow the air sample to expand. Pressure, P, is applied over a distance, ΔD. The piston has a cross-sectional area, A.

Energy conservation and the dry adiabatic process

(continued)

pushing. When an air parcel is compressed, work is therefore done on it, but when an air parcel expands, the air does work on its surroundings. Energy is either supplied (during compression) or released (during expansion).

We can calculate the work (or energy) involved in the compression or expansion of air by referring to the piston example. The energy required to compress or expand the air sample is simply the product of the required force (F) times the distance over which the force is applied. (Pressure is a force per unit area.) If we apply a pressure (P) on an air sample by moving a piston of fixed cross-sectional area (A) a variable distance (ΔD), there is a change in the volume (V) of the air sample:

$$\Delta V = A \times \Delta D$$

The energy required (ΔQ) would be:

$$\Delta Q = F \times \Delta D$$
$$\text{or } \Delta Q = P \times A \times \Delta D$$
$$\text{or } \Delta Q = P \times \Delta V$$

FIGURE 2

On a Stüve thermodynamic diagram, an air parcel at point **A** *is subjected to a dry adiabatic expansion to point* **B.**

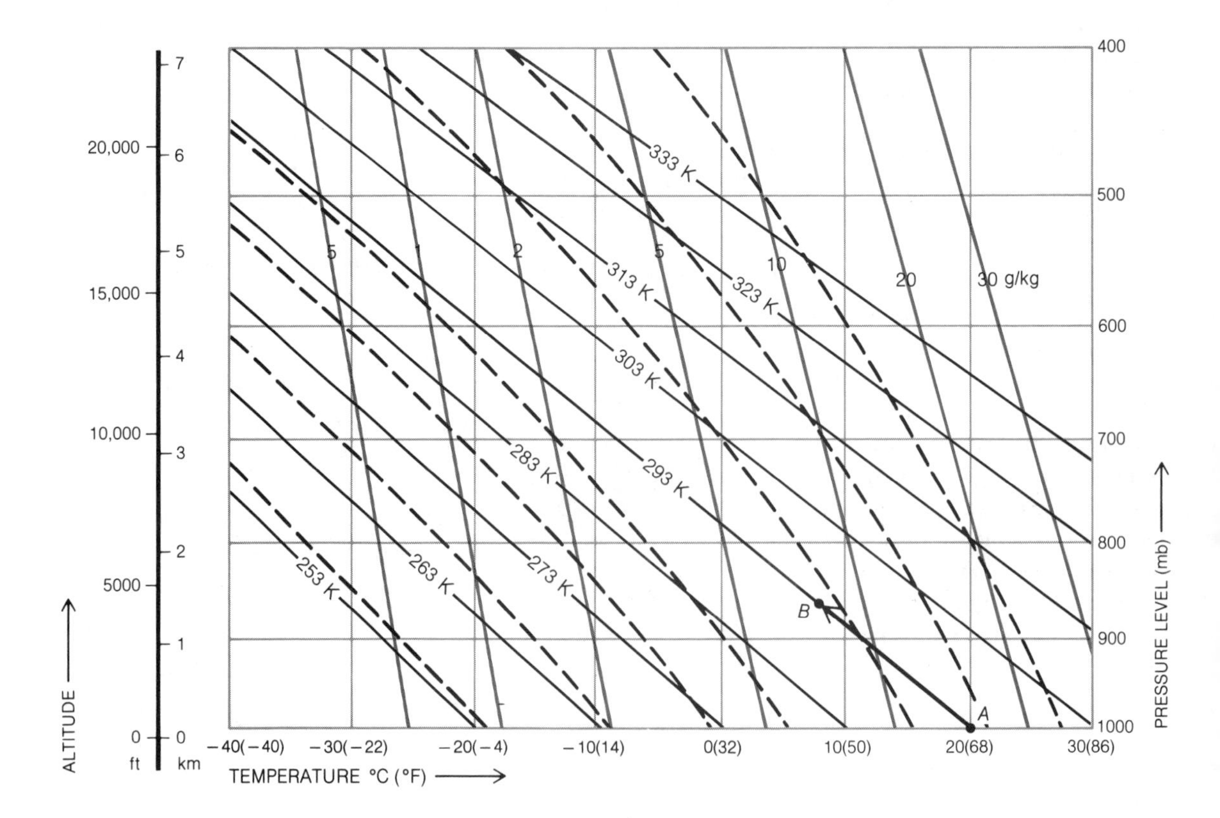

To this point, we have considered separately internal energy and the energy involved in the compression and expansion of air. We can now combine the two components in one equation:

$$\Delta Q = (C_V \times \Delta T) + (P \times \Delta V)$$

The heat flow into and out of an air parcel is thus accounted for by changes in internal energy, or by the work of expansion or compression of air, or by both.

With this statement of the law of energy conservation, we can now explore the adiabatic process. The *adiabatic assumption* is that there is no net heat flow through the imaginary walls of an air parcel, that is, in the previous equation:

$$\Delta Q = 0$$

Hence,

$$0 = (C_V \times \Delta T) + (P \times \Delta V)$$

This means that when an air parcel undergoes an adiabatic process, a temperature change (ΔT) must accompany a volume change (ΔV). The temperature drops when the volume increases, that is, when an air parcel expands as it rises in the atmosphere. Conversely, the temperature of an air parcel increases when the air volume decreases, that is, when the air parcel is compressed as it sinks within the atmosphere. Ascending unsaturated air expands and cools adiabatically at a constant rate of 10 °C per 1000 m of ascent. This is called the *dry adiabatic lapse rate*. Descending air is compressed and warms at a rate of 10 °C per 1000 m of descent.

Adiabatic processes can be displayed graphically on a *Stüve thermodynamic diagram* like the one in Figure 2. The Stüve diagram presents the relationships among several atmospheric variables. Horizontal lines are *isobars*, lines of constant air pressure, and vertical lines are *isotherms*, lines of constant temperature. Straight black sloping lines are *dry adiabats*, the temperature change of an unsaturated air parcel that is subjected to adiabatic expansion or compression. Dry adiabats are labeled in kelvins (K). Curved dashed lines are *moist adiabats*, the temperature change of a saturated air parcel that is subjected to expansional cooling. Gray sloping lines are lines of equal saturation mixing ratio in grams of water vapor per kilogram of dry air.

As an illustration of the usefulness of a Stüve diagram, consider an unsaturated air parcel that undergoes dry adiabatic expansion as a consequence of uplift (along a front, for example). The air parcel has a temperature of 20 °C (68 °F), a pressure of 1000 mb, and a relative humidity of 50 percent. These initial conditions are plotted as point *A* in Figure 2. The parcel is then lifted so that it cools dry adiabatically until it becomes saturated. At point *A*, the mixing ratio of the air parcel is 8 g per kilogram, since the relative humidity is 50 percent, and from the diagram, we see that the saturation mixing ratio is about 16 g per kilogram. The air parcel temperature thus follows the 293 K dry adiabat to point *B*, where the actual mixing ratio equals the saturation mixing ratio and the relative humidity is 100 percent. Saturation is achieved when the air parcel reaches an altitude of about 1500 m (4500 ft), where the parcel temperature is 8 °C (46.4 °F).

SUMMARY STATEMENTS

- The hydrologic cycle is the ceaseless circulation of water among the oceanic, atmospheric, and terrestrial reservoirs.

- Vapor pressure and mixing ratio are two direct measures of the water vapor concentration in air. Relative humidity indicates how close an air sample is to saturation with respect to water vapor. Relative humidity is temperature dependent.

- Saturation vapor pressure and saturation mixing ratio are uniquely related to air temperature. Both increase in value as air temperature increases. Warm air can therefore accommodate more water vapor than cold air can.

- The relative humidity of air increases when unsaturated air is cooled or when water vapor is added to the air.

- In an adiabatic process, no net heat exchange occurs between an air parcel and the surrounding air. Rising air parcels are cooled by expansion, and descending air parcels are warmed by compression.

- Atmospheric stability is determined by comparing the temperature of an air parcel moving vertically (up or down) through a layer of air with the temperature profile of the air layer. Stable air inhibits vertical motion; unstable air enhances vertical motion.

- Expansional cooling occurs through orographic lifting, frontal uplift, and convection. By these processes, the relative humidity increases, perhaps to the point of cloud development.

KEY WORDS

hydrologic cycle
evaporation
transpiration
evapotranspiration
sublimation
condensation
deposition
precipitation
water budget
drainage basin
watershed
vapor pressure
saturation vapor pressure
mixing ratio
saturation mixing ratio
relative humidity
psychrometer
wet-bulb depression
hair hygrometer
hygrograph
expansional cooling
compressional warming
adiabatic processes
air parcels
dry adiabatic cooling
dry adiabatic lapse rate
moist adiabatic lapse rate
ambient air
atmospheric stability
stable air layer
unstable air layer
absolute instability
conditional stability
isothermal
temperature inversion
absolute stability
orographic lifting
orographic precipitation
frontal uplift
warm front
cold front

REVIEW QUESTIONS

1 Distinguish among evaporation, transpiration, and sublimation.

2 Provide several examples of sublimation.

3 What is the difference between condensation and deposition?

4 Explain how heat energy is involved in the phase changes of water.

5 How does the hydrologic cycle purify water?

6 It follows from the global water budget that there must be a net flow of water from the continents to the oceans. Why?

7 Distinguish between air pressure and vapor pressure. How do they compare in magnitude?

8 How do temperature changes influence (a) the saturation vapor pressure, (b) the saturation mixing ratio, and (c) the relative humidity?

9 Under what condition is the mixing ratio equal to the saturation mixing ratio?

10 Why does the relative humidity usually fall between sunrise and early afternoon on a clear and calm day?

11 In localities where winters are cold, some central home heating systems are equipped with humidifiers. Why?

12 Describe how relative humidity is measured using a psychrometer.

13 List several ways whereby the relative humidity of air is increased.

14 Describe the change in temperature as a sample of air alternately rises and sinks within the atmosphere.

15 What is an adiabatic process and how does it occur in the atmosphere?

16 Rising parcels of saturated air do not cool as rapidly as rising parcels of unsaturated air. Please explain.

17 How does the stability of the ambient air influence upward and downward movements of unsaturated air parcels?

18 How does the topography of the land influence cloud and precipitation patterns?

19 Why are clouds and precipitation often associated with fronts?

20 Both clear air and cloudy air are associated with convection within the atmosphere. Please explain.

POINTS TO PONDER

1 A relative humidity of 25 percent measured on a cold day in January does not mean the same as a relative humidity of 25 percent measured on a hot day in July. Please explain.

2 With intense radiational cooling during the early morning hours, the air temperature is observed to drop rapidly until frost forms. Thereafter the air temperature either becomes steady or falls very slowly. Explain the change reflected by the behavior of the temperature.

3 Determine the saturation mixing ratio of an air sample having a relative humidity of 25 percent and a mixing ratio of 3 g water vapor/kg dry air.

4 Determine the relative humidity of an air sample having a temperature of 20 °C (68 °F) and a vapor pressure of 6 mb.

5 Speculate on the significance of conditional stability for convective currents.

6 How might air mass advection influence air stability?

7 Speculate on how a freshly fallen layer of snow might influence the stability of the overlying air.

PROJECTS

1 Locate some everyday examples of expansional cooling and compressional warming.

2 Check with a home appliance store to determine how a dehumidifier works.

3 For one or two full days keep track of the hourly temperature and relative humidity readings. Speculate on whether changes in relative humidity are due to local radiative effects (heating or cooling) or air mass advection.

SELECTED READINGS

Bryan, K. "The Ocean Heat Balance." *Oceanus* 21 (1978):18–26. *A detailed description of heat exchange at the ocean-atmosphere interface.*

Forrester, F. H. "An Inventory of the World's Water." *Weatherwise* 38 (1985):84–105. *A description of the reservoirs in the global hydrologic cycle.*

Leopold, L. B. *Water: a Primer.* San Francisco: W. H. Freeman, 1974. 166 pp. *Concise and well-illustrated treatment of the water cycle.*

I am the daughter of earth
and water,
And the nursling of the
sky:
I pass through the pores of the
ocean and shores;
I change, but I cannot die.
PERCY BYSSHE SHELLEY
The Cloud

Clouds take a myriad of forms. Clouds with sharp edges are composed mostly of water droplets, whereas clouds with fibrous or fuzzy edges are composed mostly of ice crystals. (Photograph by Arjen Verkaik)

SEVEN

Dew, Frost, Fog, and Clouds

CLOUDS whisk across the sky in ever-changing patterns of white and gray; fog lends an eerie silence to a dreary day; dew and frost glisten in the morning sun. Clouds, fog, dew, and frost are all products of condensation or deposition of water vapor in the atmosphere. Most clouds are the consequence of saturation brought about by uplift and expansional cooling of air, but dew, frost, and most fog develop when the lowest layers of air become saturated by other means. This chapter describes the formation of dew, frost, and fog, and the development and classification of clouds.

Low-Level Saturation Processes

Air that is in contact with the earth's surface can become saturated if its temperature is lowered sufficiently. Such cooling decreases the saturation mixing ratio and thus raises the relative humidity. As the relative humidity approaches 100 percent, dew, frost, or fog may form.

Dew and frost

The overnight emission of infrared radiation from the earth's surface lowers the temperature of the ground and, through conduction, the temperature of the air next to the ground is also lowered. If nighttime skies are clear, this **radiational cooling** may be so great that the relative humidity of the air in immediate contact with the earth's surface reaches 100 percent by dawn, which is usually the coolest time of day. If temperatures fall below the freezing point, water vapor may then be deposited on objects as **frost** (Figure 7.1), or if temperatures are above freezing, water vapor may condense as **dew** (Figure 7.2). Note that dew and frost are not forms of precipitation, because they do not "fall" from clouds, but develop in place on surfaces. A similar phenomenon occurs when water droplets appear on the outside of a glass of cold soda on a hot summer day. The "sweat" of the glass is actually dew.

The temperature to which air must be cooled to reach saturation is called the **dew point**. The dew point is also a measure of the air's water vapor content. The higher the dew point temperature is, the greater the amount of water vapor in the air. It also follows that when the difference between the actual air temperature and the dew-point temperature is small, the relative humidity will be high. Further, the dew point is an index of human comfort. Although

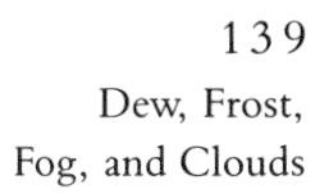

FIGURE 7.1

When air temperatures are at or below freezing, water vapor may be deposited on objects as frost, as a consequence of nocturnal radiational cooling. (Visuals Unlimited photograph by M. Gabridge)

the relative humidity is an important consideration, as a rule, many people experience discomfort when dew points rise above 20 °C (68 °F).

Note that the dew-point temperature should not be confused with the wet-bulb temperature (Chapter 6). They are not the same. Recall that the wet-bulb temperature is determined by inducing evaporative cooling. Adding water vapor to the air increases the temperature at which dew will form. Except at saturation, the wet-bulb temperature is therefore higher than the dew-point tem-

FIGURE 7.2

When air temperatures are above freezing, water vapor may condense on objects as dew, as a consequence of nocturnal radiational cooling. (Photograph by Mike Brisson)

perature because the wet bulb is adding water vapor to the air. At saturation, the dew point, the wet-bulb temperature, and the actual air temperature are the same. As shown in Appendix II, the dew point can be obtained from measurements of the dry-bulb temperature and the wet-bulb depression. For example, if the dry-bulb temperature is 20 °C (68 °F) and the wet-bulb depression is 5 °C (9 °F), the dew point is 12 °C (54 °F).

When saturation takes place at temperatures at or below 0 °C (32 °F), then frost forms, and the temperature is called the **frost point**. Frost occurs in a variety of forms: delicate feathery patterns of crystals may develop on a windowpane during a cold winter night, or fernlike crystals of **hoarfrost** may grow to a length of several centimeters on the twigs and branches of trees (Figure 7.3).

When neither cold air advection nor warm air advection is expected, the dew point (or frost point) sometimes can be used to predict the next morning's minimum air temperature. Suppose, for example, it is autumn in New England and weather forecasters are

FIGURE 7.3. **As a consequence of deposition of water vapor from humid, cold air, crystals of hoarfrost may grow to several centimeters in length on the twigs and branches of trees. (Photograph by Mike Brisson)**

calling for clear nighttime skies and calm winds—conditions ideal for rapid nocturnal radiational cooling. Gardeners know that all the ingredients are present for chilly temperatures by early morning, and they want to know whether they should go to the trouble of protecting their frost-sensitive plants. As a general rule of thumb, the late afternoon's dew point is a reasonable estimate of the next morning's low temperature: if the dew point is near or below 0 °C (32 °F), frost protection is advisable.

What is the physical basis for this rule of thumb? In response to nocturnal radiational cooling, the air temperature drops continually until the relative humidity nears 100 percent and condensation or deposition occurs (producing dew, frost, or fog). The latent heat released during condensation or deposition offsets the radiational cooling to some extent and hence, the air temperature tends to stabilize at the dew point or frost point. Many other factors, however, may complicate this simple rule. For one, the length of the night is an important control of the amount of radiational cooling. Summer nights may be too short and the air too dry (low dew point) for radiational cooling sufficient to lower the air temperature to the dew point or frost point.

It is a popular misconception that Florida citrus farmers fear a winter frost. What they actually fear is a solid *freeze*, that is, a cold snap that freezes the water in plant tissues and causes potentially lethal damage to the plant. In fact, a covering of frost may actually protect plants from such damage. (For more information on freeze and frost protection for plants, see the Special Topic.)

Fog

Fog is a visibility-restricting suspension of tiny water droplets or ice crystals in an air layer next to the earth's surface. Simply put, fog is a cloud in contact with the ground. By international convention, fog is defined as restricting visibility to 1 km (0.62 mi) or less; otherwise the suspension is called **mist**. Fog may develop when air becomes saturated through radiational cooling, advective cooling, the addition of water vapor, or expansional cooling.

With clear night skies, light winds, and an air mass that is humid near the ground and relatively dry at higher altitudes, radiational cooling may cause the layer of air at ground level to approach saturation. When these conditions occur, a cloud develops. A ground-level cloud formed in this way is called **radiation fog**

SPECIAL TOPIC

Frost and freeze control

Several measures can be taken to reduce crop damage by a freeze resulting from nocturnal radiational cooling. One of the most effective strategies to protect against late spring or early fall frosts and freezes is to avoid sites into which dense, cold air may drain. Such frost-prone areas include valley bottoms and other topographic depressions. In such sites, the growing season may be several weeks shorter than on the surrounding slopes. For this reason, orchards and vineyards are often situated on valley slopes rather than on valley floors. Even site selection on the slopes must be carefully made. Obstructions such as a hedgerow, road, or railroad embankment can impede the downslope drainage of cold air and thus damage crops immediately upslope from the barrier. To reduce this problem, growers construct channels through the barrier so that cold air can continue to drain downslope.

Based on an understanding of the conditions that contribute to radiational cooling, several other methods have been devised to reduce the incidence of radiation frosts and freezes. One factor that contributes to radiation frosts and freezes is the absence of clouds that would absorb infrared radiation emitted by the ground and vegetation, and subsequently reemit a portion of that radiation back to the crop. To protect crops, growers can create their own clouds to intercept the outgoing infrared radiation. It is widely believed that a smoke cloud emanating from smudge pots or oil burners inhibits freezing temperatures by producing a "greenhouse effect." However, smoke particles are actually nearly transparent to infrared radiation so that the "greenhouse effect" is negligible. Hence, the warming produced by smudge pots or oil burners actually consists primarily of sensible heating (that is, conductive and convective heating). On the other hand, a mist cloud produced by a fine water spray does cause a "greenhouse effect" and can protect crops. (Recall from Chapter 3 that water is strongly absorptive of infrared radiation.)

For small plants, other types of radiation screens are effective. For example, plastic "hot caps" can be placed over plants to create a protective microscale climate around them (Figure 1). During the day, the soil and the plants are warmed by the sun. If hot caps are placed over plants in the late afternoon, the heat gained during the day is better conserved at night, reducing the chances of frost formation on the plants. If the plants are somewhat larger, other radiation screens, such as wooden slats or cheesecloth, can restrict exposure to clear night skies without reducing significantly the solar energy available during the day for crop growth.

A temperature inversion often develops within the lowest air layer as a result of extreme nocturnal radiational cooling. While air temperatures at ground level may fall below the freezing point, temperatures near the top of the inversion (15 m or so above the ground) may be several degrees above freezing. The stability of the inversion layer prevents the warmer air aloft from mixing

FIGURE 1

A field of young tomato seedlings in the San Joaquin Valley, California, is protected from wind and radiational cooling by "hot caps" until the roots are established. (U.S. Bureau of Reclamation)

with the colder air at ground level. A logical strategy, then, is to mount large motor-driven fans or propellers on towers to circulate the warmer air aloft down to the ground level. Such a procedure is standard practice in citrus groves (Figure 2) where the economic value of the crop justifies the high capital investment.

Crops are sometimes protected from freezing by spraying them with a fine water mist when the plant tissues reach 0 °C (32 °F). Our initial reaction may be to question how a coating of ice can help the plants survive. Although the freezing water at the plant surface is at 0 °C, the heat released during the phase change from liquid water to ice helps stabilize plant temperatures and prevents plant damage. As the water freezes, latent heat is released to the underlying plant tissues. Furthermore, plants are typically not damaged until their tissues reach −1 °C to −2 °C (28 °F to 30 °F). Nevertheless, this procedure does require careful monitoring. As long as the sprinkling and freezing continue, the temperature of the ice remains at about 0 °C. If, however, the sprinkling is discontinued before ambient air temperatures rise enough to melt the ice, then heat will be conducted from the leaves to the ice, and the leaf temperature will be lowered to potentially lethal levels. In addition, care must be taken that the ice burden does not become so heavy that the plants are damaged by excess weight. For this reason, the sprinkling method is most suitable for low-growing vegetable crops such as cucumbers and strawberries.

Spraying reduces the threat of freeze damage by adding latent heat. Sensible heat can also be added directly through fuel combustion. Placing fuel-burning heaters at numerous sites throughout an orchard can raise both plant and air temperatures by several degrees.

(continued)

Frost and freeze control

(continued)

FIGURE 2

Ten-meter-high wind machines pull warmer air down among the orange trees during the cold nights in the San Joaquin Valley. (U.S. Department of Agriculture)

These heaters are effective because the temperature inversion tends to trap the heated air. The benefit of a heater falls only to those plants directly exposed to the warm plume of air, so using many small heaters is considerably more effective than using a few large ones.

These frost and freeze prevention techniques are not always effective, however. For three days around Christmas 1983, an arctic air mass surged into Florida dropping temperatures to well below freezing through much of the state. Damage to the citrus crop exceeded $1 billion. Again, in mid-January 1985, another cold snap of comparable intensity caused further damage to surviving citrus trees. Losses from this double blow may well mean an end to the citrus industry in the northern reaches of Florida's citrus belt.

(Figure 7.4). This type of fog is often short lived, vaporizing as the morning sun raises air temperatures and lowers the relative humidity. In winter, however, when the weak rays of the sun are reflected readily by the top of the fog layer, radiation fog may be quite persistent. For example, in California's interior valley during winter, radiation fog often lasts for several weeks.

In regions of topographic relief, air chilled by radiational cooling drains down slopes and settles in low-lying areas such as river valleys. Hilltops may thus be clear of fog, while in deep valleys, the fog is thick and persistent.

The temperature and the water vapor content of an air mass depend on the nature of the surface over which the air mass forms and travels. As an air mass moves from one place to another, termed **air mass advection**, these characteristics change, partly as a result of the modifying influence of the surfaces over which the air mass travels. When the advecting air passes over a relatively cold surface, the air mass may be cooled to saturation in its lower layers. This type of cooling is known as advective cooling and occurs, for example, in spring, when mild, humid air flows over relatively cold, snow-covered ground. Snow on the ground may chill the air to such a point that fog develops. In a more common example, warm, humid air passing over cold ocean or lake water produces persistent and dense fog (Figure 7.5). Fogs generated by advective cooling are called **advection fogs**.

Lowering the air temperature increases the relative humidity, because the capacity of air to hold water vapor is lowered as the air

FIGURE 7.4

Radiation fog develops as a consequence of extreme nocturnal radiational cooling. Radiational cooling lowers the temperature of a layer of humid air to the saturation point. (Photograph by Mike Brisson)

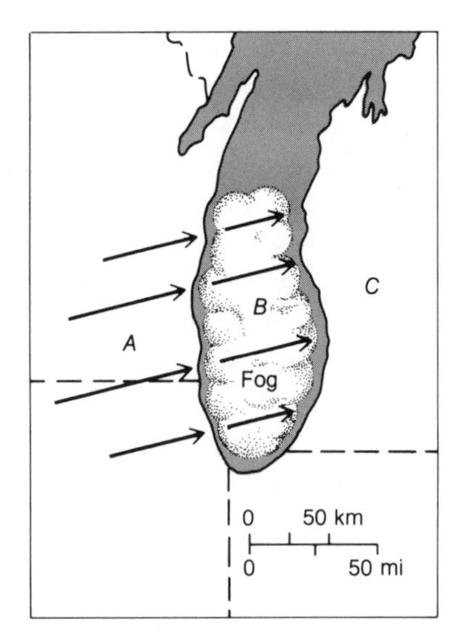

FIGURE 7.5

In summer, as warm humid air (*A*) flows over the relatively cool surface of Lake Michigan, the air is chilled to saturation and dense fog (*B*) forms. As the air continues to flow over the warmer surface of lower Michigan (*C*), air temperatures rise, the relative humidity drops, and the fog dissipates. (From Eichenlaub, V. *Weather and Climate of the Great Lakes Region*. Notre Dame, Ind.: University of Notre Dame Press, 1979, p. 109.)

temperature drops. The relative humidity of air can increase, however, without lowering air temperature. Such an increase usually occurs when a relatively dry air mass moves across an extensive water surface such as a large lake or the ocean. Water evaporates into the air, raising the mixing ratio (or vapor pressure). If evaporation is sufficient, the relative humidity may approach 100 percent. For example, **Arctic sea smoke** is actually fog that develops when extremely cold, dry air flows over a large open body of water. The lower portion of the air mass reaches saturation primarily because of an increase in the water vapor concentration of the air. Saturation occurs even though the air warms somewhat when it comes in contact with the warmer water. The increase in relative humidity caused by the rapid evaporation of the water is more important to the fog-forming process than any lowering of relative humidity caused by the water's warming of the air. Because the air is warmed from below, the fog appears as rising filaments or streamers that resemble smoke or steam. For this reason, a more general name for the fog produced when cold air comes in contact with warm water is **steam fog**. An example is the fog that develops on a cold day over a heated outdoor swimming pool.

Fog also forms on hillsides or mountain slopes as a consequence of the upslope movement of humid air (Figure 7.6). Rising humid air undergoes expansional cooling and eventually reaches saturation. Any further ascent of the saturated air produces fog. Fog formed in this way is called **upslope fog**.

FIGURE 7.6

Upslope fog forms when humid air ascends a mountainside. Expansional cooling brings the air to saturation. (Photograph by Arjen Verkaik)

Cloud Development

Water vapor is an invisible gas, but the condensation and deposition products of water vapor are visible. **Clouds** are the visible manifestations of the condensation and deposition of water vapor within the atmosphere. They are composed of tiny water droplets or ice crystals or both. In this section, we consider the mechanics of cloud development.

Laboratory studies have demonstrated that in clean air, air without dust and other aerosols, the condensation of water vapor is extremely difficult and requires supersaturated conditions (that is, a relative humidity greater than 100 percent). In clean air, the degree of supersaturation needed for cloud development increases rapidly as the radius of the droplets decreases. For example, formation of relatively small droplets with radii of 0.10 μm requires a supersaturation of nearly 340 percent. In contrast, relatively large droplets, having radii greater than 1.0 μm, need only slight supersaturations to form.

In the atmosphere, only slightly supersaturated conditions are necessary for cloud development. This is because the atmosphere contains an abundance of **nuclei**, tiny solid and liquid particles that provide relatively large surface areas on which condensation or deposition can take place. (In other words, the air in the atmosphere is *not* clean air.) Nuclei are the products of both natural and human activity. Forest fires, volcanic eruptions, wind erosion of the soil, saltwater spray, and the discharge from domestic and industrial chimneys are all continual sources.

The nuclei commonly have radii greater than 1.0 μm. This

TABLE 7.1

Some Deposition (Sublimation) Nuclei* and Their Activation Temperatures

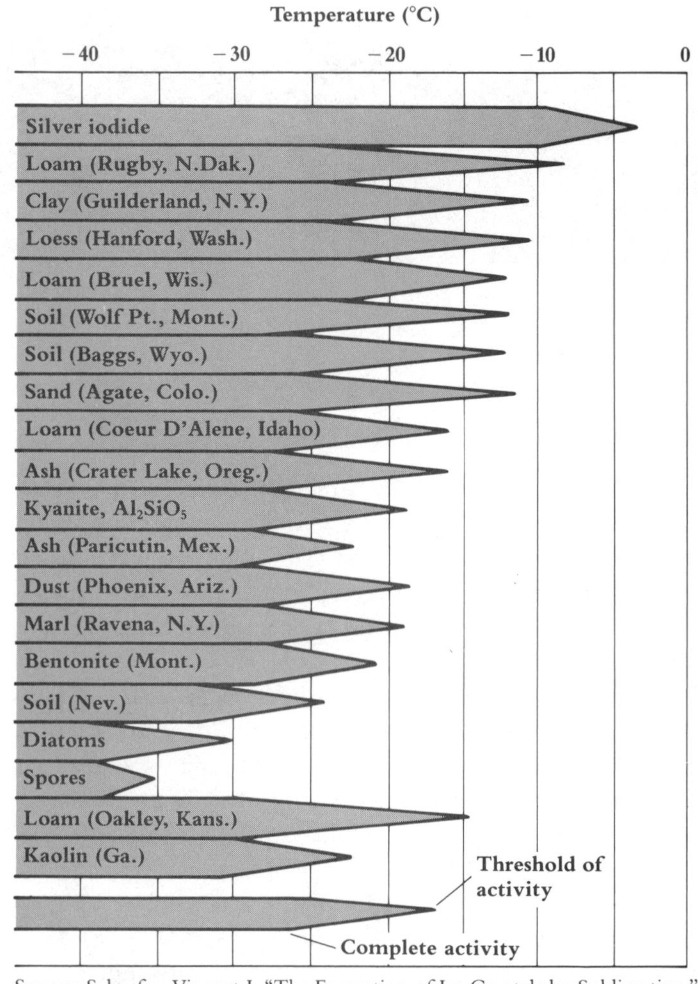

Source: Schaefer, Vincent J. "The Formation of Ice Crystals by Sublimation." *Weatherwise* 32 (1979):256

*Note that most nuclei are soil particles.

means that the nuclei are large enough to facilitate droplet condensation at relative humidities that rarely exceed 100 percent by more than a fraction of 1 percent. In addition, the air contains a plentiful supply of **hygroscopic nuclei**, that is, nuclei having a special chemical affinity for water molecules. Condensation begins on these nuclei at relative humidities under 100 percent. In fact, magnesium chloride, a constituent of sea water, is a hygroscopic substance that initiates condensation at relative humidities as low as 70 percent. Because nuclei are relatively large and because many are hygroscopic, we can expect cloud development when the relative humidity nears 100 percent.

Depending on their specific function, nuclei are classified as one of two types: **cloud condensation nuclei** and **ice-forming nuclei**. Cloud condensation nuclei are active at temperatures both above and below freezing, because water droplets condense and remain liquid even when the cloud temperature is well below 0 °C (32 °F). Such droplets are said to be *supercooled*. Ice-forming nuclei are much less abundant than cloud condensation nuclei and become active only at temperatures well below freezing. There are two types of ice-forming nuclei: (1) **freezing nuclei**, which cause liquid droplets to freeze, and (2) **deposition nuclei** (also called **sublimation nuclei**) on which water vapor deposits directly as ice. Most freezing nuclei activate only at temperatures below −9 °C (16 °F). Deposition nuclei do not become fully active until temperatures fall below −20 °C (−4 °F), as shown in Table 7.1.

Cloud Classification

A casual observer of the sky notices that clouds occur in a wide variety of forms. These forms are not arbitrary. A cloud's appearance is the consequence of processes operating in the atmosphere. In fact, observation of changes in cloud cover sometimes provides clues about future weather.

A British biologist, Luke Howard, is credited with being among the first to devise a classification of cloud types. Published shortly after 1800, the essentials of Howard's scheme are still in use today. Cloud forms are given special Latin names and are organized by appearance and by altitude of occurrence.

On the basis of *appearance*, the simplest distinction is among cirrus, stratus, and cumulus clouds. Cirrus clouds are fibrous, stratus clouds are layered, and cumulus clouds occur as heaps or puffs. On the basis of *altitude*, the most common clouds in the troposphere are grouped into four families (Table 7.2): high clouds, middle clouds, low clouds, and clouds exhibiting vertical development. Members of the first three families are produced by gentle uplift over broad areas. These clouds spread laterally to form layers. Clouds with vertical development generally cover smaller areas and are associated with much more vigorous uplift. These clouds are consequently heaped or puffy in appearance. Later chapters describe the various weather systems that trigger cloud development. Here, we describe briefly the common clouds of each family.

TABLE 7.2

Cloud Classification

GENUS	HEIGHT (IN KM) OF CLOUD BASE ABOVE GROUND	SHAPE AND APPEARANCE
HIGH CLOUDS		
Cirrus (Ci)	**6–18**	**Delicate streaks or patches**
Cirrostratus (Cs)	**6–18**	**Transparent thin white sheet or veil**
Cirrocumulus (Cc)	**6–18**	**Layer of small white puffs or ripples**
MIDDLE CLOUDS		
Altostratus (As)	**2–6**	**Uniform white or gray sheet or layer**
Altocumulus (Ac)	**2–6**	**White or gray puffs or waves in patches or layers**
LOW CLOUDS		
Stratocumulus (Sc)	**0–2**	**Patches or layer of large rolls or merged puffs**
Stratus (St)	**0–2**	**Uniform gray layer**
Nimbostratus (Ns)	**0–4**	**Uniform gray layer from which precipitation is falling**
CLOUDS WITH VERTICAL DEVELOPMENT		
Cumulus (Cu)	**0–3**	**Detached heaps or puffs with sharp outlines and flat bases, and slight or moderate vertical extent**
Cumulonimbus (Cb)	**0–3**	**Large puffy clouds of great vertical extent with smooth or flattened tops, frequently anvil shaped, from which showers fall, with thunder**

Source: Neiburger, M., J. G. Edinger, and W. D. Bonner. *Understanding Our Atmospheric Environment.* San Francisco: W. H. Freeman, 1973, p. 11.

PLATE 1

Cirrus are high, thin, wispy clouds that are composed of ice crystals. (Photo by A and J Verkaik)

PLATE 2

Cirrostratus are high clouds that form a thin, transparent veil over the sky. (Photo by A and J Verkaik)

PLATE 3

Cirrocumulus are high clouds that exhibit a wavelike or mackerel pattern of small, white puffs. (Photo by A and J Verkaik)

PLATE 4

Altostratus are middle-level clouds composed of ice crystals or water droplets or both. They form a uniform gray or white layer. (Photo by A and J Verkaik)

PLATE 5

Altocumulus are middle clouds consisting of patches arranged in a wavelike pattern. (Photo by A and J Verkaik)

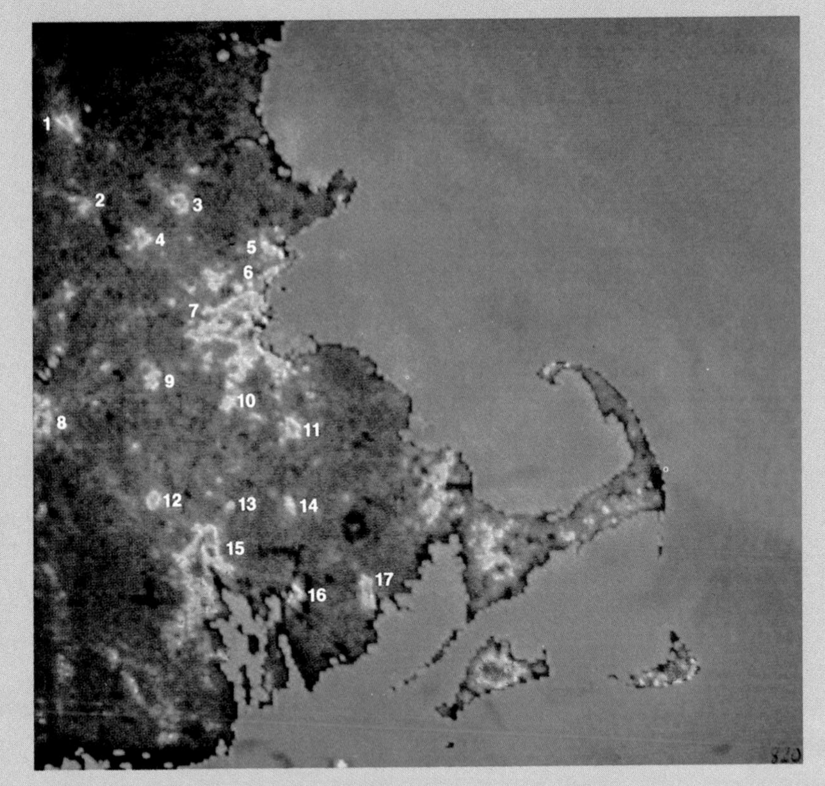

PLATE 17

A color-enhanced image shows 17 urban heat islands in southeastern New England. Number 7 is metropolitan Boston. (Temperature-color relationship: orange-yellow, 31.0–25.5 °C; light green-dark green, 25.0–16.5 °C; light blue-dark blue, 16.0–6.0 °C.) The image was detected by NOAA-5 satellite at 1400 GMT on 23 May 1978, using infrared sensors. Warm areas on Cape Cod and adjacent islands probably result from land-use patterns or soil type. (Courtesy of Michael Matson, National Environmental Satellite Service, NOAA)

PLATE 18

Computer-enhanced GOES satellite infrared photograph of the cloud of ash ejected by Mount St. Helens. The image was taken at 4:45 P.M. Pacific Daylight Time on the day of the eruption (18 May 1980). Ash cloud shows up as a dark blotch covering southern Washington, northern and eastern Idaho, southwestern Montana, and western Wyoming. The dark area to the north is high clouds. (Courtesy of Michael Matson, National Environmental Satellite Service, NOAA)

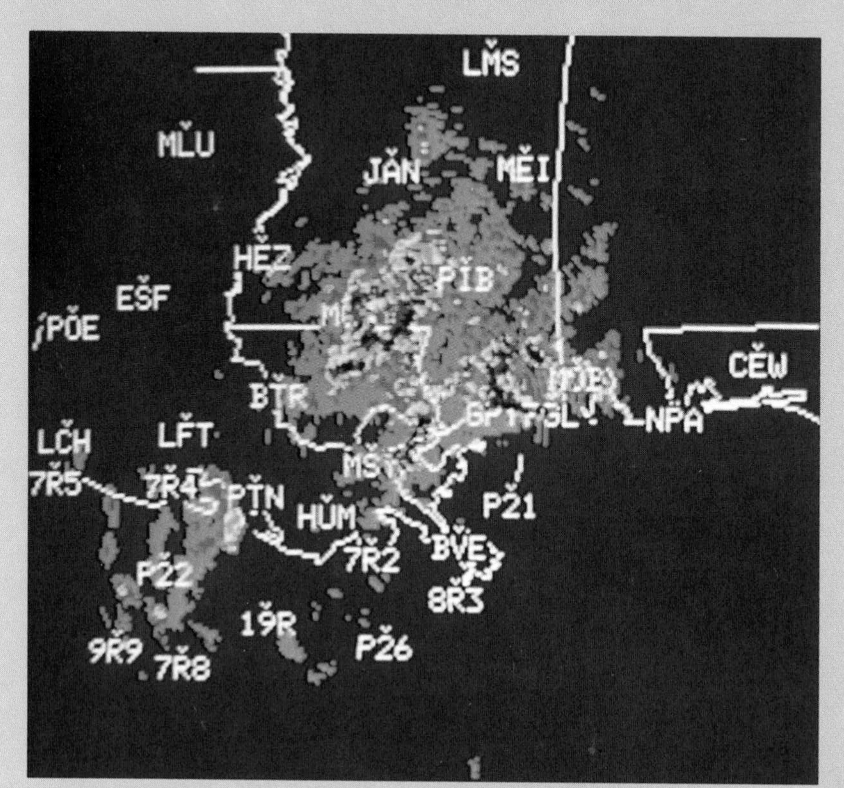

PLATE 19

A weather radar display with echo intensity graduated by color. The most intense echo (heavy rain) is indicated by the dark red, and the weakest echo (lightest rain) by light green. In this case, an area of light-to-moderate rain is occurring over southern Mississippi and southeastern Louisiana. (Courtesy of Alden Electronics)

PLATE 20

Hurricane Elena over the Gulf of Mexico as photographed from the space shuttle Discovery on 2 September 1985. Note the spiral cloud bands surrounding the central eye of the storm. The earth's curvature is visible in the background. (NASA photograph courtesy of the Johnson Space Center, Houston, Texas)

High clouds

The bases of high clouds are at altitudes above 6 km (20,000 ft). Temperatures are so low in this region of the atmosphere (below −25 °C, or −13 °F) that clouds are composed exclusively of ice crystals, a composition that gives them a fibrous or filamentous appearance. Their names include the prefix *cirro* from the Latin meaning "a curl of hair."

Clouds of the **cirrus** form are nearly transparent and occur as silky strands, sometimes called "mares tails" (Plate 1). The strands are actually streaks of falling ice crystals blown laterally by strong winds. Like cirrus clouds, **cirrostratus** clouds are also nearly transparent, so the sun or moon shines readily through them. They form a thin, white veil or sheet that partially or totally covers the sky (Plate 2). **Cirrocumulus** clouds consist of small white puffs arranged in a wavelike or mackerel pattern (Plate 3). None of these high clouds are thick enough to prevent objects on the ground from casting shadows.

Middle clouds

The bases of middle clouds are at altitudes between 2 and 6 km (6500 and 20,000 ft). (Their names include the prefix *alto* from the Latin meaning "high.") These clouds, which occur at temperatures between 0 °C and −25 °C (32 °F and −13 °F), are composed of ice crystals or supercooled water droplets or a mixture of the two. **Altostratus** clouds occur as layers that are uniformly gray or white (Plate 4). They are usually so thick that the sun is only dimly visible, as if it were viewed through frosted glass. **Altocumulus** clouds consist of roll-like patches or puffs that form in lines or waves (Plate 5). They are distinguished from the cirrocumulus form by the larger size of the cloud patches and by the clouds' sharper edge. Sharp cloud boundaries indicate the presence of water droplets rather than ice crystals.

Low clouds

The bases of low clouds range in altitude from the earth's surface (fog) up to 2 km (6500 ft). Low clouds, which form at temperatures above −5 °C (23 °F), are composed mostly of water droplets. **Stratocumulus** clouds consist of large, irregular puffs or rolls arranged in a layer (Plate 6). **Stratus** clouds occur as a uniform gray layer that stretches from horizon to horizon (Plate 7). Where stratus clouds meet the ground surface, they are known as fog. **Nimbostratus** clouds resemble the stratus form, except that nimbostratus clouds are thicker and yield more rain and snow. Typically, only drizzle falls from stratus clouds.

Vertical clouds

Air surging upward in convection currents can give rise to cumulus and cumulonimbus clouds. As unsaturated air rises, its temperature falls at the dry adiabatic rate (10 °C per 1000 m), and its relative humidity therefore increases. Eventually, with continued ascent, at some altitude known as the **lifting condensation level** (LCL), the relative humidity approaches 100 percent, the air is nearly saturated, and a cumulus cloud forms. The lifting condensation level thus coincides with the altitude of the cloud base and is usually between 1 and 2 km (3000 and 6500 ft).

Cumulus clouds are the familiar clouds that resemble puffs of cotton and dot the sky on a fair-weather day (Plate 8). They are caused by convection. Because convection is driven by solar heating, not surprisingly cumulus cloud development often follows the daily variation of solar heating. On a fair day, cumulus clouds begin forming by middle to late morning, after the sun has warmed the ground and initiated convection. Cumulus sky cover is most extensive by midafternoon—usually the warmest time of day. If the clouds are sufficiently developed, these "fair-weather" cumulus may yield a brief, light shower of rain or a light snow. As sunset approaches, convection weakens, and the cumulus clouds begin dissipating (that is, they vaporize).

The stability profile of the troposphere determines the extent of the vertical development of cumulus clouds and whether they develop into the more ominous **cumulonimbus** clouds. If the ambient air aloft is stable, vertical motion is inhibited, and cumulus clouds exhibit little vertical growth. Under these conditions we can expect the weather to remain fair. On the other hand, if the ambient air aloft is unstable for saturated air, then vertical motion is enhanced, and the tops of cumulus clouds surge upward. If the ambient air is unstable to great altitudes, adjacent cumulus merge, and the entire cloud mass takes on a cauliflower appearance as it builds into a cumulonimbus (thunderstorm) cloud (Plate 9). (Thunderstorms are discussed in greater detail in Chapter 13.)

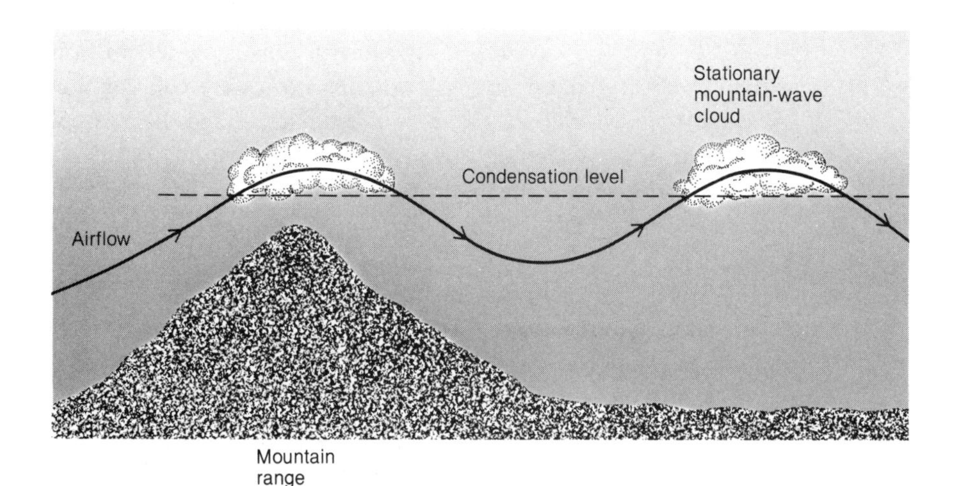

FIGURE 7.7
Mountain-wave clouds are formed when air crossing a mountain range is deflected in a wavelike pattern. Clouds develop on the ridges of the waves where the airflow is upward and expansional cooling takes place. Clouds are absent where airflow is downward and compressional warming occurs.

Unusual Clouds

Chances are that all of us have observed most of the clouds described above frequently. This is because the circumstances leading to their development are quite common. Some other, more unusual clouds are formed only by rarer atmospheric conditions.

The clouds in Plate 10 are most striking in appearance. Not surprisingly, many people have mistaken them for flying saucers. Observation of this type of cloud over a period of time reveals another strange characteristic: these clouds tend to remain stationary. They are **altocumulus lenticularis** clouds, that is, lens-shaped altocumulus clouds. These and other so-called **mountain-wave clouds** are generated by airflow that is disturbed by a mountain range.

As strong horizontal winds encounter a mountain range, the wind is deflected upward over the windward slopes and then downward over the leeward slopes (Figure 7.7). This is a common occurrence as the prevailing westerlies, flowing west to east, cross the front range of the Rocky Mountains. The disturbed winds generate a wavelike airflow pattern that extends tens of kilometers downwind of the mountain range. Within the air wave, where air flows upward, expansional cooling leads to cloud formation. Where air flows downward within the wave, compressional warming causes clouds to vaporize. Clouds thus occur on the ridges of the waves and are absent in the troughs.

Mountain-wave clouds do not move, even though the winds are

strong, because the wave itself is stationary. An analogous situation would be a waterfall on a river. If you observe a waterfall, you will note that the river's downstream flow is uninterrupted. In the pool at the base of the waterfall, however, the water is quite turbulent, but this turbulence soon dissipates downstream. The river then continues its normal flow, just like the wind over the mountain range, but the location of the irregular water turbulence of the falls or the wavelike flow of the air does not change.

Interestingly, the patterns exhibited by stratocumulus, altocumulus, and cirrocumulus are also caused by waves. These waves, however, are not linked to mountain ranges and propagate at different altitudes within the troposphere. Note that waves produce bands of clouds aligned perpendicular to the wind direction.

Because almost all water vapor is confined to the troposphere, so, too, are most clouds. One exception is the occasional intense cumulonimbus, the top of which may penetrate the tropopause and enter the lowest portion of the stratosphere. Another exception are colorful **nacreous clouds** which occur in the upper stratosphere. Temperatures at these altitudes (25 to 30 km) favor water in either the solid or supercooled state. Because of their soft, pearly luster, these rarely seen clouds are also called **mother-of-pearl clouds**. They are best viewed at high latitudes when they are illuminated by the setting sun.

Somewhat mysterious are the wavy, cirrus-like **noctilucent clouds**, which occur in the upper mesosphere, and are seen rarely, and then only at high latitudes just after sunset or before sunrise. The virtual absence of water vapor in the upper mesosphere has prompted some scientists to suggest that noctilucent clouds are composed of meteoric dust.

Conclusions

Many processes operating within the atmosphere bring air to saturation. Either through cooling or by the addition of water vapor, the relative humidity of air often approaches 100 percent. As air approaches saturation, the abundance of nuclei in the atmosphere favors deposition or condensation, that is, cloud development. Clouds may or may not yield precipitation—most do not. The special conditions required for rain and snowfall are among the topics discussed in the next chapter.

SUMMARY STATEMENTS

- Nocturnal radiational cooling reduces the air's capacity for water vapor. As the relative humidity approaches 100 percent, dew, frost, or fog may form.

- As air masses move from place to place, their temperature and water vapor content may change. Warm air passing over a cold surface may be chilled to vapor saturation, or the mixing ratio may increase when dry air is advected over a warm water surface. Fog may develop as a consequence of either situation.

- Clouds, the visible manifestations of condensation and deposition within the atmosphere, are composed of minute water droplets or ice crystals or both.

- Condensation and deposition occur on nuclei at relative humidities near 100 percent. Condensation nuclei are much more abundant than ice-forming nuclei. Most ice-forming nuclei activate at temperatures well below the freezing point. Some condensation nuclei are hygroscopic.

- Clouds are assigned special Latin names, and are classified by appearance and by altitude of occurrence.

- High, middle, and low clouds are caused by relatively gentle uplift over broad areas. These clouds therefore have a layered appearance. Clouds with vertical development are the consequence of more vigorous uplift, and are heaped or puffy in form.

- Mountain-wave clouds develop downwind of prominent mountain ranges and are stationary. Noctilucent clouds occur in the upper mesosphere, are rarely seen, and are probably composed of meteoric dust.

KEY WORDS

radiational cooling
frost
dew
dew point
frost point
hoarfrost
fog
mist
radiation fog
air mass advection
advection fogs
Arctic sea smoke
steam fog
upslope fog
clouds
nuclei
hygroscopic nuclei
cloud condensation nuclei
ice-forming nuclei
supercooled droplets
freezing nuclei
deposition nuclei
sublimation nuclei
cirrus
cirrostratus
cirrocumulus
altostratus
altocumulus
stratocumulus
stratus
nimbostratus
lifting condensation level (LCL)
cumulus
cumulonimbus
altocumulus lenticularis
mountain-wave clouds
nacreous clouds
mother-of-pearl clouds
noctilucent clouds

REVIEW QUESTIONS

1 What conditions are ideal for rapid nocturnal radiational cooling?

2 Dew and frost are not forms of precipitation. Explain this statement.

3 How is the dew-point temperature a measure of the water vapor content of air?

4 Describe the circumstances that favor the development of radiation fog. Why is this type of fog usually short lived?

5 Identify two different situations in which fog forms as a consequence of warm-air advection.

6 How does Arctic sea smoke develop?

7 What is a cloud?

8 Explain why clouds typically form without supersaturated conditions.

9 List some of the sources of cloud nuclei.

10 What is the significance of *hygroscopic* nuclei?

11 Cloud condensation nuclei are much more abundant than ice-forming nuclei. What is the implication of this for the composition of clouds?

12 Distinguish between freezing nuclei and deposition nuclei.

13 How are clouds classified?

14 How does the composition of clouds vary with altitude?

15 What is the significance of the lifting condensation level?

16 How does ambient air stability influence the vertical growth of cumulus clouds?

17 Speculate on why a cumulus or cumulonimbus cloud appears to tilt or lean with altitude.

18 What causes the ripple pattern displayed by cirrocumulus, altocumulus, and stratocumulus clouds?

19 If the temperature at the lifting condensation level is 0 °C and the temperature at the ground is 20 °C, determine the altitude of the cumulus cloud base.

20 Explain why cumulus clouds tend to vaporize toward sunset. What is the effect of a snow cover on cumulus cloud development?

POINTS TO PONDER

1 Distinguish between the dew-point temperature and the wet-bulb temperature. Under what condition are they equivalent?

2 Fog can have a catastrophic effect on a snow cover. Explain why.

3 In late spring and early summer, fog frequently develops over Lake Michigan as southwest winds push warm, humid air over the lake. Explain this phenomenon.

4 Dew formation is an *isobaric* (constant pressure) process, but expansional cooling is not an isobaric process. Explain the difference.

5 Speculate on why frost formation on fruit may actually protect the fruit.

PROJECTS

1 Using the cloud photographs in this book as a guide, identify today's clouds. Can you distinguish among low, middle, and high clouds?

2 Observe the changes in cloud type as a warm front or cold front approaches your area.

SELECTED READINGS

Schaefer, V. J., and J. A. Day. *A Field Guide to the Atmosphere.* Boston: Houghton Mifflin, 1981. 359 pp. *An exceptionally well-illustrated survey of cloud and precipitation processes.*

Scorer, R., and H. Wexler. *A Colour Guide to Clouds.* New York: Macmillan, 1963. 63 pp. *Excellent photographs of clouds plus descriptions.*

Stewart, T. R., R. W. Katz, and A. H. Murphy. "Value of Weather Information: A Descriptive Study of the Fruit-Frost Problem." *Bulletin of the American Meteorological Society* 65(1984):126–137. *Report on the use of weather information by fruit growers in the Yakima Valley, Washington, in making decisions to protect against freezing.*

Walker, J. *The Physics of Everyday Phenomena.* San Francisco: W. H. Freeman, 1979. 86 pp. *Reprints of* Scientific American *articles, including papers on the ice crystal process, raindrop shape, and fog.*

Be thou the rainbow to the
storms of life,
The evening beam that smiles
the clouds away,
And tints to-morrow with
prophetic ray!

LORD BYRON
Bride of Abydos

A special set of conditions is required for cloud particles to grow into raindrops or snowflakes. In this photo, a shaft of rain is falling from a distant cloud. (Photograph by Arjen Verkaik)

EIGHT

Precipitation, Weather Modification, and Atmospheric Optics

THIS chapter considers several closely related topics. The first portion focuses on the development of precipitation in clouds, and the forms and measurement of precipitation. This is a basis for the subsequent discussion of techniques to artificially stimulate precipitation and to dissipate fog. The chapter closes with a description of various optical phenomena produced when sunlight or moonlight interacts with clouds or with precipitation.

Precipitation Processes

The development of clouds is no guarantee that it will rain or snow. Nimbostratus and cumulonimbus clouds produce the bulk of precipitation, but most clouds—even most of those associated with a large storm system—do not yield any rain or snow. This is because a special combination of circumstances, as yet incompletely understood, is required for clouds to precipitate.

The water droplets or ice crystals composing clouds are so minute that they remain suspended indefinitely, unless they vaporize or somehow undergo considerable growth. Updrafts within clouds are usually strong enough to prevent cloud particles from falling toward the earth's surface. Even if droplets or ice crystals descend from a cloud, their descent velocities are so slow that they travel only a short distance before vaporizing in the clear, unsaturated air beneath the cloud. To precipitate, cloud particles must grow large enough to counter updrafts and to survive a descent to the earth's surface as raindrops or snowflakes. This is no minor task! It takes about 1 million cloud droplets (in the 5 to 10 μm size range) to form a single raindrop (about 1 mm in radius). How does this growth take place?

Cloud physicists have determined that condensation alone cannot result in snowflakes or raindrops. They have identified two important processes by which cloud particles grow large enough to precipitate: the collision-coalescence process and the Bergeron process.

The collision-coalescence process

The **collision-coalescence process** occurs in some **warm clouds**, that is, clouds at temperatures above the freezing point of water. These clouds are composed entirely of liquid water droplets. For precipitation to form in these clouds, some relatively large droplets (larger than 20 μm in radius), perhaps produced by "giant" sea-salt

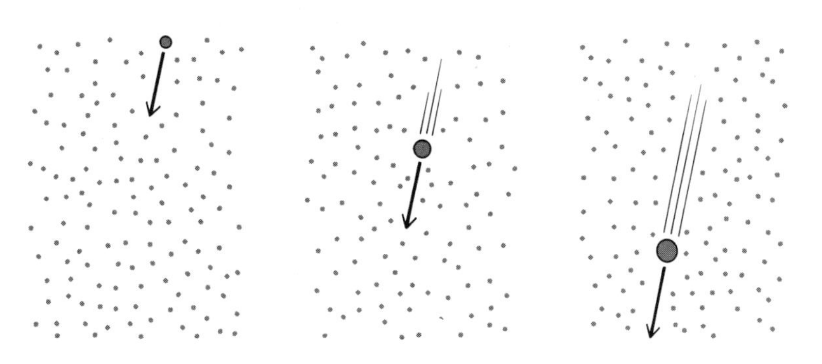

FIGURE 8.1

A relatively large water droplet falls within a cloud of much smaller droplets. The large droplet collides with the smaller droplets and thus grows by coalescence.

nuclei, must be present. A large cloud droplet falls more rapidly than the smaller cloud droplets. As it falls, a large droplet intercepts and coalesces (combines) with the smaller water droplets in its path (Figure 8.1). Collision and coalescence are repeated perhaps a million times until the droplet is large enough to fall out of the cloud as a raindrop.

It is unusual for a cloud to have droplets large enough to initiate the collision-coalescence process. In fact, most precipitation that falls on earth originates through the Bergeron process.

The Bergeron process

The **Bergeron process**, named for the Scandinavian meteorologist Tor Bergeron, who first described it in about 1930, applies to **cold clouds**, which are clouds at temperatures below 0 °C (32 °F). The process requires the coexistence of water vapor, ice crystals, and supercooled liquid water droplets.

As we noted in Chapter 7, most ice-forming nuclei are not active at temperatures above −9 °C (16 °F). Consequently, clouds at temperatures between 0 °C (32 °F) and −9 °C (16 °F) are typically composed of supercooled water droplets only. From −10 °C (14 °F) down to −20 °C (−4 °F), clouds are mixtures of supercooled water droplets and ice crystals. Below −20 °C (−4 °F), clouds usually consist of ice crystals only. Clouds with significant vertical development therefore have different components at different altitudes, depending on the temperatures involved. Cumulonimbus clouds, for example, are often composed exclusively of ice crystals aloft and of water droplets at the base. The nuclei that form water droplets far outnumber the nuclei that form ice crystals, so at cloud temperatures above −20 °C (−4 °F), the activation temperature for many deposition nuclei, supercooled water droplets are initially much more abundant than ice crystals. In fact, a single ice crystal may be surrounded by thousands to hundreds of thousands of supercooled water droplets.

The Bergeron process depends on the difference in saturation vapor pressure between air over ice and air over water. It is this difference that triggers ice crystal growth at the expense of the supercooled water droplets. Water molecules vaporize more readily from liquid water than from solid ice at temperatures below freezing. This is because water molecules are more energetic in the liquid form than in the solid phase. As a consequence, the saturation vapor pressure of air over ice is less than that of air over water (see Table 6.3). In cold clouds that are composed of a mixture of ice crystals and supercooled water droplets, a vapor pressure that is saturated for water droplets is therefore supersaturated for ice crystals. Suppose, for example, that the vapor pressure is 2.86 mb in a cloud having a temperature of −10 °C (14 °F). From Table 6.3, this vapor pressure translates into a relative humidity of 100 percent (saturation) for the air surrounding water droplets and a relative humidity of 110 percent (supersaturation) for the air surrounding ice crystals. In response to supersaturated conditions, water vapor deposits on the ice crystals, and the ice crystals grow larger. (Recall that deposition is the change in phase directly from vapor to solid state.) Deposition removes water vapor from the cloud and thereby lowers the relative humidity The relative humidity of the air surrounding the water droplets goes below 100 percent, and the droplets vaporize. Under these conditions, the ice crystals grow at the expense of the supercooled water droplets.

As the ice crystals grow larger and heavier, they begin to fall at an accelerating rate, and as the larger crystals fall, they collide and coalesce with supercooled water droplets and with the smaller ice crystals in their path, thereby growing still larger. Eventually, the ice crystals become heavy enough to fall out of the cloud. If air temperatures are below freezing most of the way to the ground, the crystals reach the earth's surface in the form of snowflakes. If the air below the cloud is above freezing, snowflakes melt and fall as rain.

Once a raindrop or a snowflake leaves a cloud, it enters a hostile environment in which either evaporation or sublimation can take place. In general, the longer the journey to the ground and the drier the air beneath the clouds, the greater the chance is that the raindrop or snowflake will return to the atmosphere as vapor. In Figure 8.2, rain falling from a distant thunderstorm is entering drier air, which causes evaporation of some of the rain during descent. This shaft of falling precipitation is called **virga**. Higher land receives more rainfall than surrounding lowlands partly because of this.

FIGURE 8.2
Some of the rain falling from the base of a distant thundershower never reaches the ground, because the precipitation vaporizes in the relatively dry air under the thundercloud. (Photograph by Don Beimborn)

Forms of Precipitation

Precipitation is defined as water in solid or liquid form that falls from clouds to the earth's surface. Besides the familiar rain and snow, precipitation also occurs as drizzle, freezing rain, ice pellets, and hail.

Drizzle consists of small raindrops from 0.2 to 0.5 mm in diameter that drift very slowly toward the ground. The small size of the droplets stems from their origin in stratus clouds. Stratus clouds are so low that droplets have only a limited opportunity to grow by coalescence. Drizzle is associated with fog and poor visibility, but never with convective storms.

Rain falls from clouds such as nimbostratus and cumulonimbus, which have upper portions at below freezing temperatures. Most raindrops originate as snowflakes, which melt on the way down as they encounter air that is above 0 °C (32 °F). Because rain originates in thicker clouds having higher bases, raindrops travel farther than drizzle and undergo more growth by coalescence. Raindrop diameters may approach 5 mm, but beyond this diameter, the drops tend to break up.

Freezing rain, or freezing drizzle, forms a coating of ice that sometimes grows thick and heavy enough, as in Figure 8.3, to bring down tree limbs, snap power lines, and totally disrupt traffic. Freezing rain develops when rain falls from a relatively mild air

FIGURE 8.3
Freezing rain can disrupt traffic and cause considerable damage to trees and power lines. (NOAA photograph)

layer aloft into a shallow layer of subfreezing air at ground level. The drops become supercooled and then freeze on contact with cold surfaces.

Snow is an assemblage of ice crystals in the form of flakes. Although it is said that no two snowflakes are identical, all snowflakes are hexagonal (six sided), just as the constituent ice crystals of the flakes are hexagonal. Snowflake form varies with water vapor concentration and with temperature, and may consist of plates, stars, columns, or needles. Snowflake size also depends in part on the availability of water vapor during the crystal growth process. At very low temperatures, water vapor concentrations are low, so the snowflakes are relatively small. Snowflake size also depends on collision efficiency as the flakes drift toward the ground. At relatively high temperatures, snowflakes are wet and stick readily together after colliding, so that flake diameters may eventually exceed 5 cm (2 in.). The appearance of such large flakes is usually a signal that the snow is about to turn to rain.

When is it too warm to snow? Surprisingly, snow is possible when the air temperature near the earth's surface is well above the freezing point. Theoretically, at least, snow can fall even when the surface air temperature is as high as 10 °C (50 °F)! The only requirement is that the wet-bulb temperature remain below 0 °C (32 °F), which also means that the relative humidity is very low. For example, if the air temperature is 5 °C (41 °F), the relative humidity must be lower than 32 percent for the wet-bulb temperature to be subfreezing (see Appendix II). Some snowflakes partially or completely vaporize as they fall through unsaturated air that is above freezing. Vaporization of snowflakes uses heat from the surrounding air, and the temperature of the air around the snowflakes thereby decreases to the wet-bulb temperature, that is, below the freezing point.

Is it ever too *cold* to snow? The answer to this question is explored in one of the Special Topics in this chapter.

Ice pellets, also called **sleet**, are actually frozen raindrops. They develop in much the same way as freezing rain with this difference: the surface layer of cold air is so deep that the raindrops freeze before striking the ground. Sleet can be distinguished readily from freezing rain, because sleet bounces when it strikes the ground and freezing rain does not. Accumulations of ice pellets or freezing rain can cause very hazardous driving conditions.

Hail is rounded or jagged lumps of ice, often characterized by concentric internal layering resembling the internal structure of an onion (Figure 8.4). Hail develops within intense thunderstorms as strong convection currents carry ice pellets upward, into the middle and upper reaches of a thundercloud (cumulonimbus). Along the way, the ice pellet grows by collecting supercooled water droplets. Hailstones eventually become too heavy to be supported by convective updrafts and fall to earth. The hailstones are large enough to survive the trip to the ground as ice, even though surface temperatures are usually well above freezing. Most hailstones are harmless granules of ice less than 1 cm in diameter, sometimes called **graupel**, but violent thunderstorms may spawn destructive hailstones the size of golf balls or larger. Hail is usually a spring and summer phenomenon that is particularly devastating to crops. (We have much more to say about hail in Chapter 13.)

FIGURE 8.4

Hailstones sometimes grow to the size of golf balls or larger (left). A hailstone consists of concentric layers of clear and opaque ice as shown in this cross section photographed in polarized light (right). (Photograph on left from NOAA; photograph on right from NCAR/NSF)

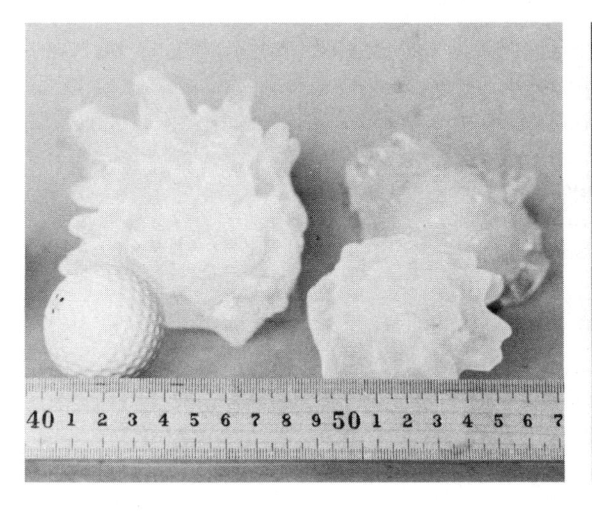

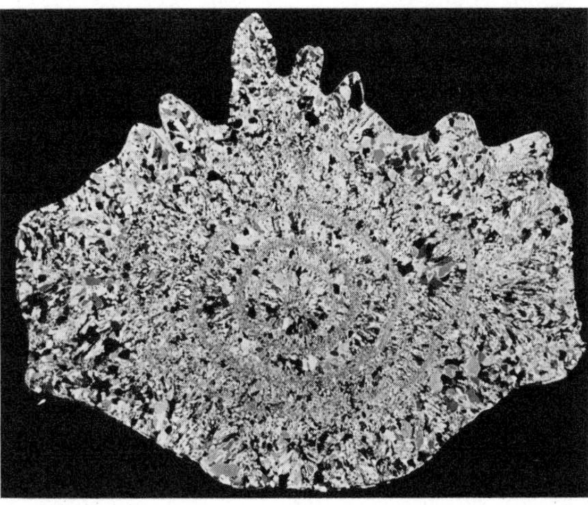

SPECIAL TOPIC

Is it ever too cold to snow?

During an episode of particularly frigid winter weather, some people will argue that it is too cold to snow. It is true that often the coldest weather is accompanied by clear skies. The climate records of the northern United States and Canada also bear out the claim. For some midwestern U.S. cities (such as Denver, Bismarck, and Minneapolis), March is both the snowiest and the mildest of the winter months (December through March). In Canada, outside of the mountains, average annual snowfall declines from more than 250 cm (100 in.) over much of the relatively mild southeast to less than 60 cm (24 in.) on the frigid shores of the Arctic Ocean.

In fact, snowfall *is* possible even at extremely low air temperatures although snowflakes are small and accumulations are usually light.

The relatively small amount of water vapor in very cold air means that comparatively little water is available for precipitation. Recall from Chapter 6 that cold air cannot hold as much water vapor as warm air, that is, the saturation vapor pressure drops rapidly as air temperature falls. For example, saturated air at −30 °C (−22 °F) contains only about 12 percent of the water vapor that is in saturated air at −5 °C (23 °F). Hence, the water available for precipitation, known as precipitable water, varies with air temperature.

Precipitable water is defined as the amount of water produced if all the water vapor in a column of air were condensed. The air column is assumed to extend from the earth's surface to the tropopause and the condensed water is described in units of depth. If all the water vapor within the troposphere were condensed, the resulting water would cover the entire earth's surface to an average depth of about 25 mm (1 in.). This depth is called the global average precipitable water depth. With the poleward decline in air temperature, precipitable water amounts also decline. Hence, average precipitable water depths decrease with latitude from more than 40 mm (1.6 in.) in the tropics to less than 5 mm (0.2 in.) near the poles.

The heaviest snowfalls typically occur when air temperatures in the lower atmosphere are within a few degrees of the freezing point because at those temperatures the potential amount of water that can precipitate in the solid form (snow) is maximum. On the other hand, moderate or heavy snowfall is very unlikely when air temperatures are below −20 °C (−4 °F) in the lower atmosphere. But even in the coldest regions of the globe where precipitable water amounts are lowest, there is some snowfall. For example, an estimated average annual 5 cm (2 in.) of snow falls on the high interior plateau of Greenland where average annual temperatures are below −30 °C (−22 °F).

Precipitation Measurement

Precipitation is collected and measured today using essentially the same device that was used as long ago as the fifteenth century: a container open to the sky. A modern **rain gage** is equipped with a funnel at the top that directs rainwater into a long, narrow cylinder, which is seated inside a larger, outer cylinder (Figure 8.5). The funnel magnifies the scale of the rainwater, so the observer can resolve rainfall in increments of 0.01 in. Total rainfall of less than 0.01 in. is recorded as a trace. Rainwater that accumulates in the inner cylinder is measured by a stick, which, in the United States, is graduated in inches. In most other nations, metric units are used.

Snowfall is measured in much the same way as rainfall, except that the funnel and inner cylinder are removed. Snow depth is usually recorded at several representative locations by using a yardstick or meterstick, and an average is then computed. It is also common practice to express snowfall in equivalent depth (inches or millimeters) of meltwater. As a general rule, 10 in. of snow melts down to 1 in. of rainwater, although this ratio varies considerably depending on the temperature at which the snow falls. "Wet" snow falling at surface air temperatures at or above freezing has a much greater water content (per inch) than "dry" snow falling at very low surface air temperatures.

Determining the rate of rainfall is often desirable. Some rain gages are therefore designed to provide a cumulative record of rainfall with time. A **weighing bucket rain gage** consists of a recording weighing scale that calibrates as water depth the weight of accumulating rainwater and records the results by pen on a clock-driven chart. A **tipping bucket rain gage** (Figure 8.6) features two small, free-swinging containers that can each collect the equivalent of 0.01 in. of rainfall. Alternating with one another, each container fills with water, tips, spills its contents, and thus trips an electric switch that marks a clock-driven chart. Both types of recording rain gage can be fitted with a propane or electric heater that melts snow as it falls into the gage, thereby providing a continuous meltwater record.

Both rainwater and snowfall are notoriously variable from one place to another, especially in convective showers. The positioning of precipitation gages is particularly important. They must be sited in representative locations, away from buildings and vegetation that might shield the gage.

FIGURE 8.5

A standard National Weather Service rain gage. (Courtesy of Weathermeasure/Qualimetrics)

FIGURE 8.6

A tipping bucket rain gage is designed to provide a continuous record of rainfall in increments of 0.01 in. (Courtesy of Weathermeasure/Qualimetrics)

Weather Modification

Weather modification is any change in weather that is induced by human activity. The activity may be either intentional or inadvertent. This section describes two principal types of intentional weather modification: cloud seeding and fog dispersal. Later, we discuss hail suppression (Chapter 13) and the impact of air pollution on weather and climate (Chapters 16 and 18).

Cloud seeding

Since World War II, considerable research has gone into methods of enhancing precipitation by cloud seeding. **Cloud seeding** is an attempt to stimulate natural precipitation processes by injecting nucleating agents into clouds—typically from aircraft (Figure 8.7). Silver iodide, a substance with crystal properties similar to those of ice, is usually injected into cold clouds that are deficient in ice

FIGURE 8.7
An airborne cloud-seeding effort. This aircraft is injecting cloud nucleating agents in an attempt to stimulate precipitation formation. (NCAR/NSF photograph)

crystal nuclei. The objective is to stimulate the Bergeron process. Silver iodide crystals are freezing nuclei that are active at −4 °C (25 °F) and below. Sometimes cold clouds are seeded with dry ice pellets instead of silver iodide crystals. Dry ice is solid carbon dioxide and has a temperature of about −80 °C (−110 °F). When introduced into a cloud, the dry ice pellets are cold enough to cause supercooled water droplets to freeze, and the frozen droplets can then function as nuclei for the Bergeron process. In warm clouds, sea-salt crystals and other hygroscopic substances are injected to stimulate the collision-coalescence process. Just how effective are all of these efforts?

Over a period of several years, in the late 1970s and early 1980s, NOAA scientists conducted a statistically rigorous experiment* designed to test the effectiveness of weather modification. The experiment was carried out over southern Florida and involved the seeding of cumulus clouds. Test days were about evenly divided between days when clouds were seeded with silver iodide crystals and days when clouds were seeded with inert (chemically inactive) sand grains. Only after the experiment was completed and the results analyzed were participating scientists made aware of the specific days when silver iodide was used. This procedure ensured an unbiased selection of clouds to be seeded and also ensured an unbiased interpretation of precipitation results.

The experiment was divided into two phases: results from an initial "exploratory" period (phase one) were to be either verified or rejected by a later "confirmatory" procedure (phase two) when the seeding was repeated. Phase one results were very encouraging, showing a 25 percent increase in rainfall on days when silver iodide was the seeding agent, as compared with days when sand was the bogus seeding agent. This finding was statistically significant at the 90 percent level, which meant that only a 10 percent probability existed that the rainfall increase was simply the result of chance. The success of phase one was, however, not repeated with phase two, when the seeding showed no statistically significant increase in rainfall.

The same type of experimental design, involving separate exploratory and confirmatory phases, was carried out in Israel from 1961 to 1967 and from 1969 to 1975. In these studies, however, statistically significant rainfall enhancement during the first phase was verified in the second phase of seeding. The contrasting findings of the Florida and Israeli experiments underscore the uncertainties of cloud seeding and the need for a better understanding of precipitation processes.

Although cloud seeding is probably successful in some instances, the actual volume of additional precipitation produced by cloud seeding and the advisability of large-scale seeding efforts are matters of considerable controversy. Some cloud seeders claim to increase precipitation by 15 to 20 percent or more, but the question remains: would the rain or snow that follows cloud seeding have

*Florida Area Cumulus Experiment (FACE).

fallen anyway? Even if seemingly successful, cloud seeding on a large geographical scale may merely redistribute a fixed supply of rain. An increase in precipitation in one area might mean a compensating reduction in another. For example, rainmaking might benefit agriculture on the high plains of eastern Colorado, but might deprive wheat farmers of rain in the adjacent downwind states of Kansas and Nebraska. Such conflicts can cause legal wrangles between adjoining counties, states, and provinces.

In some experiments, precipitation actually appears to have been reduced by seeding. An example is a well-documented seeding effort carried out over south-central Missouri during five consecutive summers in the 1950s (called **Project Whitetop**). A possible explanation for the reduced precipitation is that clouds were overseeded. Prior to seeding, cumulus clouds apparently contained just enough ice crystals for precipitation. The addition of more nuclei by seeding probably resulted in too many ice crystals competing for a limited supply of supercooled water droplets. The Project Whitetop seeding produced a large number of ice particles so small that they tended to remain suspended in the clouds, or vaporized before reaching the ground.

Fog dispersal

Fog can pose a serious hazard to both surface travel and air travel. Many auto accidents and some ship collisions and aircraft crashes have been attributed, at least in part, to visibility restrictions caused by dense fog. Fog frequently forces flight delays, reroutings, and cancellations that cost airlines millions of dollars each year and inconvenience thousands of passengers. Although the need for an effective method of **fog dispersal** is great, especially at airports, little progress has been made in this area for both technical and economic reasons.

During World War II, the British had some success in clearing warm fogs* from their airport runways. Heat from fuel burners deployed near runways raised the air temperature, thereby lowering the relative humidity to below saturation so that fog droplets vaporized. This dispersal system assisted the safe landing of an estimated 2500 Royal Air Force planes, which returned a total of 10,000 airmen from combat missions. Although this thermal approach to

*Warm fogs have temperatures above 0 °C (32 °F) and are the most common type of fog.

fog dispersal was revived from time to time after the war, fuel costs generally proved to be excessive. Today, the only warm fog dispersal systems operated routinely are at Orly and DeGaulle airports outside Paris (Figure 8.8). In this so-called Turboclair system, jet engines direct streams of warm air over runways.

FIGURE 8.8
Turboclair fog dissipation system operating at Orly Airport, Paris, France, since 1970. Jet engines lining the runway blow warm air over the runway to lower the humidity thereby causing fog droplets to evaporate. (Photograph by Jean-J. Moreau, Cliche, Aeroport de Paris)

Another approach to fog dispersal applies techniques of cloud seeding. Warm fogs are seeded with hygroscopic substances that absorb water vapor, thereby reducing the relative humidity. Fog droplets vaporize, and hygroscopic droplets grow into larger drops that fall to the surface. Cold fogs (0 °C to −20 °C, or 32 °F to −4 °F) are seeded with either dry ice pellets or with silver iodide crystals, both of which stimulate the Bergeron process. For the most part, however, fog dispersal by seeding is an experimental technology that requires further research, although very little is now taking place.

Atmospheric Optics

As the sun's rays travel through the atmosphere, they may be reflected or refracted by suspended water droplets or by ice crystals or by raindrops. The consequence is a variety of optical phenomena, including halos, rainbows, and the glory. The characteristics and origins of these optical effects are the subjects of this section. As a Special Topic, we also examine the mirage. This is another optical effect, which, however, is not due to moisture in the atmosphere.

SPECIAL TOPIC

Mirages

Appearances can be deceiving—especially in the case of mirages. Distant buildings or hills may appear higher or lower than they really are. A nonexistent pool of water may suddenly appear on the highway ahead, or a sailboat viewed from shore may appear upside down. These mirages are optical illusions that are caused by the refraction of light rays within the lower atmosphere.

Light travels at different speeds through different transparent substances. Light changes speed, for example, as it travels from air into a raindrop or from air into an ice crystal. For reasons discussed elsewhere in this chapter, a light ray is refracted at the interface between two transparent media. The speed of light also varies within a single medium if there are density variations within the medium. Light rays bend as they pass through such a substance of varying density. Indeed, this is often the case in the atmosphere.

If the density of the atmosphere were uniform throughout, then light rays would travel in straight lines at constant speed, and no refraction would occur. As described in Chapter 5, however, the atmospheric density varies considerably due to changes in temperature and pressure. Because our concern here is with optical phenomena that occur in the lowermost troposphere and involve relatively short viewing distances, we need not be concerned with the influence of vertical and horizontal pressure gradients on density. For the same reasons, we can also ignore the effect of horizontal temperature gradients on density, but we cannot ignore the effect of vertical temperature profiles on the change of air density with altitude. Indeed, light refraction within the atmosphere is primarily dependent on the air temperature lapse rate (temperature decrease with altitude).

As a general rule, light rays are refracted (bent) in the atmosphere so that the colder, denser air is on the inside (concave side) of the bend, and the warmer, less dense air is on the outside (convex side) of the bend. If the air temperature decreases with altitude at a constant rate, which is the normal condition in the troposphere, light rays reflected from a distant object bend before reaching the viewer, as shown in Figure 1A. Because human perception is based on the assumption that light travels in

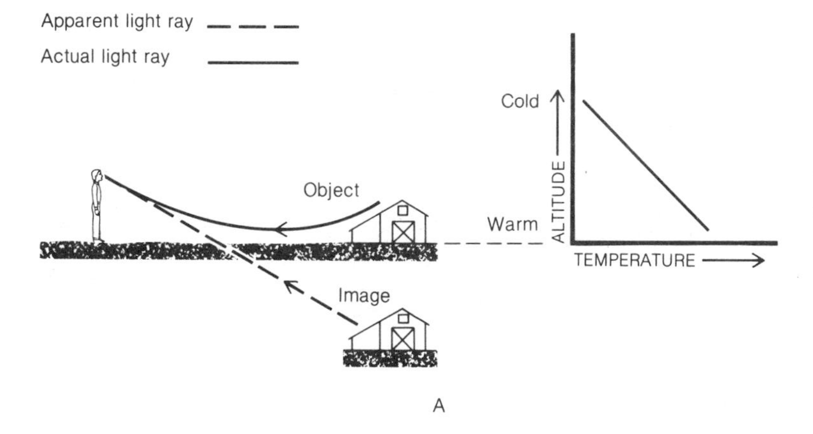

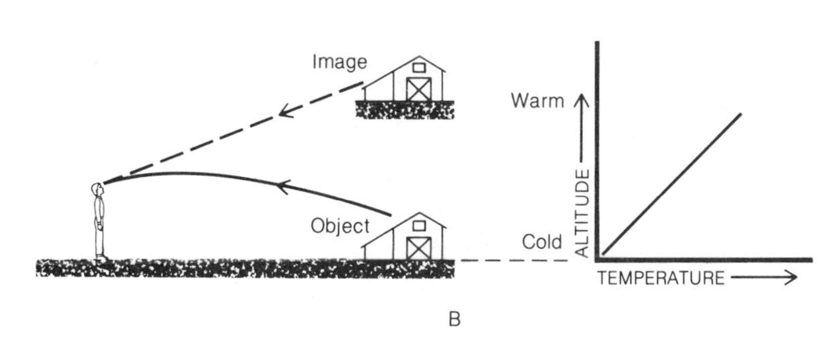

FIGURE 1

Two mirages caused by light refraction. (A) An inferior mirage in which an object appears lower than it actually is. (B) A superior mirage in which an object appears elevated from its actual position.

straight paths, to the viewer the object *appears* to be lower than it really is. This is known as an *inferior* (lower) *mirage.* If, on the other hand, the lowest air layer features a temperature inversion (that is, temperature increases with altitude), the light rays reflected from a distant object are bent in the opposite sense (Figure 1B). The object then appears to the viewer to be displaced upward. This is known as a *superior* (upper) *mirage.*

What happens when the vertical temperature gradient is not constant with altitude? As warm air is advected over a cold, snow-covered surface, the warm air mass is chilled from below. The vertical temperature gradient is greatest in the lowest layer of air and then decreases with altitude (see Figure 2A on next page). The air temperature, meanwhile, increases with altitude—a temperature inversion. A distant object appears to be displaced upward (superior mirage), but because the bottom of the object is in a steeper temperature gradient than the top of the object, the bottom appears to be lifted more than the top. The object, a building, for example, not only appears to be uplifted, it also appears shorter than it really is.

A different optical effect is observed when both the temperature and the vertical temperature gradient decrease with altitude (Figure 2B). By midday, the ground is heated intensely by the sun. The maximum heating affects the air in immediate contact with the ground so that the steepest vertical temperature gradient occurs near the surface. A distant object appears to be displaced downward (inferior mirage), but because the bottom of the object is in a steeper

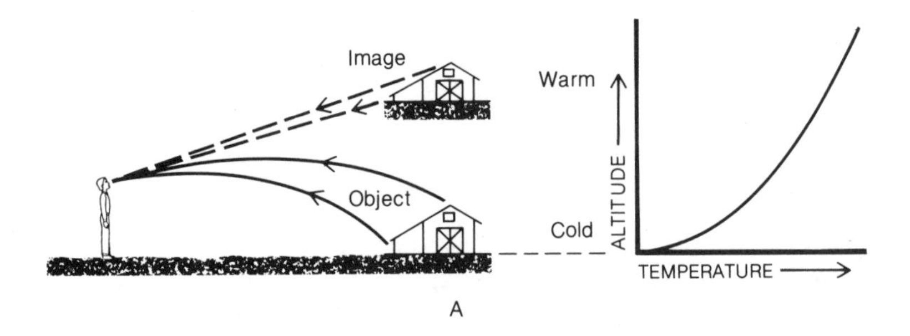

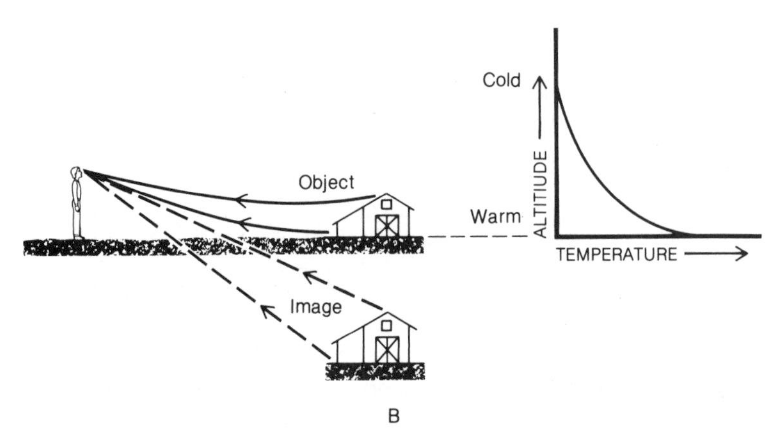

FIGURE 2

(A) A superior mirage, in which an object appears shorter than it really is, occurs when the vertical temperature gradient decreases with altitude while the temperature increases with altitude. (B) An inferior mirage, in which an object appears taller than it really is, occurs when the vertical temperature gradient decreases with altitude while the temperature decreases with altitude.

temperature gradient than the top of the object, the bottom is displaced downward more than the top. The object not only appears to sink, it also appears to be stretched vertically.

These are only a few examples of the many possible types of mirage. As the vertical air temperature gradient becomes more complex, so, too, do the types of mirage that can appear. For example, the familiar oasis mirage in a desert is an inverted image of the sky seen below the horizon caused by a complex temperature profile. All mirages, then, are displacements or distortions of something real.

Halos

A **halo** is a whitish ring of light surrounding the sun or, sometimes, the moon. It is formed when the sun's rays are refracted by the tiny ice crystals that compose high, thin clouds such as cirrostratus. **Refraction** is the bending of light as it passes from one transparent medium (such as air) into another transparent medium (such as ice or water). The light rays bend because the speed of light is greater in air than in ice or water. Refraction occurs whenever light rays strike the interface between two media at any angle other than 90 degrees (Figure 8.9). (Beams of light that are perpendicular to the water or ice surface are not refracted.)

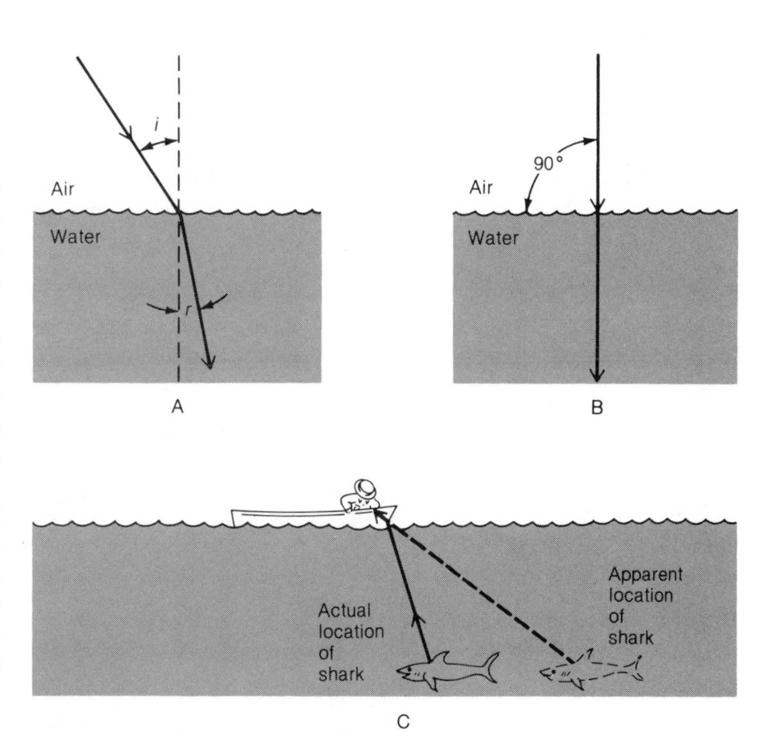

FIGURE 8.9

Light rays may be refracted (bent) as they travel from one transparent medium into another. (A) A light ray is refracted as it travels from air into water. The speed of light is less in water than in air, so the light ray is bent toward a line perpendicular to the water's surface, and angle *r* is less than angle *i*. (B) Light rays that are perpendicular to the water surface are not refracted. (C) Refracted light can be deceptive. The shark appears to be farther away from the raft than it really is. This is because of refraction and the fact that human perception assumes that light travels in straight lines.

An analogy may help to explain refraction. Suppose you are driving your auto down a highway and suddenly encounter a patch of ice along the right side of the road. The right wheels travel over the ice while the left wheels remain on dry pavement. You slam on the brakes. The left side of the car slows while the right side slips on the ice. Your auto consequently swerves to the left. This swerving of the auto is analogous to a light ray bending toward the medium in which light travels more slowly.

Light is refracted twice as it passes through an ice crystal, that is, upon entering and upon exiting. Ice crystals occur as hexagonal (six-sided) plates or columns. If crystals are very small and are oriented randomly, the light ray is refracted from side to side as it passes through the crystals (Figure 8.10). This side-to-side refraction focuses the light in a circle having a radius* of about 22 degrees (Figure 8.11). For reference, the halo's radius appears to be the same as the width of this page when held at arm's length.

*The radii of halos are usually measured as angles. Suppose, for example, that you are viewing a halo about the moon. Now visualize two lines: one line joins you with the moon, and the other line joins you with any point on the halo. Depending on the type of halo, the angle formed by the two imaginary lines will be either 22 degrees or 46 degrees.

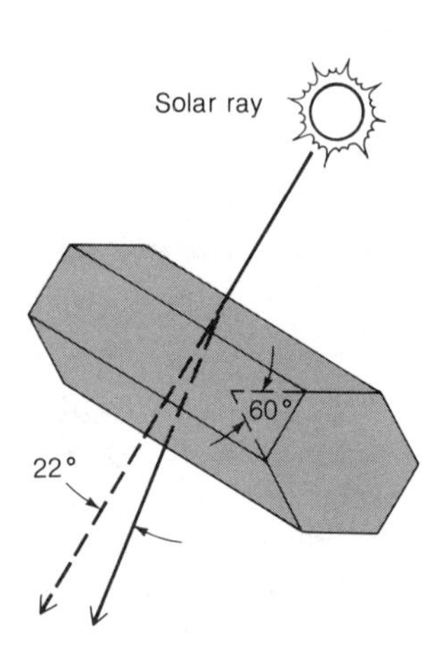

FIGURE 8.10

A ray of sunlight is refracted as it passes through an ice crystal from one side (the top side in the drawing) to another side (the bottom side showing in the drawing). The angle between the two sides of an hexagonal ice crystal is 60 degrees. This type of refraction produces a 22-degree halo about the sun (or moon).

FIGURE 8.11

A 22-degree halo about the sun is caused by refraction of sunlight by ice crystals. This photograph was taken with a wide-angle lens. (Photograph by Arjen Verkaik)

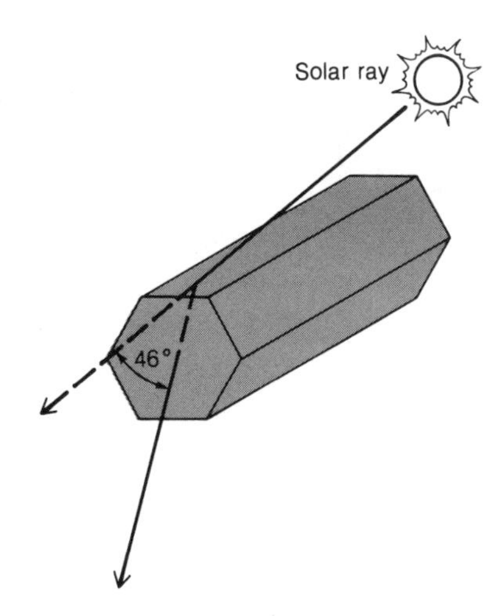

FIGURE 8.12

In a rarer situation than that shown in Figure 8.10, the sun's ray is refracted from side to top or from side to bottom as it passes through a columnar ice crystal. The angle between side and top (or bottom) is 90 degrees. This type of refraction produces a 46-degree halo about the sun or moon.

A much rarer halo has a radius of about 46 degrees about the sun (or moon). In this case, refraction is caused by columnar ice crystals with diameters in the 15 to 25 μm range. Light rays travel from side to top or from side to base of the crystals, rather than from side to side (Figure 8.12).

In some instances, light is concentrated in two bright, white spots about 22 degrees on either side of the sun (Figure 8.13). Called **sun dogs**, **mock suns**, or **parhelia**, these refractive phenomena involve larger ice crystals that are not randomly oriented.

FIGURE 8.13

Bright spots on either side of the sun, sometimes called sun dogs, are caused by refraction of sunlight by ice crystals suspended in the atmosphere. These unusually spectacular sun dogs appeared over Foam Lake, Saskatchewan. (Photothèque photograph by Daniel Comeau)

Rainbows

Rainbows are caused by a combination of refraction and reflection of sunlight (or rarely, of moonlight) by raindrops. Sunlight striking a mass of falling raindrops is refracted and internally reflected by each drop of rain. As shown in Figure 8.14, a solar ray is refracted as it enters the raindrop, then the ray is reflected by the inside back of the drop before being refracted again as it exits from the drop.

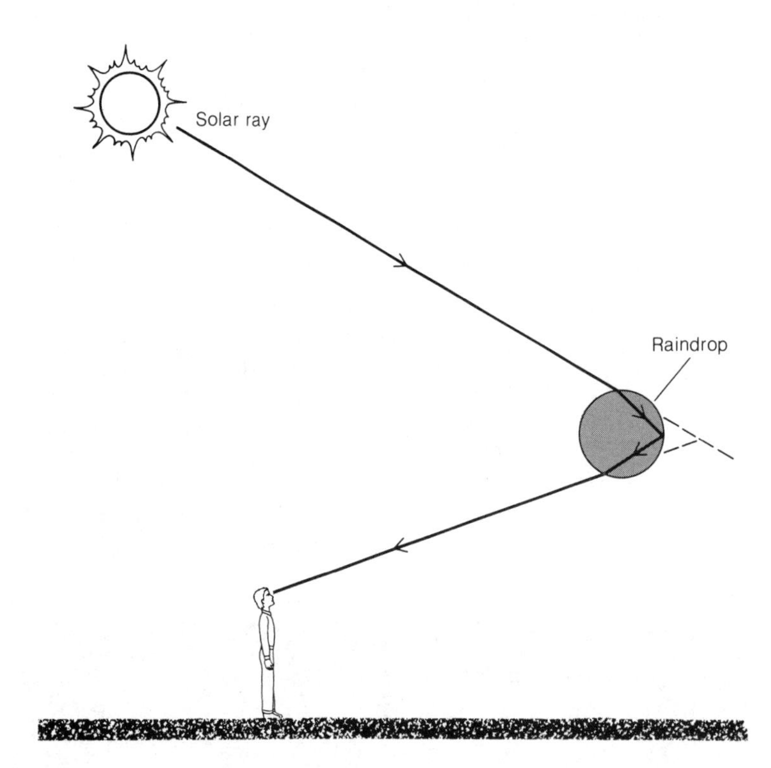

FIGURE 8.14
A solar ray is refracted and internally reflected by a raindrop. This is the basic optical process that causes a rainbow.

A rainbow appears to an observer who has his or her back to the sun and is facing a distant rain shower (Figure 8.15).* A rainbow is never visible when the sky is completely cloud covered—the sun must be shining. Because weather usually progresses from west to east, the appearance of a rainbow in the evening signals improving weather. Rain showers to the east are moving away, and clearing skies to the west, where the sun is setting, are approaching.

Raindrop refraction produces the concentric arcs of color in a rainbow. Indeed, raindrop refraction is very similar to the refraction

*You can create your own rainbow with the spray from a garden hose. Simply direct the spray so that you observe the spray with the sun at your back.

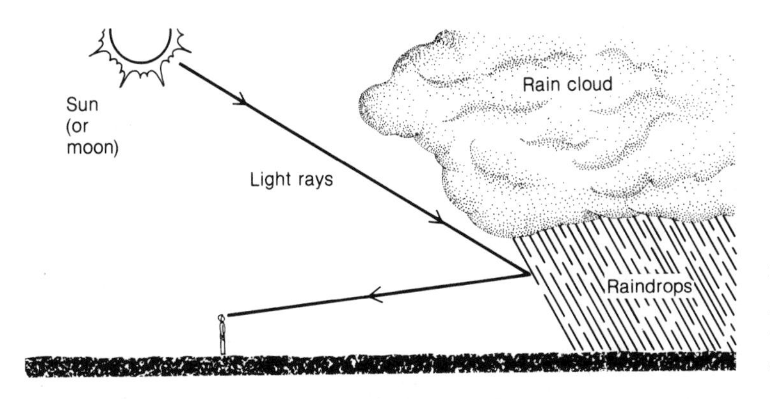

FIGURE 8.15

A rainbow appears to an observer who has his or her back to the sun and faces a distant rain shower.

of light by a glass prism: Both the prism and the raindrop disperse sunlight into its component colors (Plate 11). Violet light, at one end of the color spectrum, is slowed the most as it passes through a prism or raindrop and is refracted the most. Red light, at the other end of the color spectrum, is slowed the least and is refracted the least.

Raindrops disperse light into an arc composed of six bands of color: red, orange, yellow, green, blue,and violet, listed in order from the outer to the inner band (Plate 12). In most cases, a much dimmer, secondary rainbow appears about 8 degrees above the primary rainbow. This secondary rainbow is produced by double reflection within the raindrops (Figure 8.16). Colors in the secondary rainbow are in reverse order of colors in the primary rainbow.

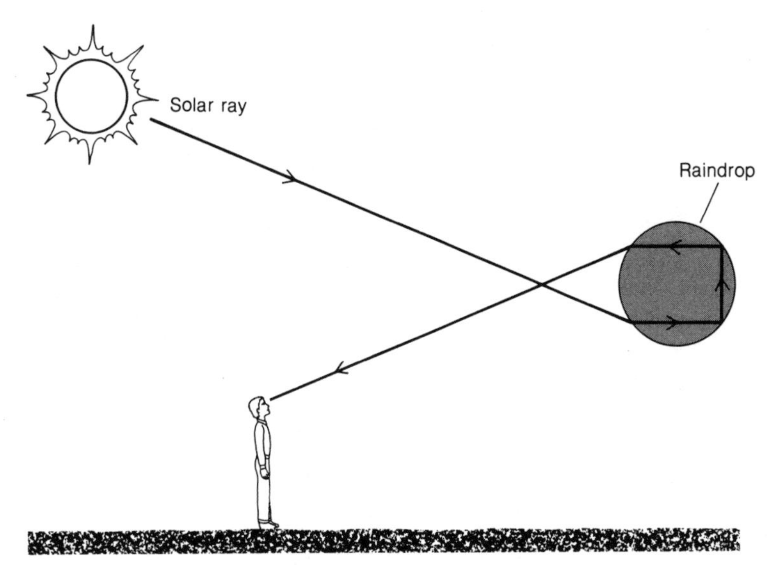

FIGURE 8.16

Refraction of sunlight by raindrops plus double reflection within raindrops produce a dimmer secondary rainbow just above the primary rainbow.

Since ice crystals refract light just as raindrops do, you may wonder why halos are not colored. In fact, ice crystals do disperse light into its component colors, but because the size and shape of ice crystals vary more than the size and shape of raindrops, the colors produced by an assemblage of ice crystals tend to overlap one another rather than form discrete bands. In halos, colors therefore wash out, although occasionally a reddish tinge is visible on the inside of a halo.

The glory

Before the age of aircraft travel, the only way to view the glory was from the vantage point of a lofty mountain peak. To see the glory, the observer must be in the bright sunshine above a cloud or fog layer, and the sun must be so situated that the observer's shadow is cast on the clouds below. The observer then sees **the glory** as concentric rings of color around the shadow of her or his head. Although less distinct, the colors of the glory are the same as those of the primary rainbow, with violet being the innermost band and red the outermost band. Today, airline pilots and observant air passengers frequently see the glory around the shadow of their aircraft on the cloud deck below (Figure 8.17).

The glory results from the same optics as the primary rainbow. The difference is in the size of the reflecting and refracting particles. Whereas rainbows occur when sunlight strikes a mass of raindrops, the glory is the consequence of sunlight interacting with a mass of smaller water droplets of uniform size that compose a cloud. In both cases, the sun's rays undergo refraction upon entering the droplet, followed by a single internal reflection, and another refraction upon exiting.

The optics of rainbows and glories have one interesting difference, however. In the glory, sunlight is refracted and reflected directly back toward the sun. As Figure 8.15 shows, this is not the case with rainbows. This explains why the glory appears about the shadow of the observer, who is situated in the direct line of the incident solar rays and of the refracted and reflected solar rays. This also explains an observation that must have fascinated ancient mountain mystics. Suppose that you and a friend are side by side high on a mountain slope at a site favorable for viewing the glory. On the cloud deck below, you see your own shadow beside that of your friend, but the glory appears about your head and not about your friend's head. Lest you presume that you have been singled

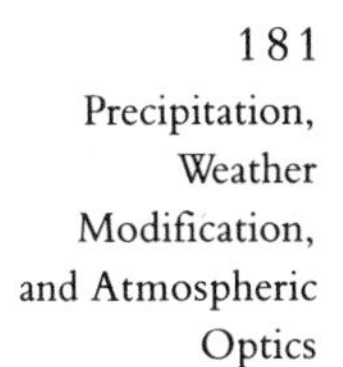

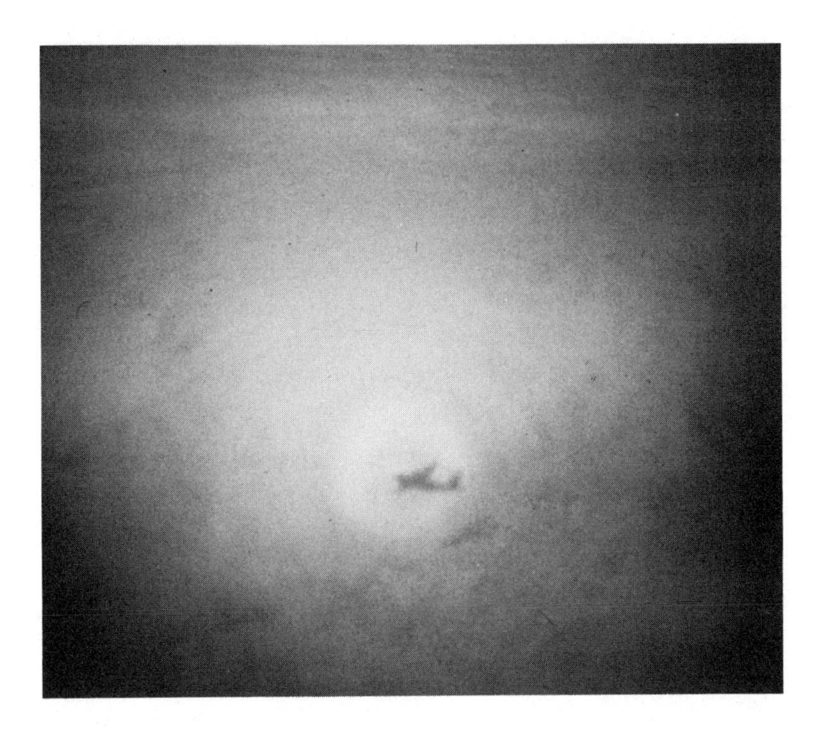

FIGURE 8.17

The glory, photographed from an aircraft, is a consequence of refraction and internal reflection of sunlight by cloud droplets of uniform size. (NCAR/NSF photograph)

out, note that your friend has the opposite observation, for your friend sees the glory only about his or her head. The fact is that each observer is in a position to view only one glory.

Conclusions

In this chapter and in previous chapters we have seen how water changes phase to produce clouds and precipitation as it is transported within the hydrologic cycle. Atmospheric circulation plays a major role in bringing air to satuation. The next several chapters describe the many atmospheric circulation systems and their associated weather conditions. We begin in Chapter 9 with a discussion of the forces that drive and shape atmospheric circulation.

SUMMARY STATEMENTS

- The collision-coalescence process and the Bergeron process account for precipitation formation. The bulk of the earth's precipitation originates in cold clouds consisting of a mixture of ice crystals and supercooled water droplets.
- The Bergeron process depends on the saturation vapor pressure difference between ice and water. As a consequence of this difference, ice crystals grow at the expense of supercooled water droplets.
- The principal forms of precipitation are rain, drizzle, freezing rain, snow, ice pellets (sleet), and hail. The form taken by precipitation depends on the type of source cloud and on the temperature profile of the troposphere.
- Cloud seeding is a type of weather modification intended to enhance rainfall or snowfall by stimulating collision and coalescence of cloud particles, or by stimulating the Bergeron process, or by affecting both processes. Cloud seeding efforts are evaluated by rigorous statistical testing and have not proven to be successful in all cases.
- Fogs can be dispersed either by raising the air temperature (thereby lowering the relative humidity) or by seeding.
- A halo is produced when the ice crystals composing cirrus clouds refract sunlight or moonlight. Rainbows develop when falling raindrops refract and internally reflect sunlight. The glory is the consequence of the same optical effect as the primary rainbow, except that the refracting and reflecting agents are cloud droplets instead of rain.

KEY WORDS

collision-coalescence process
warm clouds
Bergeron process
cold clouds
virga
precipitation
drizzle
rain
freezing rain
snow
ice pellets
sleet
hail
graupel
rain gage
weighing bucket rain gage
tipping bucket rain gage
weather modification
cloud seeding
Project Whitetop
fog dispersal
halo
refraction
sun dogs
mock suns
parhelia
rainbows
the glory

REVIEW QUESTIONS

1 Distinguish between warm clouds and cold clouds.

2 Describe the collision-coalescence process of precipitation formation.

3 Describe the Bergeron process of precipitation formation.

4 Once a raindrop or a snowflake leaves a cloud, it enters a *hostile* environment. What is meant by this statement?

5 Why are drizzle drops smaller than raindrops?

6 Distinguish between ice pellets (sleet) and freezing rain. Distinguish also between ice pellets and hail.

7 Under what conditions does freezing rain develop? Explain why freezing rain can be destructive.

8 Explain how hail may fall to the ground even when surface air temperatures are well above the freezing point.

9 Why must care be exercised in siting precipitation gages?

10 Define weather modification.

11 What is the basic objective of cloud seeding?

12 Comment on the effectiveness of cloud seeding. Does it work?

13 List some of the benefits and costs of successful cloud seeding efforts.

14 What is the principle underlying the thermal method of fog dispersal?

15 How might cloud seeding actually reduce the amount of precipitation that falls?

16 Compare and contrast the optics of rainbows with those of the glory. Explain why you can view only your own glory.

17 What does the appearance of a morning rainbow suggest about the weather later in the day?

18 Why do raindrops and ice crystals refract light rays?

19 Distinguish between a primary rainbow and a secondary rainbow.

20 What type of cloud produces a halo about the sun or moon?

POINTS TO PONDER

1 Explain why cold clouds consist more of supercooled water droplets than of ice crystals at temperatures between 0 °C and −20 °C. Explain also why most clouds colder than −20 °C are made up almost entirely of ice crystals.

2 Why is the saturation vapor pressure greater over supercooled water droplets than over ice crystals?

3 Give two reasons why rainfall is likely to be heavier on top of a mountain than in a neighboring valley.

4 Is rain possible where you are, even though you measure a relative humidity well below 100 percent? Explain your answer.

5 Discuss the factors that control the size of snowflakes.

6 As a general rule, 10 in. of snow melts down to 1 in. of rainwater. From this, determine the average density of snow.

PROJECTS

1 Design an experiment in which you create your own rainbow.

2 Assemble photographs of a variety of atmospheric optical effects. You may wish to take your own photographs. Provide an explanation for each optical phenomenon.

SELECTED READINGS

Blanchard, D. C. "Science, Success and Serendipity." *Weatherwise* 32 (1979):236–241. *Recounts the pioneering efforts of V. Schaefer, I. Langmuir, and T. Bergeron in cloud physics research.*

Cotton, W. R. "Modification of Precipitation From Warm Clouds—A Review." *Bulletin of the American Meteorological Society* 63 (1982):146–160. *An extensive summary of what is currently understood about precipitation physics.*

Davis, R. J., and L. O. Grant, eds. *Weather Modification, Technology, and Law*. Boulder, Col.: Westview Press, 1978. 124 pp. *Presentations by representatives of government, industry, and universities at the 1976 Conference on the Legal and Scientific Uncertainties of Weather Modification, sponsored by the American Association for the Advancement of Science.*

Fraser, A. B. "Chasing Rainbows." *Weatherwise* 36 (1983):280–287. *Describes a series of bright arcs observed within the primary rainbow.*

Greenler, R. *Rainbows, Halos, and Glories*. New York: Cambridge University Press, 1980. 195 pp. *Discusses the causes of a wide variety of optical phenomena in the atmosphere; includes some spectacular photographs.*

Hallett, J. "How Snow Crystals Grow." *American Scientist* 72 no. 6 (1984):582–589. *A sophisticated review of the physics of ice crystal growth.*

Kunkel, B. A. "Controlling Fog." *Weatherwise* 33 (1980):117–123. *A historical review of attempts to disperse fog.*

Walker, J. *Light From the Sky*. San Francisco: W. H. Freeman, 1980. 78 pp. *Reprints of* Scientific American *articles on various atmospheric optical phenomena, including mirages, halos, rainbows, and the glory.*

No one can tell me,
Nobody knows,
Where the wind comes from,
Where the wind goes.

A. A. MILNE
"Wind on the Hill"
Now We Are Six

Many forces interact to initiate and shape the movement of air. Ultimately, it is the sun that supplies the energy that drives the wind. (Photograph by Mike Brisson)

NINE

The Wind

SOME weather systems favor clear skies, light winds, and frosty mornings, while others bring ominous clouds, precipitation, and biting winds. Some weather systems trigger brief showers, and others are accompanied by persistent fog and drizzle. Certain weather systems are highly localized and short lived. Others dominate the weather over thousands of square kilometers for prolonged periods. Different weather systems bring different types of weather, and the type of weather depends on the circulation pattern of the air, that is, the wind.

Distinguishing between the horizontal component of the wind (east-west and north-south) and the vertical component of air motion (up-down) is useful. Although the horizontal wind is usually considerably stronger than the vertical air motion, the vertical component plays the key role in cloud and precipitation formation. As we will see, however, vertical air motion is often triggered by horizontal air motion, and vice versa.

This chapter identifies the various forces that contribute to air circulation. We show how these forces combine to initiate and modify wind, and we also examine the relationship between the horizontal and vertical motion of air.

The Forces

For convenience, imagine the wind as a continuous stream of air composed of separate volume units called **air parcels**. Assume that any force acting on an air parcel represents the influence of that same force on a stream of air parcels, in other words, on the wind. Now assume also that each air parcel consists of a unit mass of air—a single gram, for example. In considering the wind, we thus examine a force per unit mass of air, which is numerically equivalent to an acceleration, because, according to **Newton's second law of motion,***

force = mass × acceleration

For this reason, in our discussions of the wind, we sometimes use the terms "force" and "acceleration" interchangeably when we consider the motion of air parcels. Although the terms are numerically equivalent, a change in velocity is actually a response to a force. A

*This is the second of three fundamental laws of motion stated in 1687 by Sir Isaac Newton, a British mathematician and physicist. Later in this chapter, the other two laws are applied to atmospheric circulation.

force acts on an air parcel to bring about the acceleration or deceleration of that parcel.

The forces acting on air either initiate or modify motion, and are the consequence of (1) pressure gradients, (2) the Coriolis effect, (3) friction, (4) centripetal and centrifugal forces, and (5) gravity. Each of these forces is first described individually, and we then examine how they work together.

Air pressure gradients

A **gradient** is simply a change in some property with distance. A **pressure gradient** exists whenever there is a difference in air pressure from one place to another. As noted in Chapter 5, differences in the pressure exerted by air masses can arise from contrasts in air temperature, or from differing water vapor concentrations, or from both. A pressure gradient thus develops between a mass of cold, dry air and a mass of warm, humid air. In addition, diverging and converging winds can bring about air pressure changes and thereby induce a pressure gradient. This important process is covered in Chapter 11, when we examine the origin of highs and lows.

Pressure gradients develop both horizontally and vertically. A horizontal pressure gradient refers to air pressure changes along a surface of constant altitude. A vertical pressure gradient is a permanent feature of the atmosphere, since air pressure is greatest at the earth's surface and decreases rapidly with altitude.

In order to represent horizontal air pressure gradients on a weather map, the air pressure measured at each weather observation station is first reduced to sea level, as discussed in Chapter 5. Lines are then drawn joining localities that have the same air pressure reading. These lines are called **isobars**. The interval usually used between isobars is 4 mb. An isobaric analysis is used to locate centers of high and low pressure, and to determine the magnitude of the horizontal pressure gradients between weather systems. Closely spaced isobars (Figure 9.1A) mean that air pressure changes rapidly with distance, and the pressure gradient is described as steep or strong. Widely spaced isobars (Figure 9.1B) indicate that air pressure changes only slightly with distance, and the pressure gradient is weak. Note that air pressure gradients are always measured in a direction perpendicular to the isobars.

How do air pressure gradients influence the movement of air? Let us examine an analogy. Suppose a bathtub is partially filled with

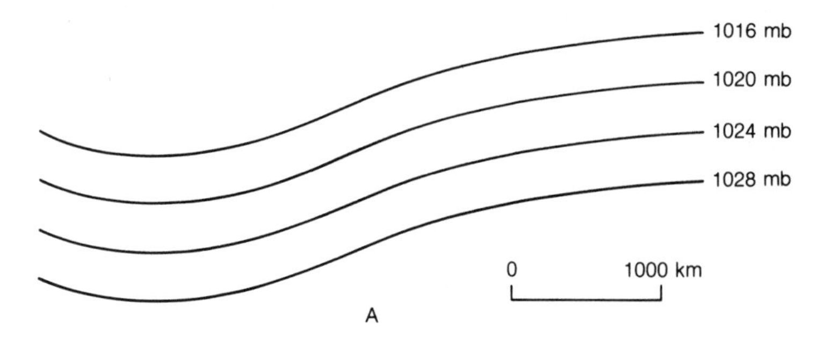

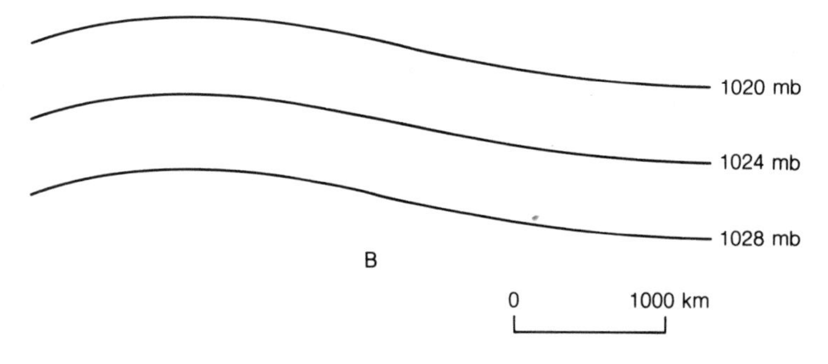

FIGURE 9.1

The horizontal air pressure gradient is relatively steep where isobars are close together (A) and weaker where isobars are farther apart (B). Isobars are lines of equal air pressure. Here, the contour interval (the difference between successive isobars) is 4 mb.

water, as shown in Figure 9.2. As we slosh the water back and forth, there is at any instant a water pressure gradient along the bottom of the tub from one end to the other. Water pressure is high where the water level is high, and low where the water level is low. In response to the water pressure gradient, water flows from one end of the tub (where the water pressure is greater) to the other end (where the water pressure is less) to eliminate the pressure gradient. If we stop agitating the water, the water level gradually returns to a horizontal surface, thus creating uniform water pressure along the tub bottom.

When an air pressure gradient develops, air tends to flow in such a way as to eliminate the pressure gradient. Hence, the wind blows

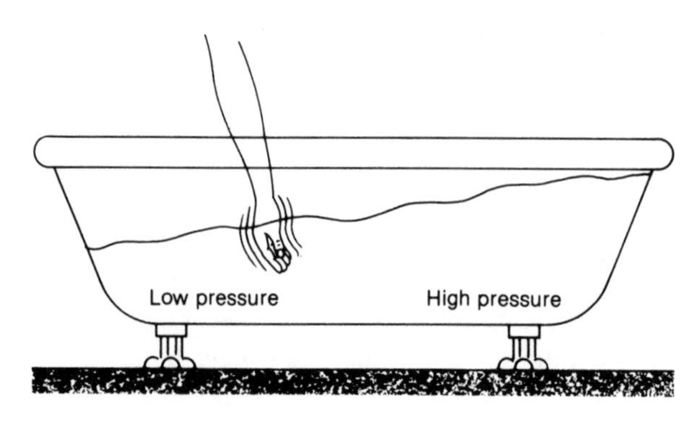

FIGURE 9.2

As we slosh water back and forth in a bathtub, we create a horizontal pressure gradient. The water pressure gradient in the tub is analogous to horizontal air pressure gradients in the atmosphere. In response to pressure gradients, the air or water flows from an area of higher pressure toward an area of lower pressure.

away from regions where air pressure is relatively high and toward locales where air pressure is relatively low. The wind is strong when the pressure gradient is steep (closely spaced isobars), and light or calm where the pressure gradient is weak (widely spaced isobars). The force that causes air parcels to move as the consequence of an air pressure gradient is the **pressure gradient force**. If this were the only force acting in the atmosphere, all pressure gradients would eventually disappear as the air flowed from high pressure centers to low pressure centers, but air motion is modified by other forces.

Coriolis effect

The pressure gradient force initiates wind, but once the air begins to move, other forces come into play. One of these forces arises from the rotation of the earth on its axis and causes a deflection of the wind. In the Northern Hemisphere, this force causes winds to swerve to the right of the initial direction of flow. In the Southern Hemisphere, the deflection is to the left of the initial direction of air motion (Figure 9.3). This deflective force is due to the **Coriolis effect**, named for Gustav Gaspard de Coriolis, who described the

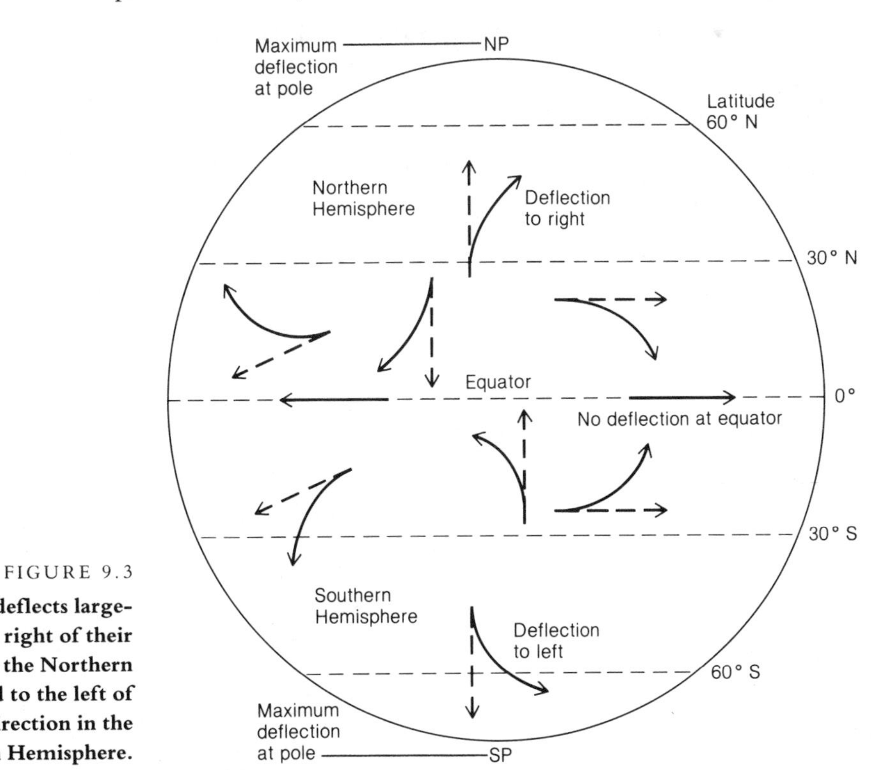

FIGURE 9.3

The Coriolis effect deflects large-scale winds to the right of their initial direction in the Northern Hemisphere and to the left of their initial direction in the Southern Hemisphere.

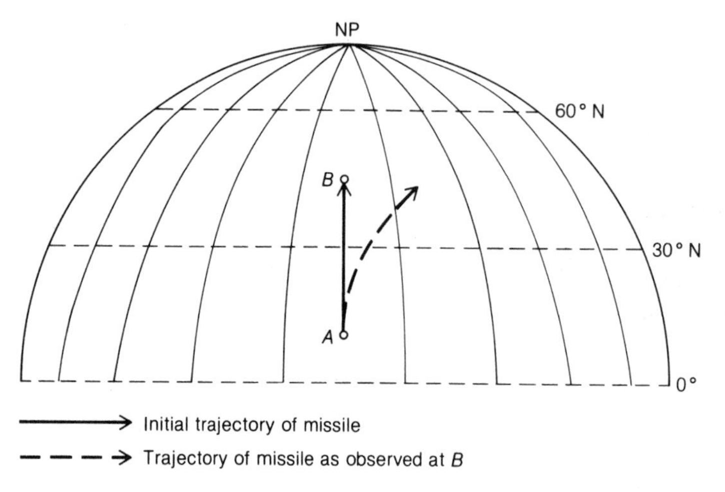

FIGURE 9.4

A missile fired straight northward from a launching pad (*A*) in a tropical latitude appears to an observer, located at the midlatitude target (*B*), to swerve toward the east. This deflection is due to the Coriolis effect.

effects of the earth's rotation on winds mathematically in 1835 An illustration helps explain the Coriolis effect.

In the Northern Hemisphere, suppose a missile is fired straight north from a launching pad in tropical latitudes toward a target in midlatitudes. As shown in Figure 9.4, the missile misses its target. In fact, to observers at the target site, the missile appears to swerve eastward. To observers watching the missile trajectory from a distant fixed point in space, however, the missile path appears straight. The deflection seen by the earthbound observers arises from the fact that their frame of reference changes as the missile travels along its path. On earth, we measure movements of objects with respect to a north-south, east-west frame of reference. From our perspective, this frame of reference does not change, because we and it rotate with the earth. Viewed from space, as in Figure 9.5, however, our

FIGURE 9.5

When viewed from a distant fixed point in space, our north-south, east-west frame of reference changes as the earth rotates on its axis. Here, a south wind in the Northern Hemisphere is deflected (relative to the frame of reference) to the right and becomes a southwest wind. This is the basic reason for the Coriolis effect. (Modified after Lutgens, F. K., and E. J. Tarbuck. *The Atmosphere*, © 1979, p. 131. Reprinted by permission of Prentice-Hall, Inc., Englewood Cliffs, N.J.)

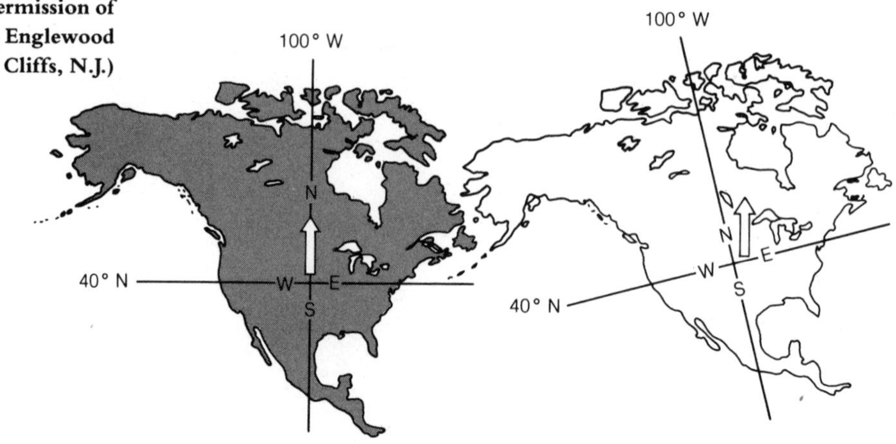

frame of reference actually shifts as the earth rotates on its axis. For the missile, or any object moving in the atmosphere, it is as if the earth twists under the object. In reality, the missile's actual direction does not change, but the perspective (frame of reference) of the observer on earth does change as the earth turns. The missile therefore appears to swerve to the right.

We also measure wind direction and speed with respect to a north-south, east-west frame of reference that rotates with the earth. We must therefore take the Coriolis effect into account in any explanation of air circulation. From our perspective on earth, the Coriolis effect gives rise to a real force that deflects the wind. In our discussion, we use the phrase "Coriolis effect" to designate this force.

Although the Coriolis effect influences the wind regardless of wind direction, the magnitude of the effect varies significantly. The Coriolis effect stems from the rotation of the earth on its axis, which imparts a rotation to our earthbound frame of reference. The rotation of our frame of reference is maximum at the poles and declines with latitude to zero at the equator. This variation can be understood by visualizing the daily rotation of towers situated at different latitudes. In a 24-hour day, the earth makes one complete rotation, as would a tower situated at the North or South Pole. In the same period, a tower at the equator would not rotate at all, but would instead turn in an end-over-end motion. For a tower located at any latitude in between, some rotation of the tower occurs as the earth rotates, but not as much rotation as occurs at the poles. The Coriolis effect is thus latitude dependent: the Coriolis effect is zero at the equator, and increases with latitude to a maximum at the poles.

The Coriolis effect also varies with wind speed—strengthening as the wind accelerates. This is because, in a given period of time, the faster air parcels cover greater distances. When compared with our north-south, east-west frame of reference, longer trajectories have greater deflections than do shorter trajectories. For practical purposes, the Coriolis effect has an important influence only on global- and synoptic-scale weather systems, and is negligible in mesoscale and microscale weather systems.

A rotational motion usually accompanies the draining of water from a sink or a bathtub. A popular misconception is that the direction of this rotation (clockwise or counterclockwise) is consistently in one direction in the Northern Hemisphere and in the opposite direction in the Southern Hemisphere—presumably because of the Coriolis effect. At this very small scale, the magnitude of the Coriolis effect is negligibly small. The drainage direction is more likely a consequence of some residual motion of the water when the sink or bathtub was first filled with water and can be clockwise or counterclockwise.

Although this book concentrates on weather systems in the Northern Hemisphere, it is useful to know why the Coriolis effect reverses direction from the Northern to the Southern Hemisphere, causing large-scale winds in the Southern Hemisphere to swerve to the left rather than to the right. This reversal is related to the difference in an observer's sense of the earth's rotation in the two hemispheres. To an observer at the North Pole, the earth rotates counterclockwise, while to an observer at the South Pole, the earth rotates clockwise. This rotation reversal translates into a reversal in deflective direction between the two hemispheres.

Friction

Like the Coriolis effect, friction is a responsive force, which does not develop until air is already moving. **Friction** is usually described as the resistance an object encounters as it moves in contact with other objects. In the lower troposphere, the horizontal wind encounters strong frictional resistance as it contacts the ground and blows against all objects on the ground, including grass, trees, buildings, and people. The rougher the surface is, the greater the frictional resistance. A forest thus offers more frictional resistance to the wind than does the relatively smooth surface of a freshly mowed lawn.

The influence of friction on the wind diminishes rapidly with altitude above the earth's surface—away from the obstacles mainly

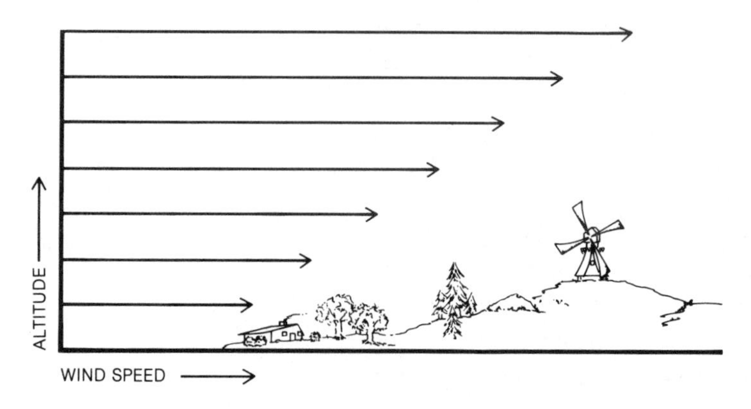

FIGURE 9.6

The horizontal wind strengthens with altitude—away from the frictional resistance offered by objects at the earth's surface.

responsible for resistance (Figure 9.6). This explains the advantage of siting a windmill at as high an altitude as possible. Above an altitude of about 1 km (0.62 mi), friction exerts an essentially negligible effect on the wind. The atmospheric zone in which frictional resistance is essentially confined is called the **friction layer**.

Centripetal and centrifugal forces

If a string tied to a rock is swung around, the rock describes a circular path of constant radius (Figure 9.7). If we cut the string, the rock flies off in a straight line. This is an illustration of **Newton's first law of motion**: An object at rest or in straight-line motion remains in that state unless acted on by a net (unbalanced) force.

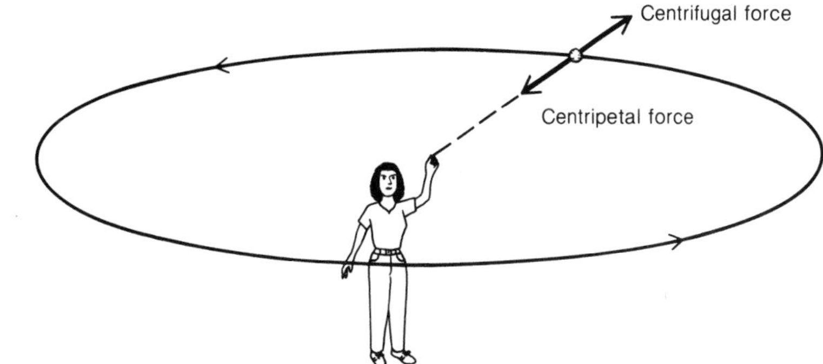

FIGURE 9.7
An object describing a circular path is acted on by a net inward directed centripetal force, which is equal in magnitude to an outward directed centrifugal force.

Before we cut the string, the string exerted a force on the rock by confining it to a circular path. This force is directed inward, toward the center of the circular path, and for this reason is known as the **centripetal force**.*

Forces bring about acceleration. We usually think of an acceleration as merely a change in the speed of an object, as when an auto speeds up or slows down. However, an acceleration may also consist of a change in the direction of an object without any change in the object's speed. This is the case in our example of the rock on a string. The centripetal force is a net force that is responsible for a continuous change in direction (a circular rather than a straight path), and not for a change in speed. Cutting the string eliminates the centripetal force, and the rock then follows a straight path.

Another of Newton's laws also applies to our orbiting rock on a string. **Newton's third law of motion** holds that for every force,

*_Centripetal_ means "center seeking."

there is an equal and opposite force, that is, for every action there is a reaction. You lean against a wall, and the wall pushes back with the same force. The centripetal force exerted on the rock by the string is opposed by an equal force exerted on the string by the rock. This outward directed force is known as the **centrifugal force**.

From the rock and string illustration, it follows that whenever the wind describes a curved path, centripetal and centrifugal forces are operating. Although the two forces are equal in magnitude, they are physically distinct, and the two terms should not be used interchangeably.

Gravity

The atmosphere is subject to the same force that holds all objects on the earth's surface: **gravity**. The force of gravity is actually the net effect of two other forces working together: (1) the force of attraction between the earth and all other objects, called **gravitation**, and (2) a centrifugal force imparted to all objects because of their spin with the earth on its axis. The two forces combine to produce the force of gravity, which accelerates a unit mass of any object downward at the rate of 9.8 m per second each second. The force of gravity always acts downward and perpendicular to the earth's horizontal surface. For this reason, gravity, unlike the Coriolis effect and frictional forces, does not modify the horizontal wind. Gravity does, however, affect air that is rising or sinking, in, for example, convection currents, and gravity is responsible for the flow of cold, dense air downhill.

We have now examined the various forces that affect horizontal or vertical air motion, and can draw the following conclusions:

1 A horizontal pressure gradient accelerates air away from regions of high air pressure and toward areas of low air pressure.

2 The Coriolis effect causes synoptic-scale and global-scale winds to swerve to the right of their initial direction in the Northern Hemisphere and to the left in the Southern Hemisphere.

3 Frictional resistance slows winds that are within 1 km of the earth's surface.

4 Curved air motion is influenced by centripetal and centrifugal forces.

5 Gravity accelerates air downward but does not modify horizontal winds.

Joining Forces

To this point, we have examined the forces operating in the atmosphere as if each force acted independently of the others. In reality, these forces interact to govern both the direction and speed of the wind. In some cases, two or more forces achieve a balance, or equilibrium. From Newton's first law of motion, it follows that when the forces acting on an air parcel are in balance, no net force occurs, and the parcel either remains stationary or continues to move along a straight path at constant speed: the net acceleration is zero.

Let us now examine how forces interact in the atmosphere to control the vertical and horizontal flow of air, that is, the wind. These interactions result in (1) hydrostatic equilibrium, (2) the geostrophic wind, (3) the gradient wind, and (4) the surface winds, which are winds within the friction layer.

Hydrostatic equilibrium

We noted earlier that the atmosphere features a vertical pressure gradient. As shown in Figure 9.8, the force due to this pressure gradient is directed upward from high pressure at the earth's surface toward the lower air pressure aloft. If this force acted alone, the vertical pressure gradient force would accelerate air away from the earth, and we would be gasping for breath. However, except in

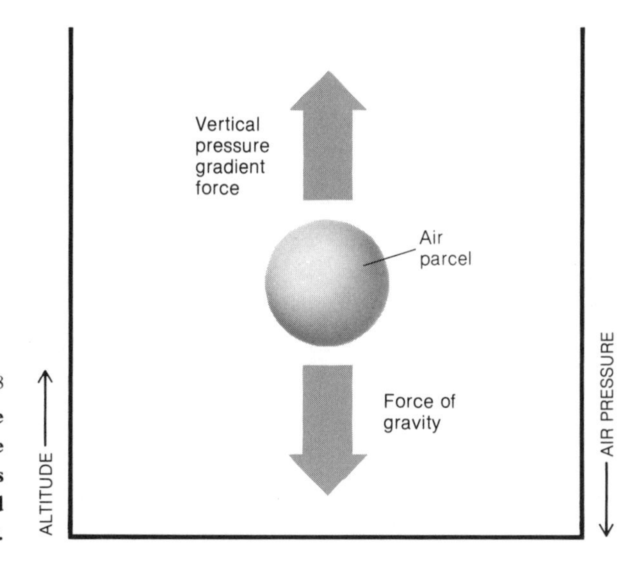

FIGURE 9.8

With hydrostatic equilibrium, the upward directed vertical pressure gradient force on an air parcel is balanced by the downward directed force of gravity.

some small-scale violent weather systems, the atmosphere's vertical pressure gradient force is exactly balanced by the equal and oppositely directed force of gravity—another example of Newton's third law of motion. This balance of forces is known as **hydrostatic equilibrium**.

Whenever forces are in balance, no net acceleration occurs, that is, there is no change in velocity. Hydrostatic equilibrium, then, does not preclude vertical (up or down) motion of air in the atmosphere. Because of the balance of forces, upward-moving air parcels continue upward at *constant* velocity, and downward-moving air parcels continue downward at *constant* velocity.

Geostrophic wind

The **geostrophic wind** is an unaccelerated horizontal wind that flows along a straight path above the friction layer. It results from a balance that develops between the horizontal pressure gradient force and the force due to the Coriolis effect. When a horizontal pressure gradient (P_H) develops, air parcels at first accelerate directly across isobars, away from high pressure and toward low pressure (Figure 9.9). As air parcels speed up, the Coriolis effect (C) strengthens, causing air parcels to swerve gradually to the right of their initial flow direction (in the Northern Hemisphere). The two forces eventually attain a balance, so that the wind blows at a constant speed in a straight path parallel to the isobars. Because the Coriolis effect is a large-scale phenomenon, the geostrophic wind can develop only in synoptic-scale or global-scale weather systems.

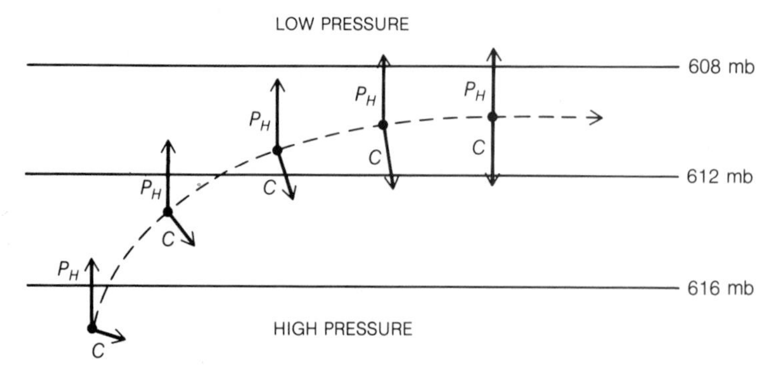

FIGURE 9.9
The horizontal air pressure gradient causes air parcels to accelerate across isobars from areas of high pressure toward low pressure. The Coriolis effect then deflects air parcels to the right in the Northern Hemisphere, and increases in magnitude until the pressure gradient force is balanced by the Coriolis effect. The result is an unaccelerated wind parallel to isobars, that is, the geostrophic wind.

Gradient wind

The **gradient wind** has many characteristics in common with the geostrophic wind. It, also, is large scale, horizontal, frictionless, and parallel to isobars. The important difference between the two winds is that the geostrophic wind travels in a straight path, while the path of the gradient wind is curved. Because a net centripetal force constrains air parcels to a curved trajectory, the gradient wind is not the consequence of balanced forces. Recall that the centripetal force changes only the direction of an air parcel, and not the parcel's speed. The horizontal pressure gradient force, the Coriolis effect, and the centripetal force thus interact in the gradient wind.

A gradient wind develops above the friction layer around a dome of high air pressure, called an **anticyclone** (or high), or around a center of low air pressure, called a **cyclone** (or low). In an anticyclone, the isobars form a series of concentric circles about the area of highest pressure, as shown in Figure 9.10. The horizontal pressure gradient force, P_H, is directed radially outward, away from

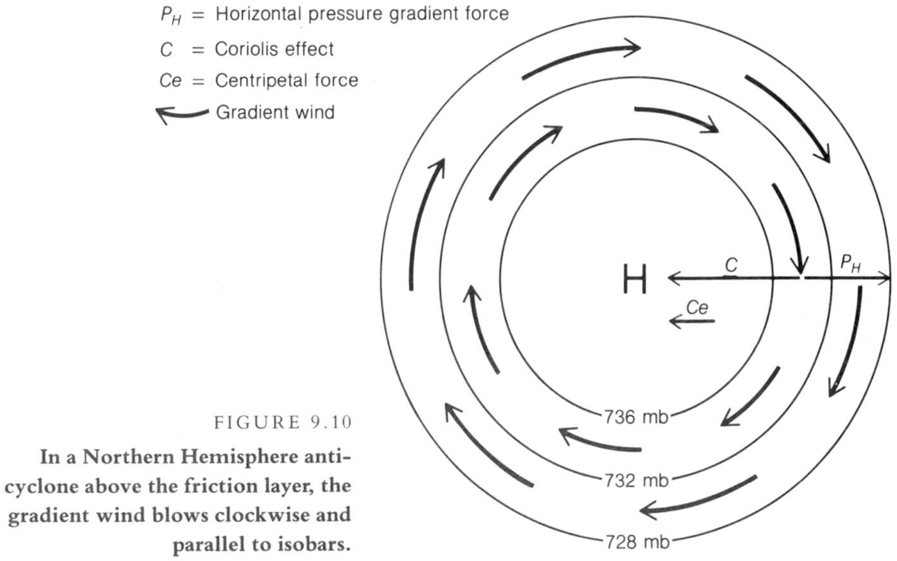

FIGURE 9.10
In a Northern Hemisphere anticyclone above the friction layer, the gradient wind blows clockwise and parallel to isobars.

the center of the high. The Coriolis effect, C, is directed inward. The Coriolis effect is slightly greater than the pressure gradient force, with the difference between the forces equal to the net inward-directed centripetal force, Ce. In a Northern Hemisphere anticyclone above the friction layer, the gradient wind consequently blows clockwise and parallel to isobars.

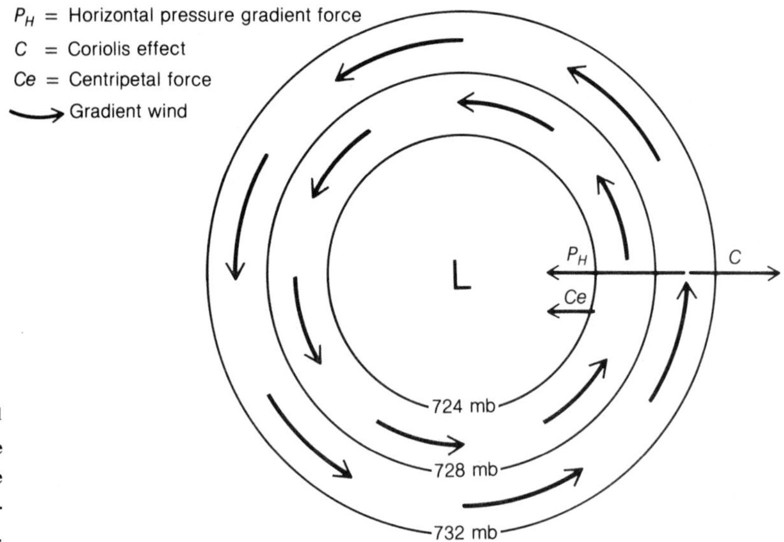

FIGURE 9.11

In a Northern Hemisphere cyclone above the friction layer, the gradient wind blows counterclockwise and parallel to isobars.

In a cyclone, the isobars form a series of concentric circles around the area of lowest air pressure. As indicated in Figure 9.11, the horizontal pressure gradient force is directed inward, toward the cyclone center, and the Coriolis effect is directed radially outward from the center of the low. The pressure gradient force is slightly greater than the Coriolis effect, with the difference equal to the net inward-directed centripetal force. In a Northern Hemisphere cyclone above the friction layer, the gradient wind consequently blows counterclockwise and parallel to isobars.

This discussion of geostrophic and gradient winds only approximates the actual behavior of winds above the friction layer. The approximation is nonetheless quite useful, and meteorologists rely on such approximations in their analysis of isobaric patterns on weather maps. A mathematical description of geostrophic and gradient winds is presented in the Mathematical Note at the end of the chapter.

Surface winds

Geostrophic winds and gradient winds are frictionless, that is, they occur at altitudes above the friction layer. What is the effect of friction on the winds within the friction layer, the surface winds? For large-scale air motion in a straight path, the frictional force (*F*) combines with the Coriolis effect (*C*) to balance the horizontal pressure gradient force (P_H), as shown in Figure 9.12. The net effect

FIGURE 9.12

Within the friction layer, the Coriolis effect and the frictional force act together to balance the horizontal pressure gradient force. This causes the wind to blow across isobars toward low air pressure. Recall that the Coriolis effect acts at right angles to the wind direction, and friction acts in a direction opposite to that of the wind.

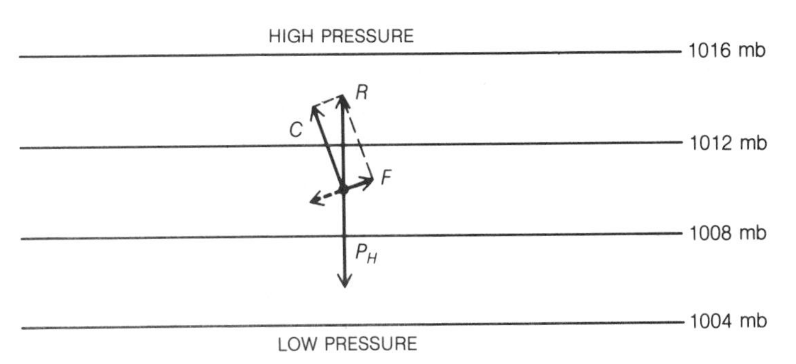

is that the wind slows down and shifts direction across isobars and toward low pressure. The deflection angle of surface winds crossing isobars varies from about 10 degrees over relatively smooth surfaces like the ocean, where friction is low, to almost 45 degrees over rough terrain, where friction is greater.

As we ascend from the earth's surface through the friction layer, the horizontal wind spirals as it strengthens and shifts direction (Figure 9.13). This is the **Ekman spiral**, named for Vagn Walfrid Ekman, who described the phenomenon mathematically in 1905.

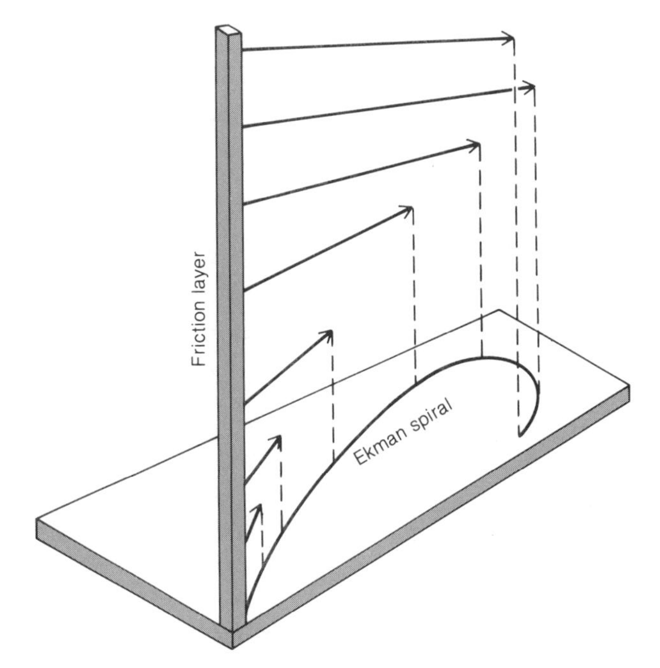

FIGURE 9.13

The horizontal wind describes an Ekman spiral as the consequence of frictional resistance. The horizontal wind strengthens and shifts direction with increasing altitude within the friction layer.

At the top of the spiral, which is above the friction layer, the wind is essentially geostrophic.

What is the effect of friction on the horizontal surface winds blowing in an anticyclone and cyclone? As with straight-line surface winds, friction slows the winds and combines with the Coriolis effect to shift winds across isobars and toward low-pressure areas. At the earth's surface, anticyclonic winds therefore blow clockwise and outward, as shown in Figure 9.14, and surface cyclonic winds blow counterclockwise and inward, as shown in Figure 9.15.

In the Southern Hemisphere, cyclonic and anticyclonic circulations are the reverse of their Northern Hemisphere counterparts. This reversal occurs because of the change in direction of the Coriolis deflection between the two hemispheres. In the Southern Hemisphere, surface winds in a cyclone blow in a clockwise and inward direction and surface winds in an anticyclone blow in a counterclockwise and outward direction. Above the friction layer, Southern Hemisphere cyclonic winds are clockwise and Southern Hemisphere anticyclonic winds are counterclockwise.

A glance at almost any national weather map (Figure 9.16) reveals that isobars seldom describe lengthy straight segments or circular patterns. Instead, isobars often form patterns of ridges and troughs.

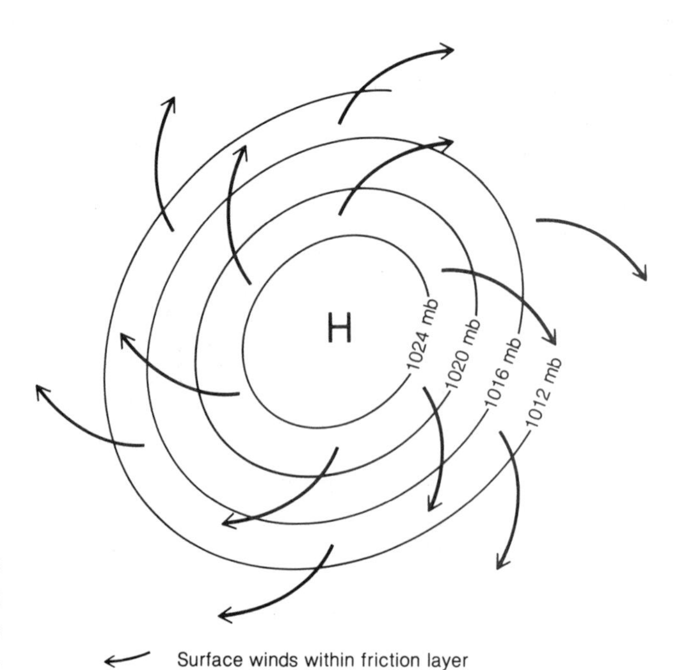

FIGURE 9.14

Suface winds blow clockwise and outward about a Northern Hemisphere anticyclone.

FIGURE 9.15

Surface winds blow counterclockwise and inward about a Northern Hemisphere cyclone.

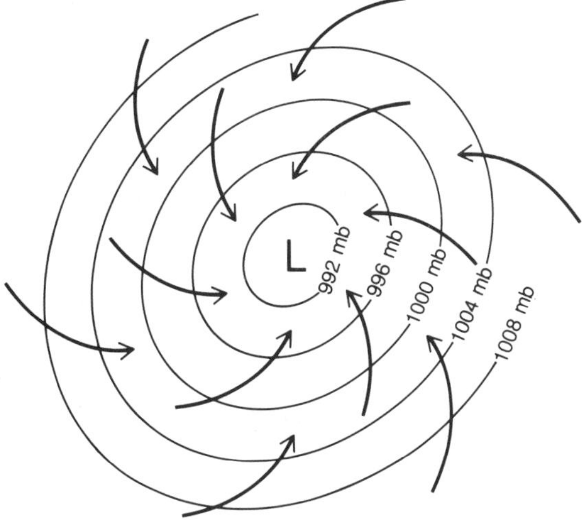

FIGURE 9.16

Note how isobars on a typical surface weather map describe trough and ridge patterns. (NOAA weather map)

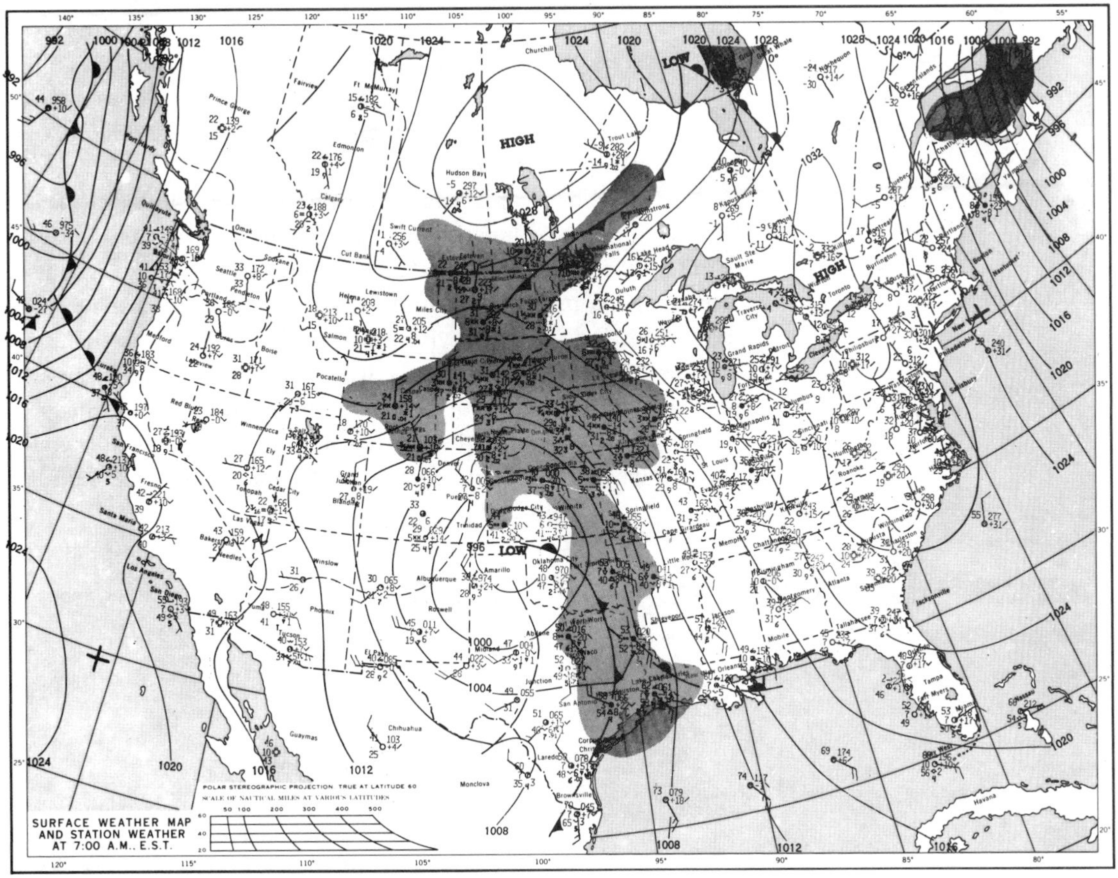

Nonetheless, in ridges and troughs, winds tend to parallel isobars above the friction layer, and tend to cross isobars toward areas of low pressure at the earth's surface.

An additional consideration in analyzing isobaric patterns for wind is the isobar spacing. As noted earlier, the stronger the air pressure gradient is, the faster the wind will be. Where isobars are closely spaced, the geostrophic and gradient winds are strong. Where isobars are widely spaced, these winds are weak. The same rule applies to surface winds.

Continuity of Wind

Air is a continuous fluid, and continuity implies a connection between the horizontal and vertical components of the wind. Horizontal winds are forced to follow the undulating landscape, ascending hills and descending into valleys. Uplift occurs along frontal surfaces as one air mass moves horizontally and either overrides or slips under another air mass (Chapter 6). Having examined the horizontal circulation of anticyclones and cyclones, we can identify other important connections between the horizontal and vertical components of the wind.

As noted in the preceding section, surface winds in an anticyclone in the Northern Hemisphere spiral clockwise and outward. Consequently, the horizontal surface winds diverge away from the center of the high. A vacuum does not, however, develop at the center. Instead, air is slowly drawn downward toward the earth's surface, replacing the air that is diverging (Figure 9.17). Now recall that descending air is adiabatically compressed, which causes the temperature to rise and the relative humidity to fall. Skies therefore tend to be fair within anticyclones, and anticyclones are appropri-

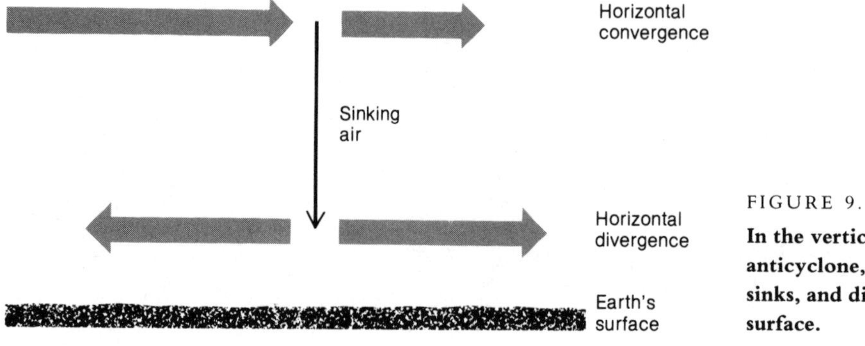

FIGURE 9.17

In the vertical cross section of an anticyclone, air converges aloft, sinks, and diverges at the earth's surface.

ately described as "fair weather" systems. Aloft, winds converge toward the center of the high, thereby compensating for the air that descends.

In contrast, surface winds in a cyclone in the Northern Hemisphere spiral counterclockwise and inward. Surface winds therefore converge toward the center of a low. Air does not simply pile up at the center, however, because the air is forced to ascend (Figure 9.18). Recall that ascending air expands and cools adiabatically, which

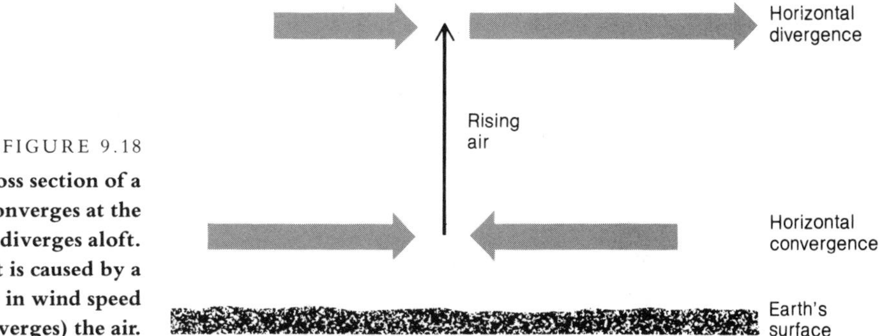

FIGURE 9.18

In the vertical cross section of a cyclone, air converges at the surface, rises, and diverges aloft. Divergence aloft is caused by a downwind increase in wind speed that stretches (diverges) the air.

causes the relative humidity to rise. Clouds and precipitation may eventually develop, thus cyclones are typically stormy weather systems. Aloft, winds diverge away from the center of the low, thereby compensating for the ascending air.

Vertical motion is also induced by downwind changes in frictional resistance. The rougher the earth's surface, the more resistance it offers to horizontal winds. When the wind blows from a rough surface to a relatively smooth surface—as when it blows from land to sea—the wind accelerates. As Figure 9.19 shows, this

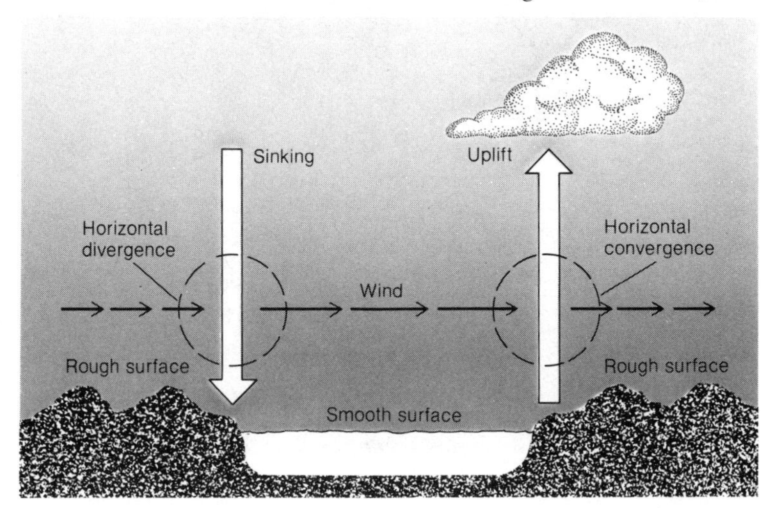

FIGURE 9.19

Surface winds undergo horizontal divergence when blowing from a rough to a smooth surface, and horizontal convergence when blowing from a smooth surface to a rough surface. Horizontal divergence of air causes air to sink, and horizontal convergence of air causes air to rise.

acceleration stretches (diverges) the wind, thereby pulling air down from aloft. In contrast, when the wind blows from a smooth to a rough surface, the wind slows, piles up (converges), and induces upward air motion. This is one reason why, along a coastline, convective cumulus clouds tend to develop with an onshore wind (from sea to land) and tend to dissipate with an offshore wind (from land to sea).

Frictionally induced convergence helps trigger **lake effect snows** on the southern and eastern shores of North America's Great Lakes (Figure 9.20). In early winter, strong north and northwest winds advect cold, dry air over the relatively warm waters of the lakes. Water evaporates from the lake surface, thus increasing the relative humidity of the lowest layer of the advecting cold air mass. The cold, now humid air converges as it approaches the shoreline, thus inducing uplift, cloudiness, and locally heavy snowfalls.

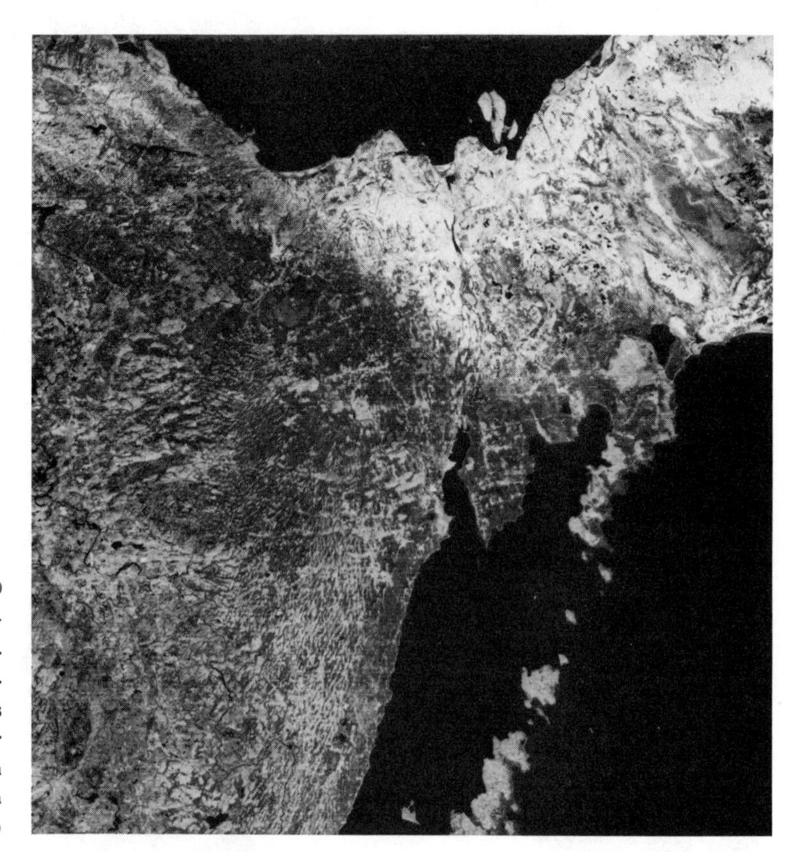

FIGURE 9.20

Landsat photograph of a sharply defined snowfall pattern on Michigan's Upper Peninsula on 20 October 1972. Cold northwest winds blowing across Lake Superior induced lake effect snows. (From U.S. Geological Survey, EROS Data Center, Sioux Falls, S.D.)

Some Practical Observations

The discussions so far in this chapter may seem quite abstract. From everyday experience, we are aware of the end product of the forces described, namely, the weather. We probably were not aware of the forces themselves. In learning about atmospheric forces and how they interact, we see that each force is bound by certain constraints. For example, friction is important only at the bottom of the troposphere, and the Coriolis effect always acts to the right of the initial wind direction in the Northern Hemisphere. In the following chapters, your awareness of these and other constraints will aid in understanding the characteristics of the various weather systems—for example, why hurricanes do not form at the equator, and why winds in a tornado may blow in either a clockwise or counterclockwise direction.

In spite of the emphasis on theoretical considerations, some practical observations can be drawn from our discussion. For example, we learned that surface winds spiral in a counterclockwise and inward direction around a low. With this information, we can formulate a useful rule of thumb for locating storm centers. If you stand with your back to the wind and then turn approximately 45 degrees to your right, the storm center will be located to your left. This rule is a modification of an observation first stated in 1857 by the Dutch meteorologist Christopher H. D. Buys-Ballot. It must be applied with caution, however, because surface winds can be modified by local effects such as sea breezes or other mesoscale air circulation.

Another practical observation stems from our discussions of air circulation in an anticyclone. As we saw, descending air currents associated with highs favor fair skies. In addition, the horizontal air pressure gradient is typically very weak over a large area in the middle of a high. The resulting light winds coupled with clear nighttime skies allow intense radiational cooling to occur. In an anticyclone, the air adjacent to the ground may thus be chilled to the point that dew, frost, or even radiation fog develops.

Wind Measurement

Scientists are interested in monitoring both the speed and the direction of wind. Most wind monitoring instruments are designed to measure only the horizontal component of the wind since it is

SPECIAL TOPIC

Wind chill

At low air temperatures, high winds heighten the danger of *frostbite*—the freezing of body tissue—by increasing the rate of sensible heat loss from the body. As wind speed increases, the thickness of a still air layer (called the *boundary layer*) adjacent to the body diminishes. Because the boundary layer insulates the body from heat loss, any reduction in its thickness expedites heat loss from the body. As winds strengthen up to approximately 50 km (30 mi) per hour, sensible heat transport from the body increases. Further increases in wind speed do not effectively increase heat loss, however, because other factors become limiting.

Because of the danger of frostbite, weather reports during winter in northern localities and in mountainous regions report the *wind chill equivalent temperature*, as presented in Tables 1 and 2. This index, first introduced by P. A. Siple and C. F. Passel in 1945 and refined by A. Court in 1948, is now reported regularly by the National Weather Service. When the ambient air temperature is −1 °C (30 °F) and the air is calm, the wind chill equivalent temperature will be the same (−1 °C or 30 °F) as the ambient temperature. If, however, the wind speed picks up to 32 km (20 mi) per hour, then the wind chill equivalent temperature drops to −15 °C (4 °F). What is the significance of this value? Contrary to popular opinion, this value does not mean that skin

TABLE 1

Wind Chill Equivalent Temperature (°C)

WIND SPEED (m/sec)	AIR TEMPERATURE (°C)															
	6	3	0	−3	−6	−9	−12	−15	−18	−21	−24	−27	−30	−33	−36	−39
3	**3**	**−1**	**−4**	**−7**	**−11**	**−14**	**−18**	**−21**	**−24**	**−28**	**−31**	**−34**	**−38**	**−41**	**−45**	**−48**
6	**12**	**−6**	**−10**	**−14**	**−18**	**−22**	**−26**	**−30**	**−34**	**−38**	**−42**	**−46**	**−50**	**−54**	**−58**	**−62**
9	**−6**	**−10**	**−14**	**−18**	**−23**	**−27**	**−31**	**−35**	**−40**	**−44**	**−48**	**−53**	**−57**	**−61**	**−65**	**−70**
12	**−8**	**−12**	**−17**	**−21**	**−26**	**−30**	**−35**	**−39**	**−44**	**−48**	**−53**	**−57**	**−62**	**−66**	**−71**	**−75**
15	**−9**	**−14**	**−18**	**−23**	**−27**	**−32**	**−37**	**−41**	**−46**	**−51**	**−55**	**−60**	**−65**	**−69**	**−74**	**−79**
18	**−10**	**−14**	**−19**	**−24**	**−29**	**−33**	**−38**	**−43**	**−48**	**−52**	**−57**	**−62**	**−67**	**−71**	**−76**	**−81**
21	**−10**	**−15**	**−20**	**−25**	**−29**	**−34**	**−39**	**−44**	**−49**	**−53**	**−58**	**−63**	**−68**	**−73**	**−77**	**−82**
24	**−10**	**−15**	**−20**	**−25**	**−30**	**−35**	**−39**	**−44**	**−49**	**−54**	**−59**	**−63**	**−68**	**−73**	**−78**	**−83**

temperature actually drops to −15 °C. Through sensible heat transfer, skin temperature can drop no lower than the temperature of the ambient air, which is −1 °C in this example. In fact, any exposed body parts lose heat at a *rate* equivalent to conditions induced by calm winds at −15 °C.

Air in motion (wind) is more effective in removing heat from the body than temperature alone would imply. We must therefore counteract the cooling effect of wind by adding more layers of clothing to slow the rate of sensible heat loss from our body. Cold temperatures and high winds are especially hazardous to those body parts that are usually exposed and have a high surface-to-mass ratio, such as the ears, nose, and fingers. These parts are especially susceptible to frostbite.

In summary, wind chill equivalent temperature is a measure of the effectiveness of moving air as a heat sink. When the wind chill equivalent temperature is significantly lower than the ambient air temperature, we are wise to dress more warmly than the actual ambient temperature would normally dictate. Especially at low wind chill equivalent temperatures, all body parts, including the face and ears, should be protected if a person expects to be exposed to the wind for more than a few minutes at a time.

TABLE 2

Wind Chill Equivalent Temperature (°F)

WIND SPEED	AIR TEMPERATURE (°F)																		
(mph)	45	40	35	30	25	20	15	10	5	0	−5	−10	−15	−20	−25	−30	−35	−40	−45
5	43	37	32	27	22	16	11	6	1	−5	−10	−15	−20	−26	−31	−36	−41	−47	−52
10	34	28	22	16	10	4	−3	−9	−15	−21	−27	−33	−40	−46	−52	−58	−64	−70	−76
15	29	22	16	9	2	−5	−11	−18	−25	−32	−38	−45	−52	−58	−65	−72	−79	−85	−92
20	25	18	11	4	−3	−10	−17	−25	−32	−39	−46	−53	−60	−67	−74	−82	−89	−96	−103
25	23	15	8	0	−7	−15	−22	−29	−37	−44	−52	−59	−66	−74	−81	−89	−96	−104	−111
30	21	13	5	−2	−10	−18	−25	−33	−41	−48	−56	−63	−71	−79	−86	−94	−102	−109	−117
35	19	11	3	−4	−12	−20	−28	−35	−43	−51	−59	−67	−74	−82	−90	−98	−106	−113	−121
40	18	10	2	−6	−14	−22	−29	−37	−45	−53	−61	−69	−77	−85	−93	−101	−108	−116	−124
45	17	9	1	−7	−15	−23	−31	−39	−47	−55	−62	−70	−78	−86	−94	−102	−110	−118	−126

usually considerably stronger than the vertical component. For some specialized research purposes, very sensitive instruments are available to measure vertical wind speeds or to measure a combination of vertical and horizontal wind components.

A **wind vane**, shown in Figure 9.21, is the most common directional indicator and always points into the wind. A modification of this design, shown in Figure 9.22, is the airport **wind sock**, which points downwind. Wind direction is always designated as the direction *from which* the wind blows. For example, a wind blowing from the east toward the west is described as an "east wind," and a wind blowing from the northwest toward the southeast is designated as a "northwest wind." A wind vane may be linked electronically or mechanically to a dial that is calibrated to read in points of the compass or in degrees. An east wind is specified as 90 degrees, a south wind as 180 degrees, a west wind as 270 degrees, and a north wind as 360 degrees. The wind is recorded as 0 degrees only under calm conditions.

Wind speed can be estimated by observing the wind's effect on lake or ocean surfaces or on flexible objects such as trees. Such observation is the basis of the Beaufort scale, devised by Sir Francis Beaufort in 1805 and presented here in Table 9.1. A considerably more accurate measure of horizontal wind speed is provided by a

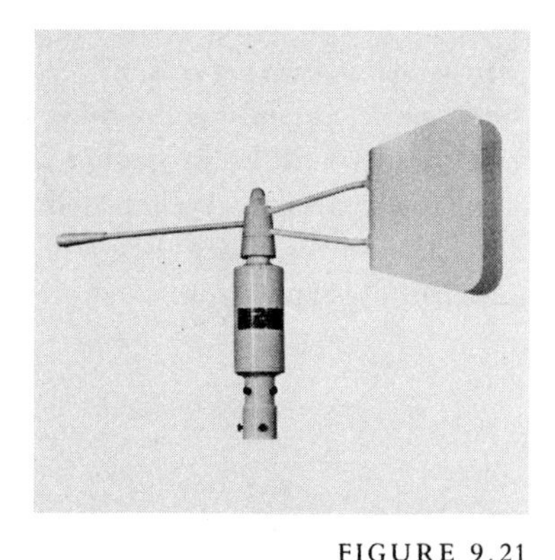

FIGURE 9.21
Horizontal wind direction is monitored by a wind vane. The instrument arm points in the direction from which the wind blows. (Courtesy of Belfort Instrument Company)

FIGURE 9.22
An airport wind sock gives wind direction and a general indication of wind speed. (Photograph by J. M. Moran)

TABLE 9.1

Beaufort Scale of Wind Force*

BEAUFORT NUMBER	GENERAL DESCRIPTION	LAND AND SEA OBSERVATIONS FOR ESTIMATING WIND SPEEDS	WIND SPEED 10 M ABOVE GROUND (KM/HR)
0	Calm	Smoke rises vertically. Sea like mirror.	Less than 1
1	Light air	Smoke, but not wind vane, shows direction of wind. Slight ripples at sea.	1–5
2	Light breeze	Wind felt on face, leaves rustle, wind vanes move. Small, short wavelets.	6–11
3	Gentle breeze	Leaves and small twigs moving constantly, small flags extended. Large wavelets, scattered whitecaps.	12–19
4	Moderate breeze	Dust and loose paper raised, small branches moved. Small waves, frequent whitecaps.	20–28
5	Fresh breeze	Small leafy trees swayed. Moderate waves.	29–38
6	Strong breeze	Large branches in motion, whistling heard in utility wires. Large waves, some spray.	39–49
7	Near gale	Whole trees in motion. White foam from breaking waves.	50–61
8	Gale	Twigs break off trees. Moderately high waves of great length.	62–74
9	Strong gale	Slight structural damage occurs. Crests of waves begin to roll over. Spray may impede visibility.	75–88
10	Storm	Trees uprooted, considerable structural damage. Sea white with foam, heavy tumbling of sea.	89–102
11	Violent storm	Very rare; widespread damage. Unusually high waves.	103–117
12	Hurricane	Very rare; much foam and spray greatly reduce visibility.	118 and over

*Developed in 1805 by Irish hydrographer Sir Francis Beaufort.

cup anemometer, as shown in Figure 9.23. This device works on the same principle as a bicycle or an automobile speedometer. The wind spins the cups, thus generating an electric current, which is calibrated on a dial in meters per second, kilometers per hour, miles per hour, or knots.* Several other types of anemometer are available, including the very sensitive **hot-wire anemometer**. In this instrument, the wind blows past a heated wire, or wires, and the heat lost to the air is then calibrated in terms of wind speed.

Recording a continuous trace of wind speed and direction is sometimes useful. Wind vanes and anemometers can be linked to

*A knot is 1 nautical mile (1.85 km) per hour.

pens that record on a clock-driven drum. As shown in Figure 9.24, the trace indicates a considerable variation in both wind direction and wind speed with time.

One reason for monitoring the wind is its influence on human comfort. A breeze can make a hot, muggy day more tolerable. On the other hand, the wind exacerbates the chilling effect of cold air. For more information about wind chill, see this chapter's Special Topic.

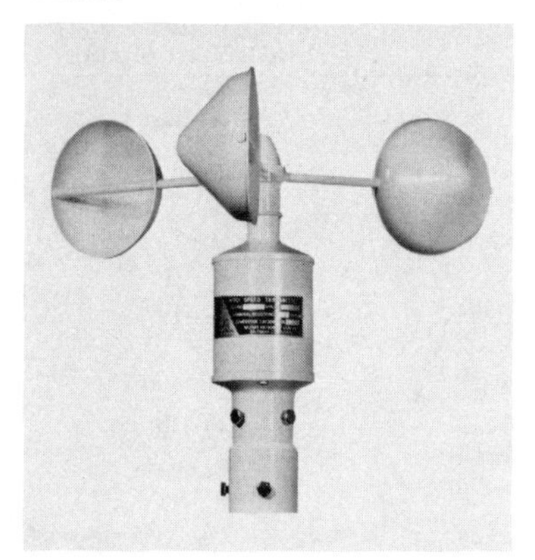

FIGURE 9.23

Wind speed is measured by a cup anemometer. The faster the wind speed is, the faster the spin of the cups. (Courtesy of Belfort Instrument Company)

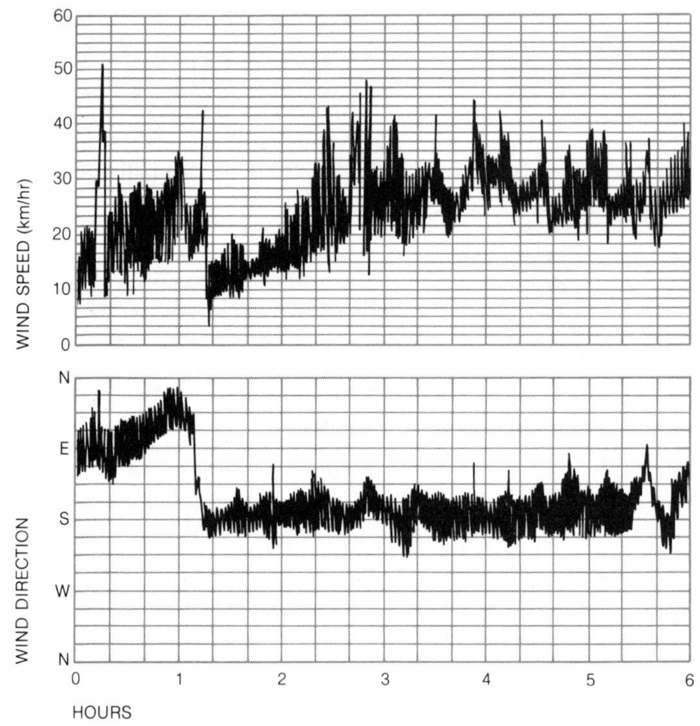

FIGURE 9.24

Continuous trace of time variations in wind speed and direction.

Conclusions

In Chapter 4, we discussed air circulation and redistribution of heat from areas of excess radiational heating to areas of excess radiational cooling. In this chapter, we identified and described the various forces that work together to shape air circulation on global and synoptic scales. Weather systems differ on the basis of the air circulation pattern that characterizes each system. The pattern of horizontal and vertical winds distinguishes fair weather systems (anticyclones) from stormy weather systems (cyclones). We are now ready to examine the characteristics of global-scale weather systems.

MATHEMATICAL NOTE

Geostrophic and gradient winds

The geostrophic wind is the result of a balance between the horizontal pressure gradient force and the force due to the Coriolis effect. Here we examine the geostrophic motion of a unit mass of air (1 g, for example). The acceleration imparted to this air parcel by a horizontal pressure gradient is given by:

$$\frac{1}{\rho}\left(\frac{\Delta P}{\Delta N}\right)$$

where ρ is the density of the air, and ΔP is the change in air pressure over a horizontal distance ΔN measured perpendicular to isobars. (Pressure gradients are always measured perpendicular to isobars.)

The acceleration imparted to the air parcel by the Coriolis effect is given by:

$$(2\Omega \sin \phi)V$$

where V is the speed of the air parcel, Ω is the angular velocity of the earth as it rotates on its axis ($\Omega = 7.29 \times 10^{-5}$ radians per second), and ϕ is the latitude.

For the geostrophic wind,

$$\frac{1}{\rho}\left(\frac{\Delta P}{\Delta N}\right) = (2\Omega \sin \phi)V$$

Solving for V, the speed of the geostrophic wind, we have:

$$V = \frac{1}{2\Omega \sin \phi\, \rho}\left(\frac{\Delta P}{\Delta N}\right)$$

The geostrophic wind thus strengthens with an increasing pressure gradient (i.e., closer spacing of isobars) and a decreasing latitude. Recall that the geostrophic wind blows parallel to isobars and at right angles to the oppositely directed pressure gradient force and Coriolis effect.

The gradient wind results from an interaction of the horizontal pressure gradient, the Coriolis effect, and the centripetal force. Again, assume that we are examining the motion of a unit mass of air. The acceleration imparted to the air parcel by the centripetal force is:

$$\frac{V^2}{r}$$

where V is the speed of the air parcel, and r is the radius of curvature of the path described by the air parcel. The centripetal force is the force that constrains the air parcel to a curved trajectory. This force is strong where the curvature is sharp (small r), and weak where the curvature is gradual (large r).

The three forces (pressure gradient, Coriolis, and centripetal) interact so that the sum of the accelerations is zero, that is, the speed of the gradient wind is constant (unaccelerated). Hence,

$$\frac{V^2}{r} + (2\Omega \sin \phi)V - \frac{1}{\rho}\left(\frac{\Delta P}{\Delta N}\right) = 0$$

The net force, which is the centripetal force, operates in the gradient wind only to change the direction of the wind as it follows a curved path, and not to change the wind's speed. We could solve the above equation for V to determine the speed of the gradient wind for some radius of curvature, latitude, and horizontal pressure gradient (measured perpendicular to isobars). Note that for wind blowing in a straight line,

$$\frac{V^2}{r} = 0$$

and the equation reduces to that presented earlier for geostrophic flow.

SUMMARY STATEMENTS

■ Wind is the consequence of an interaction of forces generated by air pressure gradients, the Coriolis effect, friction, centripetal and centrifugal forces, and gravity.

■ The pressure gradient force initiates air motion and arises in part from spatial variations in air temperature and water vapor concentration.

■ The Coriolis effect, which is due to the earth's rotation on its axis, deflects the wind to the right in the Northern Hemisphere and to the left in the Southern Hemisphere. The deflective force is zero at the equator and increases by latitude to a maximum at the poles.

■ Friction has an important influence on winds blowing within 1 km of the earth's surface. Friction slows the wind and, in combination with the Coriolis effect, deflects surface winds across isobars toward lower air pressure.

■ The force of gravity always acts downward and perpendicular to the earth's surface. Hydrostatic equilibrium is the balance between the upward directed pressure gradient force and the downward directed force of gravity.

■ The geostrophic wind is an unaccelerated horizontal wind that blows along straight paths above the friction layer and results from a balance between the horizontal pressure gradient force and the Coriolis effect. The geostrophic wind always blows parallel to *straight* isobars.

■ The gradient wind is a horizontal, frictionless air flow that parallels curved isobars. Above the friction layer in the Northern Hemisphere, the gradient wind blows clockwise about anticyclones and counterclockwise about cyclones.

■ In large-scale surface winds (global and synoptic scales), friction slows the wind and combines with the Coriolis effect to shift the wind direction across isobars toward low-pressure areas.

■ Within the friction layer, horizontal winds blow clockwise and outward in Northern Hemisphere anticyclones, and counterclockwise and inward in Northern Hemisphere cyclones.

■ The concept of wind *continuity* implies a connection between the horizontal and vertical components of the wind. In an anticyclone, horizontal winds converge aloft and diverge at the surface. Continuity of air means that air descends near the high center, the relative humidity drops, and the weather is fair. In a cyclone, horizontal winds diverge aloft and converge at the surface. Continuity of air means that air ascends near the center of the low, the relative humidity increases, and the weather is stormy.

KEY WORDS

air parcels
Newton's second law of motion
gradient
pressure gradient
isobars
pressure gradient force
Coriolis effect
friction
friction layer
Newton's first law of motion
centripetal force
Newton's third law of motion
centrifugal force
gravity
gravitation
hydrostatic equilibrium
geostrophic wind
gradient wind
anticyclone
cyclone
Ekman spiral
lake effect snows
wind vane
wind sock
cup anemometer
hot-wire anemometer

QUESTIONS

1 Why does the vertical component of the wind play the key role in cloud and precipitation formation?

2 What causes horizontal gradients in air pressure?

3 A weak horizontal air pressure gradient is indicated by what type of pattern of isobars?

4 How do air pressure gradients alone influence the motion of air?

5 Why is it appropriate to refer to the Coriolis effect and friction as "responsive" forces?

6 Why does the Coriolis deflection reverse direction between the Northern and Southern Hemispheres?

7 Explain how the Coriolis deflection arises from a change in our frame of reference.

8 The Coriolis deflection is latitude dependent. Explain why.

9 How does wind speed change with altitude within the friction layer?

10 State Newton's three laws of motion.

11 Distinguish between the centrifugal force and the centripetal force.

12 What is hydrostatic equilibrium?

13 What is the difference between the geostrophic wind and the gradient wind?

14 Describe the horizontal air circulation about a low pressure system (a) within the friction layer and (b) above the friction layer.

15 Describe the horizontal air circulation about a high pressure system (a) within the friction layer and (b) above the friction layer.

16 Describe how forces interact to shape the horizontal air circulation within (a) cyclones and (b) anticyclones.

17 Explain why the horizontal wind parallels isobars above the friction layer and crosses isobars within the friction layer.

18 How does isobar spacing affect horizontal wind speeds?

19 Provide several examples of how horizontal winds are linked to the vertical motion of air.

20 Explain why cyclones produce stormy weather and anticyclones are fair weather systems.

POINTS TO PONDER

1 How does the magnitude of the typical horizontal air pressure gradient compare with that of the vertical air pressure gradient?

2 How might air mass advection influence vertical air pressure gradients?

3 Speculate on why vertical wind speeds typically are considerably weaker than horizontal wind speeds.

4 In view of Newton's first law of motion, is gradient wind the consequence of "balanced" forces? Explain your response.

5 What type of isobaric pattern is associated with a high pressure system?

6 What is meant by the statement that the Coriolis effect is more "apparent" than real?"

7 Why is the Coriolis effect important only in global- and synoptic-scale weather systems?

8 Explain why radiation fog might be associated with an anticyclone.

PROJECTS

1 From your observation of televised or newspaper national weather maps predict the wind direction at various locations.

2 Predict how wind direction and speed change as a cyclone approaches your location.

SELECTED READINGS

Crutcher, H. L. "Winds, Numbers, and Beaufort." *Weatherwise* 28 (1975):260–271. *Discusses the Beaufort scale of wind force.*

Eagleman, J. R. *Meteorology*. New York: D. Van Nostrand, 1980. 384 pp. *Includes a general description of forces operating in the atmosphere that shape air circulation.*

The fair breeze blew, the
white foam flew,
The furrow followed free;
We were the first that ever
burst
Into that silent sea.

Down dropt the breeze, the
sails dropt down,
'Twas sad as sad could be;
And we did speak only to
break
The silence of the sea!

All in a hot and copper sky,
The bloody Sun, at noon,
Right up above the mast did
stand,
No bigger than the Moon.

Day after day, day after day,
We stuck, nor breath nor
motion;
As idle as a painted ship
Upon a painted ocean.

SAMUEL TAYLOR COLERIDGE
The Rime of the Ancient Mariner

Global-scale circulation encompasses wind belts that encircle the globe and pressure systems that cover huge areas of the earth's oceans. (NOAA photograph)

TEN

Global-Scale Circulation

Idealized Pattern

Pressure Systems and Wind Belts

Winds Aloft

Seasonal Shifts

Singularities

Upper-Air Westerlies

Long-Wave Patterns

Blocking Systems

Jet Stream

Short Waves

GLOBAL-SCALE air circulation is ultimately responsible for the development and displacement of most smaller scale weather systems. Beginning a description of weather systems with those that are planetary in scope is therefore appropriate.

Idealized Pattern

Global-scale air circulation is a complex pattern of winds and pressure systems. To better understand how this circulation is shaped, we start with an idealized model of the earth. Picture the earth as a nonrotating sphere with a uniform solid surface. As on the real earth, the sun heats the equatorial regions more intensely than the polar regions. In response to the equator-to-pole temperature gradient, two huge convection currents form—one in each hemisphere (Figure 10.1A). Warm, light air rises at the equator and flows aloft toward the poles. Along the way, the air cools radiationally. At the poles, the now cold, dense air sinks, and flows at the surface toward the equator, thus completing the convective circulation.

As our idealized planet begins to rotate, we expect the Coriolis effect to come into play (Figure 10.1B). In the Northern Hemisphere, surface winds would shift to the northeast (that is, northeast to southwest), and in the Southern Hemisphere, surface winds would become southeasterly (that is, southeast to northwest). On our hypothetical planet, surface winds would thus blow counter to the planet's rotation from east to west. This is an impossible situation, because the rotating earth would have a breaking effect on the air circulation. The kinetic energy of the winds would be converted to frictional heat, and the winds would slow.

Circulation is maintained when the winds divide into three zones in each hemisphere, so that some winds blow with the rotational direction and other winds blow counter to the rotational direction of the earth (Figure 10.1C). In the Northern Hemisphere, surface winds are northeasterly from 0 degrees to 30 degrees, southwesterly from 30 degrees to 60 degrees, and northeasterly from 60 degrees to 90 degrees. In the middle latitudes of both hemispheres (30 degrees to 60 degrees), surface winds thus blow with the rotating earth. In the process, these winds acquire kinetic energy from the moving earth, and some of this kinetic energy is transported to tropical and high latitudes to help sustain winds that blow against the earth's rotation.

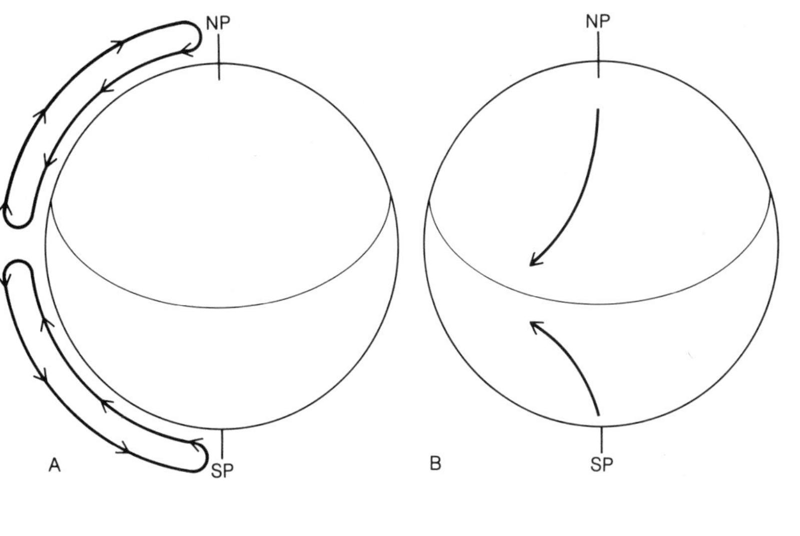

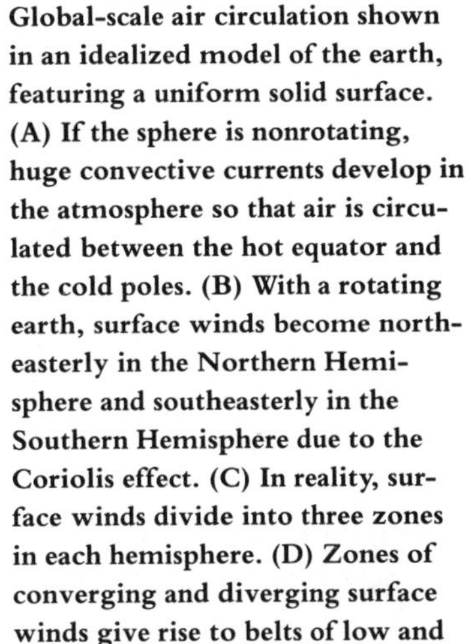

FIGURE 10.1

Global-scale air circulation shown in an idealized model of the earth, featuring a uniform solid surface. (A) If the sphere is nonrotating, huge convective currents develop in the atmosphere so that air is circulated between the hot equator and the cold poles. (B) With a rotating earth, surface winds become northeasterly in the Northern Hemisphere and southeasterly in the Southern Hemisphere due to the Coriolis effect. (C) In reality, surface winds divide into three zones in each hemisphere. (D) Zones of converging and diverging surface winds give rise to belts of low and high air pressure.

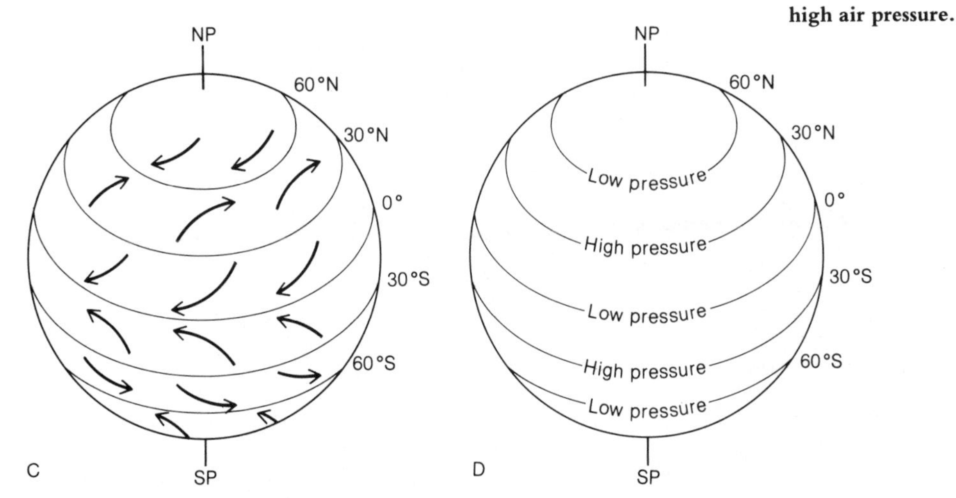

Surface winds converge along the 0-degree and 60-degree latitude circles. Convergence causes air to rise, which leads to expansional cooling, cloud development, and possible precipitation. These convergence zones are therefore belts of low pressure (Figure 10.1D). Surface winds diverge at the poles and along the 30-degree latitude circles. In these regions, divergence causes air to descend, which leads to compressional warming and fair weather. These are zones of high pressure.

If the continents and oceans are added to our idealized model, the thermal characteristics of the earth's surface become more complex and more realistic, and so does the global-scale circulation. Some of the pressure belts break into separate cells, and important changes in air pressure occur over land versus sea. In the next section, we describe the principal features of global-scale circulation as these features are presently understood.

Pressure Systems and Wind Belts

Northern Hemisphere maps of the average air pressure at sea level for January and July (Figure 10.2) reveal several areas of relatively high and low atmospheric pressure. These areas are the

FIGURE 10.2

Mean Northern Hemisphere sea-level air pressure for (A) January and (B) July. Solid and dashed lines are isobars in mb. (Source: NOAA)

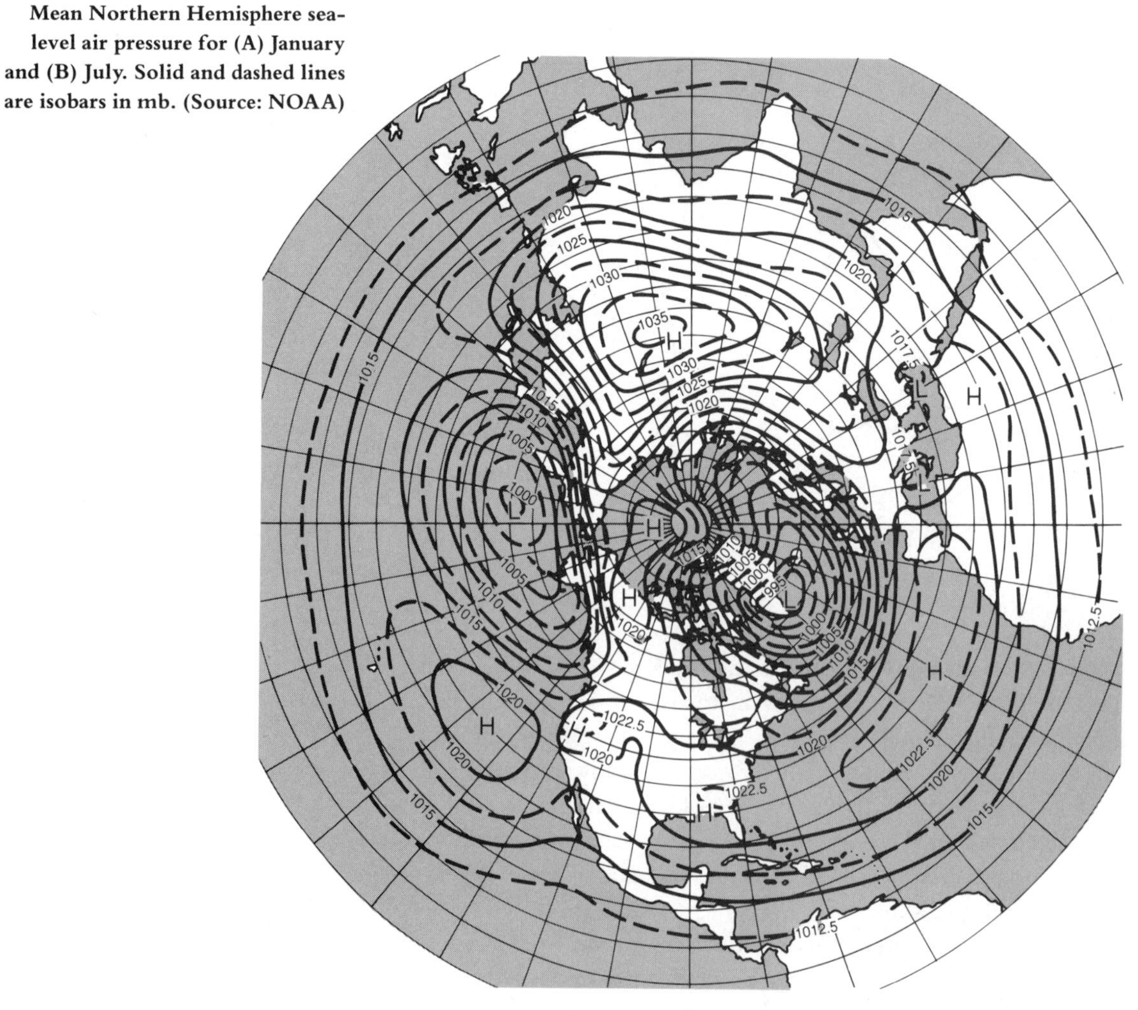

semipermanent pressure systems. Although these systems are persistent features of global circulation, they do exhibit important seasonal changes in both location and strength, hence the modifier *semipermanent*. The pressure systems include the subtropical anticyclones, the equatorial trough, and the subpolar lows.

The **subtropical anticyclones** are imposing features of global circulation that are centered over subtropical latitudes (near 30 degrees N and S) of the North and South Atlantic, the North and South Pacific, and the Indian Ocean. These highs extend vertically from the ocean surface to the tropopause, and exert a strong influence on weather and climate. Stretching from the center of the high, out over its eastern flanks, are extensive areas of subsiding

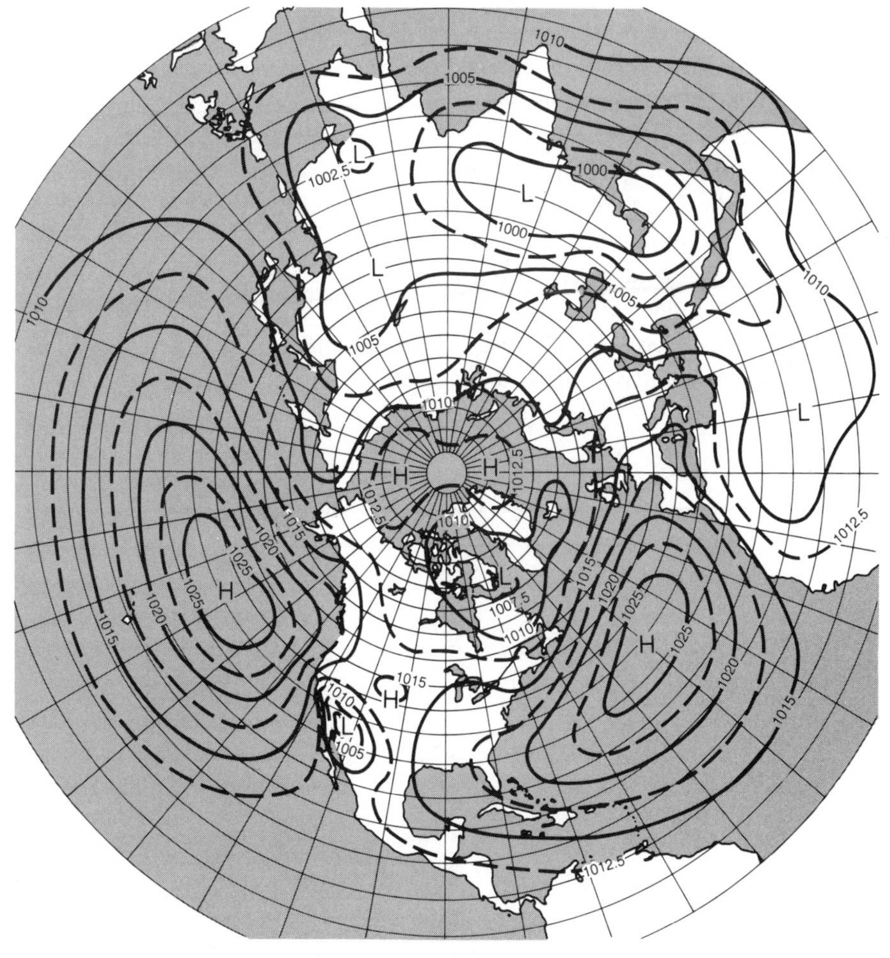

stable air, which cause compressional warming and low relative humidities and sunny skies. The world's major deserts, including the Sahara and the Sonoran, are situated under the eastern sides of the subtropical anticyclones. On the far-western flanks of the subtropical highs, however, there is less subsidence and stability. As a result, episodes of cloudy, stormy weather are more frequent in these regions. This contrast in weather from one flank of a subtropical high to the other is particularly apparent across southern North America. The weather of the American Southwest (on the eastern side of the North Pacific high, or the Hawaiian high) is considerably drier than the weather of the American Southeast (on the western side of the North Atlantic high, which is also called the Azores, or the Bermuda, anticyclone).

As is typical of anticyclones, subtropical highs have very weak horizontal pressure gradients at their center. Surface winds are consequently very light or even calm over extensive areas of the subtropical oceans. This situation played havoc with ancient sailing ships, which were becalmed for days or even weeks at a time. Ships setting sail from Spain to the New World were often caught in this predicament, and crews were forced to jettison their cargo of horses when supplies of food and water ran low. Early mariners therefore referred to this area of calm winds as the **horse latitudes**, a name now applied to all latitudes under subtropical highs. The stanzas from Coleridge's *Rime of the Ancient Mariner* at the beginning of this chapter aptly describe the weather conditions of horse latitudes.

TABLE 10.1

Features of Global-Scale Circulation

LATITUDE	GLOBAL-SCALE SYSTEMS	
90° N	**High**	**Polar anticyclones**
	↙↙	**Polar northeasterlies**
60° N	**Low**	**Subpolar cyclones**
	↗↗	**Westerlies**
30° N	**High**	**Subtropical anticyclones**
	↙↙	**Northeast trades**
0°	**Low**	**Equatorial trough (ITCZ)**
	↖↖	**Southeast trades**
30° S	**High**	**Subtropical anticyclones**
	↘↘	**Westerlies**
60° S	**Low**	**Subpolar cyclones**
	↖↖	**Polar southeasterlies**
90° S	**High**	**Polar anticyclones**

In the Northern Hemisphere, surface winds flow clockwise and outward, away from the centers of the subtropical highs (Table 10.1). Surface winds north of the horse latitudes constitute the highly variable **midlatitude westerlies** (actually southwesterlies). The surface winds blowing out of the southern flanks of the anticyclones are known as the northeast **trade winds**. Analogous winds develop in the Southern Hemisphere. Recall, however, that the Coriolis effect is reversed in the Southern Hemisphere. A counterclockwise and outward surface air flow thus causes southeast trade winds on the northern flanks of the Southern Hemisphere subtropical highs, and a belt of northwest winds on the southern flanks.

The trade winds of the two hemispheres flow into a broad east-west belt of low pressure near 0 degrees latitude, called the **equatorial trough**, where rising air motion induces cloudiness and rainfall. The most active weather develops along the **intertropical convergence zone** (ITCZ), a discontinuous belt of thundershowers paralleling the equator (Figure 10.3).

On the poleward side of the subtropical highs, surface westerlies flow into regions of low pressure. In the Northern Hemisphere, there are typically two separate **subpolar lows**—the **Aleutian low** over the North Pacific Ocean, and the **Icelandic low** over the

FIGURE 10.3

This satellite photograph shows a discontinuous east-west band of cumulonimbus clouds (white blotches) near the equator. The cloud band marks the convergence of the trade winds of the two hemispheres and is known as the intertropical convergence zone (ITCZ). (NOAA photograph)

North Atlantic. Because of the counterclockwise and inward circulation of surface winds in these pressure cells, the midlatitude southwesterlies converge with the polar northeasterlies. In the Southern Hemisphere, in contrast, the midlatitude northwesterlies and the polar southeasterlies converge along a nearly continuous belt of low pressure around the Antarctic continent.

The surface westerlies meet and override the polar easterlies along the **polar front**. A **front** is a narrow zone of transition between air masses of contrasting density, that is, air masses of differing temperature, or differing water vapor content, or both. In this case, dense, cold air masses flowing toward the equator meet milder, lighter midlatitude air masses moving toward the pole. The polar front is not continuous around the globe, but is instead well defined in some areas and not in others, depending on the magnitude of the temperature contrast across the front. Where the north-south temperature gradient is great, the front is well defined and is a potential site for storm development. Where the temperature contrast is minimal, the front is poorly developed and inactive.

This brief description gives a generalized view of the major components of global-scale circulation at the earth's surface. The distribution of these surface circulation features is portrayed schematically in Figure 10.4.

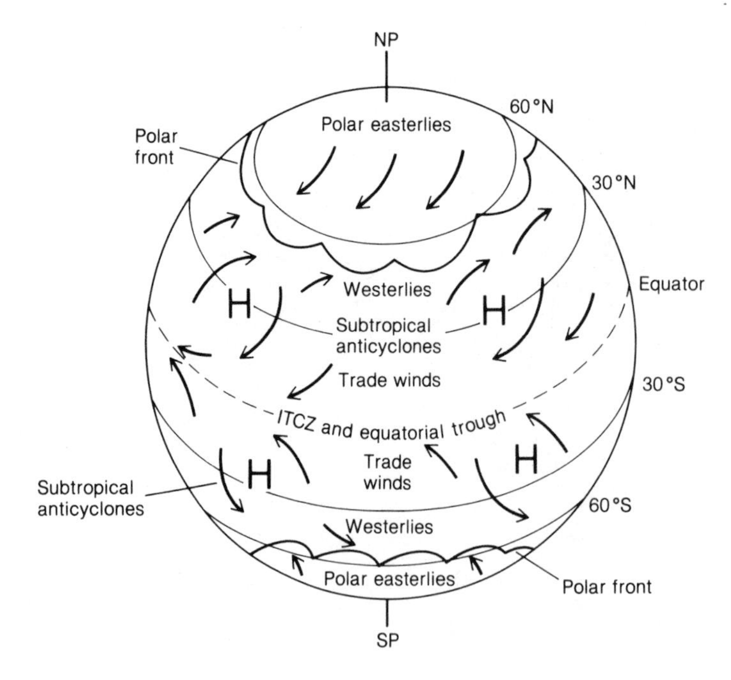

FIGURE 10.4

Schematic representation of the global-scale surface circulation of the atmosphere.

Winds aloft

What is the pattern of the global winds aloft, in the middle and upper troposphere?

As noted, air subsides in subtropical anticyclones, sweeps toward the equator as the surface trade winds, and then rises in the ITCZ. Aloft, in the middle troposphere, air flows poleward, away from the equatorial trough and into the subtropical highs. The Coriolis effect shifts these upper-level winds toward the right in the Northern Hemisphere (southwest winds), and toward the left in the Southern Hemisphere (northwest winds). In the tropics, therefore, the winds aloft actually blow in the opposite direction from the winds at the surface. The vertical profile of this circulation, shown schematically in Figure 10.5, resembles a huge convective cell and is known as a **Hadley cell**, named after the scientist who first proposed its existence in 1735. Hadley cells are thus situated on either side of the ITCZ and extend poleward to the subtropical highs.

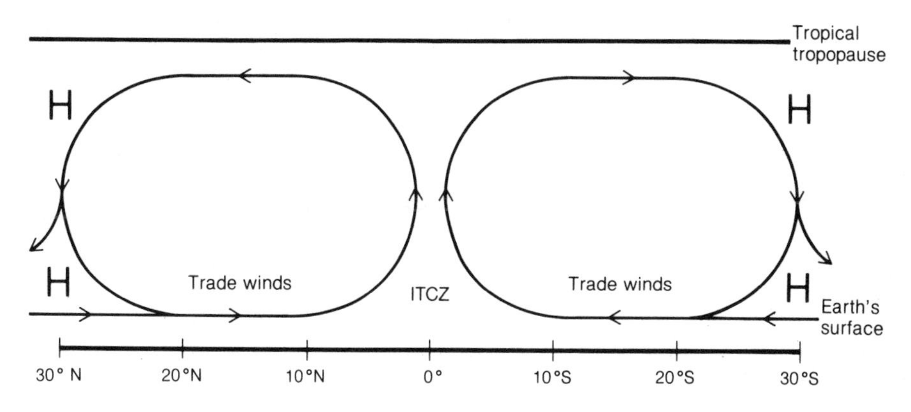

FIGURE 10.5
Schematic representation of the Hadley cell circulations in tropical latitudes of the Southern and Northern hemispheres. Air rises over the intertropical convergence zone (ITCZ), and sinks in the subtropical anticyclones.

It was once assumed that separate cells, similar to the Hadley cells, occurred in both midlatitudes and polar latitudes. Detailed upper-air monitoring has shown, however, that no such cells exist. Instead, midlatitude winds aloft blow from west to east in a wavelike pattern of ridges and troughs, as shown in Figure 10.6. These winds are responsible for the development and displacement of synoptic weather systems (highs, lows, and air masses), and for the poleward heat transport described in Chapter 4. Because these winds are so important to our midlatitude weather, we examine the

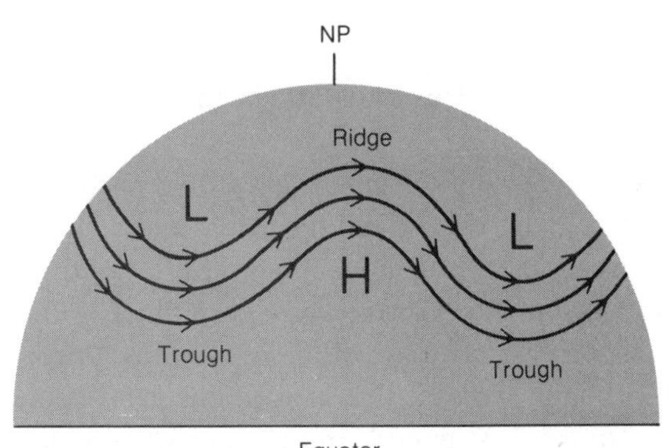

FIGURE 10.6

Aloft, in the middle and upper troposphere, the westerlies flow in a wavelike pattern of ridges and troughs.

upper-air westerlies in more detail in a separate section of this chapter.

In polar regions, air subsides and flows away at the surface from shallow, cold anticyclones. In the Northern Hemisphere, these highs are well developed only in winter over the continental interiors. In the Southern Hemisphere, cold highs persist over the Antarctic continent year-round. Aloft, polar winds are westerly.

Figure 10.7 shows the vertical profile of Northern Hemisphere winds in the troposphere from the equator to the poles. In this

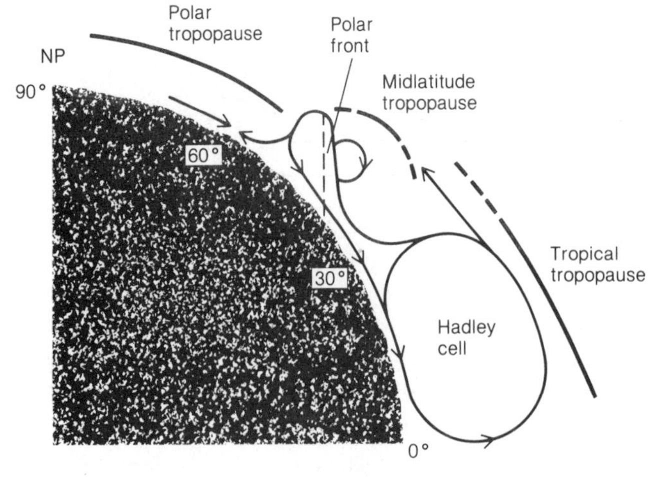

FIGURE 10.7

A vertical cross section showing the north-south (meridional) winds in the Northern Hemisphere troposphere. Note that the tropopause occurs in three segments and that the vertical scale is considerably exaggerated.

perspective, we are viewing only the north-south and up-down components of the wind and are neglecting the west-east component. The tropopause is not continuous from pole to equator, but usually occurs in segments. The polar tropopause is at a lower altitude than the midlatitude tropopause, which, in turn, occurs at a lower altitude than the tropical tropopause.

Seasonal shifts

Important changes take place in planetary circulation between winter and summer. Pressure systems, the polar front, and the global wind belts follow the sun, shifting toward the poles in spring and toward the equator in autumn. Because seasons are reversed in the Northern and Southern hemispheres, global-scale systems of both hemispheres move north and south in tandem. In addition, the strength of pressure cells varies between seasons. Subtropical anticyclones exert lower surface pressures in winter than in summer. The Icelandic low deepens in winter and greatly weakens in summer, and although well developed in winter, the Aleutian low disappears in summer.

Seasonal reversals in surface air pressure also occur over the continents. These pressure shifts stem from the contrast in solar heating between land and sea. For reasons detailed in Chapter 2, the sea surface exhibits smaller temperature variations over the course of a year than does the earth's land surface. This land-sea contrast is displayed vividly by the maps of global surface temperature in Plate 15. Temperatures were derived from radiation measurements obtained by sensors aboard weather satellites.* Parts A and B are color-enhanced images of January and July temperatures, respectively. In the Northern Hemisphere, these are usually the coldest and warmest months of the year. The final image (part C) is the temperature difference between the two months. Note that the temperature contrast is considerably greater over the continents (up to 30 °C, or 54 °F) than over the oceans (typically 8 to 10 °C, or 14 to 18 °F).

As a consequence of the thermal contrast, the continents are sites of relatively high pressure in winter and of relatively low pressure in summer. In winter, cold anticyclones appear over northwestern North America and over the interior of Eurasia. In summer, prominent belts of low pressure form across North Africa and from Arabia eastward into Southeast Asia. Warm low-pressure cells also develop in summer over the desert country of southwestern North America.

In terms of weather and climate, the importance of the ocean with its special thermal properties cannot be overstated. Even rela-

*Recall from Chapter 3 that the radiation emitted by any object is temperature dependent, so temperature can be measured remotely by infrared sensors on earth-orbiting satellites.

tively small departures from normal sea surface temperatures can have serious global ramifications. An example is the El Niño phenomenon, which was particularly well developed in 1982–1983, as described in the Special Topic.

Seasonal shifts in the location of the global-scale wind belts and in the location of pressure systems leave their mark on the world's climates. For example, northward migration of the ITCZ triggers

SPECIAL TOPIC

The unprecedented El Niño of 1982–1983

El Niño **refers to a warm surface ocean current that occasionally appears along the coast of Ecuador and Peru. This warm current cuts off the usual upwelling of cold, nutrient-rich bottom water that supports the economically important fish population. El Niño typically occurs toward the end of the year, hence the name El Niño (the child), for the Christmas season. El Niño is usually weak and short lived, but not so in 1982–1983.**

El Niño is associated with the ***southern oscillation,*** **a seesaw pattern of air pressure variation between the eastern and western tropical Pacific. Surface air pressures are normally lower in the region of northern Australia and Indonesia than over the Southeast Pacific. Consequently, the trade winds of the equatorial Pacific have an east to west component. This persistent easterly flow literally drags warm surface waters toward the west. Along the coast of tropical South America, cold bottom water wells up to the surface to replace the warm water that is dragged westward.**

One of the first signs of an El Niño is a drop in air pressure over large areas of the Southeast Pacific, while pressures rise over the western Pacific (the southern oscillation). This shift in pressure pattern causes the equatorial easterlies to diminish and eventually to reverse direction. Westerly winds then drag warm surface waters toward the east. When the warm waters reach the South American coast, they are deflected southward along the coastlines of Peru and Ecuador.

In 1982, El Niño began in May, and by December, the warming of surface waters was unprecedented in magnitude and extent (see Plate 16). The unusual warming spread westward along the equatorial Pacific to near the international dateline (180 degrees W), and in places, sea-surface temperature anomalies (departures from normal) reached +6 °C (+10.8 °F). This anomalously warm

summer monsoon rains in Central America, North Africa, India, and Southeast Asia (see Chapter 12). As a subtropical anticyclone shifts north and south with the sun, its dry eastern flank influences some localities in winter and other localities in summer, and thus brings a pronounced seasonality to rainfall. (We have more to say about the influence of the planetary circulation on world climates in Chapter 17).

---

water over vast areas led to major changes in atmospheric circulation patterns over tropical and middle latitudes. Circulation changes, in turn, led to weather extremes in many areas.

The winter storm track was displaced hundreds of kilometers southeast of its usual position, which resulted in destructive high winds and heavy rains in California. Excessive rains between mid-November 1982 and late January 1983 caused the worst flooding of the century in Ecuador. French Polynesia was struck by no less than five hurricanes, and at the other extreme, severe drought parched Australia, Indonesia, and southern Africa. Australia suffered a $2 billion loss in crops, and millions of sheep and cattle died.

El Niño also had a devastating impact on marine life in the eastern tropical Pacific, particularly off the coast of South America. El Niño brought warm surface waters eastward, thus preventing the upwelling of nutrient-rich cold water from the ocean depths. As a consequence, the growth of microscopic algae that normally flourish in the upwelling areas diminished sharply. The 1982–1983 El Niño was so strong that it triggered a 20-fold decrease in amount of algae along the South American coast. This decline reduced the numbers of anchovy, which feed heavily on the algae, to a record low. Other fish, such as the jack mackerel, that feed on tiny animals, which in turn feed on the algae, also decreased in number as algae production declined. With the decline of the fish population, marine birds (such as frigate birds and terns) and marine animals (such as fur seals and sea lions) also suffered large population declines because of breeding failures and the disappearance of their food sources.

Even the far-removed wildlife on and near Christmas Island (2 degrees N 157 degrees W) in the Central Pacific suffered from the effects of the 1982–1983 El Niño. Hundreds of thousands of sea birds, which feed on fish and squid, normally nest on this island. In November of 1982, however, when scientists arrived on the island, they discovered that virtually all the adult birds had deserted the site, leaving their young behind to starve. In addition to reduced food sources, heavy rains and flooding probably contributed to the reproductive failure of the birds. By July 1983, with the return of more normal atmospheric and oceanic conditions near Christmas Island, the few surviving birds began to nest again.

The 1982–1983 El Niño provides a remarkable example of the vast complex of interconnections between the atmosphere, the ocean, and the many forms of life, not only in the equatorial Pacific, but around the entire globe. Scientists will be studying these interconnections for many years in an attempt to understand and predict the origin and impact of future El Niños.

Singularities

A **singularity** is a weather event that occurs on or near a certain date with unusual regularity. The frequency of these events in the climatic record is greater than would be expected on the basis of chance alone. Most singularities are linked to regular changes in features of the global-scale atmospheric circulation. For example, in regions of seasonal precipitation, the onset and ending of the rainy season may constitute singularities. Consider some North American examples.

The **January thaw** is the most widely recognized and perhaps the only real singularity in a statistically rigorous sense. It is a period of relatively mild weather around January 20 to 23 and occurs primarily in the New England states. The thaw is caused by a flow of warm air on the back (west) side of the Bermuda-Azores anticyclone, which, for some unexplained reason, temporarily shifts north of its usual midwinter location.

Regular weather episodes that do not fit precisely the definition of a singularity, but are fairly predictable, are the July rainfall maximum in Arizona and the so-called Indian summer weather in the northeastern United States. In late June, the North Pacific subtropical anticyclone shifts abruptly northward, allowing the Bermuda-Azores high to extend its influence westward, across subtropical North America. The anticyclone's clockwise circulation pumps warm, humid air into the southwestern United States and brings a rainy end to Arizona's dry spring. For example, in Phoenix, Arizona, the mean monthly rainfall for April, May, and June is only 8 mm (0.30 in.), 3 mm (0.12 in.), and 2 mm (0.08 in.) respectively. For July, however, the mean monthly rainfall jumps abruptly to 20 mm (0.20 in.).

Indian summer usually develops in October, but it may also occur in November—there is no exact date. Large, warm anticyclones stagnate over the eastern United States, displacing the principal storm track northward along the Saint Lawrence river valley. Typical Indian summer weather consists of persistently warm, sunny days with hazy skies, cool nights, and frosty mornings.

Upper-Air Westerlies

The midlatitude westerlies of the Northern Hemisphere merit special attention here, because they govern the weather in the United States and Canada. As we have noted, in the middle and upper troposphere, the westerlies flow about the hemisphere in wavelike patterns of ridges and troughs (see Figure 10.6). Winds exhibit a clockwise (anticyclonic) curvature in the ridges, and a counterclockwise (cyclonic) curvature in the troughs. Between two and five waves typically encircle the hemisphere at any one time. These **long waves** are called **Rossby waves**, after Carl G. Rossby, the Swedish-American meteorologist who described and explained them in the late 1930s. The winds' wavelike configuration allows us to describe the westerlies by wavelength (distance between successive troughs or successive ridges), amplitude (north-south extent), and the number of waves encircling the hemisphere. The westerlies exhibit changes in all three of these measures and, as a direct consequence, the weather changes.

The westerlies are more vigorous in winter than in summer. In winter, they strengthen and have more waves of shorter length and greater amplitude. This seasonal difference stems from the north-south pressure gradient, which is steeper in winter because of the greater temperature contrast between north and south at that time of year. In summer, north-south temperature differences are typically minimal, pressure gradients are weak, and as a consequence, so are the westerlies.

Long-wave patterns

The "weaving westerlies" consist of a north-south wind superimposed on a west-east wind. We refer to the north-south airflow as the westerlies' **meridional component**, and the west-east airflow as the **zonal component**. The meridional component of Rossby waves brings about a north-south exchange of air masses and the poleward transport of heat. In the Northern Hemisphere, winds from the south carry warm air masses northward, and winds from the north transport cold air masses southward. Cold air is thus exchanged for warm air, and heat is transported poleward. As Rossby waves change in length, amplitude, and number, however, concurrent changes take place in the advection of air masses. Consider some examples.

Occasionally, the westerlies flow almost directly from west to east, nearly parallel to latitude circles, with a weak meridional component (Figure 10.8). This is a **zonal flow pattern** in which the north-south exchange of air masses is minimal. Cold air stays to the north, and warm air remains in the south. At the same time, the United States and southern Canada are flooded by air that originated over the Pacific Ocean. The Pacific air dries out to some extent as it passes through the western mountains, and it then warms adiabatically as it descends onto the Great Plains—spreading uniformly mild and generally fair weather east of the Rocky Mountains.

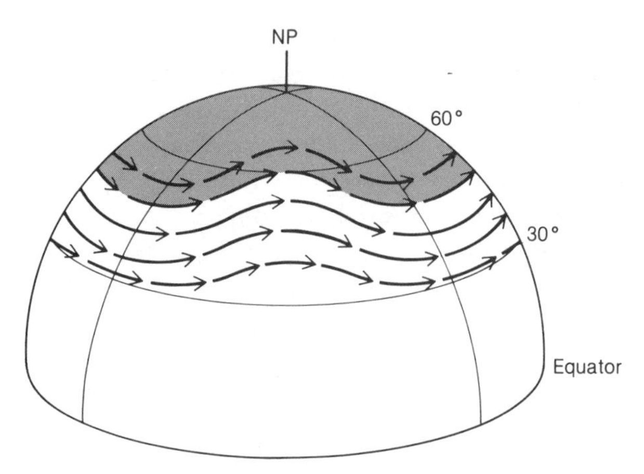

FIGURE 10.8

The midlatitude westerlies exhibit a zonal flow pattern when winds are almost directly west to east with only a small meridional (north-south) component.

At other times, the westerlies exhibit considerable amplitude, flowing in a pattern of deep troughs and sharp ridges (Figure 10.9). In this **meridional flow pattern**, masses of cold air surge southward, and warm air streams northward. The collision of contrasting air masses causes warm air to override cold air and sets the stage for the development of storms that are then swept along by the westerlies.

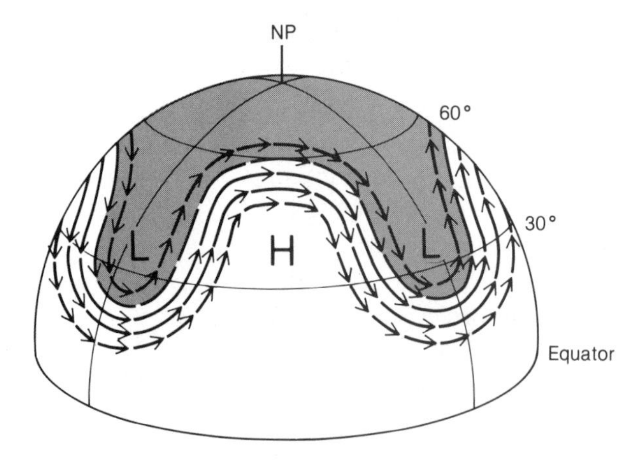

FIGURE 10.9

The midlatitude westerlies exhibit a meridional flow pattern when west to east winds have a strong meridional (north-south) component.

These two illustrations of Rossby wave configurations are the opposite extremes of a wide range of possible westerly wind patterns, each featuring varying combinations of meridional and zonal flow. The situation may be further complicated by a **split flow pattern**, in which westerlies to the north (over Canada, for example) have a wave configuration that differs from that of westerlies to the south (over the lower United States).

The westerly wind pattern typically shifts back and forth between dominantly zonal and dominantly meridional flow. For example, zonal flow might persist for a week and then give way to a more meridional flow for a few weeks, and then return to zonal flow again. The transition from one wave pattern to another is usually abrupt, sometimes taking place within a day. This abruptness poses a challenge to weather forecasters, because a sudden shift in the upper-air winds may divert a storm toward or away from a locality, or may cause an abrupt influx of colder air, which would change rain to snow.

Unfortunately for the long-range weather forecaster, the shifts between wave patterns have no regularity. No predictable zonal-meridional cycle has been discerned. The only observation useful to forecasters is that meridional patterns tend to persist for longer periods than zonal patterns. During the winter of 1976–1977, for example, the strong meridional wave patterns shown in Figure 10.10 persisted through much of the winter. Northwesterly flow

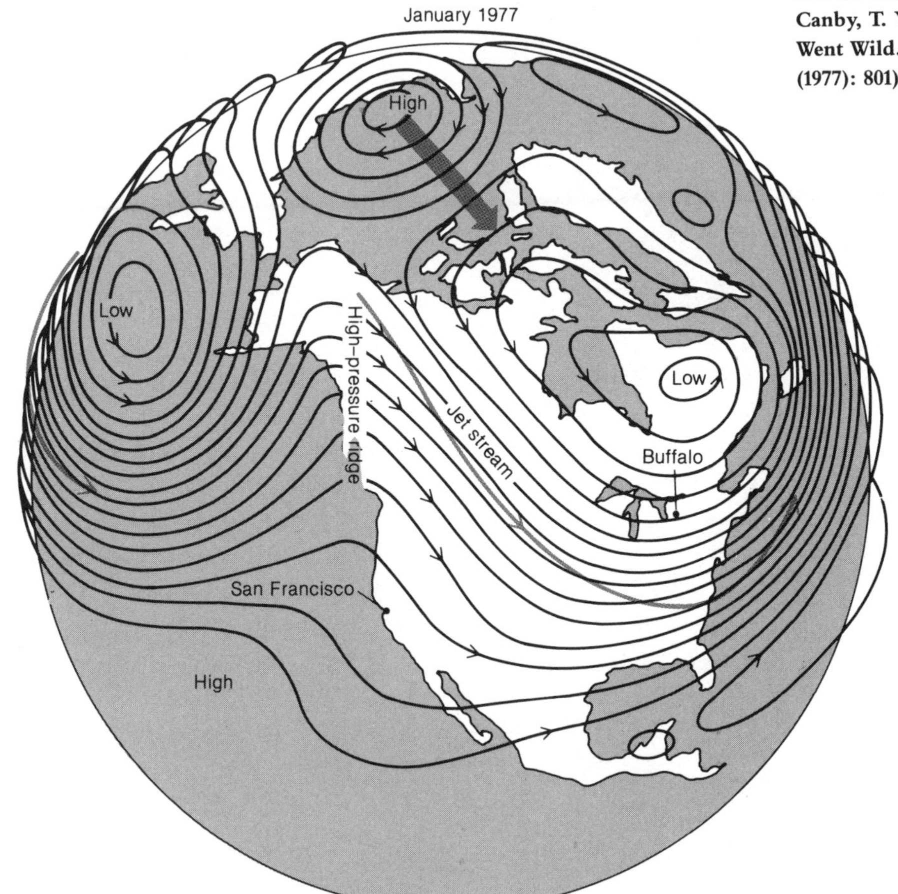

FIGURE 10.10

The strong meridional westerly wind pattern that prevailed over North America during the winter of 1976–1977 brought record cold to the East, and record drought and heat to the West. (Modified after Canby, T. Y. "The Year the Weather Went Wild." *National Geographic* 152 (1977): 801)

aloft brought surge after surge of bitter arctic air into the midsection of the United States, resulting in one of the coldest winters of this century for that area. Meanwhile, southwesterly winds aloft brought unusually mild air to the far western United States.

Blocking systems

For the continent as a whole, the weather in North America is more dramatic when the westerly wave pattern is strongly meridional. Sometimes undulations of the westerlies become so extreme that large, whirling masses of air actually separate from the main westerly air flow. This situation, shown in Figure 10.11, is analogous to the whirlpools that form in rapidly flowing rivers. In the atmosphere, cutoff masses of air whirl in either a cyclonic or an anticyclonic direction. A **cutoff low** or a **cutoff high** that blocks the usual west-to-east progression of weather systems is referred to as a **blocking system**. Because of the persistent nature of these systems, extreme weather usually results.

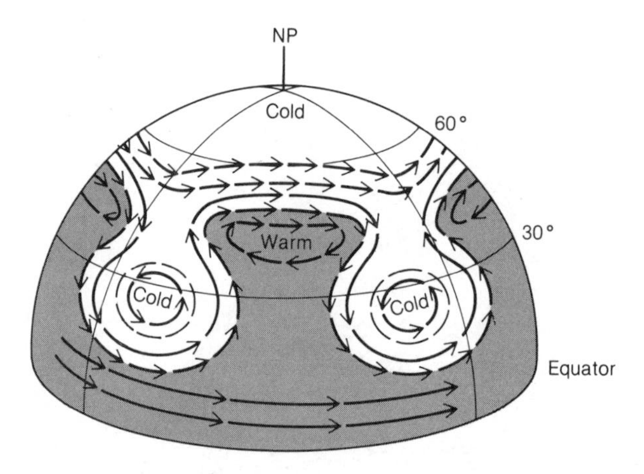

FIGURE 10.11

The midlatitude westerlies sometimes exhibit an extreme meridional flow pattern in which rotating pools of cold and warm air are cut off from the main circulation. The pool of cold air rotating counterclockwise is a blocking cyclone, and the pool of warm air rotating clockwise is a blocking anticyclone.

In July and August of 1975, a large blocking high, stationed to the west of the British Isles, deflected the flow of cool humid oceanic air far to the north of its usual path. As a consequence, residents of Great Britain and of most of Europe experienced one of the hottest, driest summers on record. Fires raged through the parched forests of Germany and Italy, and drought cut severely the yields of many food crops.

Blocking highs were also responsible for the serious 1972 shortfall in the wheat harvest of the Soviet Union. This event had worldwide economic ramifications, because it forced the Soviets to enter the international grain market. As a result, the price of wheat was sharply elevated, and the reserves of the grain-exporting nations were depleted. (In the United States, this increased demand caused the price of wheat to more than triple between early 1972 and early 1974.) Persistent anticyclones over Soviet wheatlands adversely affected yields of both winter and spring wheat. In winter, stationary anticyclones favored prolonged periods of fair, cold weather. Because of an unusually thin snow cover, frost penetrated deep into the soil and killed much of the winter wheat. The wheat replanted in the spring did not fare much better because of drought. Through much of the summer of 1972, a blocking high diverted storm systems away from the wheat-growing region, so rainfall totals for May, July, and August were less than half of normal. This drought, following a relatively dry winter, meant that much of the wheat withered in dessicated soil.

In summer, stagnant anticyclones bring extended periods of searing heat that can take many human lives. In the summer of 1980, more than 1200 people in the southern Mississippi River Valley died from heat-related stress. A stalled, warm anticyclone caused air temperatures to hover near 38 °C (100 °F) each day for over a month. A similar episode in the Midwest two years later contributed to an estimated 200 deaths.

Jet stream

Embedded in the upper-level westerly winds are relatively narrow ribbons of very strong wind called **jet streams**. In midlatitudes, a jet stream is situated in the upper troposphere between the midlatitude tropopause and the polar tropopause, and directly over the polar front (Figure 10.12). For this reason, it is known as the **polar front jet stream**. This jet stream follows the meandering path of the planetary westerly waves, and attains wind speeds that frequently exceed 125 km (80 mi) per hour. Westbound aircraft understandably avoid the jet stream because it is a head wind, whereas eastbound flights seek it because it is a tail wind.

Why is a jet stream associated with the polar front? Recall that air temperature influences air density and hence air pressure. Where temperature contrasts are great, pressure gradients are steep and winds are strong. The polar front, situated between the cold polar easterlies and the mild westerlies, is a zone of great thermal contrast, which causes a relatively steep pressure gradient aloft. The strongest winds (that is, the jet stream) would be anticipated over

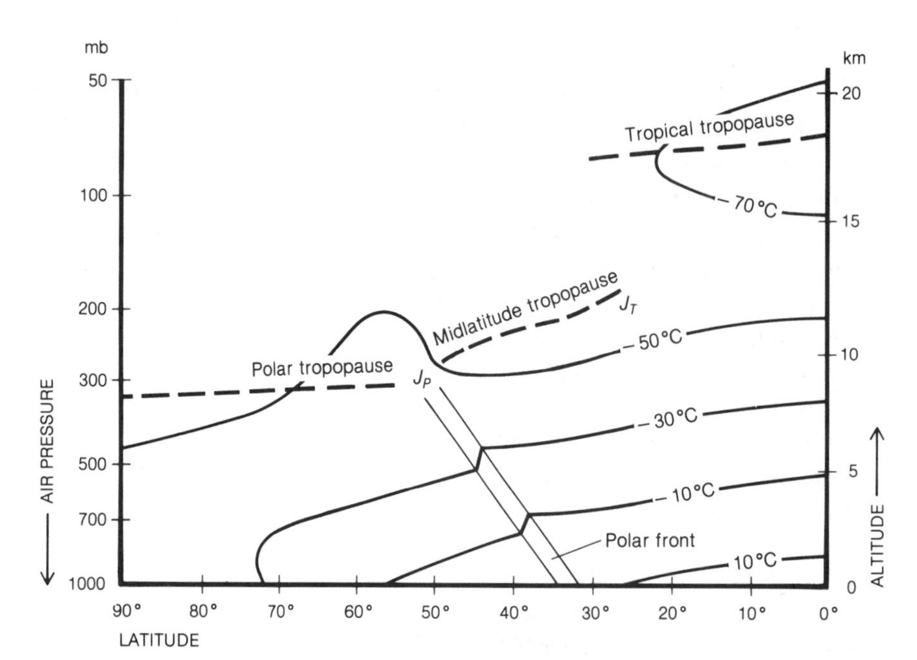

FIGURE 10.12

A vertical cross section (meridional profile) through the Northern Hemisphere troposphere shows the location of the polar front jet stream (J_P) above the polar front and at the midlatitude tropopause, and the location of the subtropical jet stream (J_T). In this perspective, the jet streams blow perpendicular to the page. Solid lines are isotherms in °C. Note that there is considerable vertical exaggeration.

the zone of greatest temperature contrast, the polar front. The relationship between the jet stream and the polar front is examined in more detail in the Mathematical Note at the end of the chapter.

Like the polar front, the jet stream is not uniformly well defined around the globe. Where the polar front is well defined, that is, where surface temperature gradients are particularly steep, jet stream winds accelerate. Such a segment, in which the wind may accelerate by as much as 100 km (62 mi) per hour, is known as a **jet maximum**. A typical jet maximum is 160 km (100 mi) wide, 2 to 3 km (1 to 2 mi) thick, and several hundred kilometers in length.

What, then, is the role of the polar front jet stream in the generation and maintenance of synoptic-scale storms? Air flowing through a jet maximum changes in both speed and direction, and these changes induce a complex pattern of horizontal divergence and horizontal convergence aloft. In Figure 10.13, a jet maximum

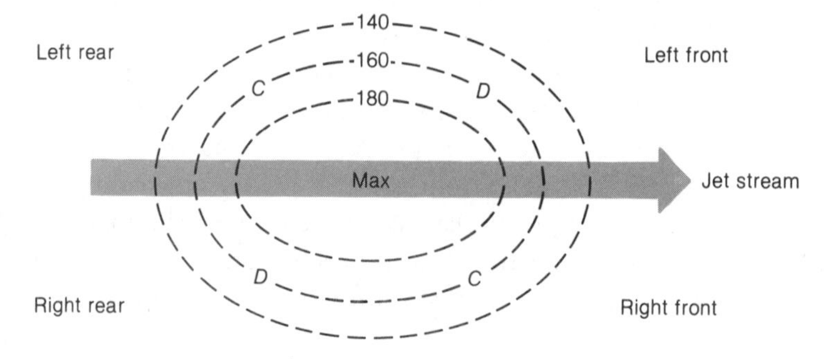

FIGURE 10.13

A jet stream maximum is divided into quadrants. The jet maximum is outlined by isotachs (dashed lines) of equal wind speed (in km per hour). Air movement into and out of the jet maximum induces areas of divergence (D) and areas of convergence (C).

(viewed from above) is divided into four quadrants: left rear, right rear, left front, and right front. Horizontal divergence occurs in the left front and right rear quadrants, and horizontal convergence occurs in the right front and left rear quadrants. Now recall from Chapter 9 that a synoptic-scale cyclone is characterized by the convergence of air at the surface and by divergent airflow aloft. Where the jet stream triggers upper-air horizontal divergence, the jet stream therefore contributes to the development and maintenance of cyclones that form and travel along the polar front. The strongest horizontal divergence has been found in the left front quadrant, so it is under this sector of the jet stream maximum that a cyclone typically develops.

Other characteristics of the polar front jet stream add to our understanding of weather. Although the jet stream is a component of the westerly winds and weaves with the westerlies, a jet maximum typically progresses from west to east at a faster rate than the west to east displacement of the troughs and ridges in Rossby waves (Figure 10.14). The jet stream is also not always a single stream of air, but may splinter into separate ribbons.

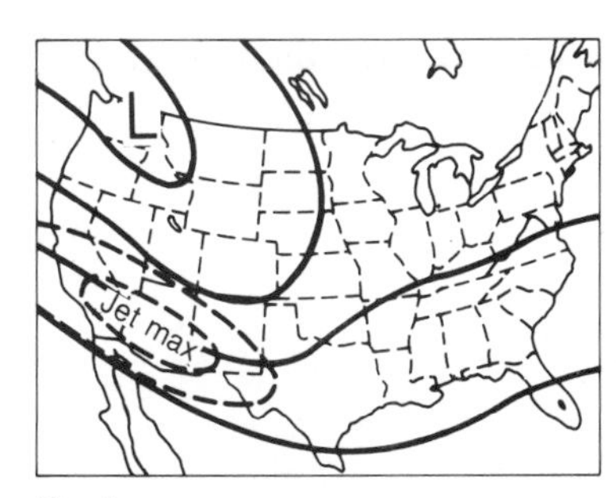

Monday

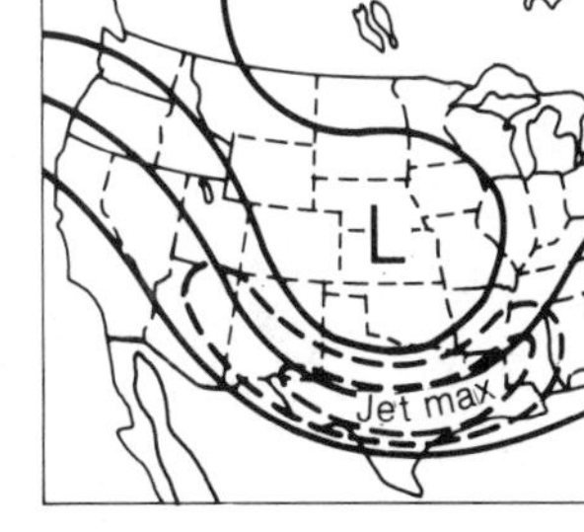

Tuesday

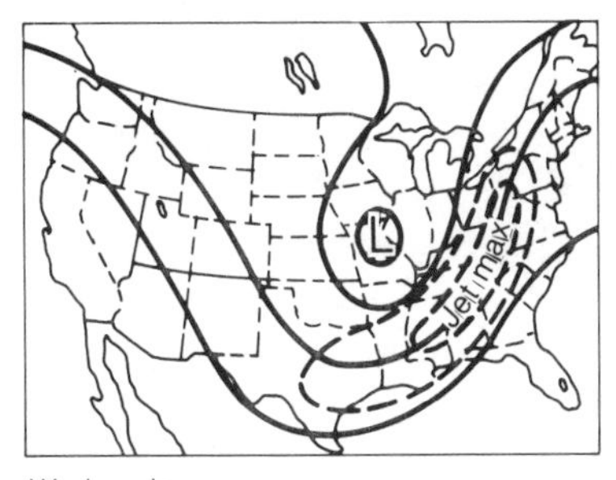

Wednesday

FIGURE 10.14

An upper-level trough is shown progressing from west to east across the United States. Meanwhile, a jet maximum (dashed outline) travels eastward at a more rapid rate than the trough. Dashed lines are isotachs, lines of equal wind speed. (From Sechrist, F. S., and E. J. Hopkins. *Meteorology: Weather and Climate*. Madison, Wis.: University of Wisconsin Extension, 1974, p. 146.)

Like the polar front, the jet stream undergoes important seasonal shifts. The jet stream strengthens in winter, when north-south temperature contrasts are great, and weakens in summer, when temperature contrasts are minimal. As shown in Figure 10.15, the average summer location of the jet stream is across southern Canada, and the average winter position is across the southern United States. These locations represent long-term averages: The jet stream actually exhibits considerable range in latitude from week to week, and even from day to day. As a general rule, when the polar front jet stream is south of us, the weather tends to be winterlike, and when the polar front jet stream is north of us, the weather tends to be summerlike.

The polar front jet stream is not the only jet stream. The **subtropical jet stream** occurs near the break in the tropopause between the tropical and middle latitudes (Figure 10.12). It is less energetic and less variable in latitude than its northerly counterpart. Jet streams also occur in the Southern Hemisphere, but so far have not received as much study as those to the north.

Short waves

Short waves are another feature of the upper-level westerlies. These waves, which are ripples superimposed on Rossby long waves, are important to surface weather systems. While Rossby

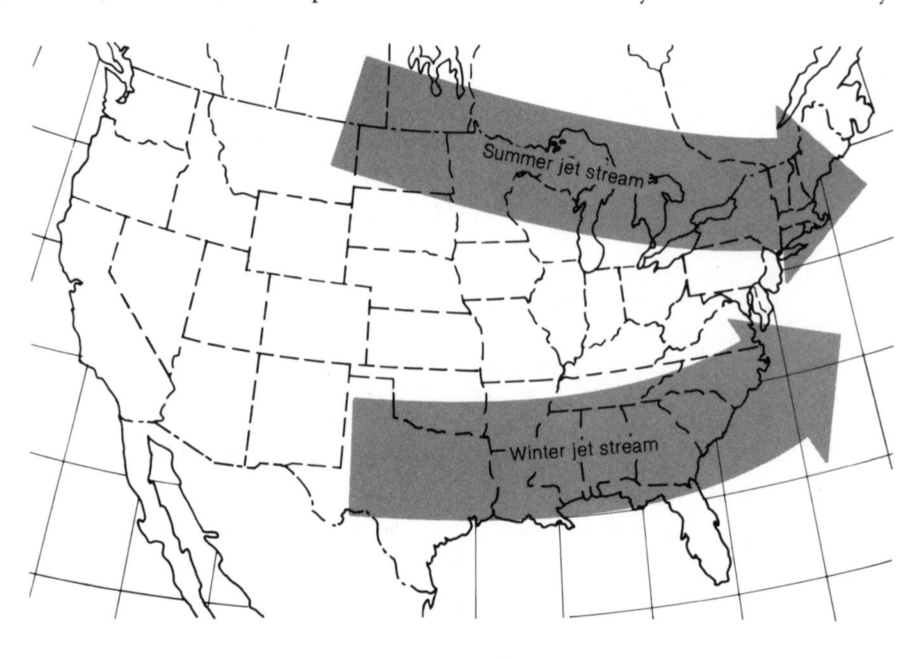

FIGURE 10.15
Approximate average position of the polar front jet stream in winter (December through March) and in summer (June through October).

waves drift very slowly eastward, short waves travel rapidly through the Rossby waves. As five or fewer long waves encircle the hemisphere, a dozen or more short waves may be encircling.

Both short waves and long waves in the westerlies can contribute to cyclone development. For the same isobar spacing (pressure gradient), gradient and geostrophic wind speeds are not equal. Anticyclonic gradient winds are stronger than geostrophic winds, and cyclonic gradient winds are weaker than geostrophic winds. Hence, as shown in Figure 10.16, westerly winds tend to strengthen in a ridge and weaken in a trough. The result is horizontal divergence of

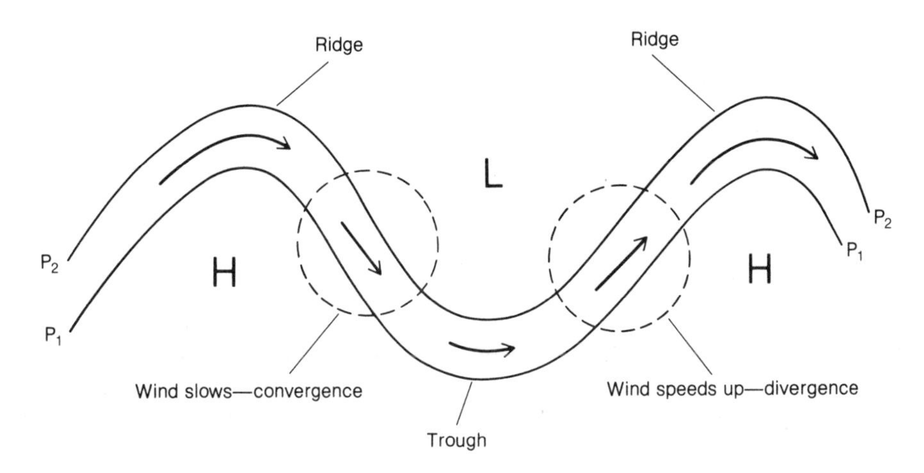

FIGURE 10.16

Westerly winds speed up in ridges and slow down in troughs. This induces convergence aloft ahead of the ridge, and divergence aloft ahead of the trough. Solid lines are isobars (P_1 is greater than P_2).

air to the east of the trough (and west of the ridge). Short and long waves thus favor storm development in the same way as the polar front jet stream—by inducing divergent airflow aloft. Conditions aloft are consequently most favorable for storm development when the jet maximum appears on the east side of a trough (and west of a ridge).

Conclusions

Features of the global-scale circulation are important controls of weather and climate. This chapter traced the linkages between the westerly winds aloft and the weather at the earth's surface. Our examination continues in the next chapter with a focus on synoptic-scale weather makers.

MATHEMATICAL NOTE

The polar front and the midlatitude jet stream

Why is a jet stream found over the polar front? To understand the reasons for this important association, we must return to the concept of hydrostatic equilibrium. We learned in Chapter 9 that hydrostatic equilibrium is the balance between the upward directed pressure gradient force and the downward directed force of gravity. We now express this relationship mathematically.

The air parcel in Figure 1 has a density, ρ, a cross-sectional area, A, and a volume, V. The air pressure acting on the top of the parcel, P_2 (at altitude Z_2), arises from the weight of the total atmospheric column above Z_2. The air pressure acting on the bottom of the parcel, P_1 (at altitude Z_1), is due to the weight of the total atmospheric column above Z_1. The pressure difference between the parcel top and the parcel bottom is the vertical pressure gradient and depends on the weight of the parcel itself. The parcel weight, W, is equal to its mass, M, times the acceleration of gravity, g, and is distributed over the area, A. Hence:

$$P_2 - P_1 = W/A$$

but since $$W = Mg = \rho Vg$$

and $$V = A(Z_2 - Z_1)$$

then $$P_2 - P_1 = \rho g(Z_2 - Z_1)$$

that is, $$\Delta P = -\rho g \Delta Z$$

This is known as the *hydrostatic equation*. The minus sign appears on the right side of the equation to account for the in-

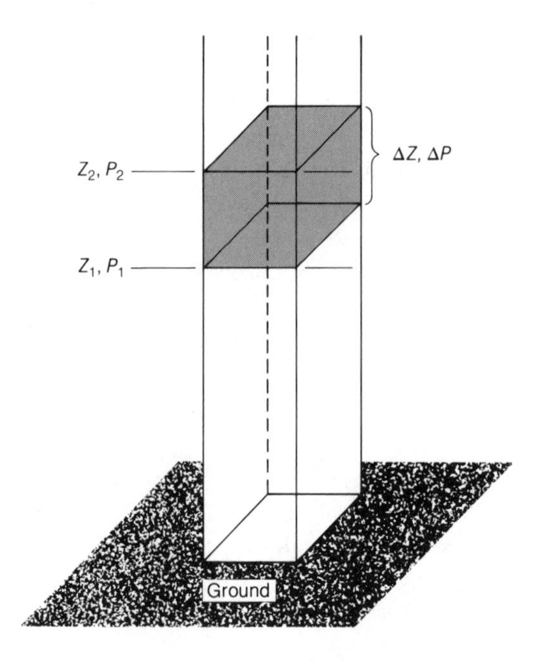

FIGURE 1

In a column of air, a change in pressure (ΔP) is associated with a change in altitude (ΔZ).

verse relationship between air pressure and altitude, that is, pressure drops as altitude increases.

According to the hydrostatic equation, the pressure change (ΔP) that accompanies an altitude change (ΔZ) depends on the density of air and on the acceleration of gravity. For practical purposes, we may assume that gravity is constant with altitude, at least for several kilometers. Air density is thus the ultimate determinant of the rate at which air pressure declines with altitude. Because cold air is denser than warm air, air pressure drops more rapidly with altitude in cold air than in warm air. This is a very useful conclusion.

Now suppose that a sharp horizontal air temperature gradient exists at the earth's surface, as across the polar front. The hydrostatic equation tells us that on the warm side of the front the air pressure drops less rapidly with altitude than it does on the cold side of the front. In the vertical, pressure surfaces consequently slope as shown in Figure 2. Notice that the horizontal pressure gradient strengthens with altitude, so the horizontal wind must also strengthen with altitude. The wind reaches its maximum speed just below the tropopause and over the area of maximum temperature change (the polar front), constituting the jet stream. Above this level, in the stratosphere, the westerlies weaken because the horizontal pressure gradient weakens. A jet stream is thus associated with steep horizontal air temperature gradients at the earth's surface. Where the temperature contrast across a front is particularly great, we would expect to encounter a jet maximum aloft.

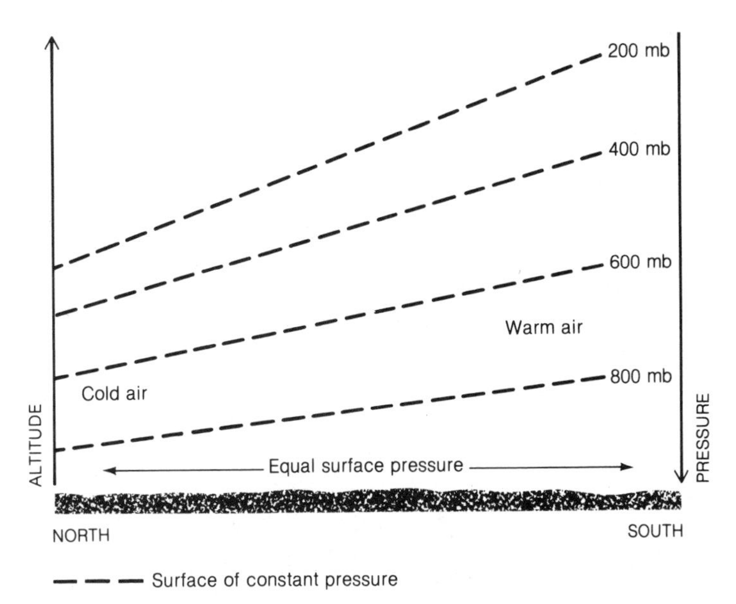

FIGURE 2

Pressures fall more rapidly with altitude in cold air than in warm air. The horizontal pressure gradient consequently strengthens with altitude when cold air is situated next to warm air. A strengthening of the horizontal pressure gradient in turn means stronger winds aloft.

SUMMARY STATEMENTS

- The principal features of global-scale circulation are semipermanent pressure cells, polar easterlies, westerlies, trade winds, the intertropical convergence zone, and the polar front. These features shift poleward during spring and toward the equator during autumn.

- Aloft, global-scale winds form Hadley cells in the tropics, westerly waves in the midlatitudes, and westerly winds over polar regions.

- Contrasts in solar heating favor relatively high pressure over continents and low pressure over seas in winter, and relatively low pressure over continents and high pressure over seas in summer.

- Upper-air westerlies describe a wavelike pattern that, with time, undergoes changes in length, amplitude, and wave number. These changes in turn influence the degree of meridional and zonal flow and the north-south exchange of air masses.

- A cutoff anticyclone or a cutoff cyclone blocks the usual west to east progression of weather systems and may lead to extreme weather conditions.

- The midlatitude jet stream is a corridor of very strong winds within the westerlies and is situated at the top of the troposphere over the polar front. The jet stream and waves in the westerlies cause divergence of air aloft, which favors storm development.

KEY WORDS

semipermanent pressure systems
subtropical anticyclones
horse latitudes
midlatitude westerlies
trade winds
equatorial trough
intertropical convergence zone (ITCZ)
subpolar lows
Aleutian low
Icelandic low
polar front
front
Hadley cell
singularity
January thaw
Indian summer
long waves
Rossby waves
meridional component
zonal component
zonal flow pattern
meridional flow pattern
split flow pattern
cutoff low
cutoff high
blocking system
jet streams
polar front jet stream
jet maximum
subtropical jet stream
short waves

REVIEW QUESTIONS

1 Why are global-scale pressure systems described as "semipermanent"?

2 Why do some of the world's major deserts occur on the eastern side of the subtropical anticyclones?

3 Explain the origin of the name "horse latitude."

4 How are the trade winds linked to the subtropical anticyclones?

5 How are the westerlies linked to the subtropical anticyclones?

6 Describe the weather along the intertropical convergence zone (ITCZ).

7 What is a front?

8 How is the Hadley cell like a giant convective current?

9 How is the tropopause segmented?

10 The major features of a global-scale circulation "follow the sun" through the course of a year. Explain this statement.

11 Give an example of a singularity.

12 Is Indian summer a true singularity?

13 Distinguish between the meridional component and the zonal component of the midlatitude westerly waves. Which component is responsible for poleward heat transport?

14 Distinguish between a zonal flow pattern and a meridional flow pattern.

15 Why is the weather more dramatic when the westerly wave pattern is strongly meridional?

16 What is the relationship between the polar front jet stream and the midlatitude westerlies?

17 Define jet stream maximum.

18 Describe the seasonal changes in the polar front jet stream.

19 What is the relationship between short waves and Rossby long waves?

20 In westerly waves, why does divergence of air occur east of a trough and convergence of air occur east of a ridge?

POINTS TO PONDER

1 Speculate on why the tropical tropopause is at a greater altitude than the polar tropopause.

2 Explain how the jet stream and short waves contribute to cyclone development.

3 Suppose that the prevailing westerly circulation pattern shifts from dominantly zonal to dominantly meridional. How will this change influence north-south air mass exchange? How will the polar front be affected?

4 Describe the relationship of blocking systems to extremes in weather.

5 Speculate on how a split flow pattern might influence the weather over North America.

6 Why is a jet stream associated with the polar front?

PROJECTS

1 From your viewing of televised weather summaries for the nation, speculate on whether the westerly wave pattern is dominantly zonal or dominantly meridional. Support your choice.

2 From your analysis of surface temperatures across the nation, locate the polar front and the polar front jet stream.

SELECTED READINGS

Chang, J. *Atmospheric Circulation Systems and Climates.* Honolulu, Hawaii: The Oriental Publishing Company, 1972. 328 pp. *An excellent, detailed discussion of atmospheric circulation systems.*

Harman, J. R. "Tropospheric Waves, Jet Streams, and United States Weather Patterns." *Association of American Geographers Resource Paper*, no. 11 (1971). 37 pp. *Detailed description of the relationships of westerly waves, jet streams, and the weather.*

Lorenz, E. N. "A History of Prevailing Ideas About the General Circulation of the Atmosphere." *Bulletin of the American Meteorological Society* 64 (1983):730–733. *A brief survey of the principal people and ideas involved in research on global-scale circulation.*

The Westerly Wind asserting his sway from the southwest quarter is often like a monarch gone mad, driving forth with wild imprecations the most faithful of his courtiers to shipwreck, disaster, and death.
JOSEPH CONRAD
The Mirror of the Sea

Synoptic-scale circulation is a subdivision of the global-scale circulation and includes weather systems that influence continental or oceanic areas. Air masses, fronts, cyclones, and anticyclones are principal features of synoptic-scale weather. (NOAA photograph)

ELEVEN

Synoptic-Scale Weather

Air Masses

North American Types

Modification

Frontal Weather

Cold Front

Warm Front

Cyclones

Life Cycle

Cold and Warm Cores

Anticyclones

Cold and Warm Cores

Anticyclonic Weather

THE ceaseless succession of synoptic-scale weather systems is directly responsible for the day-to-day variability of our weather. This chapter examines the weather conditions that accompany the major weather systems of the midlatitudes: (1) air masses, (2) fronts, (3) cyclones, and (4) anticyclones.

Air Masses

An **air mass** is a huge volume of air that is relatively uniform in temperature and water vapor content. The properties of an air mass are dictated by the type of surface over which it develops, that is, its source region. In order for an air mass to become homogeneous, the surface must be homogeneous, and the air mass must reside there for at least several days. Air masses that form over the cold surfaces of the far north (polar) are relatively cold, and those that develop over the warm surfaces of the far south (tropical) are relatively warm. Air masses that form over land (continental) tend to be relatively dry, and those that develop over the oceans (maritime) are relatively humid. There are thus four basic types of air mass: cold and dry, cold and humid, warm and dry, and warm and humid. It follows that a convenient way to classify air masses is on the basis of source region.

North American types

Over North America, the major air mass types are designated continental tropical (cT), maritime tropical (mT), maritime polar (mP), continental polar (cP), and arctic (A). Note that both cP and A air are cold and dry. Figure 11.1 shows the locations of the principal source regions of each of these air mass types.

Continental tropical air, which forms over the subtropical deserts of the southwestern United States, develops primarily in summer, and is hot and dry. **Maritime tropical air** is very warm and humid, having source regions over tropical and subtropical oceans. It retains these properties year-round and is responsible for oppressive summer heat and humidity east of the Rocky Mountains. The source region for **maritime polar air** is the cold ocean waters of the North Pacific and North Atlantic. Along the West Coast, mP air brings heavy winter rains (snows in the mountains) and persistent coastal fogs in summer. Dry **continental polar air** develops over the northern interior of North America. In winter, cP air is

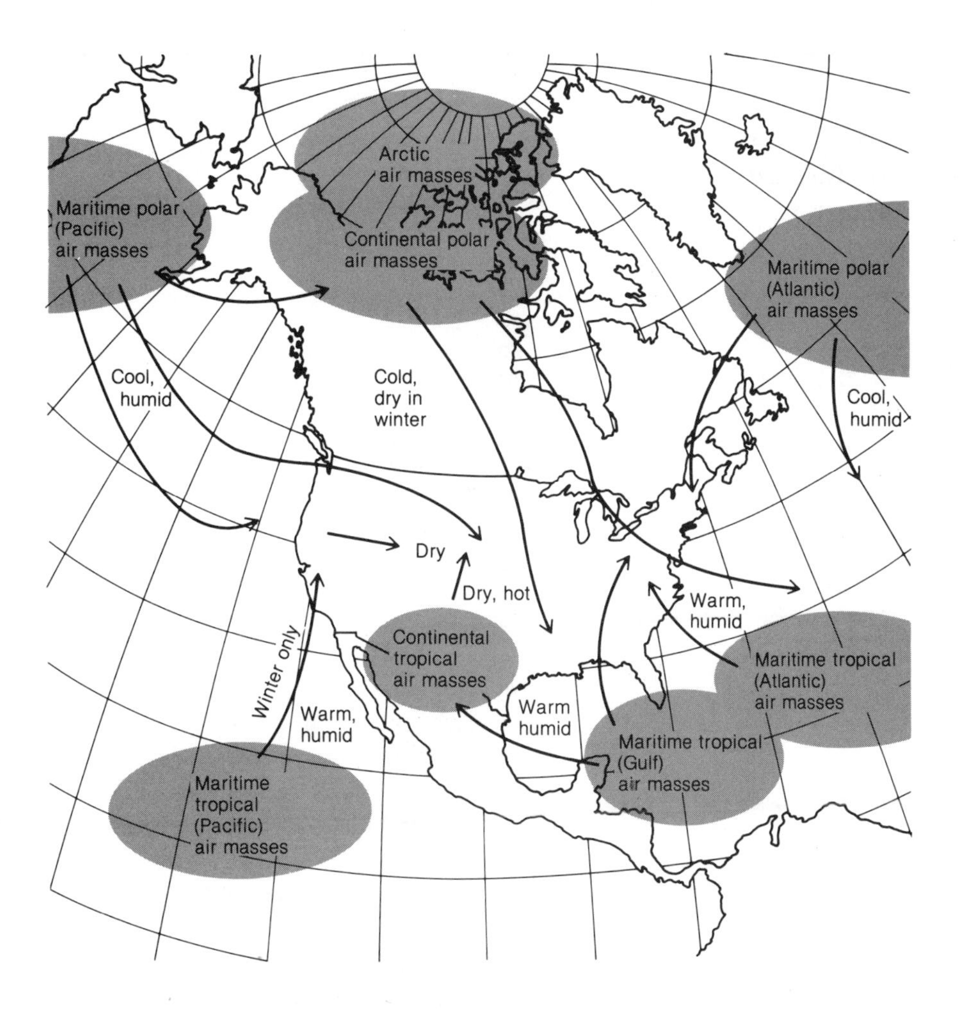

FIGURE 11.1

Air mass source regions for North America.

typically very cold, because the source region is snow covered and subject to extreme radiational cooling. In summer, when the source region warms in response to lengthy days of sunshine, cP air is quite mild. **Arctic air** forms primarily in winter over the Arctic Basin, Greenland, and interior portions of North America north of about 55 degrees latitude. It is exceptionally cold and dry, and is responsible for the bone-numbing winter cold waves that sweep across the Great Plains.

In addition to differing in temperature and humidity, air masses also differ in stability. As we noted in Chapter 6, stability is an important property of air because it influences vertical motion and the consequent development of cloudiness and precipitation. Table 11.1 lists the usual stability, temperature, and humidity characteristics of the various air masses within their source region.

TABLE 11.1

Stability, Temperature, and Humidity Characteristics of North American Air Masses

AIR MASS TYPE	SOURCE REGION STABILITY		CHARACTERISTICS	
	WINTER	SUMMER	WINTER	SUMMER
A	**Stable**		**Bitter cold, dry**	
cP	**Stable**	**Stable**	**Very cold, dry**	**Cool, dry**
cT	**Unstable**	**Unstable**	**Hot, dry**	**Hot, dry**
mP (*Pacific*)	**Unstable**	**Unstable**	**Mild, humid**	**Mild, humid**
mP (*Atlantic*)	**Unstable**	**Stable**	**Cold, humid**	**Cool, humid**
mT (*Pacific*)	**Stable**	**Stable**	**Warm, humid**	**Warm, humid**
mT (*Atlantic*)	**Unstable**	**Unstable**	**Warm, humid**	**Warm, humid**

Modification

Air masses do not remain in their source regions indefinitely but move from place to place. As they travel, their properties are modified—those of some air masses changing more than those of others. Changes may occur in temperature, mixing ratio (or vapor pressure), and stability. Air masses are modified primarily in two ways: (1) by exchanging heat or moisture or both with the terrain over which the air mass travels and (2) by undergoing large-scale adiabatic compression or expansion. Let us examine examples of each of these mechanisms.

In winter, as a mass of cP air plunges southeastward from Canada into the United States, its temperatures are usually modified quite rapidly. While temperatures in the upper plains might dip below −15 °C (5 °F), by the time the polar air reaches the southern United States, temperatures may not drop much below the freezing point. This rapid air mass modification occurs because, outside of its source region, polar air is colder than the ground over which it travels. The warmer ground heats the bottom of the air mass, destabilizing it and causing convection currents that distribute heat throughout the air mass. When cP air travels over snow-covered ground, however, modification is much less rapid. A cold surface stabilizes the air and reduces convective heating.

Tropical air masses are not modified as readily as polar air masses, because tropical air is usually warmer than the ground over which it travels. The bottom of the air mass cools by contact with the ground, but this cooling stabilizes the air and suppresses convection currents. As a result, cooling is restricted to the very bottom of the air mass. A cold wave loses its intensity as it pushes southward, but a summer heat wave can retain much of its intensity from the Gulf of Mexico well into Canada.

As a consequence of orographic lifting, air masses undergo significant changes in temperature and mixing ratio (or vapor pressure). When cool, moist mP air sweeps inland off the Pacific Ocean, the air is forced up the windward side of the coastal mountain ranges and thereby cools adiabatically. Clouds and precipitation develop, lowering the mixing ratio and releasing latent heat. The air then warms adiabatically as it descends down the mountains' leeward slopes into the Great Basin. These processes are repeated as the air mass flows through the Rockies and emerges on the Great Plains—milder and drier than the air mass was when it began flowing inland. The net warming of the air is due to the conversion of the latent heat of vaporization to sensible heat when water vapor condenses on the windward slopes of the mountains.

Frontal Weather

Air is uplifted where contrasting air masses meet. The uplift often cools the air sufficiently for clouds and precipitation to develop along the frontal surface. Depending on the slope of the front, frontal weather may be confined to a very narrow band, or it may extend over a broad region.

Cold front

When denser air displaces lighter air, the heavier air advances along the ground, where frictional resistance steepens the frontal slope (Figure 11.2). This happens, for example, when cP air replaces mT air. The zone of transition between such air masses is commonly called a **cold front**, although sometimes—especially in summer—the density contrast between air masses is due more to differences in vapor pressure than to differences in temperature. Because of the steep frontal slope, uplift is restricted to a very narrow zone along the front's leading edge.

If the cold front advances slowly but steadily, the type of frontal weather depends on the stability of the warmer air. If the warm air is relatively stable, then nimbostratus and altostratus clouds may

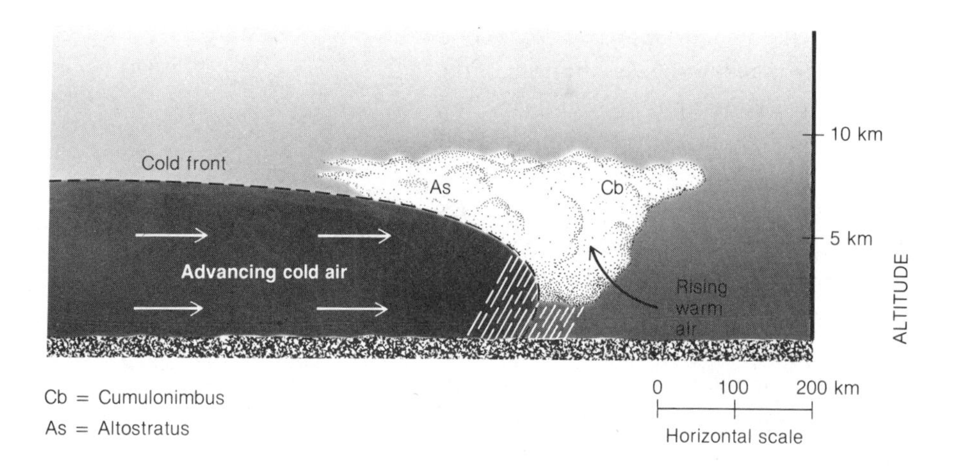

FIGURE 11.2
Storm clouds are caused by uplift along a cold front.

form. Any precipitation is likely to be showery and brief. If the warm air is unstable, the uplift is vigorous, giving rise to cumulonimbus clouds towering above nimbostratus. Thunderstorms typically develop, perhaps accompanied by strong, gusty winds or other violent weather.

If the cold front moves along at a rapid pace, then a **squall line**, a band of vigorous thunderstorms, may develop ahead of the front. (Squall lines are discussed in detail in Chapter 13, when we consider severe weather systems.)

FIGURE 11.3
A sequence of clouds forms along an approaching warm front where stable warm air is replacing cooler air. Note the vertical exaggeration in this cross section.

Warm front

Relatively light, warm air replaces denser, colder air by overrunning, as shown in Figure 11.3. The consequence is a broad area of

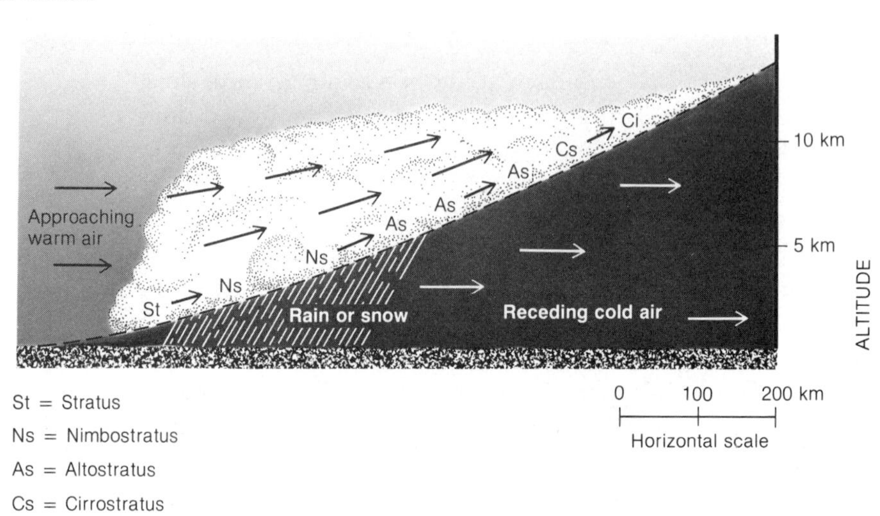

FIGURE 11.4
Thin, wispy cirrus clouds may appear more than 1000 km (620 mi) in advance of a surface warm front. (NOAA photograph)

gradual uplift along a gently sloping frontal surface that usually stretches many hundreds of kilometers ahead of the surface **warm front**. As a warm front advances, clouds develop and gradually lower and thicken in the following sequence: cirrus, cirrostratus, altostratus, nimbostratus, and stratus.

The initial wispy cirrus clouds (Figure 11.4) may appear more than 1000 km (620 mi) in advance of the surface warm front. Slowly the clouds spread laterally, forming thin sheets of cirrostratus that turn the sky a bright milky white. The tiny ice crystals composing these high clouds (bases above 6 km) may reflect and refract sunlight to produce various images such as halos, sun pillars, and sun dogs (Chapter 8). The appearance of these optical phenomena may herald the coming of stormy weather a few days in advance.

In time, cirrostratus give way to altostratus, thin clouds with bases at altitudes of 2 to 6 km. Soon after altostratus clouds thicken enough to block out the sun, light rain or snow begins. Steady precipitation falls from low, gray nimbostratus clouds and persists until the warm front finally passes—a period that may exceed 24 hours. Copious amounts of rain may therefore fall ahead of the warm front, and since the precipitation intensity is usually only light to moderate, much of the water is likely to infiltrate the soil. This is the type of soaking rain that farmers appreciate. If it is cold enough for the precipitation to fall in the form of snow, accumulations may be substantial.

Just ahead of the warm front, steady precipitation gives way first to drizzle falling from low stratus clouds (bases below 2 km) and then to thick fog. After the warm front finally passes, skies usually clear, since the zone of overrunning has also passed, and the weather turns warm and muggy.

The cloud and precipitation sequence just described for a warm front applies when the advecting warm air is relatively stable. The sequence changes somewhat when the warm air is relatively unstable. In that case, uplift is more vigorous and gives rise to cumulonimbus clouds (thunderstorms) ahead of the surface warm front, and produces scattered areas of brief but intense rainfall or perhaps snowfall.

Clouds and precipitation develop along fronts only when there is a significant contrast between air masses. If the air masses are nearly identical in temperature and humidity, the front may pass unnoticed except for a shift in wind direction.

Frontal weather does not occur in isolation from cyclones. We explore this relationship next.

Cyclones

The **cyclone**, or low, is the principal weather maker of the midlatitudes. The counterclockwise and inward surface flow of air about a center of low pressure brings together contrasting air masses to form fronts and attendant clouds and precipitation. Most of the ascending air that characterizes a low actually occurs along frontal surfaces.

Life cycle

If conditions are favorable in the middle and upper troposphere, a synoptic-scale cyclone may develop. **Cyclogenesis**, the birth of a cyclone, usually takes place along the polar front directly under an area of upper-level divergence. As noted in Chapter 10, divergence aloft occurs to the east of a trough and under the left front quadrant of a jet stream maximum. If the divergence of air aloft exceeds the convergence of air at the surface, the surface air pressure drops, a horizontal pressure gradient develops, and a cyclonic circulation begins. A storm is born. The westerly flow aloft then steers the cyclone across the continent while the storm passes through its life cycle (Figure 11.5).

Initially, the polar front is stationary, and surface winds are directed parallel to the front. As surface winds converge and the

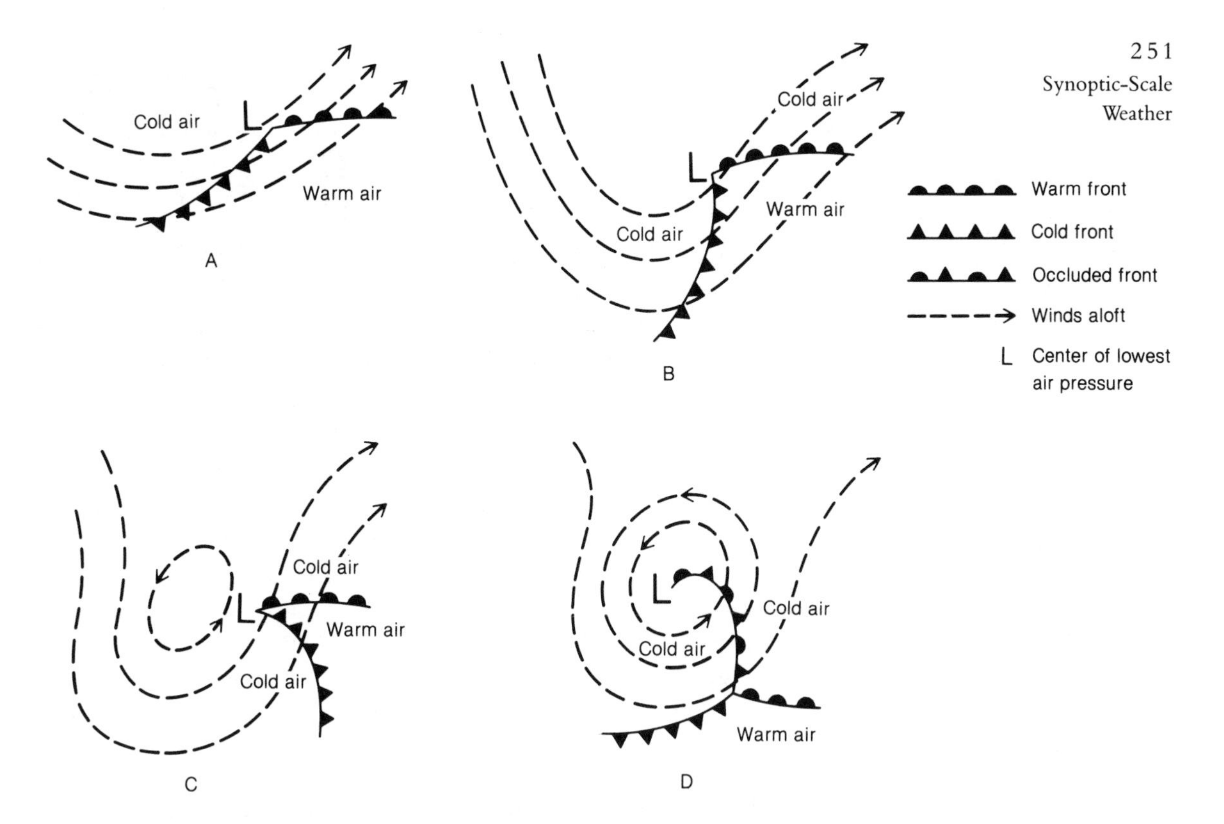

FIGURE 11.5

A midlatitude cyclone passes through its life cycle: (A) initial stage, (B) mature stage, (C) late mature stage, and (D) occluded stage. At maturity, the storm is east of the upper-level trough, and at occlusion, the storm is under the upper-level trough. (From Trewartha, G.T., and L.H. Horn. *An Introduction to Climate*, 5th edition. New York: McGraw-Hill, 1980, p. 165. Used with permission.)

surface air pressure drops, the front begins to move (Figure 11.5A). West of the low center,* the front advances toward the southeast as a cold front. East of the low center, the front moves northward as a warm front. At this early stage in the storm's life cycle, the front forms a wave pattern, hence the descriptive name, **wave cyclone**.

As the cyclone matures, the central pressure continues to drop, the horizontal pressure gradient steepens, and the counterclockwise winds strengthen. Meanwhile, the warm and cold fronts become very active and travel along with the counterclockwise airflow of the storm system. Because the cold front typically moves faster than the warm front, however, the cold front forms almost a right angle with the warm front by the time the storm achieves full maturity (Figure 11.5B).

We are most interested in the fully mature cyclone, because the storm is most intense at this stage of its life cycle. At late maturity, the central pressure of a cyclone is approaching its lowest value, wind speeds are beginning to peak, and the precipitation is wide-

*The low center is the point of lowest surface air pressure.

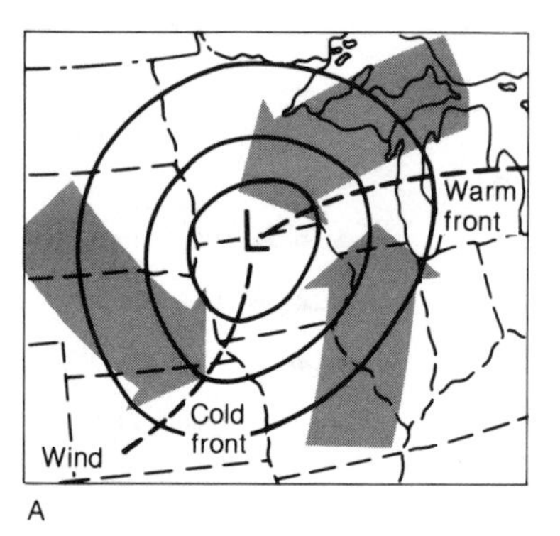

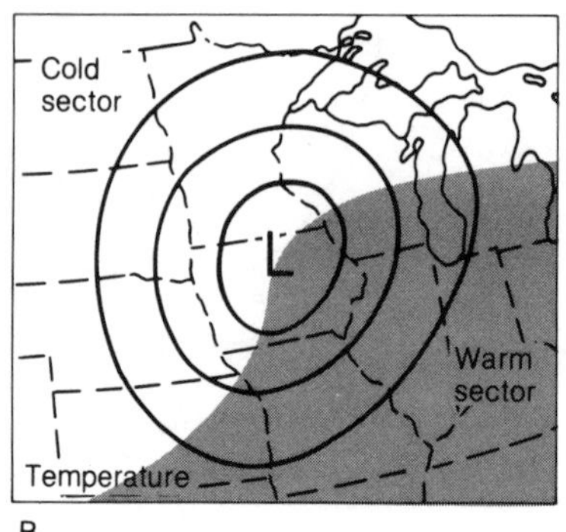

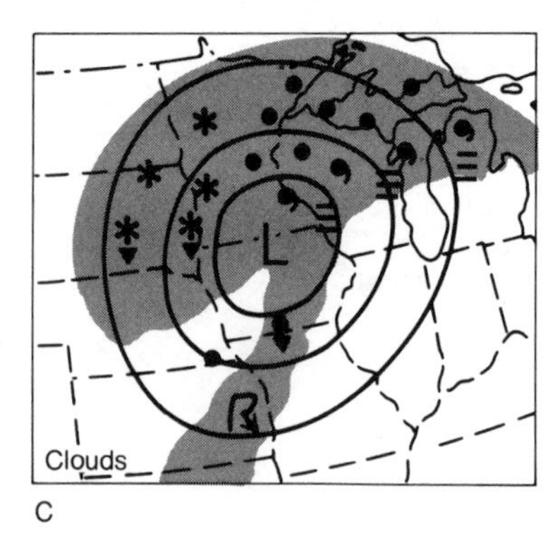

FIGURE 11.6

A midlatitude wave cyclone at maturity showing distribution of (A) surface winds, (B) surface air temperatures, and (C) clouds and precipitation.

spread and heaviest. Figure 11.6 shows the winds, temperature distribution, cloud shield, and precipitation patterns associated with a typical midlatitude low at full maturity. Figure 11.7 is a satellite view of the whirling cloud shield around a mature low. These features are consistent with our earlier description of cold and warm frontal weather. The stormy weather associated with cyclones is thus due to uplift of air along the associated fronts.

The typical weather pattern of a mature midlatitude cyclone shows that storms have a warm side and a cold side. Note in Figure 11.6 that the coldest air is located to the west of the low center (where winds are northwesterly) and the warmest air is to the

FIGURE 11.7

Satellite view of the cloud pattern associated with a mature midlatitude cyclone centered over the southeastern United States. (NOAA photograph)

southeast of the low center (where winds are southerly). Air temperatures to the northeast of the low center (where winds are northeasterly) are between the temperatures of the cold and warm sectors of the storm. As a cyclone tracks across the continent, the weather to the left (cold side) of the path is quite different from the weather to the right (warm side) of the storm's trajectory. Consider an illustration. A winter storm develops in eastern Colorado. As it matures, it moves northeastward toward the Great Lakes region and takes either track *A* or track *B* (Figure 11.8). The storm center passes within 150 km (90 mi) of Chicago, but track *A* takes the

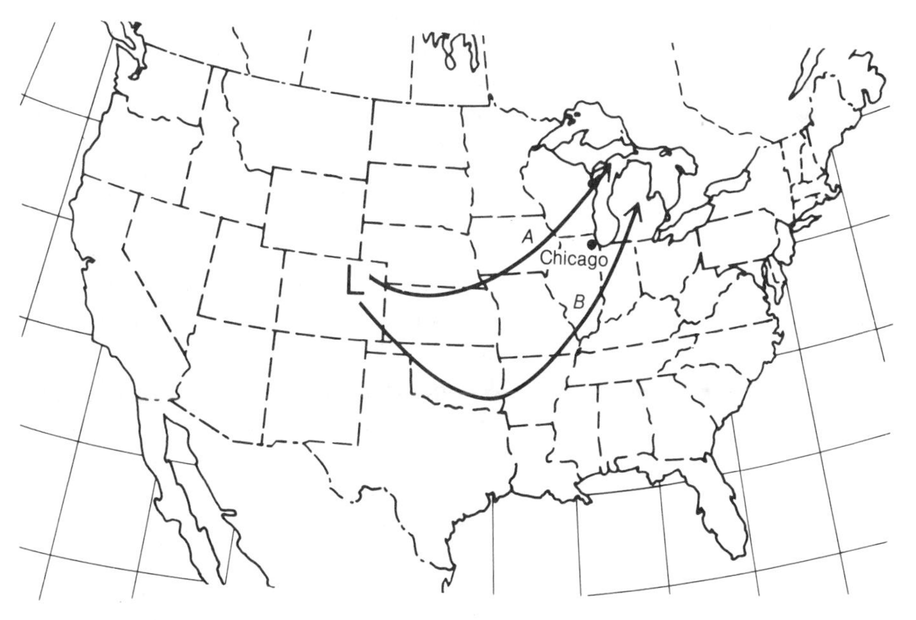

FIGURE 11.8

A storm that developed in eastern Colorado follows a path either to the west (track *A*) or to the east (track *B*) of Chicago. For track *A*, Chicago is on the warm side of the storm, and for track *B*, Chicago is on the storm's cold side.

storm west of Chicago and track *B* takes it east of Chicago. The storm track thus has a major impact on Chicago's weather, as summarized briefly in Table 11.2.

If the storm takes track *A*, Chicago residents experience the warm side of the storm. Steady rain (or perhaps snow, briefly, at the onset) tapers off to drizzle and fog after 12 to 24 hours. As the warm front passes, skies clear partially, and winds freshen from the south, advecting warm and humid (mT) air at the surface. Clearing is short lived, however, as scattered showers and thunderstorms herald the arrival of colder air. As the cold front passes, wind direction **veers** (turns clockwise), blowing first from the southwest, then west, and finally northwest. Skies clear again, and the air temperature falls.

TABLE 11.2

Sequence of Surface Weather Conditions in Chicago as Mature Winter Storm Tracks West (Track *A*) and East (Track *B*) of City

WF = Warm front
CF = Cold front
F = Pressure falling
R = Pressure rising

	TRACK *A*					
Wind direction	E	SE	S	SW	W	NW
Frontal passage	—	WF	—	CF	—	—
Advection	—	Warm	Warm	Cold	Cold	Cold
Air pressure tendency	F	F	F	R	R	R
	TIME →					

	TRACK *B*			
Wind direction	E	NE	N	NW
Frontal passage	—	—	—	—
Advection	—	—	Cold	Cold
Air pressure tendency	F	F	R	R
	TIME →			

If the storm takes track *B*, Chicago residents experience the cold side of the storm and no frontal passages. Gusty east and northeast winds drive steady snow or rain (depending on how cold the air is) for 12 hours or longer. Then winds **back** (turn counterclockwise), blowing from the north. Precipitation tapers off to snow flurries or showers, and temperatures begin to drop. Finally, winds shift into the northwest, skies clear, and temperatures continue to drop.

The specific path taken by a cyclone (track *A* or track *B* in the example) depends on the direction of the upper-level westerlies in which the storm is embedded. Figure 11.9 shows the principal storm tracks across 48 states of the United States. All storms tend to converge toward the northeast; their ultimate destination is the Icelandic low of the North Atlantic, or western Europe. Although many storm tracks appear to begin just east of the Rocky Mountains, in reality they originate over the Pacific Ocean. As they travel through the mountains, cyclones temporarily lose their identity, but redevelop on the Great Plains. (What acually happens to these cyclones is explained in more detail in the Special Topic, The Case of the Missing Storms.)

The notion that storms in middle latitudes generally move from west to east was first suggested in 1703 by Daniel Defoe, the English journalist and novelist. Defoe drew this conclusion from his study of a great storm that lashed the British Isles in late November of that year and reports he received that several days earlier the same storm had ravaged the east coast of North America. In the United

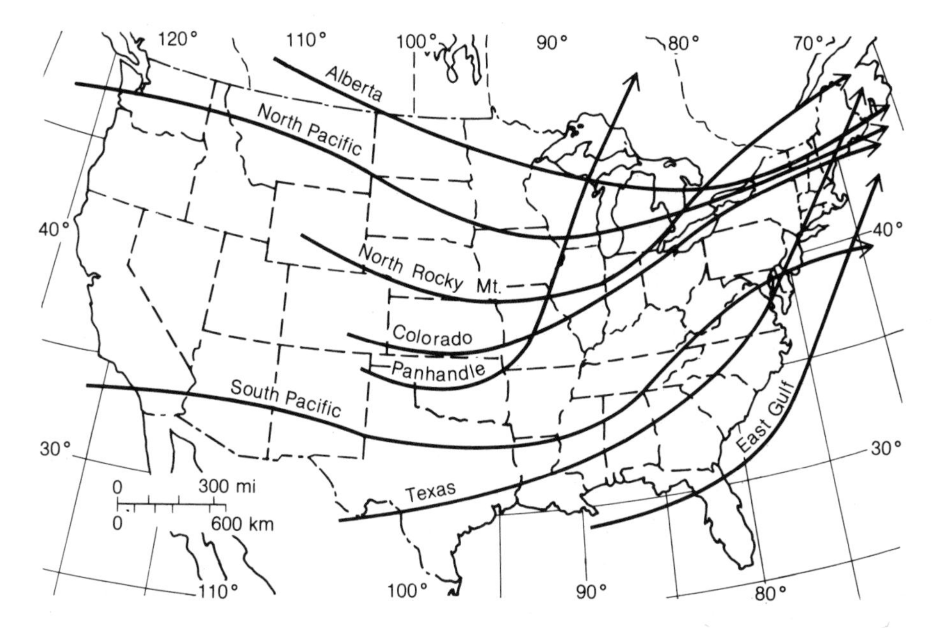

FIGURE 11.9

Principal storm tracks across North America. (From Trewartha, G.T. *An Introduction to Climate*. New York: McGraw-Hill, 1968, p. 219.)

States, Benjamin Franklin is generally credited with discovering that storms usually move in an easterly or northeasterly direction. Franklin, a resident of Philadelphia, reached this conclusion in 1743 from correspondence with his brother who lived in Boston, 500 km (300 mi) to the northeast. Franklin learned that cloudiness associated with a coastal storm first reached Philadelphia and many hours later, according to his brother, reached Boston.

It may be confusing to think of a storm as moving *toward* the northeast while the winds in the northeast portion of the storm blow *from* the northeast. In fact, New Englanders long assumed that the powerful "nor'easters" that pounded their shores moved down the coast from the northeast because the storm winds were northeasterly. Actually, "nor'easters" develop along the South Atlantic coast and then travel up the coast. In effect, there are two motions: the movement of the storm center up the coast and the counterclockwise circulation of the winds around the storm center. The circulation of a storm is quite independent of the storm's path much as the spin of a Frisbee is independent of its trajectory.

As a general rule, storms that form in the South yield more precipitation than those that develop in the North, because southern storms are closer to the prime source of moisture and energy, maritime tropical air. For example, Alberta track cyclones typically yield only light amounts of rain or snow, but East Gulf track storms usually bring heavy accumulations of rain or snow.

SPECIAL TOPIC

The case of the missing storms

After cyclones sweep ashore along the West Coast of North America, they seem to disappear as they move inland over the mountainous West. Later, they redevelop east of the front range, typically on the plains of Alberta or eastern Colorado. What actually happens to these storms as they track through the mountains?

Visualize a storm as it moves ashore—a huge cylinder of air spinning about a vertical axis in a cyclonic direction (Figure 1). As the cylinder moves up the windward slope of a mountain range, it shortens and widens, so the column of air converges vertically and diverges horizontally. As these changes take place, the spin of the cylinder slows so that, in effect, the storm's circulation weakens. As the cylinder of air then descends the leeward slope, the air column diverges vertically (stretches) and converges horizontally, strengthening the cyclonic spin. This weakening of cyclonic circulation upslope and the strengthening of the circulation downslope account for the seeming disappearance and reappearance of a storm as it crosses mountainous terrain.

The situation is analogous to that of an ice skater performing a pirouette. The skater changes the rate of spin by extending or drawing in the arms. When arms are extended (analogous to horizontal divergence), the spin rate slows. When arms are brought as close as possible to the skater's body (analogous to horizontal convergence), the spin rate increases. In effect, the skater with arms extended and the storm passing through mountainous terrain both conserve angular momentum. Simply put, the conservation of angular momentum means that a change in the radius of a rotating mass is balanced by a change in its rotational speed. An increase in radius (horizontal divergence) is accompanied by a reduction in rotational speed, and a decrease in radius (horizontal convergence) is accompanied by an increase in rotational speed.

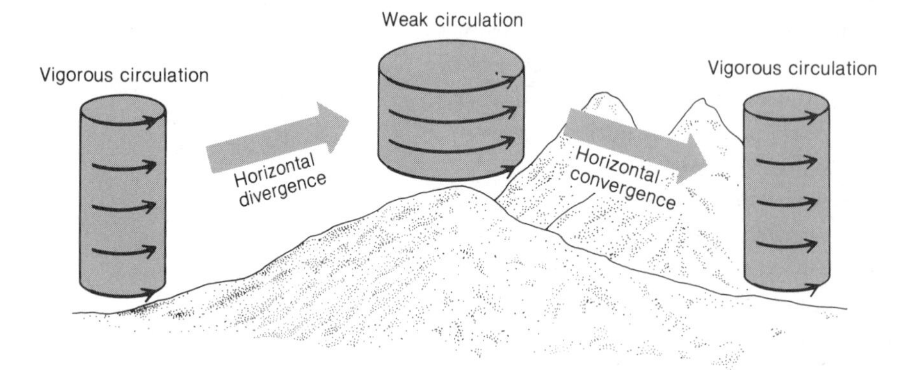

FIGURE 1

The cyclonic spin (circulation) of an air cylinder slows with horizontal divergence and speeds up with horizontal convergence. This depicts what happens to a cyclone as it travels through mountainous terrain.

Eventually, usually after 3 to 5 days, a storm completes its life cycle by occluding. An **occlusion** occurs when the faster moving surface cold front catches up with the surface warm front—forcing the warm air aloft and cutting off the supply of warm, humid air to the storm center (Figure 11.5D). The merging of the cold front and the warm front forms an **occluded front.** Skies remain cloud covered, but the steady precipitation gives way to drizzle, and the low begins to "fill" as the central pressure rises. At occlusion, the upper-level trough is directly over the surface cyclone. Because the storm is no longer embedded in the main westerly flow aloft, it stalls.

The occlusion just described takes place when the air behind the advancing cold front is colder than the cool air ahead of the warm front. The advecting cold air slides under and lifts both the warm air and the cool air. A second and less common type of occlusion occurs when the air behind the advancing cold front is not as cold as the air ahead of the warm front. This is the situation, for example, when an oceanic cyclone occludes along the Pacific Northwest coast. In this case, the air behind the cold front is relatively mild—having traversed ocean waters—whereas the air ahead of the warm front is relatively cold because it has traveled over land. With the storm's occlusion, the advecting mild air slides under the warm air but overrides the cold air.

This idealized description of the life cycle of a cyclone does not apply to all synoptic-scale disturbances that affect our weather. Sometimes storms develop with meager upper-air support and are therefore weak and poorly defined. At other times, deteriorating weather is linked to an upper-air or surface trough, and not to a closed cyclonic circulation. Nevertheless, our cyclone weather model provides a useful approximation of the sequence of weather events that accompanies midlatitude synoptic-scale storms.

The basic features of this cyclone model were first formulated during World War I by researchers at the Norwegian School of Meteorology in Bergen. The model is therefore referred to as the **Norwegian cyclone model**. Major advances in weather monitoring, especially those involving remote sensing by satellite, have verified the Norwegian model and it remains a remarkably close approximation of our current understanding of midlatitude cyclones.

From earlier discussions of the relationships between winds aloft and migratory synoptic-scale cyclones, we can deduce that storms should exhibit seasonal variability. In summer, when the mean positions of the polar front and jet stream lie across southern Canada, very few well-organized cyclones occur in the United States, and the Alberta storm track shifts northward across central Canada. In winter, however, when the mean positions of the polar front and jet stream shift southward, vigorous cyclogenesis is more frequent in the United States. In general, Alberta track storms are the most common, since they occur year-round, whereas storms with other tracks develop primarily in winter.

Cold and warm cores

Mature and occluding cyclones are cold-core systems, which means that they occupy relatively cold columns of air. In vertical cross section, pressure surfaces within **cold-core lows** are concave upward (Figure 11.10), and the circulation strengthens with altitude.

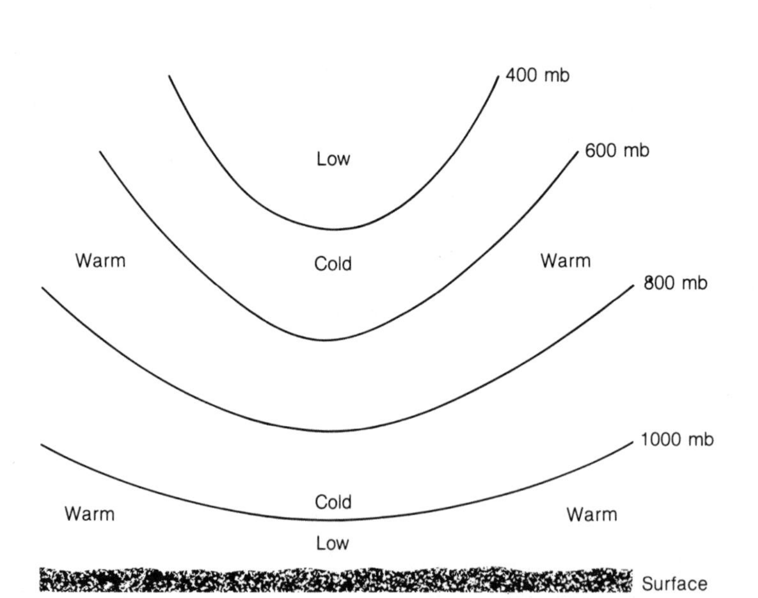

FIGURE 11.10

A vertical cross section shows sloping isobaric (constant pressure) surfaces within a cold-core low pressure system.

On the other hand, the vertical axis of a mature cyclone slopes toward the upper-air cold trough (Figure 11.11), that is, the surface cyclone becomes a trough aloft. Recall, however, that when a storm occludes, the surface cyclone is situated directly under the trough.

A second type of low that sometimes appears on midlatitude surface weather maps has characteristics markedly different from those of cold-core lows. These cyclones of the second type are stationary, have no fronts, and are associated with fair weather.

They develop over arid or semiarid deserts, including the interior of the southwestern United States, when the hot summer sun heats the ground, which in turn heats the overlying air. Intense heating lowers the air density over an area wide enough for a synoptic-scale low to appear. This **warm-core low**, or **thermal low**, is very shallow, and its circulation weakens rapidly with altitude, that is, away from the heat source. In fact, the surface counterclockwise circulation frequently reverses at some altitude, and the thermal low is then overlain by an anticyclone (Figure 11.12).

The preceding sections discussed the general weather patterns caused by lows. In the daily march of weather across the mid-latitudes, lows are usually followed by highs. Let us now examine these fair-weather pressure systems.

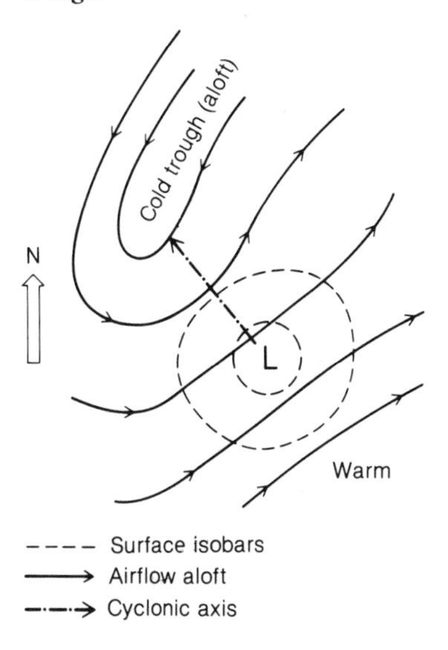

FIGURE 11.11

A surface cyclone slopes with altitude toward an upper-level cold trough.

Anticyclones

An anticyclone is the opposite of a cyclone. Cyclonic circulation favors the convergence of contrasting air masses and the development of fronts. In anticyclones, subsiding air and diverging surface airflow favor formation of a uniform air mass and fair skies. Like cyclones, however, anticyclones can have either cold or warm cores.

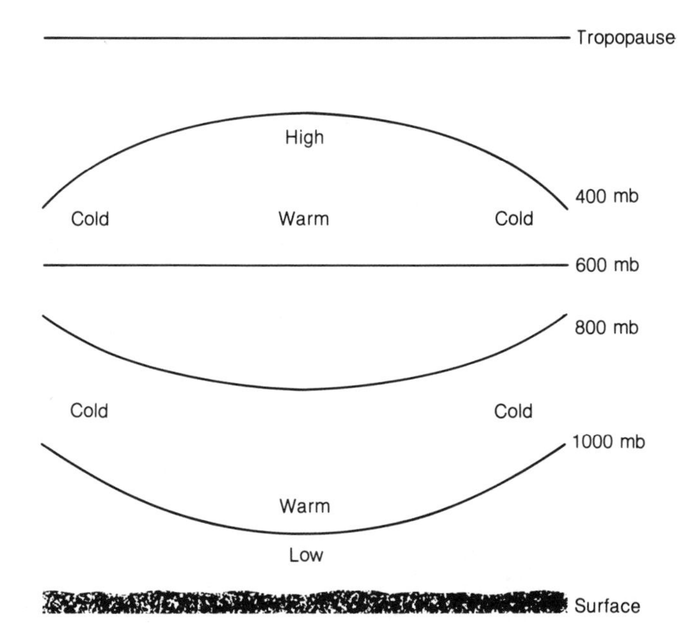

FIGURE 11.12

A vertical cross section shows sloping isobaric surfaces within a warm-core low pressure system. The shallow near-surface cyclonic circulation weakens and then reverses with altitude.

Cold and warm cores

Cold-core anticyclones coincide with domes of cP or A air and, depending on the specific air mass, are labeled **polar highs** or **arctic highs**. They are the product of extreme radiational cooling over the continental interiors of North America north of the polar front. Cold anticyclones are shallow systems in which the clockwise circulation weakens with altitude and often reverses. A cold trough situated over a cold anticyclone is therefore not unusual.

Like cP and A air, cold anticyclones are most intense (that is, they exhibit their highest surface temperatures) in winter. Although arctic or polar anticyclones with very high central pressures tend to remain stationary over their source regions in winter, lobes of cold air (smaller cold highs) often break out of the source region and slide southeasterly across Canada and into the United States. These cold air surges are then pulled into the circulation of migrating cyclones. This explains why winter storms are usually followed by clearing skies and colder temperatures.

On rare occasions in the winter, a particularly strong arctic high brings a surge of bitter cold air that sweeps as far south as southern Florida. The resulting subfreezing temperatures can spell disaster for citrus growers. As an illustration, the synoptic weather pattern that caused the costly Florida freeze of December 1983 is shown in Figure 11.13.

FIGURE 11.13 (facing page)
Satellite view of much of North America on 24 December 1983 and the corresponding pattern of surface air pressure. The heavy solid lines are isobars labeled in millibars. An unusually extensive arctic high pressure system is centered over eastern Montana where surface pressures exceed 1060 mb. The frigid air mass associated with the high covers almost the entire United States except the Southwest. Early that morning, temperatures plunged to record low levels over the Plains, Midwest, and Deep South. Subfreezing temperatures in the citrus and vegetable growing areas of south Texas and south Florida caused more than a billion dollars of crop damage. In the satellite photograph, much of the white in the northern United States is snow cover rather than clouds. The bright white over the Gulf is cloudiness caused by cold air streaming over the relatively warm ocean water. (Dashed line off East Coast is a trough.) (Photograph from NOAA/Satellite Data Services; map from NOAA).

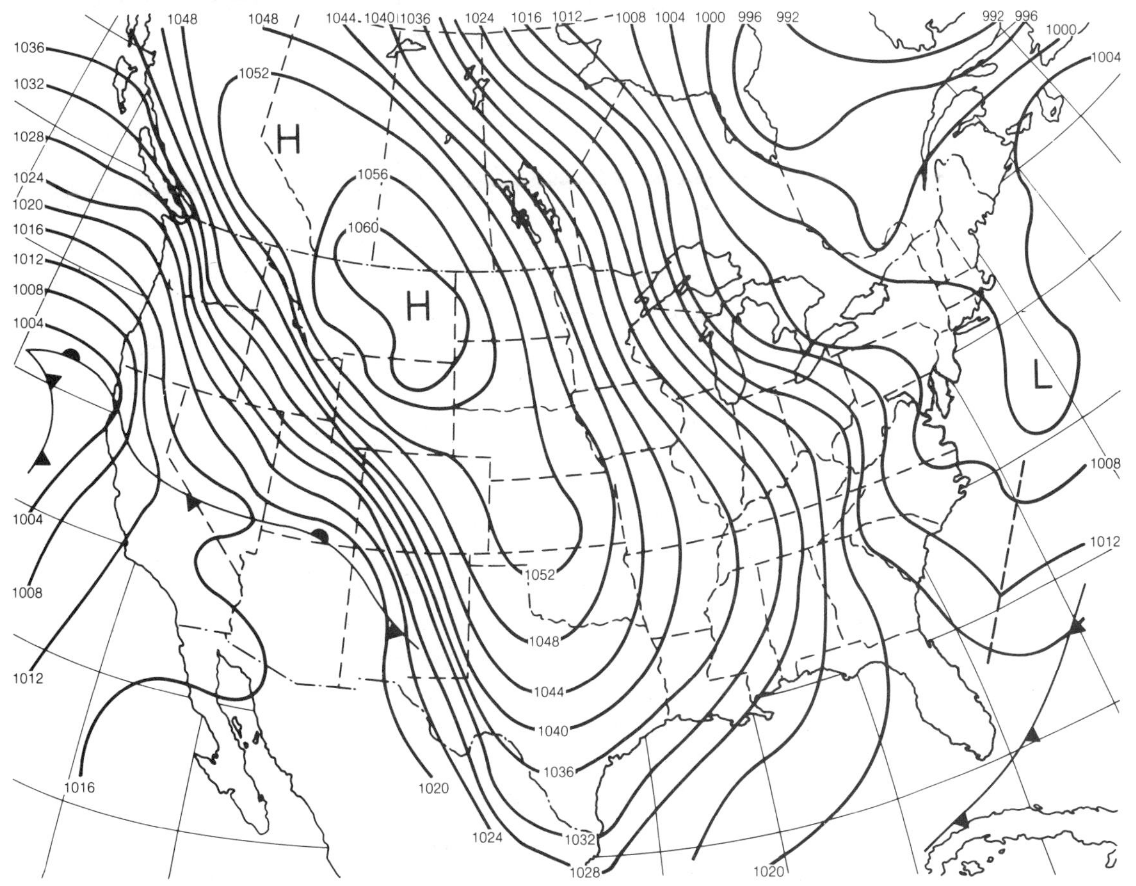
1048
1048
1044
1040
1036
1024
1016
1012
1008
1004
1000
996
992
992
996
1000
1004
1036
1032
1028
1024
1020
1016
1012
1008
1004
1052
H
1056
1060
H
L
1008
1012
1004
1008
1012
1052
1048
1044
1040
1036
1032
1028
1024
1020
1016
1020

Warm-core anticyclones form south of the polar front and consist of extensive areas of subsiding warm, dry air. They are massive systems with circulation extending from the earth's surface to the tropopause. Semipermanent subtropical anticyclones are examples of warm-core highs, but other warm-core anticyclones may occur over the interior of the United States in summer. Cold-core anticyclones eventually become warm-core systems. As a cold-core anticyclone traverses the land, the air mass moderates considerably. By the time the high leaves the continent over the extreme southeast, it has merged with the subtropical high as a warm-core system.

A cold anticyclone produces high surface air pressures because cold air is relatively dense, but how does a warm anticyclone produce high surface pressures? After all, in equal volumes, warm air is lighter than cold air. The high surface pressure of warm-core anticyclones, such as the subtropical highs, is attributed to the system's circulation. Horizontal convergence (inflow) of air aloft exceeds horizontal divergence of air (outflow) at the surface, thereby maintaining a relatively high surface air pressure.

Anticyclonic weather

As noted in Chapter 9, an anticyclone is a fair-weather system. This is because surface winds blowing in a clockwise and outward pattern* induce subsidence of air over a wide area under a high pressure system. Subsiding air is compressionally warmed, so relative humidities drop and clouds vaporize or fail to develop. Also, as noted earlier, the horizontal pressure gradient is weak over a wide region near the center of an anticyclone, so prevailing winds are very light or calm. At night, clear skies and light winds favor rapid radiational cooling and, in some instances, the development of dew, frost, or fog (Chapter 7).

The horizontal pressure gradient strengthens away from the central region of an anticyclone and so do the winds. With strengthening winds, air mass advection then occurs. Typically, well to the east of the high center, northwest winds advect cold air southward, while to the west of the high center, south and southeasterly winds advect warm air northward. This air mass advection brings about significant changes in air temperature in winter when the contrasts between air masses are greatest.

An understanding of the basic circulation characteristics of an

*In the Northern Hemisphere, this is the pattern of surface winds. In the Southern Hemisphere, surface winds in a high blow counterclockwise and outward.

anticyclone helps us predict the sequence of weather events as an anticyclone travels into and out of a midlatitude location. Consider what happens in winter as a cold anticyclone moves southeastward out of southern Canada and into the northeastern United States. Ahead of the high, strong northwest winds bring a surge of cold continental polar or arctic air. Strong winds and falling temperatures mean low wind chill equivalent temperatures and more work for furnaces. To the lee of lakes Erie and Ontario, heavy lake effect snow showers break out. Even hundreds of kilometers downwind of the lakes, instability showers* bring scattered light snow falls. However, as the high approaches an area, winds slacken, skies clear, and nighttime radiational cooling brings the lowest temperatures associated with the pressure system. Then, as the anticyclone center drifts away toward the southeast, winds again strengthen, but this time from the south, and warm air advection begins. The first sign of warm air advection is the appearance of high, thin cirrus clouds in the western sky. The entire sequence just described may take several days to a week, depending on the anticyclone's forward speed.

In summer, a Canadian high pressure system causes the same pattern of air mass advection as in winter except that the temperature contrast between air masses is considerably less. Air advected ahead of the high by northwesterly winds is not likely to be much cooler than the air advected behind the high by southwesterly winds. The most noticeable difference between the advecting air masses often is the contrast in the water vapor content. Air advected ahead of the high is often less humid, and therefore more comfortable, than the air advected behind the high.

The pattern of air mass advection associated with an anticyclone also applies to a ridge. Cold air advection occurs ahead (to the east) of a ridge, and warm air advection occurs behind (to the west) of a ridge.

The circulation pattern of an anticyclone (or ridge) does not occur in isolation from that of a cyclone (or trough). The atmosphere is after all a continuous fluid, and anticyclones follow cyclones and cyclones follow anticyclones. In our illustration, therefore, northwest winds develop ahead of the high and on the back (west) side of a retreating low. The southerly airflow behind the high develops to the east of a low, which is approaching from the west. In both cases, winds are caused by pressure gradients set up between traveling anticyclones and cyclones.

*As the cold air travels over the warmer ground, the air is destabilized, giving rise to convective showers.

Conclusions

We have now examined the features of the principal weathermakers of the midlatitudes: air masses, fronts, cyclones, and anticyclones. We have seen how these synoptic-scale systems are linked to the planetary-scale circulation. Chapter 12 analyzes some special circulation patterns, most of which operate at the mesoscale.

SUMMARY STATEMENTS

- Air masses are classified on the basis of source region characteristics. As an air mass flows from one place to another, it is modified. The extent of modification depends on its stability and the nature of the surface over which it travels.
- Weather along a cold front or ahead of a cold front usually consists of a narrow band of clouds and brief rain or snow showers. Weather associated with warm fronts typically consists of a broad cloud and precipitation shield that extends hundreds of kilometers ahead of the surface warm front.
- As a midlatitude cyclone progresses through its life cycle, it is transported by upper-level winds toward the northeast. The storm typically begins as a wave along the polar front, and enters maturity as surface pressures continue to drop, winds strengthen, and frontal weather develops. The storm finally occludes as the cold front catches up with the warm front. The track followed by a cyclone is critical to the type of weather experienced at a given locality.
- Thermal lows are stationary, have no fronts, and are associated with very hot, dry weather.
- Cold-core anticyclones are shallow systems that coincide with domes of continental polar or continental arctic air. Warm-core anticyclones, such as the subtropical highs, extend deep into the troposphere and are accompanied by wide areas of subsiding, warm, dry air on their eastern flanks.

KEY WORDS

air mass
continental tropical (cT) air
maritime tropical (mT) air
maritime polar (mP) air
continental polar (cP) air
arctic (A) air
cold front
squall line
warm front
cyclone
cyclogenesis
wave cyclone
veer
back
occlusion
occluded front
Norwegian cyclone model
cold-core lows
warm-core low
thermal low
cold-core anticyclones
polar highs
arctic highs
warm-core anticyclones

REVIEW QUESTIONS

1 What determines the properties of an air mass?

2 Distinguish among the four basic types of air masses.

3 How and why do the properties of continental polar air change from winter to summer?

4 What causes air mass modification and how do air masses modify?

5 Explain why continental polar air modifies more rapidly than does maritime tropical air. What is the significance of this difference for cold waves and heat waves?

6 What types of clouds are associated with a cold front?

7 Describe the cloud sequence as a warm front approaches your locality.

8 Explain why the surface wind flow about a center of low pressure favors the development of fronts, whereas surface anticyclonic airflow does not.

9 Explain why the appearance of a halo about the sun or moon may signal the approach of stormy weather.

10 Describe the stages in the life cycle of a midlatitude cyclone.

11 A mature midlatitude cyclone features a warm side and a cold side. Please explain.

12 A mature midlatitude cyclone tracks northeastward along the East Coast and passes out to sea to the east of Boston. Describe the wind shifts and changes in air mass advection at Boston.

13 Distinguish between winds that back and winds that veer.

14 An Alberta track storm usually produces much less precipitation than a storm that tracks out of eastern Colorado. Why?

15 Describe what happens when a cyclone occludes.

16 Explain why the site of principal cyclone activity shifts from the United States into Canada from winter to summer.

17 Distinguish between a warm-core high and a cold-core high.

18 Distinguish between a warm-core low and a cold-core low.

19 Why is frontal weather not generated by a thermal low?

20 A warm anticyclone produces relatively high surface air pressures and yet it is composed of a column of warm (light) air. Explain.

POINTS TO PONDER

1 Speculate on the lowest air temperatures likely to be associated with maritime polar air in its source region.

2 Describe how Pacific air modifies as it travels from west to east across the United States.

3 Why is the slope of a cold front steeper than that of a warm front?

4 In summer we may experience passage of a cold front, and yet the air ahead of the cold front has about the same afternoon temperatures as the air behind the cold front. Explain.

5 Why is air mass stability an important factor in the type of weather that occurs along a front?

6 Under what circumstances will a frontal passage not be accompanied by cloudiness?

7 List the conditions at the surface and aloft required for cyclogenesis.

PROJECTS

1 Determine the tracks of winter cyclones that influence the weather of your locality. Are you usually on the warm side or cold side of these storms?

2 As a major winter storm approaches your area, keep an hourly log of changes in (a) cloud cover, (b) cloud type, (c) temperature, (d) wind speed, (e) wind direction, (f) air pressure, and (g) precipitation.

SELECTED READINGS

Eagleman, J. R. "Conceptual Models of Frontal Cyclones." *Journal of Geography* 80 (1981):87–91. Distinguishes among three basic types of frontal cyclones.

Namias, J. "The History of Polar Front and Air Mass Concepts in the United States—An Eyewitness Account." *Bulletin of the American Meteorological Society* 64 (1983):734–755. *Provides a historical perspective on a key concept in understanding midlatitude weather.*

Salmon, E. M., and P. J. Smith. "A Synoptic Analysis of the 25–26 January 1978 Blizzard Cyclone in the Central United States." *Bulletin of the American Meteorological Society* 61 (1980):453–460. *Detailed analysis of a severe winter storm.*

Let me snuff thee up, sea
breeze!
and whinny in thy spray.
HERMAN MELVILLE
White Jacket

Along a coastline such as this, we can often anticipate a refreshingly cool sea breeze on a summer afternoon. This is just one of a wide variety of local and regional winds. (Photograph by Joseph M. Moran)

TWELVE

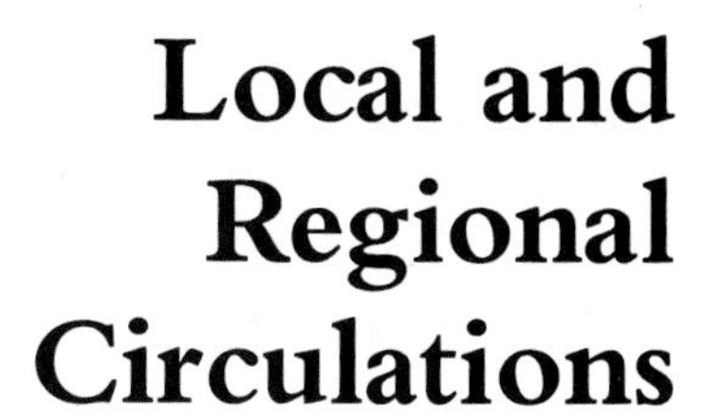

Local and Regional Circulations

Monsoon Circulation

Land and Sea (or Lake) Breezes

Heat Island Circulation

Katabatic Winds

Chinook Winds

Desert Winds

Mountain and Valley Breezes

IN THIS chapter, we describe several specialized types of atmospheric circulation systems. Although these systems cover varying scales of size and persist for varying periods, each is strongly influenced by topography or other features of the earth's surface. We also discuss wind power in the Special Topic for this chapter.

Monsoon Circulation

A **monsoon circulation** characterizes regions where seasonal reversals in winds cause wet summers and dry winters. Although a weak monsoon develops over central and eastern North America, we are concerned here with the much more vigorous monsoon circulation over Africa and Asia, where 2 billion people depend on monsoon rains for their drinking water and agriculture.

What causes monsoons? A complete explanation is not yet available. Monsoons are certainly linked to seasonal shifts in global-scale circulation features, specifically, the north-south shifts of the intertropical convergence zone (ITCZ). However, as proposed in 1686 by Edmund Halley, monsoons also depend on seasonal differences in the heating of the land and sea. As we have seen, land warms up more than the sea in response to the same insolation. Beginning in spring, relatively cool air over the ocean and relatively warm air over the land gives rise to a horizontal air pressure gradient directed from sea to land,* initiating a flow of humid air inland. Over the land, humid air is heated and rises, and consequent adiabatic expansional cooling leads to condensation, clouds, and rain. The release of latent heat intensifies the buoyant uplift—triggering even more rainfall. Aloft, the air spreads seaward and subsides over the relatively cool ocean surface, thus completing the monsoon circulation. Beginning in autumn, radiational cooling chills the land more than the adjacent sea, which sets up a horizontal air pressure gradient directed from land to sea. Air subsides over the land, and dry surface winds sweep seaward. Air rises over the relatively warm sea and, aloft, drifts landward—completing the winter monsoon circulation. The summer monsoon is therefore wet, and the winter monsoon is dry.

Monsoon winds cover sufficiently long trajectories and persist long enough to be influenced by the Coriolis effect. As shown in

*Recall that air pressure gradients are always directed from areas of high pressure toward areas of low pressure.

Figure 12.1, January and July surface monsoon winds are deflected to the right in the Northern Hemisphere and to the left in the Southern Hemisphere.

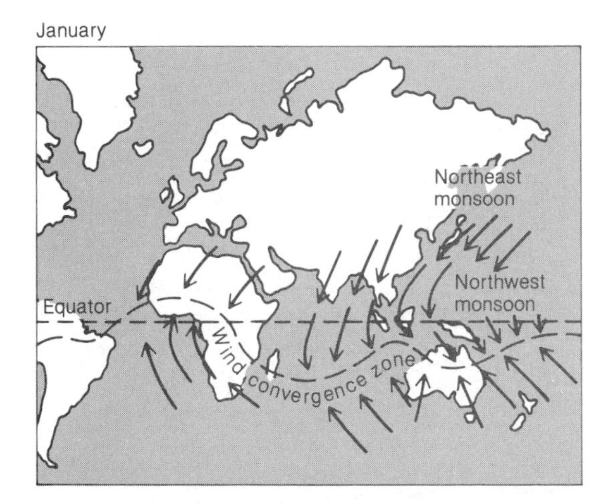

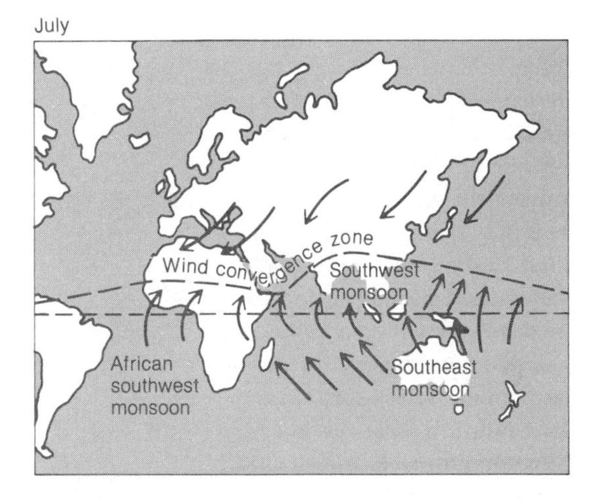

FIGURE 12.1

Surface air streams during monsoons of January and July are deflected to the right in the Northern Hemisphere and to the left in the Southern Hemisphere. (After Webster, P. J. "Monsoons." *Scientific American* 245, no. 2 (1981):112. Copyright © 1981 by Scientific American, Inc. All rights reserved.)

Monsoon rains usually commence abruptly a few weeks before the summer solstice and cease just after the equinox, but rainfall is not continuous. On the contrary, the rainy season typically consists of a sequence of active and dormant phases. During **monsoon active phases**, the weather is cloudy with frequent deluges of rain, but during **monsoon dormant phases**, the weather is sunny and hot. The shift between active and dormant phases reflects the tendency of heavy rainfall to surge inland. Heavy rains first strike coastal areas and soak the ground. Because the ground becomes moist, more of the available solar radiation is used for evaporation and less for sensible heating. Coastal areas cool as a consequence, and eventually the uplift weakens, and skies clear partially. Meanwhile, the area of maximum heating, vigorous uplift, and heavy rain shifts inland. Back in coastal areas, however, the hot sun eventually dries the soil, sensible heating intensifies, uplift strengthens, and the rains resume (monsoon active phase). This sequence of active and dormant phases is repeated about every 15 to 20 days in summer.

Insolation, land and water distribution, and topography thus impose some regularity on the monsoon circulation, that is, summers are wet and winters are dry. The global-scale circulation (especially shifts of the ITCZ) and the strength and distribution of convective activity, however, vary from year to year. These variations mean that monsoon rains change in intensity and duration from one year to the next.

SPECIAL TOPIC

Wind power

Today scientists are using the methods and materials of space age technology to design and construct giant windmills to convert the wind's kinetic energy into electrical energy. The Wind Energy Systems Act of 1980 allocated $900 million for the development of a cost-effective wind power system in the United States. Large, modern windmills, like the one shown in Figure 1, are capable of generating 200 kilowatts of electricity. If such machines were to sustain this level of power production continuously, each would provide enough power for nearly 100 average American homes. By the year 2000, however, an estimated 30,000 large wind-driven

FIGURE 1

This 200-kilowatt wind turbine generator supplies part of the electric power used by residents of Clayton, New Mexico. (Photograph courtesy of the U.S. Department of Energy)

turbines and thousands of small windmills would be required to meet one tenth of our electrical needs.

In Chapter 4, we saw that the sun drives the atmosphere. Actually, only about 2 percent of the solar radiational energy that reaches the earth is ultimately converted to the kinetic energy of wind. This is nevertheless a tremendous quantity of energy. Theoretically, windmill blades can convert 60 percent of the wind's energy into mechanical energy. Typically, however, wind-generating systems extract only about 35 percent of the wind's energy, and average wind speeds must be at least 19 km (12 mi) per hour before most wind-powered electrical generating systems can operate economically.

The power that a windmill can extract from the wind is directly proportional to the following: air density, the area swept out by the windmill blade, and the cube of the instantaneous wind speed (V^3). Wind speed is by far the most important consideration in evaluating wind energy potential at any locality, for even small changes in wind speed translate into important changes in energy generation. For example, doubling the wind speed multiplies the available wind power by a factor of eight. Unfortunately, both wind speed and direction vary continuously with time, and wind speed also varies with exposure, roughness of the terrain, elevation above the earth's surface, and the season. As a general rule, a minimum of several years of detailed wind monitoring is needed for a preliminary evaluation of wind power potential at any site. The long-term climatic record should also be consulted to check for the frequency of destructive winds. Wind data from nearby weather observing stations can be very useful, as long as care is taken in extrapolating wind data from one locality to another and from one elevation to another.

The most formidable obstacles to the development of wind power potential stem from the inherent variability of the wind. The electrical output of wind turbines varies in consequence, and a wind power system must include a means of storing the energy generated during gusty periods for use during lulls. Several thousand dollars' worth of storage batteries are required to provide a reasonable storage capacity in a wind power system for a typical home. A 3 to 5 kilowatt wind turbine is needed to meet the total electrical requirements, including heating, of a typical household. Today, such systems are commercially available, but the cost of materials and construction—including a tower, storage batteries, and generator—ranges from about $5000 to $20,000.

Economy of scale suggests that centralized "wind farms" serving entire communities are preferable to individual household wind turbines. These "wind farms" would consist of 50 or more super wind power generators, each capable of producing one or more megawatts of electricity. Indeed, multimegawatt wind systems are currently under development. In late 1982, the Pacific Gas and Electric Company of California announced plans to construct an array of 36 wind turbines in Solano County northeast of San Francisco. Each turbine will generate 3500 kilowatts of power.

A wind power expert, William Heronemus, has proposed that rows of windmills be anchored offshore to tap the energy of strong oceanic winds

(Figure 2), but the idea may not be practical because of the high cost of underwater cables (needed to carry electricity to shore) and the cost of huge support platforms. Electricity generated by "wind farms" would be fed through existing electrical distribution systems to avoid high storage costs.

Given current technical and economic limitations, wind power has its greatest immediate potential in those regions where winds are consistent in direction and average a relatively high speed. In North America, such regions include the western High Plains, the Pacific northwest coast, the eastern Great Lakes, the south coast of Texas, and the exposed summits and passes in the Rockies and Appalachians. Low-powered wind systems have great potential in small, isolated communities and on individual farms and ranches. The impact of wind systems on the environment is minimal. They may be somewhat noisy, detract some from the beauty of the landscape, and kill some birds.

FIGURE 2

An artist's conception of a proposed multi-unit, offshore wind-power system. (Westinghouse photograph provided by U.S. Department of Energy)

With all of the interest in wind power today, it is ironic to note that in the 1930s and 1940s, the Rural Electrification Administration eliminated about 50,000 wind-powered pumping and electric-generating systems on midwestern farms.

Land and Sea (or Lake) Breezes

For those lucky enough to live near the ocean or a large lake, sea or lake breezes bring welcome respite from the oppressive heat of a summer afternoon. On warm days, a weak regional (synoptic-scale) pressure gradient allows a cool wind to sweep inland from the sea or from a large lake. Depending on the source, this refreshing wind is called either a **sea breeze** or a **lake breeze**. Both breezes are caused by differential heating between land and water.

When both land and water are exposed to the same intensity of solar radiation, the land warms more than the water. The relatively warm land heats the overlying air, thereby lowering air density. Compared with the land, the water remains relatively cool, as does the air overlying the water. Consequently, as shown in Figure 12.2A, a local horizontal air pressure gradient develops between land and water, with the highest pressure near the surface over the water. In response to this gradient, cool air sweeps inland—in some cases tens of kilometers, and in other cases only a few hundred meters. Continuity then requires that there be a return airflow aloft from the land to the water body, with air sinking over the water and rising over the land.

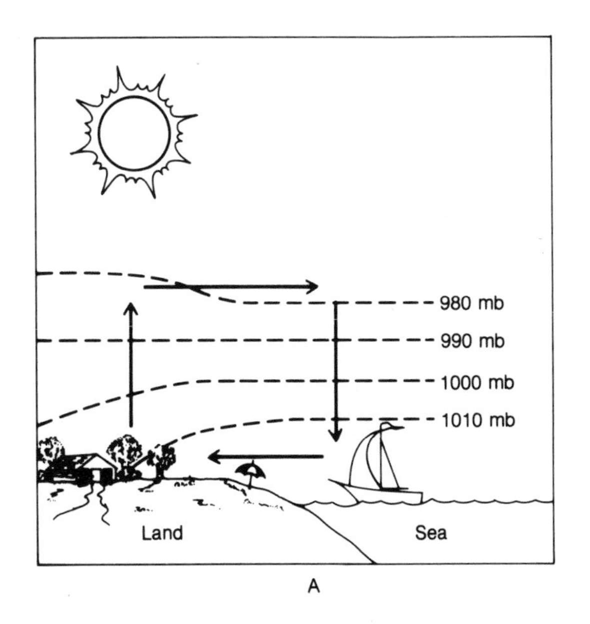

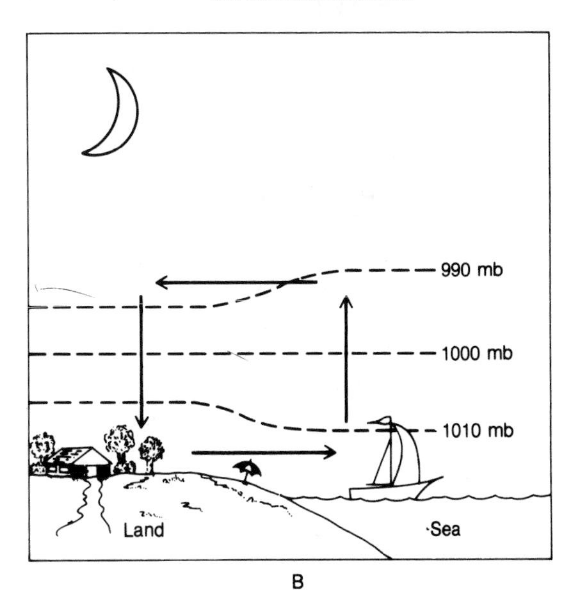

FIGURE 12.2

Vertical cross section of a sea (lake) breeze (A) and vertical cross section of a land breeze (B). Arrows indicate wind direction and dashed lines are isobaric surfaces.

At night, a weak regional pressure gradient causes sea and lake breezes to shift to **land breezes**. The change in circulation direction is caused by a reversal in heat differential between land and water. At night, radiational cooling chills the land (and the air over the land) more than the water (and the air over the water). The land thus becomes cool relative to the water surface. Horizontal density differences in air density give rise to a horizontal pressure gradient directed from land to sea or lake. A cool offshore breeze develops, along with a return flow aloft, and air subsides over the land and rises over the water (Figure 12.2B). Land and sea or lake breezes typically develop and diminish so rapidly that they are not significantly influenced by the rotational effect of the earth.

When the surface of Lake Michigan is cooler than the adjacent land (May through August), a shallow mesoscale high pressure system may develop over the lake on days when synoptic-scale winds are weak. As shown in Figure 12.3, surface winds diverge from the high center toward the shoreline. The leading edge of this surge of cool air, the lake breeze, forms a miniature cold front, which, by midafternoon, may move inland many kilometers. As with all lake, sea, and land breezes, this circulation is vulnerable to changes in synoptic-scale pressure gradients. Steepening pressure gradients mean stronger regional winds that will overwhelm and dominate local circulations.

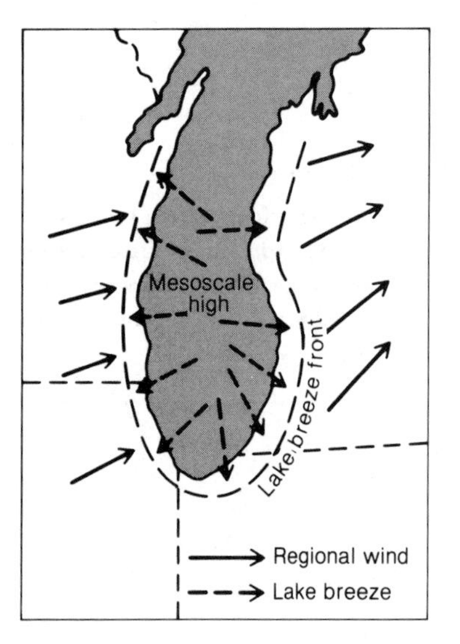

FIGURE 12.3

When synoptic-scale (gradient) winds are weak and the Lake Michigan waters are cooler than the adjacent land, a mesoscale high develops over the lake. Outward surface airflow gives rise to lake breezes during the day. (From Lyons, W.A. "Some Effects of Lake Michigan Upon Squall Lines and Summertime Convection." ***Proceedings, Ninth Conference on Great Lakes Research.*** **Ann Arbor: The University of Michigan, 1966, p. 262.)**

Heat Island Circulation

Travelers approaching large cities are often greeted by an unsightly veil of dust, smoke, and haze hanging over the city. This dome of air pollutants is the result of a convective circulation of air that concentrates pollutants over the city. The circulation is in turn related to a temperature contrast between the city and the surrounding rural areas.

The average annual temperature of a city is typically slightly warmer than that of the surrounding countryside, although on some days the thermal contrast may be as great as 10 °C (18 °F) or more. Snow consequently melts faster and flowers bloom earlier in a city. This climatic effect is known as the **urban heat island**. Figure 12.4 illustrates the urban heat island of Washington, D.C., and Plate 17 is a satellite view of heat islands in southeastern New England. One of the factors that contributes to the development of an urban heat island is the relatively high concentration of heat sources (for example, people, cars, industry, air conditioners, and furnaces) in cities. Because all of the heat from every source even-

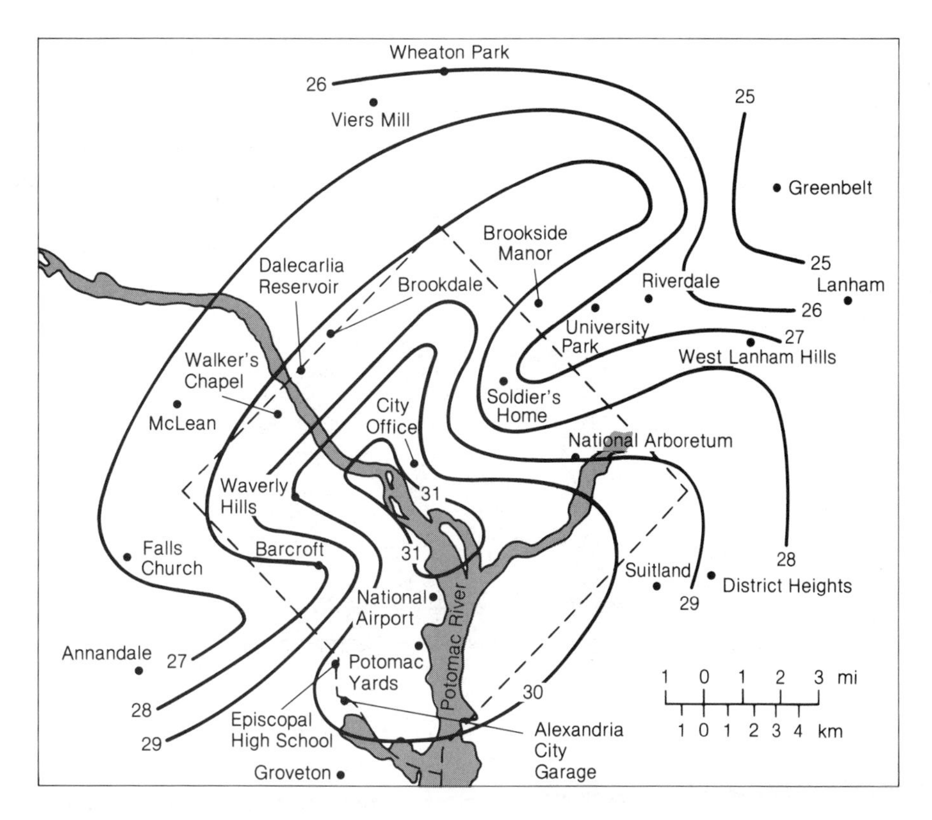

FIGURE 12.4

Average winter low temperatures (in °F) in Washington, D.C., illustrate the urban heat island effect. (From Woolum, C.A. "Notes From a Study of the Microclimatology of the Washington, D.C. Area for the Winter and Spring Seasons." *Weatherwise* 17, no. 6 (1964):6.)

tually reaches the atmosphere, the air of a large city receives a considerable input of waste heat. In fact, on a very cold winter day in New York City, heat from urban sources may approach 100 watts per m^2—about 8 percent of the solar constant.

The daytime absorption of solar radiation and the emission of heat into city air is also facilitated by the thermal properties of urban building materials. Concrete, asphalt, and brick conduct heat more readily than the soil and vegetative cover of rural areas. The heat loss at night by infrared radiation to the atmosphere and to space is thus partially compensated for in cities by a release of heat from the buildings and streets and by the obstruction of the sky by tall buildings. The temperature contrast between city and country is further accentuated by a city's typically low evapotranspiration rate. Urban drainage systems (sewers) quickly and efficiently remove most runoff from rain and snowmelt, so less of the available solar energy is used for evaporation (latent heating). More solar energy is therefore available to heat the ground and air directly (sensible heating).

When regional winds are weak, the relative warmth of a large city compared with its surroundings can promote a convective circulation of air, as shown in Figure 12.5. Warm air at the city's center rises and is replaced by cooler, denser air flowing in from the countryside. The rising columns of air gather aerosols into a **dust dome** over the city. In this way, dust may become a thousand times more concentrated over urban-industrial areas than in the air of the

FIGURE 12.5

Heat island of a large urban area may give rise to a convective circulation pattern that transports air pollutants to a dust dome over the city. (Modified after Lowry, W.P. "The Climate of Cities." *Scientific American* 217, no. 2 (1967):20. Copyright © 1967 by Scientific American, Inc. All rights reserved.)

rural countryside. If regional winds strengthen to more than about 15 km (10 mi) per hour, the dust dome elongates downwind in the form of a **dust plume**, and spreads the city's pollutants over the countryside. The Chicago dust plume, for example, is sometimes visible 240 km (150 mi) from its source.

Katabatic Winds

Under the influence of gravity, a shallow mass of cold, dense air slides downhill. This **katabatic wind** usually originates over extensive snow-covered plateaus or over other highlands in winter. Although adiabatic compression warms the air to some extent, the air is so cold to start with that these winds are still quite cold when they reach the lowlands.

Among the best known katabatic winds are the **mistral**, which descends from the snow-capped Alps down the Rhone River Valley of France and into the Gulf of Lyons along the Mediterranean coast, and the **bora**, which originates in the high plateau region of Yugoslavia and cascades onto the narrow Dalmatian coastal plain along the Adriatic Sea. Both the mistral and the bora are winter phenomena.

In some places, such as inlets of the coastal range of British Columbia, katabatic winds are channeled by mountain valleys, and this constricted flow sometimes accelerates the wind speeds to potentially destructive proportions. Steep slopes can also accelerate katabatic flow. Along the edge of the massive Greenland and antarctic ice sheets, for example, katabatic winds frequently exceed 100 km (62 mi) per hour.

Chinook Winds

Like the katabatic wind, the chinook is a downslope breeze. While the katabatic wind is cold and dry, however, the chinook wind is warm and dry.

Chinook winds develop when relatively mild air aloft is adiabatically compressed as it descends the leeward slopes of mountain ranges. For every 1000 m of descent, the air temperature rises about 10 °C (Chapter 6). Air that flows down the slopes of high mountain ranges such as the Rockies thus undergoes considerable warming.

Because of the warmth of the air, it does not flow downslope under the influence of gravity, as does a katabatic wind; rather, the air must be drawn downslope. In the United States and Canada, a chinook wind may develop when a cyclone or an anticyclone is situated on the Great Plains to the east of the Rocky Mountains (Figure 12.6). West winds associated with these pressure systems pull the air downslope for thousands of meters.

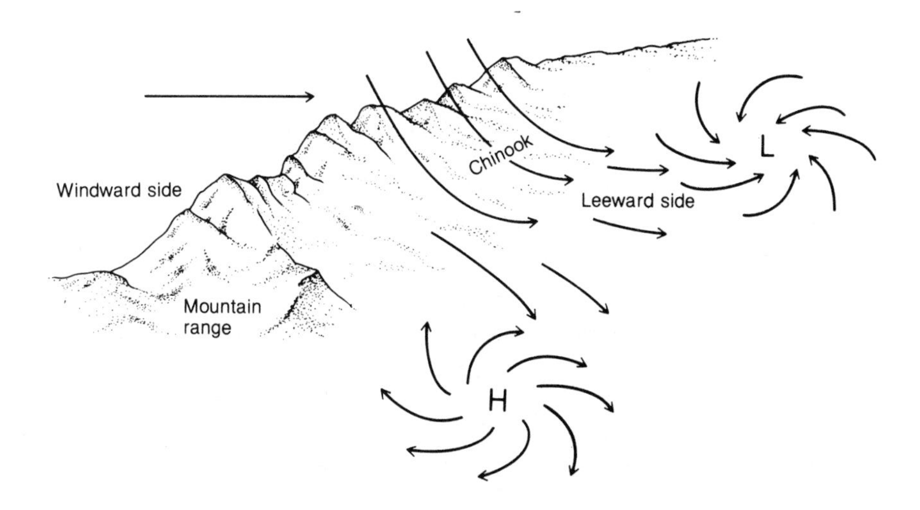

FIGURE 12.6

A chinook wind develops when the circulation about an anticyclone or cyclone (situated to the lee of a mountain range) pulls air down the leeward side of the range.

At the onset of a chinook wind, surface air temperatures often climb abruptly tens of degrees in response to compressional warming. For example, on 6 January 1966, at Pincher Creek, Alberta, a chinook sent temperatures soaring 21 °C (38 °F) in only 4 minutes. An even more dramatic temperature surge was recorded at Spearfish, South Dakota, on 22 January 1943: the temperature rose from −30 °C (−4 °F) at 7:30 A.M. to 7 °C (45 °F) at 7:32 A.M.—37 °C (49 °F) in only 2 minutes! The sudden springlike warmth can, however, just as quickly give way to severely cold conditions. For example, at the foot of the Rockies, a wind shift from west to north brings an abrupt end to the chinook and the return of polar or arctic air.

Chinook is a Native American word that means "snoweater." The term is appropriate because of the wind's catastrophic effect on a snow cover. As the chinook is compressionally warmed, the relative humidity drops dramatically. Because the chinook is both warm and very dry, a snow cover melts and vaporizes rapidly. It is not unusual, for example, for a half meter of snow to disappear in this way in only a few hours.

Chinook winds are not restricted to the leeward side of the Rockies. A similar wind develops in the Alpine valleys of Austria and Germany. There, the warm dry wind is known as the **foehn wind**. In southern California, the notorious **Santa Ana wind** is actually a chinook wind. A strong high pressure system over the Great Basin sends northeast winds over the southwestern United States, driving air downslope from the desert plateaus of Utah and Nevada and around the Sierra Nevada mountains as far west as coastal southern California. These hot, dry, gusty winds desiccate vegetation and contribute to the outbreak of brush fires and forest fires (Figure 12.7).

Desert Winds

Deserts typically are very windy places primarily because of the intense solar heating of the ground. The ground absorbs most of the solar radiation in deserts, because little is used to evaporate the scarce water. In some deserts the mid-day temperature of the sandy surface exceeds 55 °C (130 °F). This hot surface produces a steep temperature lapse rate in the air layer next to the ground. A steep temperature lapse rate, in turn, means great instability (Chapter 6), vigorous convection, and gusty surface winds. The strength and gustiness of the wind varies with the intensity of solar radiation so that wind speeds and gustiness usually peak in the early afternoon and during the warmest months.

FIGURE 12.7

Hot, dry Santa Ana winds fan brush and forest fires such as this one in Mandeville Canyon, Los Angeles County. (Courtesy of FOTO-HETZEL)

Variations in the surface characteristics of a desert (albedo, moisture, topography) cause some spots to become hotter than others. If regional (synoptic-scale) winds are relatively weak, local surface winds converge toward hot spots replacing the hot air rising over the hot spots. And, just as water begins to rotate as it converges toward a drain in a sink, the converging surface winds rotate about a vertical axis. In the process, dust is lifted from the ground and the circulation is visible as a whirling mass of dust-laden air known as a **dust devil.** Typically dust devils are very small systems that average about 10 m (30 ft) in diameter and less than 2000 m (6000 ft) in altitude with a life expectancy of only several minutes. Normally, dust devil winds are not very strong and cause no property damage.

Larger scale winds are produced in deserts by thunderstorms or migrating cyclones. Surface winds associated with these weather systems can give rise to either dust storms or sand storms, the difference between the two hinging on the size range of the loose surface sediment that is eroded by the wind. Dust consists of very small particles (less than 0.06 mm in diameter) that can be carried by the wind to great altitudes. Sand, which typically covers only a small fraction of desert terrain, consists of large particles (0.06 to 2.0 mm in diameter) that are transported by the winds within about a meter of the ground.

One of the most spectacular dust storms, known as a **haboob,** is generated by the strong, gusty downdraft of a thunderstorm. In a desert, thunderstorm rains often completely vaporize in the dry air beneath cloud level. The thunderstorm downdraft, however, exits the cloud base, rushes toward the ground, and then surges ahead of the thunderstorm as a mass of cool, gusty air. Dust picked up from the ground fills the air, severely restricting visibility, and the mass rolls along the ground as a huge ominous black cloud. A haboob may be more than 100 km (62 mi) wide and reach altitudes of several kilometers. They are most common in the Sudan of North Africa but also occur in the American Southwest desert.

Under some conditions, winds blow out of deserts and into quite different climatic regions. For example, in spring when the semipermanent subtropical anticyclones shift poleward, hot, dry winds stream northward out of the Sahara Desert of North Africa over the Mediterranean Sea and into southern Europe. These hot, dry winds are little changed after crossing the Mediterranean except in the western Mediterranean where winds approaching Spain have traveled over water for a longer distance and become humid.

Mountain and Valley Breezes

Along mountain slopes that are exposed to intense solar heating, a localized air circulation may develop that reverses direction from day to night (Figure 12.8). After the winter snows have melted, bare

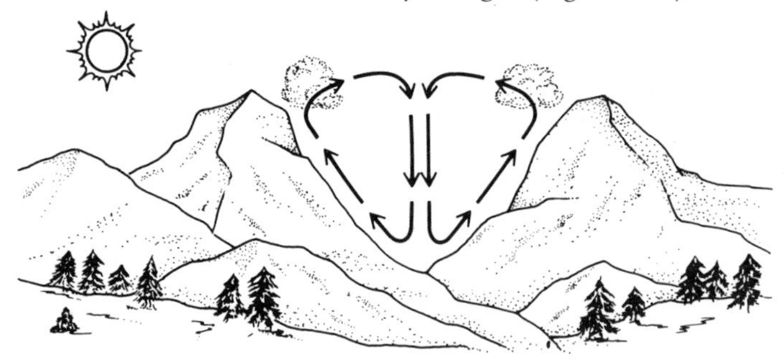

Valley breeze

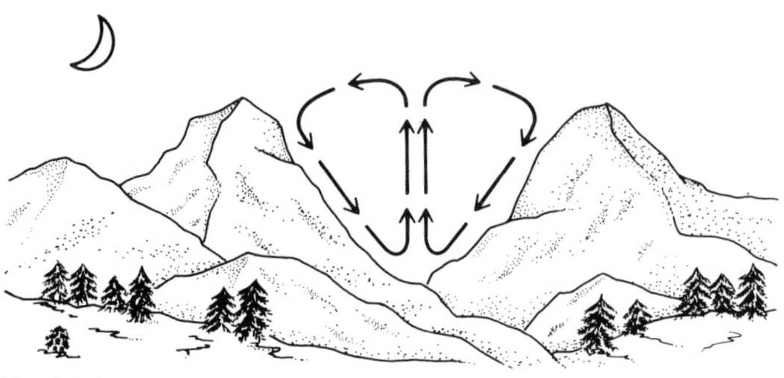

Mountain breeze

FIGURE 12.8

Valley and mountain breeze circulation.

valley walls facing the sun absorb solar radiation, and the air in contact with the walls is heated. This warmed, light air flows upslope as a **valley breeze** and may even cause cumulus clouds to develop near the mountain top. At night, with infrared radiational cooling, the mountain slopes cool rapidly and the air in contact with the slopes is chilled. This cold, dense air then flows downslope as a gusty **mountain breeze**.

Conclusions

This chapter described several local and regional circulation systems, and in subsequent chapters we will consider others. The dominance of these circulation systems by larger scale circulation is apparent, that is, the global and synoptic-scale patterns set boundary conditions for any smaller scale circulation. In some cases, synoptic-scale winds reinforce mesoscale winds, as when regional winds blow in the same direction as a sea or lake breeze. In other cases, however, synoptic-scale winds negate or overwhelm mesoscale winds, as when northerly winds sweep along the edge of the Rocky Mountains, thus eliminating the possibility of chinook winds. Even the large-scale monsoon winds are influenced by shifts of the ITCZ and the subtropical anticyclones.

The discussion of weather systems is continued in the next chapter with a focus on thunderstorms.

SUMMARY STATEMENTS

- Monsoon circulations in tropical latitudes illustrate the interplay of climate controls. Differences in solar heating between land and sea, and seasonal shifts in global-scale circulation play important roles in monsoon development. Although summers are predictably wet and winters are predictably dry, the intensity and duration of monsoon rains vary from one year to the next.

- Human activities influence the climate of large cities by altering the local radiation balance. This gives rise to an urban heat island and, in some cases, to a convective circulation pattern that concentrates pollutants in a dust dome over the city.

- A variety of mesoscale circulations are favored by certain synoptic-scale conditions and are triggered by localized pressure gradients. These include land and sea or lake breezes, and mountain and valley breezes.

- Katabatic winds are gravity-induced flows of dense, cold air. Chinook winds, on the other hand, are compressionally warmed air pulled down the leeward slope of mountains by synoptic-scale winds.

KEY WORDS

monsoon circulation
monsoon active phases
monsoon dormant phases
sea breeze

lake breeze
land breezes
urban heat island
dust dome
dust plume
katabatic wind
mistral
bora
chinook winds
foehn wind
Santa Ana wind
dust devil
haboob
valley breeze
mountain breeze

REVIEW QUESTIONS

1 Define the monsoon circulation.

2 In what areas of the globe are monsoon circulations dominant features of the climate?

3 Why does land warm up more than the sea in response to the same intensity of solar radiation?

4 Explain the shifts between active and dormant phases of a monsoon.

5 How is the monsoon circulation linked to seasonal shifts of the intertropical convergence zone and the subtropical anticyclones?

6 The Coriolis effect influences the monsoon circulation but not land and sea breezes. Why the difference?

7 Why does a sea breeze develop on some summer days but not on others?

8 Speculate on how topography affects the inland extent of sea breezes or lake breezes.

9 Sea breezes, lake breezes, and land breezes develop in response to a horizontal air pressure gradient. What causes this pressure gradient to develop?

10 What is meant by the continuity of the atmosphere?

11 Describe the surface air circulation about a mesoscale high.

12 Explain why a lake breeze develops along the Lake Michigan shoreline in summer but not during the winter.

13 What factors contribute to the development of an urban heat island?

14 Speculate on steps that city planners might take to reduce the intensity of urban heat islands, thereby decreasing the likelihood of dust dome development.

15 How is an urban dust dome linked to an urban heat island?

16 Compare and contrast the thermal properties of vegetation with the thermal properties of materials composing a city (brick, asphalt, concrete).

17 What force drives katabatic winds?

18 A katabatic wind is not a warm wind in spite of adiabatic compressional heating. Explain why.

19 How do chinook winds differ from katabatic winds?

20 What special conditions are needed for a valley breeze to develop?

POINTS TO PONDER

1 Describe in detail the mechanics of the dry and wet monsoons.

2 In portions of India that are influenced by the monsoon circulation, the warmest month of the year is usually May. Explain.

3 In view of your understanding of atmospheric stability, explain the following observation: In spring, in the Great Lakes region, convective cumulus clouds tend to develop over land but not over lake waters.

4 Explain why a mesoscale anticyclone sometimes develops over Lake Michigan. Is this a cold-core high or a warm-core high?

5 Speculate on the type of weather that might develop along a sea breeze front.

6 How does the Bowen ratio of an urban area compare with that of a rural countryside? Elaborate on your response.

PROJECT

1 Determine whether your area is affected by any local or regional air circulation patterns. You may wish to consult the NOAA Climatic Summary for your area.

SELECTED READINGS

Eichenlaub, V. *Weather and Climate of the Great Lakes Region.* Notre Dame, Ind.: The University of Notre Dame Press, 1979. 335 pp. *Describes local circulation phenomena in the Great Lakes region.*

Flavin, C. "Wind Power: A Turning Point." *Worldwatch Paper* 45 (1981). 56 pp. *An assessment of the wind's energy potential.*

Idso, S. B. "Arizona Weather Watchers: Past and Present." *Weatherwise* 28 (1975):56–60. Includes descriptions of desert weather systems.

Matson, M., et al. "Satellite Detection of Urban Heat Islands." *Monthly Weather Review* 106 (1978): 1725–1734. *Discusses heat island detection using sensors on NOAA-5 satellite.*

Monteverdi, J. P. "The Santa Ana Weather Type and Extreme Fire Hazard in the Oakland-Berkeley Hills." *Weatherwise* 26 (1973):118–121. *Describes synoptic weather pattern associated with Santa Ana winds.*

Sorensen, B. "Turning to the Wind." *American Scientist* 69 (1981):500–508. *Gives an evaluation of the potential of wind-power technology.*

Webster, P. J. "Monsoons." *Scientific American* 245, no. 2 (1981):109–118. *Discusses in some detail the energetics involved in monsoon circulations.*

Zipser, E. J., and A. J. Bedard, Jr. "Front Range Windstorms Revisited: Small Scale Differences Amid Large Scale Similarities." *Weatherwise* 35 (1982):82–85. *Details characteristics of a destructive local wind.*

Blow, winds, and crack your cheeks; rage, blow.
You cataracts and hurricanes, spout
Till you have drench'd our steeples, drown'd the cocks.
You sulph'rous and thought-executing fires,
Vaunt-couriers of oak-cleaving thunderbolts,
Singe my white head. And thou, all-shaking thunder,
Strike flat the thick rotundity o' th' world;
Crack nature's moulds, all germens spill at once,
That makes ingrateful man.

WILLIAM SHAKESPEARE
King Lear

Lightning is one of many hazards associated with thunderstorms. Hail, torrential rains, and strong winds are other hazards that can also accompany thunderstorms. (Photograph by Arjen Verkaik)

THIRTEEN

Thunder-storms

Thunderstorm Life Cycle

Thunderstorm Genesis

Severe Thunderstorms

Thunderstorm Hazards

Lightning

Microbursts

Flash Floods

Hail

MOST OF US are familiar with typical thunderstorm weather—the blackening sky and abrupt freshening of wind followed by bursts of torrential rain, flashes of lightning, and rumbles of thunder. Often the cool breezes and rains bring us welcome—albeit temporary—relief on hot, muggy summer afternoons. For a farmer whose crops are wilting under the summer sun, the heavy rains may be an economic lifesaver. Some thunderstorms become violent, however, and wreak havoc. Lightning starts fires, strong winds level trees and buildings, heavy rains cause flooding, and, in some cases, thunderstorms spawn destructive hail or tornadoes. At this very moment, an estimated 2000 thunderstorms are in progress somewhere on earth—most of them in tropical and subtropical latitudes. In this chapter, we discuss the life cycle and characteristics of thunderstorms and the hazards posed by severe thunderstorms.

Thunderstorm Life Cycle

A **thunderstorm** is a small-scale system and thus affects a relatively small area and is short lived. It is the product of very strong convection air currents that extend deep into the troposphere, perhaps reaching to the tropopause or higher. Upward-surging air currents are made visible by billowing cauliflower-shaped cumuliform clouds, as shown in Figure 13.1.

Recall from Chapter 2 that convection currents begin when heat is conducted from the relatively warm earth's surface to cooler air immediately above the surface. We can visualize the upward moving branch of a convection current as consisting of a continuous stream of bubbles (or parcels) of warm, unsaturated air. The rising

FIGURE 13.1

(A) As convection currents surge upward into the atmosphere, cumulus clouds billow upward. In some cases (B), the cumulus clouds may build both vertically and horizontally into a thunderstorm cloud (cumulonimbus). Photograph B was taken approximately one hour after photograph A. (Photographs by Arjen Verkaik)

A

B

bubbles cool at the dry adiabatic lapse rate (10 °C per 1000 m) until they reach the lifting condensation level (LCL) where water vapor condenses and cumulus clouds begin to form. The more humid the air is to start with, the less the expansional cooling that is needed for the bubbles to achieve saturation and the lower is the base of cumulus clouds. Hence, the base of cumulus clouds usually is lower in Florida where relative humidities are high than in New Mexico where relative humidities are low.

The latent heat that is released during condensation adds to the buoyancy of the rising convective bubbles and they surge upward while cooling at the moist adiabatic lapse rate (averaging 7 °C per 1000 m). Convective bubbles continue ascending for as long as they are warmer, and thus lighter, than the surrounding air, that is, as long as the ambient air is unstable. Some of the saturated bubbles surge through the cloud top and vaporize in the relatively dry air above the cloud. As a consequence, the water vapor content of the air above the cloud increases. Because the air above the cloud is more humid, subsequent bubbles are able to rise higher before vaporizing. As this process is repeated, the cumulus cloud billows upward. This billowing character of cumulus cloud growth is evident in Plate 13; at this state of vertical development, the cloud is called **cumulus congestus**. If vertical growth continues, the cumulus congestus cloud builds into a **cumulonimbus cloud**, a thunderstorm cloud (Plate 9).

A thunderstorm is typically composed of five to eight convection cells, each a few kilometers in diameter. The life cycle of a thunderstorm cell may be divided into three stages: cumulus, mature, and dissipating (Figure 13.2).

The **cumulus stage** is initiated when towering cumulus clouds merge to form a thunderstorm cell. Air streams upward throughout the cell as an **updraft**, which frequently reaches altitudes of 8000 m (26,000 ft) or higher. The updraft is usually strong enough to keep precipitation suspended in the upper reaches of the cloud.

The weight of rain droplets and of ice crystals eventually becomes sufficient to allow them to fall faster than the updraft rises, and precipitation begins to descend through the cloud, ushering in the **mature stage** of the thunderstorm cell. Falling rain and snow drag the adjacent air downward, creating a strong **downdraft** alongside the updraft. The downdraft leaves the base of the cloud and spreads out along the ground, well in advance of the parent cell, as a mass of relatively cool, gusty air. (The air is cool because the chilling effect of the cold rain and snow offsets to some extent the

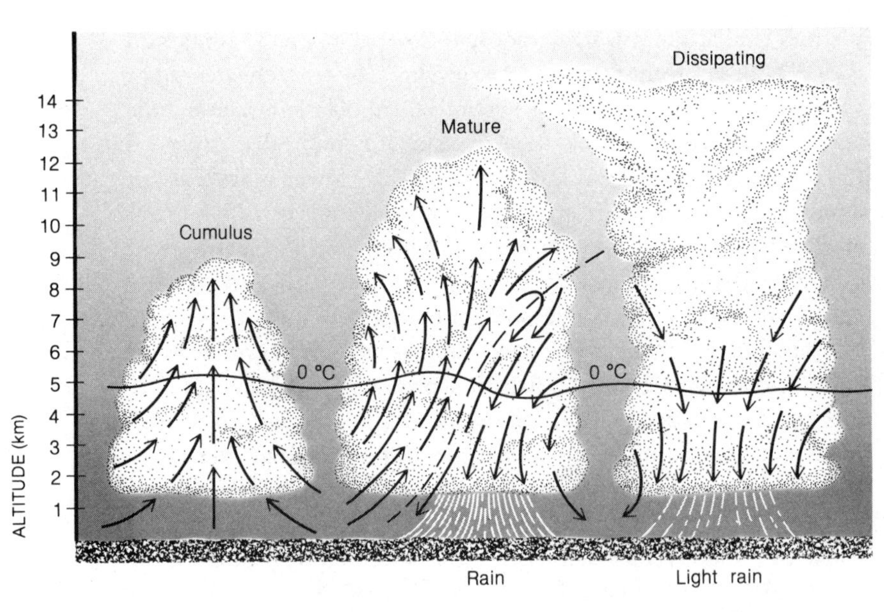

FIGURE 13.2

A thunderstorm's life cycle consists of cumulus, mature, and dissipating stages.

downdraft's compressional warming.) The leading edge of this localized air mass resembles a miniature cold front and is called a **gust front**. Convective clouds sometimes develop as a consequence of uplift along the gust front and may initiate the formation of secondary thunderstorm cells ahead of the parent cell.

Ominous-appearing low clouds are sometimes associated with thunderstorm gust fronts. A **roll cloud** is an elongated cylindrical dark cloud that appears to rotate slowly about its horizontal axis. The roll cloud occurs behind the gust front and beneath, but detached from, a cumulonimbus cloud. Why this cloud forms is not fully understood. Although appearances might suggest otherwise, roll clouds are in fact seldom accompanied by severe weather. This is *not* the case with shelf clouds. A **shelf cloud** is a low elongated cloud that is wedge-shaped with a flat base (Figure 13.3). This cloud occurs at the edge of a gust front, and beneath and attached to a cumulonimbus cloud. The shelf cloud is thought to develop as a consequence of uplift of stable, warm humid air along the gust front. Damaging surface winds may occur under the shelf cloud, and sometimes this cloud precedes a severe thunderstorm.

The thunderstorm cell reaches its maximum intensity during the late mature stage when cloud tops can exceed altitudes of 18,000 m (60,000 ft) and exhibit the anvil shape shown in Figure 13.4. Strong winds at these high altitudes are responsible for distorting the cloud top into the anvil shape. Also, temperatures within the upper por-

FIGURE 13.3

A shelf cloud such as this one may develop along a thunderstorm gust front. Often shelf clouds are accompanied by strong and gusty surface winds and may be associated with a severe thunderstorm. (Photograph by Arjen Verkaik)

FIGURE 13.4

When the billowing cumulonimbus cloud reaches the tropopause, it spreads out and forms an anvil top. (NCAR/NSF photograph)

tion of the cloud are so low that the anvil is composed solely of ice crystals which give it a fibrous appearance. During this phase, the heaviest rains fall and the darkened sky is streaked with lightning. In addition, hail, strong surface winds, and even tornadoes may develop.

Viewed from space by satellite, clusters of mature thunderstorm cells appear as bright white blotches, such as those over Texas and northeast Oklahoma in Figure 13.5. The brightness of these clusters is due to the high albedo (reflectivity) of the cloud tops. Solar radiation that does penetrate must travel through great thicknesses of cloud before emerging at the cloud base. A considerable amount of solar radiation is absorbed by the clouds, so the sunlight emerging at the cloud bases is very weak. For this reason, from our perspective on the earth's surface, the sky darkens as thunderstorm clouds approach.

FIGURE 13.5
In this satellite photograph, clusters of thunderstorm cells appear as bright white blotches over Texas and northeast Oklahoma. (NOAA photograph)

As precipitation spreads throughout the thunderstorm cell, so does the downdraft, heralding the demise of the cell. During the **dissipating stage**, upward air motion is replaced by sinking air that is adiabatically warmed by compression. Relative humidity decreases, precipitation tapers off and ends, and convective clouds gradually vaporize.

Typically, a thunderstorm cell completes its life cycle in less than an hour, but on some occasions lightning, thunder, and bursts of heavy rain may persist for many hours. This is because a thunderstorm usually consists of a cluster of cells, as shown schematically in Figure 13.6. Each cell may be at a different stage in its life cycle, and new cells form and old cells dissipate continuously. A succession of many cells is thus responsible for prolonged periods

FIGURE 13.6

These individual thunderstorm cells are traveling at an angle of about 20 degrees to the direction of movement of a multicellular thunderstorm. As they move, the individual cells go through their life cycle. (Modified after Browning, K.A., and F.H. Ludlam. "Radar Analysis of a Hailstorm." *Technical Note*, no. 5, Meteorology Department, Imperial College, London.)

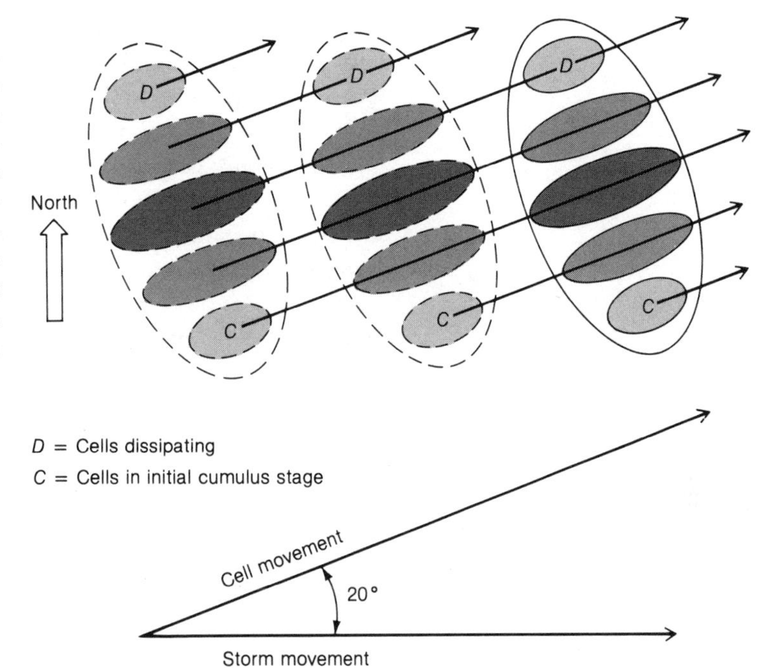

of thunderstorm weather. Although a locality may be in the direct path of a distant, intense cell, the relatively brief life span of an individual cell means that the storm may dissipate long before reaching that locality.

The multicellular character of many thunderstorms complicates the motion of the weather system. A thunderstorm cluster usually tracks at an angle of about 20 degrees to the right of the paths of the constituent cells. This phenomenon is illustrated in Figure 13.6, which shows a thunderstorm cell cluster tracking from west to east, while its five component cells head off toward the northeast. In this idealized case, new cells form in the southern sector of the storm, and old cells dissipate in the northern sector.

Thunderstorm Genesis

Most thunderstorms develop within masses of warm, humid maritime tropical (mT) air when the air mass destabilizes. Uplift is the key to destabilizing mT air. Because mT air is usually conditionally stable,* it becomes unstable and surges upward only when it is

*Recall from Chapter 6 that conditionally stable air is *stable* for unsaturated air parcels and *unstable* for saturated air parcels.

lifted to the condensation level. If the air is very humid to begin with, then minimal lifting (that is, cooling) is needed for the mT air to achieve saturation and for cumuliform clouds to develop. The uplift of air is initiated by (1) frontal activity, (2) orographic effects, (3) convergence of air at the surface, or (4) intense solar heating of the ground. Any one or a combination of these mechanisms may be sufficient to induce the uplift necessary to destabilize mT air. In addition, the potential instability of mT air is enhanced by cold air advection aloft or by warm air advection at the surface. Either one or both of these processes can steepen the air mass lapse rate, as shown in Figure 13.7, and reduce the stability of the ambient air.

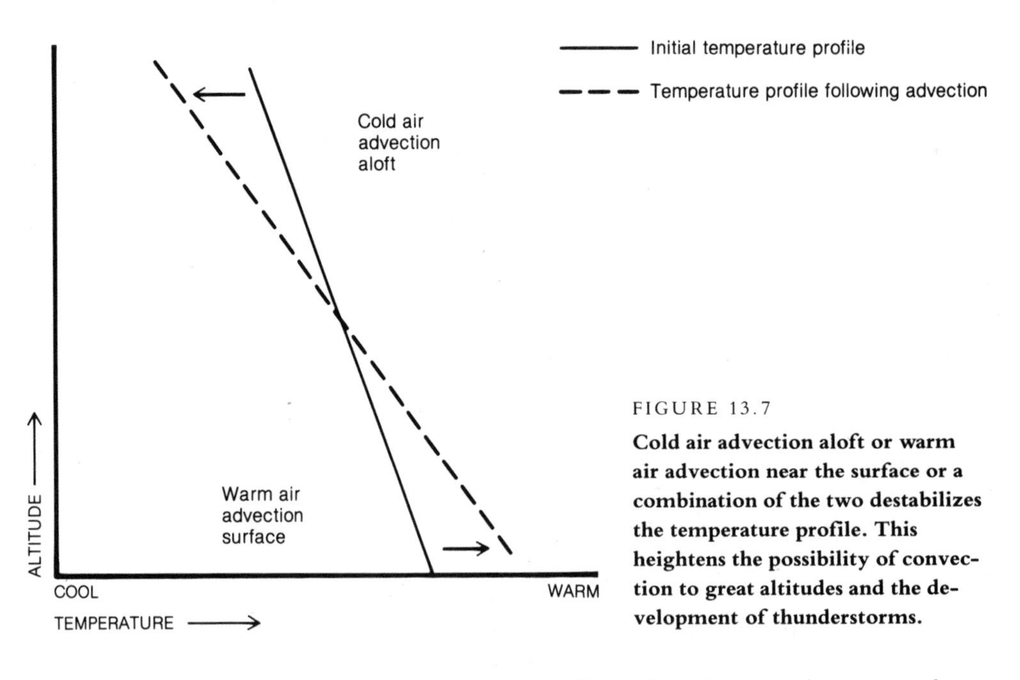

FIGURE 13.7

Cold air advection aloft or warm air advection near the surface or a combination of the two destabilizes the temperature profile. This heightens the possibility of convection to great altitudes and the development of thunderstorms.

Depending on the specific triggering mechanisms, thunderstorms are categorized as either **air mass thunderstorms** or **frontal thunderstorms**. Air mass thunderstorms pop up almost randomly anywhere within a mass of mT air. Because they are caused by convection currents driven by intense solar heating, they develop in the afternoon, during the warmest hours of the day. A noteworthy exception to this general rule occurs in the central United States over the Missouri River Valley and the adjacent portion of the upper Mississippi River Valley where air mass thunderstorms are more frequent at night than during the day. Several explanations have been proposed for this. One possible explanation centers on the low-level jet stream of maritime tropical air that streams northward up the Mississippi River Valley. This jet stream strengthens at night and causes warm air advection at low levels that destabilizes the air and spurs convection.

As the name implies, frontal thunderstorms are associated with uplift along frontal surfaces. Most are triggered by the vigorous uplift of mT air along or ahead of a well-defined cold front, but sometimes thunderstorms break out in the overrunning zone ahead of a surface warm front. Frontal thunderstorms occur parallel to the surface front, and because frontal activity typically persists for many days, these thunderstorms may develop at any time of the day or night. In some instances, a line of thunderstorms forms well in advance of a sharply defined cold front—perhaps 30 to 180 km (20 to 110 mi) distant. These thunderstorms are aligned as a **squall line** and are often severe.

Both air mass and frontal thunderstorms are usually associated with the mT air occupying the southeast sector of a mature mid-latitude cyclone. Convergent surface airflow and the consequent upward air motion associated with the low contribute to convection currents and thunderstorm genesis.

In recent years, study of satellite images has revealed a third pattern of thunderstorm occurrence in addition to the air mass and frontal types. A **mesoscale convective complex** (MCC) is a nearly circular cluster of many thunderstorms that covers an area that may be a thousand times larger than that of an individual air mass thunderstorm. In fact, it is not unusual for a single MCC to cover an area equal to that of Iowa. New thunderstorms develop continuously within a mesoscale convective complex so that the life expectancy of the system is typically 12 to 24 hours. The longevity of an MCC coupled with its typically slow movement (15 to 30 km per hour) means that rainfall is widespread and substantial. These weather systems are primarily a warm season (March through September) phenomenon and occur chiefly in the eastern two-thirds of the United States where more than 50 may be expected in a single season.

In subtropical and tropical latitudes, intense solar heating combined with convergent surface airflow triggers thunderstorm development. As we saw in Chapter 10, this combination characterizes the ITCZ, which is actually a discontinuous line of thunderstorms. Another example of thunderstorm development comes from the interior of the Florida peninsula, where converging coastal sea breezes induce upward motion of mT air and frequent thunderstorms. In mountainous areas, orographic lifting and solar heating combine to induce thunderstorm activity. This mechanism is at work in the Rocky Mountain region, which ranks second only to Florida in frequency of thunderstorms in the United States (Figure 13.8).

An average of more than 60 thunderstorm days occur per year along the Front Range of the Rocky Mountains in a band from southeastern Wyoming southward through central Colorado and into north central New Mexico. This high thunderstorm frequency is linked to differences in heating arising from variations in topography. Recall from our discussion of mountain breezes in Chapter 12 that mountain slopes facing the sun absorb the direct rays of the sun and become relatively warm. The warm slopes, in turn, heat the air in immediate contact with the slopes. At the same time, air at the same altitude but located to the east of the mountains, out over the relatively flat terrain of the western Great Plains, is much

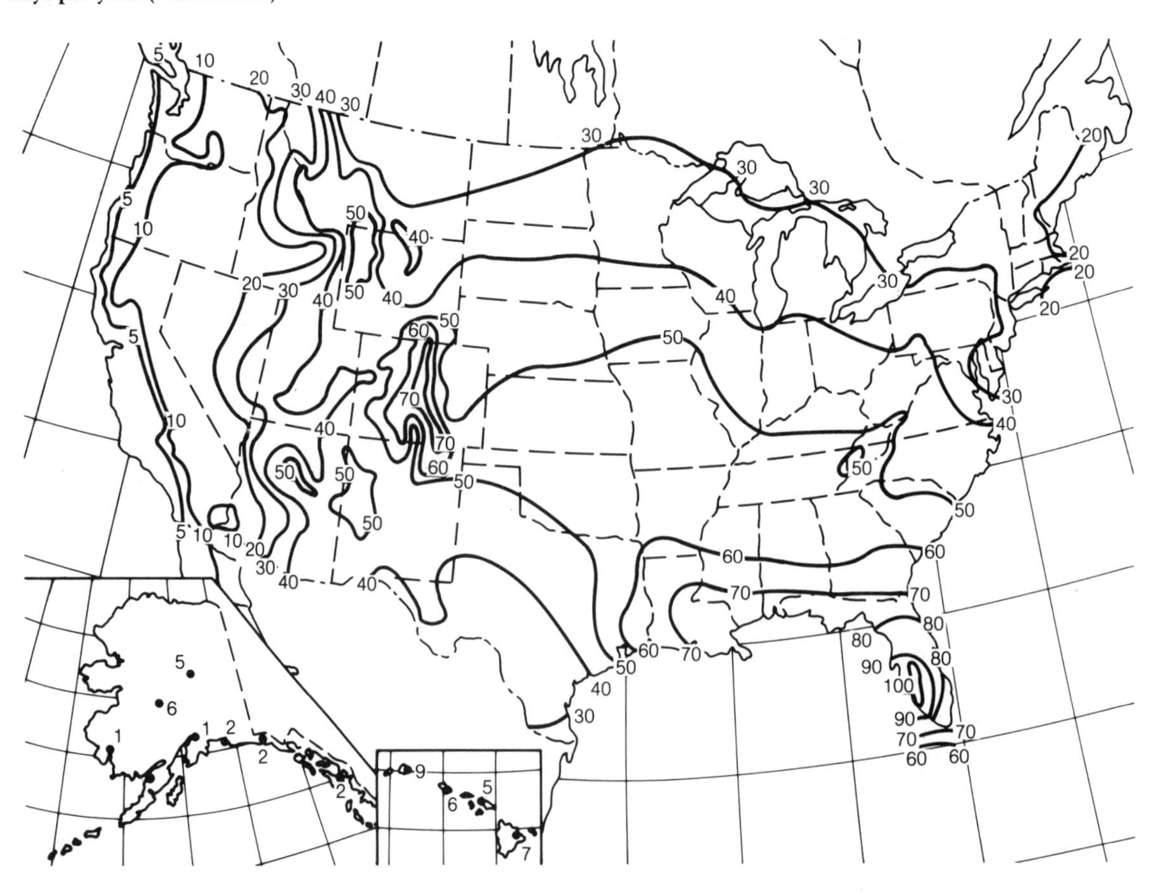

FIGURE 13.8

Thunderstorm frequency across the United States in average number of days per year. (NOAA data)

cooler. The cooler, denser air over the Great Plains sweeps westward and replaces the warm, light air that rises over the mountain slopes. Warm air rising over the mountain produces convective clouds that frequently billow upward to form thunderstorms. This process of thunderstorm development is enhanced whenever the synoptic-scale pressure pattern favors east winds over the western Great Plains.

The distinction is sometimes made between forced convection and free convection. In the case just described, topography helps to force air upward to form convective currents. This situation is one of **forced convection**. On the other hand, convection that is triggered by intense solar heating of relatively flat terrain is known as **free convection**.

To this point we have considered those conditions conducive to deep convection and thunderstorm development. Under other circumstances, however, convection is inhibited and thunderstorms cannot form. This situation usually occurs when air masses reside or travel over relatively cold surfaces and are thereby stabilized. Snow-covered ground cools and stabilizes the overlying air. Convection is thus suppressed, so thunderstorms are rare in middle and high latitudes during winter. Because large bodies of water do not warm as much as adjacent land masses, thunderstorms are less frequent over the oceans than over the continents. In localities such as coastal California, where a shallow layer of relatively cool mP air is usually advected inland from the Pacific, thunderstorms are infrequent.

Because convective activity is dependent on intense solar heating, thunderstorms occur with the greatest frequency in tropical latitudes and in continental interiors. The steamy interiors of Brazil, equatorial Africa, and the island of Indonesia experience the greatest number of thunderstorms in the world. In these localities, thunderstorm activity can be expected just about every day.

Severe Thunderstorms

By convention, a **severe thunderstorm** is accompanied by locally damaging winds, frequent lightning, or large hail. As a general rule, the greater the altitude of a thunderstorm top, the more likely the thunderstorm is to produce severe weather. Why, then, do some thunderstorm cells surge to great altitudes and trigger severe weather, while others do not? One possible explanation is that in severe thunderstorm cells the updraft is tilted (Figure 13.9). The tilt deflects much of the precipitation away from the updraft, so much of the precipitation falls alongside the updraft rather than through

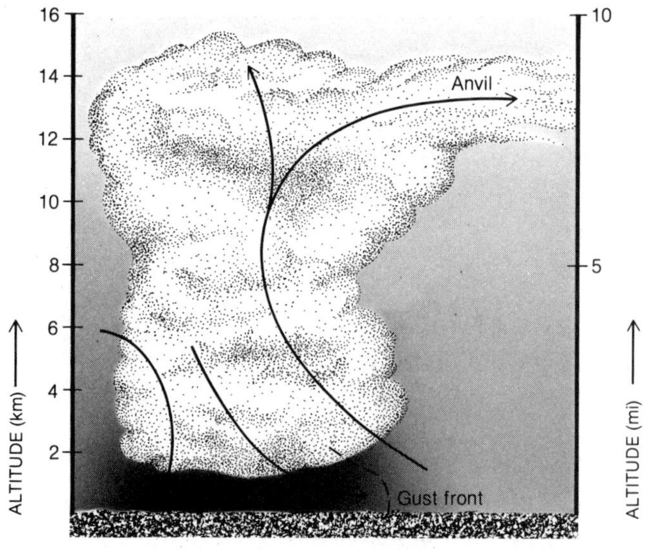

FIGURE 13.9

When the updraft is tilted within a thunderstorm cell, the precipitation does not fall against the updraft. The cell can therefore billow upward to great altitudes. (From Trewartha, G.T., and L.H. Horn. *An Introduction to Climate*, 5th edition. New York: McGraw-Hill, 1980, p. 180. Used with permission.)

it. Consequently, precipitation does not drag against the updraft, and the updraft continues to build the cell to greater altitudes. The reason for a tilted updraft may be linked to the special synoptic weather pattern favorable to severe thunderstorm development.

In the United States and Canada, most severe thunderstorms break out over the Great Plains and are associated with mature synoptic-scale cyclones. Severe thunderstorm cells usually form a squall line within the cyclone's warm sector, ahead of and parallel to a fast-moving, well-defined cold front. The squall line appears as an ominous, rolling and twisting mass of low, black clouds (Figure 13.10), often hundreds of kilometers long. The squall line moves very rapidly at speeds that may approach 80 km (50 mi) per hour.

The midlatitude jet stream appears to be an important ingredient in the development of severe thunderstorm cells—sometimes called **supercells**. First, the jet is responsible for tilting the updraft within the cell. In addition, as we saw in Chapter 10, the midlatitude jet stream maximum induces both divergence and convergence of air aloft. Divergent airflow triggers cyclone development, while convergent airflow causes weak subsidence of air over the warm sector of the cyclone. The subsiding air warms compressionally (and dries), but it is prevented from reaching the surface by a shallow

FIGURE 13.10

An ominous, rolling and twisting mass of low, black clouds marks a squall line. (Photograph by Donn Quigley)

layer of mT air. In North America, the mT air surges rapidly northward, out of the Gulf of Mexico, as a tongue—perhaps 3000 m (10,000 ft) deep—and is often described as a **low-level jet stream**. The warm, humid air is pumped northward by the circulation on the western flank of the Bermuda-Azores subtropical anticyclone. This synoptic situation is shown in Figure 13.11. Note that the air mass between the cold front and the moist mT air is usually cT air.

FIGURE 13.11

The synoptic situation favorable for development of severe thunderstorm cells consists of a mature synoptic-scale cyclone with a low-level mT jet stream underlying the midlatitude high tropospheric jet stream in the southeast sector of the cyclone.

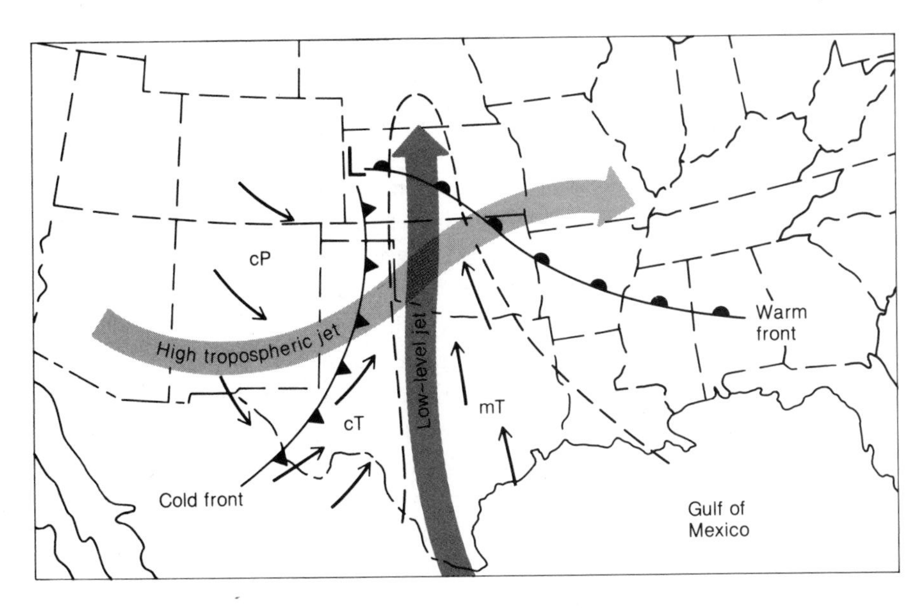

The two jet streams produce a special stratification of air that is ideal for the eruption of severe thunderstorm cells. As a consequence of compressional warming, the air subsiding from aloft becomes warmer than the underlying layer of mT air. A zone of transition develops between the two air masses as a temperature inversion (Figure 13.12). As we learned in Chapter 6, a temperature inversion is extremely stable, so the two air masses do not mix and

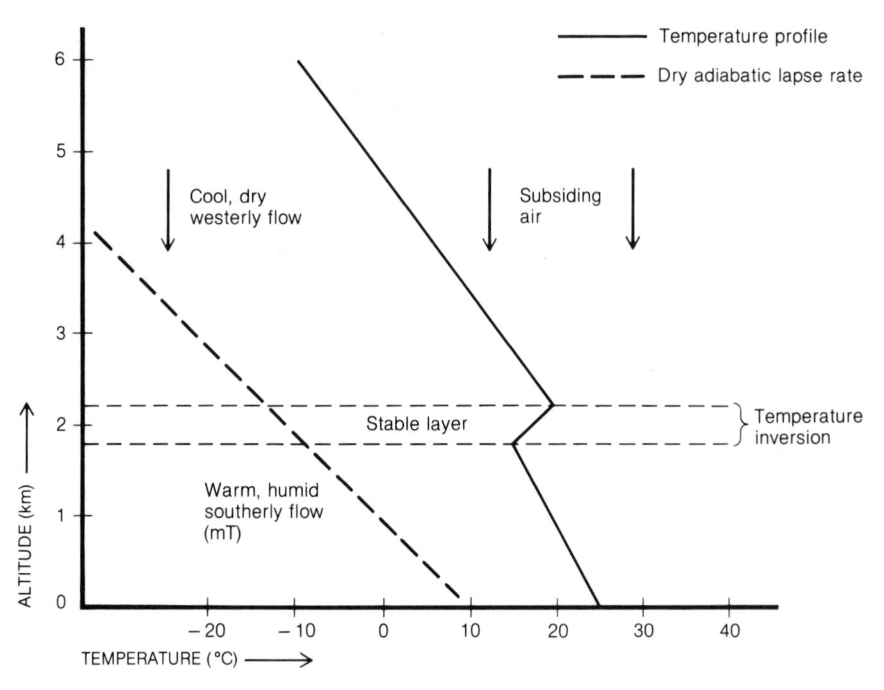

FIGURE 13.12
The temperature profile that is most favorable for the sudden eruption of severe thunderstorm cells. An elevated temperature-inversion layer separates subsiding dry air aloft from warm, humid mT air at the surface.

convection is confined to the surface mT air layer. As long as this synoptic situation persists, the contrast between the two air masses mounts. The subsiding air becomes drier, and the underlying air becomes more humid. The potential for severe weather continues to grow, and all that is needed is a trigger that will force the convection currents to penetrate the elevated temperature inversion. The needed upward impetus may be supplied by the intense solar heating of midafternoon or by the lifting of air associated with the approaching cold front. By either mechanism, convection currents eventually break through the temperature inversion, and cumulus clouds billow upward at speeds that may exceed 100 km (60 mi) per hour. Such explosive updrafts can even penetrate the tropopause and surge into the lower stratosphere.

Occasionally, ominous pouchlike mammatus clouds will appear on the underside of the anvil top of a cumulonimbus cloud (Plate

14). Contrary to popular belief, **mammatus clouds** do not always indicate a severe thunderstorm. In fact, in the mountainous western United States, they often are associated with relatively weak convective showers. Their unusual appearance is attributed to blobs of cold dense saturated air that sink down from the anvil into the dryer air beneath the anvil.

Thunderstorm Hazards

Thunderstorm hazards include lightning, microbursts, torrential rains, hail, and the spawning of tornadoes. Because tornadoes are an especially severe hazard, they are considered in detail in the next chapter.

Lightning

Lightning is a weather phenomenon that is directly hazardous to human life. Each year lightning kills perhaps a hundred people in the United States and Canada. This number of fatalities may seem surprisingly high. Fatal lightning strikes are typically isolated events that do not make headlines. A single, fatal stroke of lightning is not as newsworthy as a single disastrous tornado, but so many lightning bolts occur daily throughout North America that injuries and fatalities quickly add up. For some tips on lightning safety, refer to the Special Topic.

Lightning is feared not only because it may strike people but also because it ignites forest and brush fires. In the Rocky Mountain region, lightning is the most common cause of forest fires, starting more than 9000 fires per year. One might think that the heavy rains associated with thunderstorms would quickly quench a lightning-induced fire. In the western basins, however, the base of a cumulonimbus cloud is usually so far above the ground, and the lower air layer is so dry that much of the rain evaporates before reaching the fire.

What is lightning? **Lightning** is a brilliant flash of light produced by an electrical discharge of about 100 million volts. The potential for an electrical discharge exists whenever a charge difference develops between two objects. A normally neutral objective becomes negatively charged when it gains electrons (negatively charged subatomic particles) and positively charged when it loses electrons. When differences in electrical charge develop within a cloud or

SPECIAL TOPIC

Lightning safety

Lightning kills and injures, and a blinding flash of lightning followed by a crash of thunder terrorizes many people. In the United States between 1959 and 1983, an average of 100 people died each year as a result of lightning, and 250 were injured.

While the danger of lightning cannot be ignored, some simple safeguards will minimize the risk of injury when a thunderstorm threatens.

The odds of being struck and killed by lightning are slim, actually—about 350,000:1—and these odds improve when precautions are taken. By comparison, the odds of being struck and killed by an auto are 50 times greater. Although no place is absolutely safe, the risk can be minimized in even the most lightning-prone area of the nation, south and central Florida, where each square kilometer of land is struck by an estimated 10 lightning bolts yearly.

When a thunderstorm approaches, the best action is to seek shelter in a house or other building, avoiding contact with conductors of electricity that provide pathways for lightning. These include pipes (don't shower), stoves (don't cook), and wires (don't use the telephone). Electrical appliances pose no hazard if properly grounded, but why tempt fate by using them?

Some confusion surrounds the safety of motor vehicles during a lightning storm. A metal car or truck is a good shelter. Cars with cloth convertible tops and the backs of pickup trucks, on the other hand, are not. In Texas, in 1979, people riding in the back of a pickup truck were struck and killed by lightning while people riding in the cab of the truck were unhurt.

If a building or auto is not accessible when a thunderstorm approaches, find shelter under a cliff, in a cave, or in a low area, such as a ravine, a valley, or even a roadside ditch. Avoid: (1) tall, isolated objects, such as telephone poles and flagpoles, (2) metallic objects, such as wire fencing, rails, clotheslines, bicycles, and golf clubs, (3) high areas, such as hilltops and rooftops, and (4) bodies of water, such as swimming pools and lakes.

Individual trees in open spaces are hazardous, but a thick grove of small trees may offer safe haven. A group of people in the open should spread out. Lightning may be about to strike if your hair stands on end. In this unlikely event, immediately drop to your knees and, placing your hands on your knees, bend forward. One should not lie flat on the ground.

Contrary to popular belief, lightning can strike the same place more than once. New York City's Empire State Building typically is struck more than 20 times a year and on one occasion was hit 15 times in only 15 minutes. Such sites should be avoided during a thunderstorm.

Fortunately, two of every three persons "struck" by lightning recover fully. Most survivors are jolted by a nearby lightning bolt and not hit directly. Immediate mouth-to-mouth resuscitation or cardiopulmonary resuscitation may revive victims who are not breathing. Those who appear merely stunned may require treatment for burns or shock. Victims of lightning bolts carry no electrical charge and can be handled safely.

between a cloud and the ground, the stage is set for lightning. On a clear day, the earth's surface is negatively charged and the upper troposphere is positively charged. As a cumulonimbus cloud forms, however, a charge separation takes place within the cloud such that the upper region of the cloud becomes positively charged and the cloud base becomes negatively charged. The negatively charged cloud base then induces a positive charge on the portion of the ground underlying the cloud. Air is a very good insulator, and so, as a thunderstorm cell forms, electrical charges build and a tremendous potential soon develops for an electrical discharge, lightning. When the thunderstorm cell reaches its mature stage, the electrical resistance of the air breaks down and lightning occurs, thereby neutralizing the electrical charges. Lightning may take a path between the positive and the negative portions of a cloud, or between clouds, or between a cloud and the ground.

The cause of charge separation in cumulonimbus clouds is not well understood. But based on recent field studies (involving aircraft measurements of electric fields within thunderstorms) and laboratory simulations, a promising explanation is emerging. John Hallet of the University of Nevada's Desert Research Institute proposes that charge separation is a consequence of in-cloud collisions between ice crystals and graupel. **Graupel** (German for "soft hail") consists of pea-sized ice particles formed when supercooled water droplets collide and immediately freeze together. As ice crystals strike and bounce off graupel, opposite charges develop on both the ice crystals and graupel. Updrafts and downdrafts then separate the two types of charged particles and carry them to different portions of the cloud. In this way, the cloud is divided into positively and negatively charged regions.

We are most concerned about lightning discharged between a cloud and the ground, because this path poses the greatest hazard, although this discharge represents only about 20 percent of all lightning bolts. Using high-speed photography, scientists have determined that a lightning flash consists of a regular sequence of events. Initially, surges of negative electrical charge, called **stepped leaders**, travel from the cloud base to within 50 m (150 ft) of the ground. Stepped leaders are met by a positively charged **return stroke** from the ground. The return stroke forms a narrow conductive path, about 10 cm (4 in.) in diameter, between the cloud and the ground. Electrons flow, neutralization occurs, and the path is illuminated from ground to cloud by a lightning flash. Note that this return stroke is contrary to the common perception that a lightning flash progresses from cloud to ground. After this initial electrical discharge, subsequent surges of negative electrical charge (from the cloud) called **dart leaders**, follow the same conductive path and each dart leader is met by a return stroke (from the ground) and the conductive path is again illuminated. Typically, a single lighhning strike consists of two to four dart leaders plus return strokes. Electricity flows at the astonishing rate of nearly 50,000 km per second, so the entire lightning sequence just described occurs in less than two-tenths of a second. The human eye has difficulty in separating the individual flashes of light that constitute a single lightning bolt. Hence, we perceive a lightning flash as a flickering light.

In some cases, a dart leader is met by a return stroke that forges a new conductive path from the ground. The result is a forked lightning bolt that strikes the ground in more than one place.

Cloud-to-ground lightning appears as streaks or bolts (Figure 13.13). **Sheet lightning** consists of bright flashes across the sky and indicates cloud-to-cloud discharges. **Heat lightning** is simply light reflected by clouds from thunderstorms occurring beyond the horizon.

Where there is lightning, there is **thunder**, although sometimes we may see distant lightning and not hear the thunder. Lightning heats the air along the narrow conducting path to temperatures that may exceed 30,000 °C (54,000 °F). For this reason, people can be burned severely by a lightning bolt that strikes nearby. Such intense

FIGURE 13.13

Cloud-to-ground lightning appears as bright streaks or bolts. (Photograph by Arjen Verkaik)

heating expands the air violently and initiates a sound wave that we hear as thunder.

Because light travels about a million times faster than sound, we see the lightning almost instantaneously, but we hear the thunder later. The closer we are to the thunderstorm cell, the shorter the time interval will be between lightning flash and thunder. As a rule of thumb, thunder takes about 3 seconds to travel 1 km (and 5 seconds to travel 1 mi). If you must wait 9 seconds between lightning flash and thunderclap, the thunderstorm cell is about 3 km (1.8 mi) away.

Microbursts

Severe, and sometimes not so severe, thunderstorms can produce **microbursts**, also called **downbursts**. These are very intense downdrafts, lasting no more than 15 minutes, that can cause considerable damage over a small area, typically several square kilometers or less. In a microburst, the air surges downward, striking the ground at speeds near 100 km (62 mi) per hour, and then bursts radially outward along the ground. Microbursts are particularly dangerous for aircraft on takeoff or landing, because they trigger **wind shear**, an abrupt change in wind speed or direction over a short distance. Wind shear plays havoc with the aerodynamics of the aircraft and may cause an abrupt loss of altitude.

The existence of microbursts was first proposed by T. T. Fujita of the University of Chicago. Fujita has conducted an extensive investigation of the phenomenon, and he suggests that microbursts may have contributed to two commercial aircraft accidents in 1975. Apparently, the pilots unwittingly flew their jet planes through the center of a microburst cell. One pilot did so on takeoff; the other while attempting to land. Both planes abruptly lost altitude and crashed. The 8 July 1982 crash of a jet aircraft on takeoff from New Orleans may also have been caused by wind shear associated with a microburst. In retrospect, perhaps 27 or more civil airline accidents since 1964 have stemmed from encounters with microbursts.

An intensive field study of microburst activity was conducted during the summer of 1982 over a 1600 km^2 (615 mi^2) area near Denver's Stapleton International Airport. Using experimental aircraft and a dense grid of meteorological sensors, this **Joint Airport Weather Studies** (JAWS) project detected about 100 downbursts in only 86 observation days. One of the most important findings of JAWS is the apparent inadequacy of the **Low-Level Wind Shear**

Alert System (LLWSAS), an array of ground-level anemometers currently operated by the Federal Aviation Administration at 59 airports.* Many of Denver's downbursts were either too small or too far off the ground to be detected by LLWSAS. Indeed, microbursts may affect only a small portion of a runway. A promising alternative for detection of a microburst is a British airborne system that uses a laser-beam radar to provide the pilot with a 3-second warning of an approaching microburst. The most effective microburst detection system, however, would probably be an airport-based Doppler radar (described in Chapter 14).

Flash floods

Torrential rains often accompany severe thunderstorms. Prolonged heavy rains (more than 7.5 mm, or 0.3 in., per hour) can greatly exceed the infiltration capacity of the ground. Simply put, the ground cannot absorb all of the rainwater. Excess water runs off to creeks, streams, rivers, or sewers, or collects in low-lying areas. If a drainage system cannot accommodate the sudden input of huge quantities of water, **flash flooding** is the consequence.

Flash flooding is a special hazard in mountainous terrain where steep slopes channel runoff into narrow stream and river valleys. The water level of streams and rivers may rise so rapidly that campers and other visitors are caught by surprise and trapped. On 31 July 1976, in the Big Thompson Canyon of Colorado, torrential rains from a thunderstorm caused a sudden rise in water level that claimed 130 lives; most of the victims were campers. Indeed, the recent increase in flood fatalities in the United States can be attributed to the fact that more people are visiting remote areas prone to flash flooding. During the 1970s, flash floods took an average of 200 lives annually, twice the flood fatalities of the 1960s, and three times those of the 1940s.

Because of their design and composition, urban areas are also prone to flash floods during intense downpours. Concrete and asphalt render a city surface virtually impervious to water, so elaborate sewer systems must be installed to transport runoff to nearby natural drainageways. These systems have a limited capacity for water, however, and may be unable to handle the excess water produced during a torrential rainfall. Water backs up and collects under viaducts and in other low-lying areas. Sometimes water lev-

*Installation of 51 additional systems was planned by 1985.

els rise so abruptly in these areas that motorists are trapped in their vehicles (Figure 13.14). For example, on the evening of 1 August 1985, a slow-moving severe thunderstorm drenched Cheyenne, Wyoming, with 15.4 cm (6.06 in) of rain in less than four hours. Flash flooding filled city streets with up to 2 m (6 ft) of water, 12 people lost their lives, and early estimates of property damage were well in excess of $25 million.

FIGURE 13.14
Motor vehicles and occupants may be trapped by the sudden rise of water accompanying flash floods such as this one in Montgomery County, Maryland, 15 July 1975. (NOAA photograph)

Hail

Hail, or **hailstones**, is precipitation in the form of balls or lumps of ice. Hail falls from intense thunderstorm cells that are characterized by strong updrafts, great vertical development, and an abundant supply of supercooled water droplets. Hailstones range from pea size to the size of a grapefruit or even larger. The nation's largest hailstone on record was collected at Coffeyville, Kansas, on 3 September 1970. It weighed 758 g (1.67 lb) and measured 45 cm (17.5 in.) in circumference, and 14 cm (5.6 in.) in diameter—about the size of a softball.

On rare occasions, the fall of hail is so great that snowplows must be used to clear highways. Up to 45 cm (18 in.) of hail blanketed a 150 km^2 (54 mi^2) area in northwest Kansas on 3 June 1959. On 6

August 1980 at Orient, Iowa, drifts of hail were reported to be 1.8 m (6 ft) deep.

A hailstone forms when an ice pellet is transported through portions of a cumulonimbus cloud in which there are variable concentrations of supercooled water droplets. The ice pellet may descend gradually through the entire cloud, or it may follow a more complex pattern of ascent and descent as it is caught alternately in updrafts and downdrafts. In the process, the ice pellet grows by accretion (addition) of freezing water droplets, and eventually becomes so large and heavy that it falls out of the cloud base and reaches the ground as hail.

When an ice pellet enters portions of the cloud containing a relatively large concentration of supercooled water droplets, water collects on the ice pellet as a liquid film, which freezes slowly to form a transparent layer, or glaze. When the ice pellet travels through portions of the cloud where the concentration of water droplets is relatively low, droplets freeze immediately on contact with the ice pellet. As the droplets freeze, many tiny air bubbles are trapped within the ice, producing an opaque whitish layer of granular ice, or **rime**. The result is alternating lamina of clear and opaque ice, which, in cross section, resembles the internal structure of an onion. This concentric layering is evident when a hailstone is sliced open (Figure 8.4).

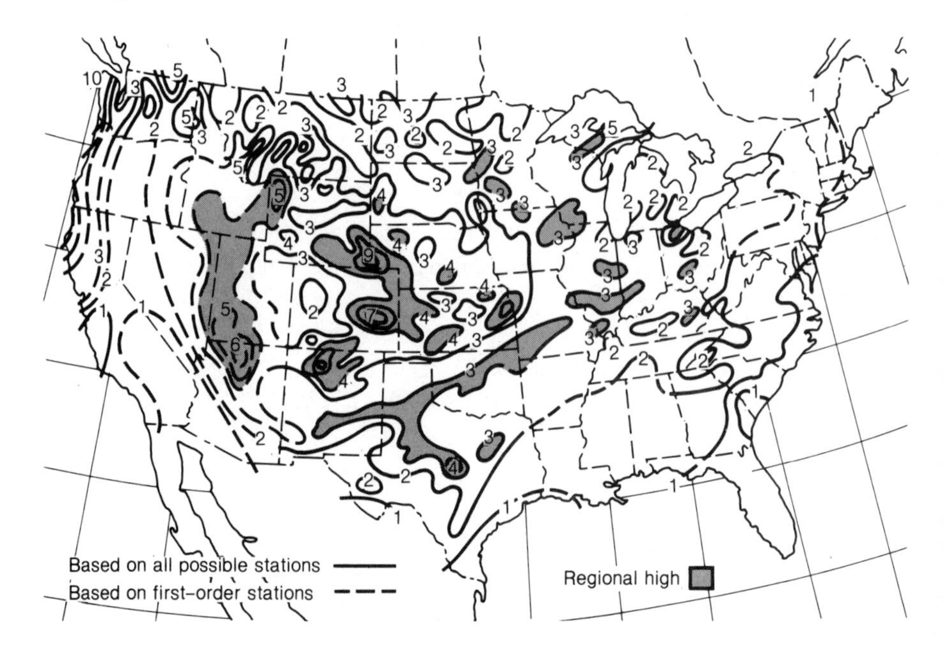

FIGURE 13.15

Hail frequency across the United States in average number of days per year. (Data from Illinois State Water Survey)

Perhaps surprisingly, hail frequency is not necessarily related to thunderstorm frequency. Although Florida experiences the greatest frequency of thunderstorms in the United States, hail is very unusual in Florida. In North America, hail is most likely on the High Plains just east of the Rockies, where it can be expected to fall from about 10 percent of all thunderstorms. Figure 13.15 is a map of hail frequency in the United States.

Because hail usually falls during the growing season and because large hailstones can cause considerable damage to crops, some efforts have been make to suppress or prevent hail. A brief historical sketch of these activities is presented in the Special Topic on hail suppression.

SPECIAL TOPIC

Hail suppression*

Efforts to suppress hail have deep historical roots. Indeed, in fourteenth century Europe, church bells were rung and cannons fired in the belief that the attendant noise would somehow ward off hail. A period of particularly intense hail suppression activity took place in the grape-growing regions of Austria, France, and Italy during the late nineteenth century. M. Albert Stiger, a wine grower and burgomaster of Windisch-Feistritz, Austria, designed and built a special funnel-shaped hail suppression cannon (Figure 1). Stiger be-

FIGURE 1

A hail suppression cannon popular in the grape-growing regions of Austria, France, and Italy during the late nineteenth century. (Photograph courtesy of S.A. Changnon and J.L. Ivens, Illinois State Water Survey, and copyright by the American Meteorological Society. Appeared in Changnon and Ivens, "History Repeated: The Forgotten Hail Cannons of Europe."* Bulletin AMS *62, no. 3 (1981):372.)

*This discussion is based on Changnon, S. A., Jr., and J. L. Ivens. "History Repeated: The Forgotten Hail Cannons of Europe." *Bulletin of the American Meteorological Society* 62 (1981):368–375.

(continued)

lieved that smoke particles in the cannon fire would inhibit hailstone development. Amazingly, in experimental firings in 1896 and 1897 at Windisch-Feistritz, Stiger reported no hail, although severe hail damage occurred in neighboring areas. Word of Stiger's apparent success spread throughout the vineyard regions of Europe, and hail cannons soon became commonplace. In fact, there were so many cannons that the accidental shooting of people became a problem in some localities. After Stiger's much heralded success was not duplicated elsewhere, however, the interest in hail cannons waned rapidly, and by 1905 this early attempt at weather modification ended.

The modern era of hail suppression experimentation began after World War II. Although founded on a much better, albeit not yet complete, understanding of cloud physics, the new techniques shared some similarities with earlier efforts. For example, until the practice was outlawed in the early 1970s, Italian farmers regularly fired explosive rockets into threatening clouds in an attempt to shatter developing hailstones. In the Soviet Union today, scientists fire silver iodide crystals, a cloud-seeding agent (see Chapter 8), into thunderclouds. They theorize that the silver iodide crystals will stimulate the formation of large numbers of small hailstones, which will melt long before they reach the ground, instead of the normal development of small numbers of larger hailstones, which could devastate crops. Among western scientists, much skepticism surrounds the Soviet technique. In the United States, where annual agricultural losses due to hail exceed $700 million (Figure 2), ground-based and aircraft cloud seeding were used in hail suppression experiments. In an interesting parallel with events in nineteenth century Europe, decreased public confidence in the effectiveness of modern hail suppression techniques brought an end in the late 1970s to the federal funding of hail suppression research.

FIGURE 2
Large hailstones devastated this cornfield in a matter of minutes. (NCAR/NSF photograph)

Conclusions

Thunderstorms are the products of convective currents that surge to great altitudes within the troposphere. As such, thunderstorms channel excess heat at the earth's surface into the atmosphere. They may also be accompanied by damaging lightning, winds, torrential rains, and hail. These, however, are not the only hazardous progeny of thunderstorms. The storms also spawn tornadoes, a subject of the next chapter. In Chapter 14, we also discuss hurricanes, the most violent of tropical storms, which often begin as a cluster of thunderstorms over tropical oceans.

SUMMARY STATEMENTS

- A thunderstorm is a mesoscale weather system produced by strong convective air currents that surge high into the troposphere.

- The life cycle of a thunderstorm cell consists of a three-stage sequence: cumulus, mature, and dissipating. The storm's maximum intensity is reached during the mature stage.

- A thunderstorm usually consists of a cluster of cells, each of which may be at a different stage in its life cycle. Thunderstorms typically track at a small angle to the path of the cell cluster.

- Thunderstorms usually develop in maritime tropical air as a consequence of uplift by one or more of the following: (1) frontal activity, (2) orographic effects, (3) surface convergence, or (4) intense solar heating of the earth's surface. They are classified as air mass thunderstorms, frontal thunderstorms, or mesoscale convective complexes.

- Severe thunderstorm cells typically form a squall line ahead of a fast-moving, well-defined cold front trailing a mature midlatitude cyclone. The midlatitude jet stream causes dry air to subside over a surface layer of mT air. This produces a layering of air that can lead to explosive convection and the development of severe thunderstorms.

- Lightning is light produced by the discharge of electricity within a cloud, between clouds, or between a cloud and the ground.

- Some severe thunderstorm cells may produce microbursts, very intense and potentially destructive downdrafts.

- Flash flooding is a special hazard in mountainous terrain where steep slopes channel excess runoff into narrow stream and river valleys.

- Hail develops in intense thunderstorm cells characterized by strong updrafts, great vertical development, and an abundant supply of supercooled water droplets.

KEY WORDS

thunderstorm
cumulus congestus
cumulonimbus
cumulus stage
updraft
mature stage
downdraft
gust front
roll cloud
shelf cloud
dissipating stage
air mass thunderstorm
frontal thunderstorm
squall line
mesoscale convective complex (MCC)
forced convection
free convection
severe thunderstorm
supercell
low-level jet stream
mammatus clouds
lightning
graupel
stepped leaders
return stroke
dart leaders
sheet lightning
heat lightning
thunder
microburst
downbursts
wind shear
Joint Airport Weather Studies (JAWS)
Low-Level Wind Shear Alert System (LLWSAS)
flash flooding
hail
hailstones
rime

REVIEW QUESTIONS

1 Describe the characteristics of each stage in the life cycle of a thunderstorm cell.

2 What is a gust front?

3 What is the significance of the relatively short life span of thunderstorms for local weather forecasting?

4 Distinguish between air mass thunderstorms and frontal thunderstorms. Which is likely to be more intense and why?

5 Why do air mass thunderstorms usually develop in the afternoon, during the warmest hours of the day?

6 What is a squall line?

7 Describe how thunderstorms develop over the interior of the Florida peninsula.

8 Why are thunderstorms most frequent in tropical latitudes and in continental interiors?

9 Under what conditions is a thunderstorm considered to be severe?

10 In what sector of a mature cyclone are thunderstorms most likely to develop?

11 Describe the synoptic situation in middle latitudes that is most favorable for the formation of severe thunderstorms.

12 What causes lightning? Why is it dangerous?

13 What causes thunder?

14 What is a microburst, and how might it pose a hazard for aircraft?

15 How might the flash flood hazard vary with the season in middle and high latitudes?

16 Why is flash flooding a particular hazard in mountainous terrain? In urban areas?

17 What type of thunderstorm may produce hail?

18 Explain the internal concentric layering of hailstones.

19 Where in the United States is hail most frequent?

20 Today, little research is directed at hail suppression. Explain.

POINTS TO PONDER

1 Explain how secondary thunderstorms might develop along a gust front.

2 What causes the anvil shape of a thunderstorm top?

3 Thunderstorms usually track at an angle to the paths of their constituent cells. What is the significance of this behavior for thunderstorm forecasting?

4 Maritime tropical air is conditionally stable. How is this significant for convection and thunderstorm development?

5 Explain the diurnal and seasonal variations in thunderstorm occurrence. Explain also why thunderstorms are rare along the coast of southern California and over snow-covered terrain.

6 Identify the roles of the midlatitude jet stream and the low-level mT jet stream in the formation of severe thunderstorms.

7 Why does a tilted updraft prolong the vertical development of a thunderstorm cell?

PROJECT

1 Review the available climatic summary data for your area for the frequency of (a) thunderstorms, (b) hail, and (c) flash flooding.

SELECTED READINGS

Changnon, S. A., Jr., and J. L. Ivens. "History Repeated: The Forgotten Hail Cannons of Europe." *Bulletin of the American Meteorological Society* 62 (1981):368–375. *A fascinating discussion of the parallels between modern and early weather modification efforts.*

Few, A. A. "Thunder." *Scientific American* 233, no. 1 (1975):80–90. *Discussion of the relationship between thunder and lightning flash.*

Fujita, T. T., and F. Caracena. "An Analysis of Three Weather-Related Aircraft Accidents." *Bulletin of the American Meteorological Society* 58 (1977):1164–1181. *Possible role of microbursts in aircraft accidents.*

Maddox, R. A., and J. M. Fritsch. "A New Understanding of Thunderstorms: The Mesoscale Convective Complex." *Weatherwise* 37 (1984):128–135. *Discusses the recent discovery of large circular clusters of thunderstorms.*

Marrero, J. "Danger: Flash Floods." *Weatherwise* 32 (1979):34–37. *In 1978, floods claimed more than 100 lives in the United States and caused damage that exceeded $1 billion.*

McCarthy, J., and R. Serafin. "The Microburst Hazard to Aircraft." *Weatherwise* 37 (1984):120–127. *Reviews what is currently understood about the genesis and characteristics of microbursts.*

Schulz, L. W. "The Central Kansas Flash Floods of June 1981." *Bulletin of the American Meteorological Society* 65 (1984):228–234. *Describes the synoptic weather conditions that led to two flash floods only 8 days apart.*

Wheeling, the careening
winds arrive with lariats
and tambourines of rain.
Torn-to-pieces, mud-dark
flounces of Caribbean

cumulus keep passing,
keep passing. By afternoon
rinsed transparencies begin
to open overhead, Mediterra-
nean
windowpanes of clearness

AMY CLAMPITT
The Kingfisher, "The Edge of the Hurricane"

Hurricanes are ocean storms. Those that strike coastal areas are often accompanied by a destructive surge of ocean water. (Photograph by John D. Cunningham)

FOURTEEN

Tornadoes and Hurricanes

THIS CHAPTER covers the genesis and characteristics of two special types of weather systems: the tornado and the hurricane. They are singled out because of their great threat to human life and the tremendous property damage they can cause. Actually, the two systems are quite different. Tornadoes are small-scale, short-lived disturbances that usually occur over certain continental localities at midlatitudes. Hurricanes are much larger and much longer-lived storms that spend much of their life cycle over tropical seas.

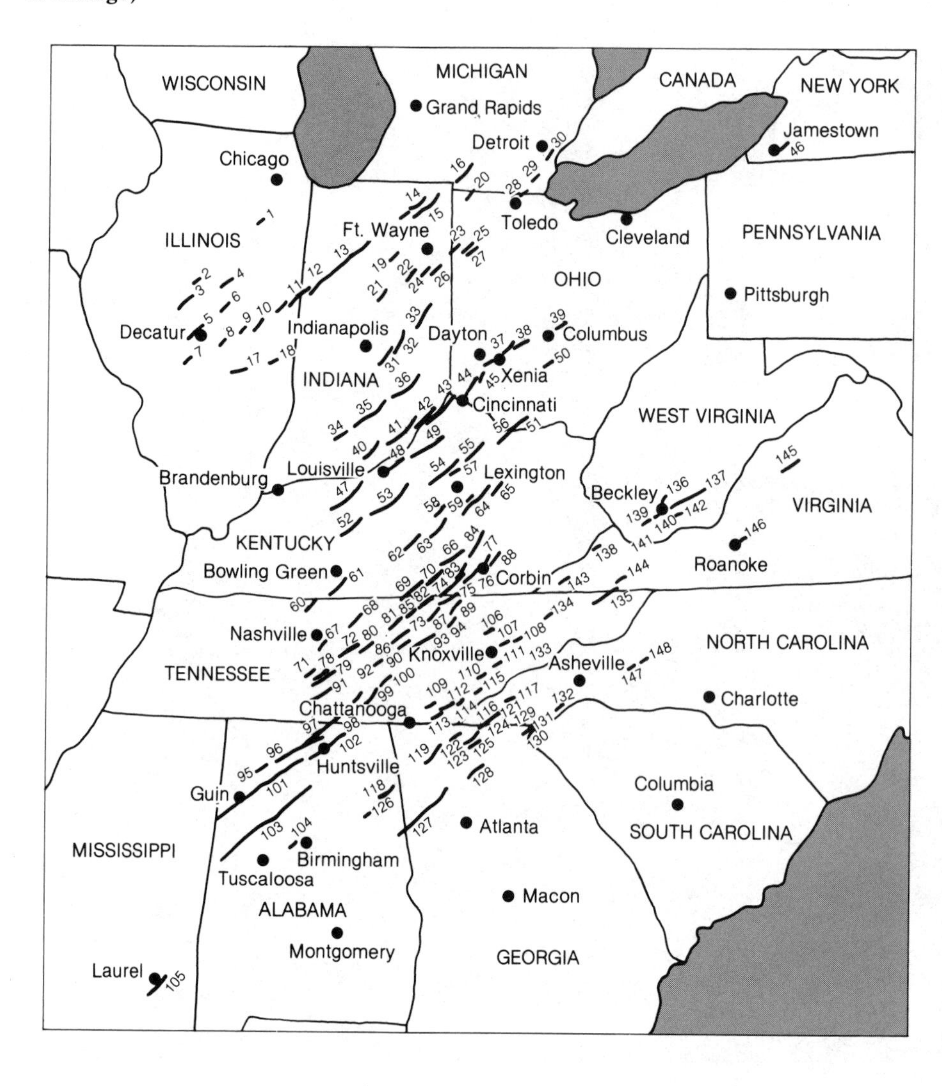

FIGURE 14.1
Tracks of the 148 tornadoes that hit a 13 state area on 3–4 April 1974. (From T. T. Fujita, The University of Chicago)

Tornado Characteristics

Tornadoes are by far the most violent of weather systems. They are fortunately small and short-lived storms that usually affect sparsely populated regions. Occasionally, however, a major tornado outbreak causes incredible devastation, death, and injury. On 3 and 4 of April 1974, 148 tornadoes left 315 people dead and 5484 people injured, and caused extensive property damage in 13 states (Figure 14.1). Major tornado outbreaks are listed in Table 14.1.

A **tornado** is a small mass of air that whirls rapidly about an almost vertical axis. It is made visible by clouds and by dust and debris sucked into the system. Tornadoes are approximately funnel shaped, although a variety of forms have been observed, ranging from cylindrical masses of nearly uniform lateral dimensions (Figure 14.2A) to long, slender, ropelike pendants (Figure 14.2B). By convention, when the circulation remains aloft, it is termed a **funnel cloud**, but when it touches down on the ground, it is called a tornado.

A typical tornado cuts a path on the ground 3 km (1.9 mi) long and about 50 m (150 ft) wide. The average affected area is therefore

TABLE 14.1

Five Greatest Tornado Outbreaks (in Number of Tornadoes Per Outbreak) From 1950–1978

	DATE	TORNADOES	DEATHS	INJURIES	AREA
1	**April 3–4, 1974**	**148**	**315**	**5484**	**Area between Mississippi River and Appalachian Mts.: Illinois to New York; Mississippi to Virginia**
2	**April 11, 1965**	**51**	**256**	**over 1500**	**Southern Great Lakes; Iowa to Ohio**
3	**May 4, 1959**	**46**	**0**	**2 (minor)**	**Great Plains: Oklahoma to Minnesota and Wisconsin**
4	**April 21, 1967**	**43**	**58**	**1068**	**Central Mississippi Valley to southern Great Lakes: Missouri to Michigan**
5	**January 9–10, 1975**	**42**	**11**	**287**	**Southern Plains, lower Ohio Valley and southeastern states: Texas to Alabama; Oklahoma to southern Indiana**

Source: Modified after NSSFC, "Tornadoes." *Weatherwise* 33 (1980):55

A

B

FIGURE 14.2

A tornado may appear as (A) a cylindrical mass of relatively uniform lateral dimensions or (B) a long, slender ropelike pendant. (NCAR photographs)

only about 0.15 km^2 (0.06 mi^2). Some notable departures from this average have, however, occurred. The most deadly tornado in North American history, the Tri-State tornado of 18 March 1925, left a 352-km (217-mi) long path of devastation stretching from southeastern Missouri through the southern tip of Illinois and into southwest Indiana. Along the path, fatalities totaled 689, injuries approached 2000, and 11,000 people were left homeless.

Tornadoes are formed by and travel with intense thunderstorm cells. Tornadoes and their parent thunderstorm cells usually (about 90 percent of the time) travel from southwest to northeast. Trajectories are often erratic, however, with many tornadoes exhibiting a hopscotch pattern of destruction as they alternately touch down and lift off the ground (Figure 14.3). The average forward speed is around 55 km (34 mi) per hour, although there are reports of tornadoes racing along at speeds approaching 240 km (149 mi) per hour.

An extremely steep horizontal air pressure gradient between the tornado center and edge is the force ultimately responsible for a tornado's violence. The pressure drop in only 100 m (300 ft) or so may be equivalent to the normal pressure drop between sea level and an altitude of 1 km, that is, about a 10 percent reduction. The Coriolis effect is also at work, but the system is so small that the effect is negligible. This means that the winds in a tornado may rotate in either a clockwise or a counterclockwise direction, although the latter dominates by far in Northern Hemisphere tornadoes. In the tornadic winds whirling about a vertical axis, the

FIGURE 14.3
The pattern of destruction produced by a tornado is often erratic, as shown in this aerial photograph of a residential area in Birmingham, Alabama, hit by a tornado on 4 April 1977. Note how some houses are completely destroyed while neighboring houses are still standing. (NOAA photograph)

inward-directed pressure gradient force is countered by the outward-directed centrifugal force.

When a tornado strikes, damage is caused by (1) very strong winds, (2) an abrupt pressure drop, and, to a lesser extent, (3) suction vortices. Winds sometimes estimated at approaching 500 km (310 mi) per hour blow down trees, power poles, buildings, and other structures. Flying debris has caused much of the death and injury associated with tornadoes.

It was once widely believed that a tornado can cause a building to explode. The air pressure within the building supposedly could not adjust rapidly enough to the abrupt pressure drop associated with the tornado. People were therefore advised, in the event of a tornado sighting, to open windows to help equalize the internal and external air pressure. However, most buildings have sufficient air leaks so that a potentially explosive pressure differential never develops. The destruction of buildings is due instead to very strong currents of air that blow over roofs and cause the structure to lift, much as air induces lifting as it flows over the curved upper surface of an airplane wing.

Suction vortices swirl about within a tornado and act like vacuum cleaners, sucking debris into the circulation. These vortices are sometimes strong enough to lift freight cars off their tracks.

In addition to reports of the terrible devastation caused by a tornado strike, unusual and even bizarre events associated with tornadoes are sometimes recounted. These include observations of straws driven deeply into trees, a rain of fish or frogs (presumably pulled out of a nearby pond or lake), and autos and other objects lifted, carried a kilometer or more, and redeposited on the ground. Among the oddest reports is the deplumation of chickens and other fowl. No one knows why fowl lose their feathers during a tornado. Speculation is that the phenomenon is due to the pressure drop within the tornado, to strong winds (perhaps aided by natural molting), or to the relaxation of the feather follicles—a reaction brought on by nervous stress.

T. T. Fujita of the University of Chicago has devised a six-point intensity scale for evaluating tornado strength. Called the F-Scale and presented as Table 14.2, it is based on estimated wind speeds and categorizes tornadoes as weak, strong, or violent. Using this scheme, Fujita reports that of the tornadoes that occurred in the United States between 1950 and 1978, about 63 percent were rated as weak and only 2 percent were violent. The violent systems were responsible for 68 percent of the total fatalities, however. These statistics are updated through the early 1980s in Figure 14.4. (An increase in tornadoes recorded in the United States can be attributed to a better observation network rather than to any real increase in tornado frequency.)

Tornado Genesis and Distribution

The previous chapter considered the special synoptic-scale weather pattern favorable for the development of severe thunderstorms. To this pattern must be added another requirement for tornado de-

TABLE 14.2

The Fujita Tornado Intensity Scale

		ESTIMATED WIND SPEED	
F-SCALE	CATEGORY	IN KM PER HOUR	IN MI PER HOUR
0	**Weak**	**65–118**	**40–73**
1		**119–181**	**74–112**
2	**Strong**	**182–253**	**113–157**
3		**254–332**	**158–206**
4	**Violent**	**333–419**	**207–260**
5		**420–513**	**261–318**

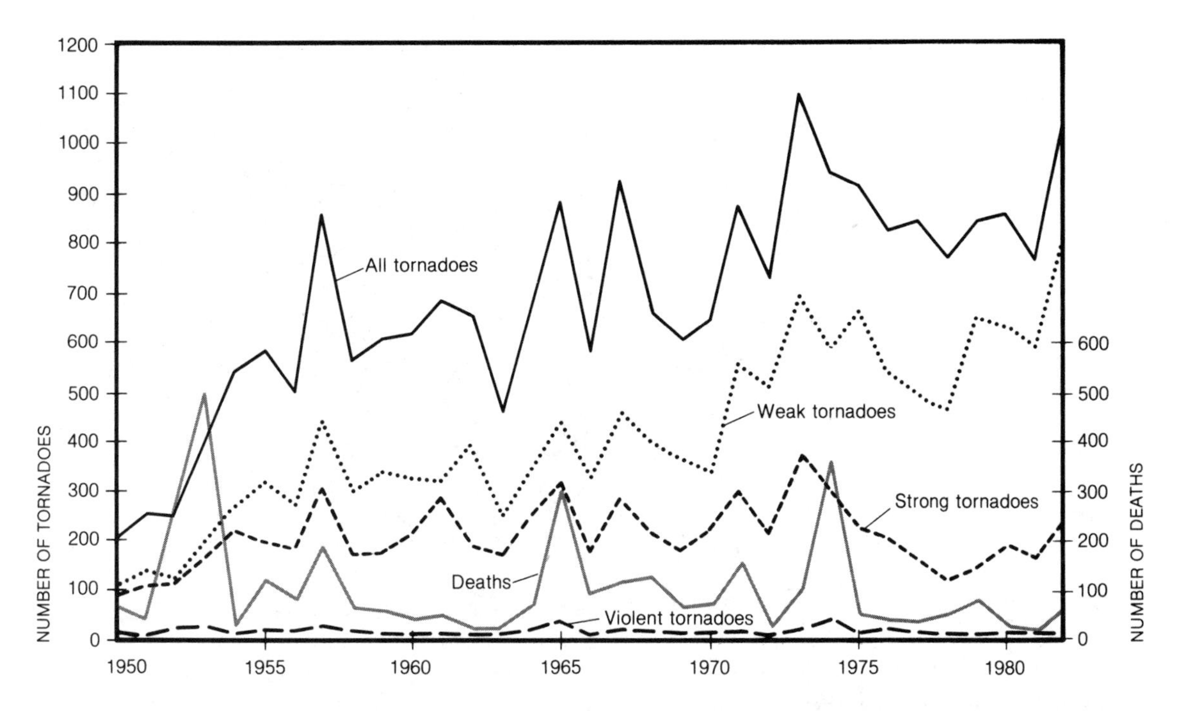

FIGURE 14.4

Annual totals of tornadoes by intensity category and tornado-related deaths in the United States. (From Snow, J. T. "The Tornado." *Scientific American* 250, no. 4 (1984):95. Copyright © 1984 by Scientific American, Inc. All rights reserved.)

velopment: tornadoes are most likely to form over relatively flat, dry terrain. They are rare in areas of great topographic relief such as the Rocky Mountain region. Flat terrain offers a minimum of frictional resistance, and dry conditions mean that most of the absorbed solar radiation is channeled into sensible heating, thereby spurring deep convection.

Central and southeastern portions of the United States are among the few places in the world where synoptic weather conditions and terrain are ideal for tornado development; the interior of Australia is another. Although tornadoes have been reported in all 50 of the states and throughout southern Canada, most occur in **tornado alley**, a corridor stretching from the Texas Panhandle northeast to Missouri and including Oklahoma, Kansas, and portions of Nebraska. Central Oklahoma has the maximum annual incidence of tornadoes (Figure 14.5).

The spring maximum

Almost two thirds of all tornadoes develop during the warmest hours of the day, and almost three quarters of tornadoes in the

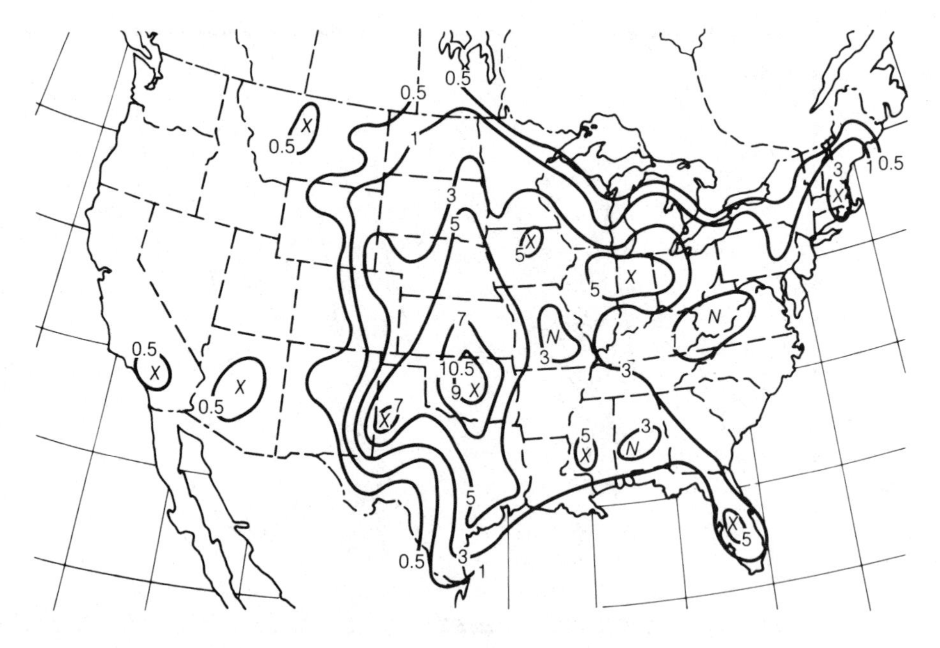

FIGURE 14.5

Tornado frequency in number per year within areas defined by 91-km-radius (56.5-mi-radius) circles, based on 29 years of data collected since 1 January 1950. An *X* indicates a relative maximum, and an *N* denotes a local minimum. (From NSSFC. "Tornadoes." *Weatherwise* 33, no. 2 (1980):54.)

United States occur from March to July (Figure 14.6). The peak months of tornado frequency are April (15 percent), May (22 percent), and June (20 percent). During these times, the weather conditions are optimal for spurring deep convection and the severe thunderstorms that spawn tornadoes.

One factor that contributes to the spring peak in tornado frequency is the relative instability of the lower atmosphere during that time of year. During the transition from winter to summer, days lengthen and insolation increases, thereby warming the ground. Heat is transported from the ground into the troposphere (Chapter 4), but it takes time for the entire troposphere to adjust to the heating from below. The upper troposphere, in fact, usually retains its winter coldness well into spring. The result is a relatively steep air temperature lapse rate that is favorable for deep convective activity and severe thunderstorm development.

Another factor that contributes to the spring tornado maximum is the greater likelihood that ideal synoptic weather conditions will occur at that time of year. Recall from Chapter 13 that severe thunderstorms develop in the southeast (warm) sector of a strong mid-latitude cyclone. Such cyclones achieve their greatest intensity when there are sharp temperature contrasts across the nation, that is, in spring when the polar front is best defined.

During spring there is a steady northward progression of tornado outbreaks. In effect, the center of maximum tornado frequency

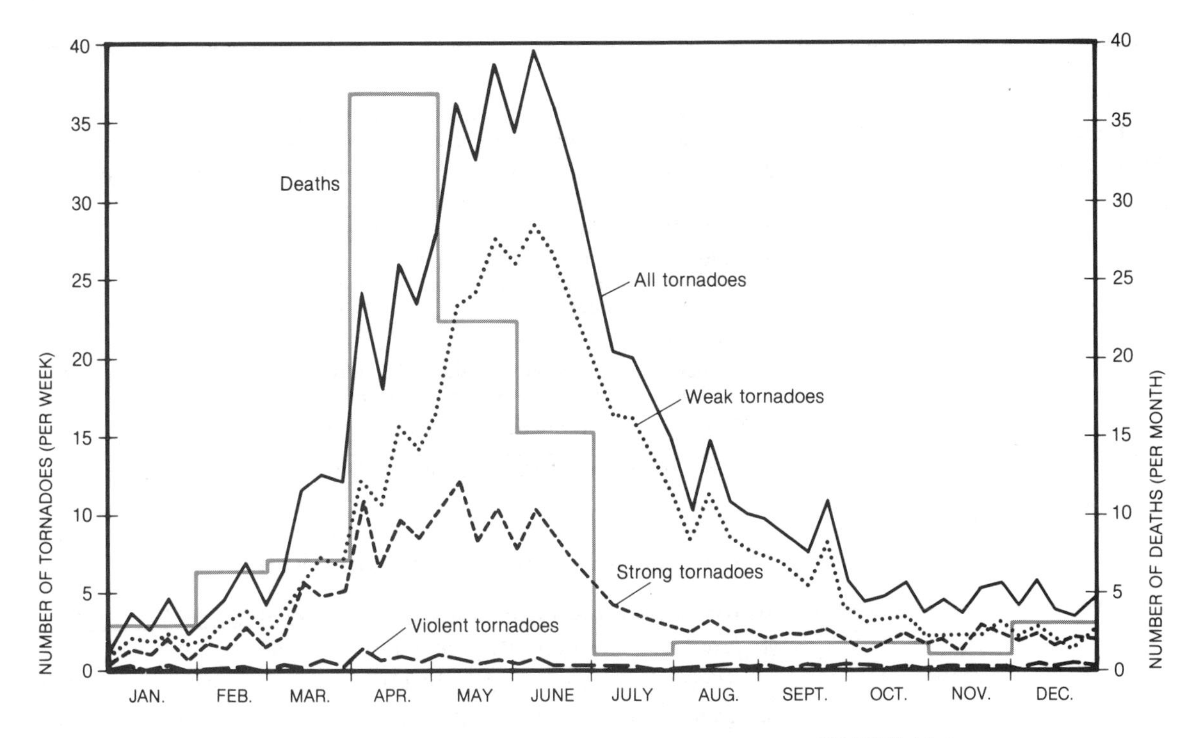

FIGURE 14.6

Average weekly tornado occurrence in the United States from 1950 to 1982 and monthly death totals from tornadoes. (From Snow, J. T. "The Tornado." *Scientific American* 250, no. 4 (1984):95. Copyright © 1984 by Scientific American, Inc. All rights reserved.)

follows the sun, as do the midlatitude jet stream, the principal storm tracks, and the northward incursion of maritime tropical (mT) air. In late winter the maximum tornado frequency is therefore along the Gulf Coast states. By May the maximum frequency reaches the southern Great Plains, and by June the highest tornado incidence is in the northern plains and the prairie provinces east of the Rockies.

Perhaps 80 percent of all North American tornadoes are linked to midlatitude cyclones. Most of the others are the product of convective instability triggered by hurricanes. In fact, most hurricanes that strike the southeastern United States are accompanied by tornadoes. The tornadoes usually develop on the northwest flank of a hurricane, after the system has turned toward the northeast.

What are your chances of experiencing a tornado? Very slim. Even in the most tornado-prone regions of North America, a tornado is likely to strike only once every 250 years. There are, of course, exceptions to this rule. Tornadoes have hit Oklahoma City no less than 26 times since 1892. To be prepared in the unlikely event that a tornado should strike your area, study the recommendations in Table 14.3. These precautions will reduce the hazard posed by tornadoes.

TABLE 14.3

What to Do if a Tornado Is Approaching*

1	**Seek shelter in a tornado cellar, an underground excavation, or a steel-framed or substantial reinforced-concrete building.**
2	**Avoid auditoriums, gymnasiums, supermarkets, or other structures having wide, free-span roofs.**
3	**In an office building, go to an interior hallway on the lowest floor.**
4	**At home, go to the basement. If there is no basement, go to a small room (closet or bathroom) in the center of the house on the lowest floor. Seek shelter under a mattress or a sturdy piece of furniture.**
5	**Stay clear of all windows.**
6	**In open country, travel at right angles to the tornado track. If this is not possible, lie flat in a ravine, creek bed, or open ditch.**
7	**Do not seek shelter in mobile homes, automobiles, or trucks.**

*Modified slightly after NOAA recommendations.

The tornado-thunderstorm connection

Tornadoes develop in the strong updraft of severe thunderstorms (Figure 14.7), although the precise physical relationship between the two systems is not completely understood. Tornadic circulation apparently stems from an interaction between the updraft and the horizontal wind. The horizontal wind must exhibit strong vertical shear in both speed and direction, that is, wind speed must increase with altitude and wind direction must veer from southeast at the surface to west aloft. The shear in wind speed produces air rotation about a horizontal axis. When this rotation interacts with the updraft, the region of rotating air is tilted to a vertical position. This rotation of air about a vertical axis is added to by the shear in the horizontal wind direction. As a consequence, the entire updraft spins as a cylinder 10 to 20 km (6.2 to 12.4 mi) in diameter. This circulation system, known as a **mesocyclone**, may or may not give rise to a tornado.

The mesocyclone circulation actually begins in the mid-troposphere and builds upward and downward from there. Meanwhile, the updraft strengthens as more air converges toward the base of the thunderstorm. The updraft may eventually become so strong that it overshoots the top of the thunderstorm and produces a cloud bulge on top of the thunderstorm anvil (Figure 14.7).

For reasons again not well understood, the mesocyclone in a tornadic thunderstorm narrows and spirals downward toward the ground as a funnel. As the spinning mass of air narrows, its speed increases, perhaps to violent proportions. This increased speed is analogous to that of an ice skater performing a pirouette: the skater's rate of spin increases as the arms are brought closer to the body.

One major reason why this description of tornado genesis is tentative is the rarity of direct instrument measurements of tornadoes. Much of what is known about the circulation of tornadoes and tornadic thunderstorms has been derived from the analysis of motion picture footage and from laboratory and computer simulations. The future is promising, however, in terms of tornado monitoring. A new type of radar, called Doppler radar (discussed later in this chapter), can better resolve the circulation inside thunderstorms than the traditional radar. Doppler radar is now undergoing extensive field testing. Since 1981, scientists at the University of Oklahoma have had some success in obtaining direct data on tornadoes by dropping a special tornado-resistant instrument package (called TOTO*) from a speeding truck in the path of oncoming storms. The traditional weather instruments are usually destroyed by tornadoes.

*TOTO is the acronym for Totable Tornado Observatory, as well as the name of Dorothy's dog in *The Wizard of Oz*.

FIGURE 14.7

Components of a severe thunderstorm that spawns a tornado. (From Snow, J. T. "The Tornado." *Scientific American* 250, no. 4 (1984):91. Copyright © 1984 by Scientific American, Inc. All rights reserved.)

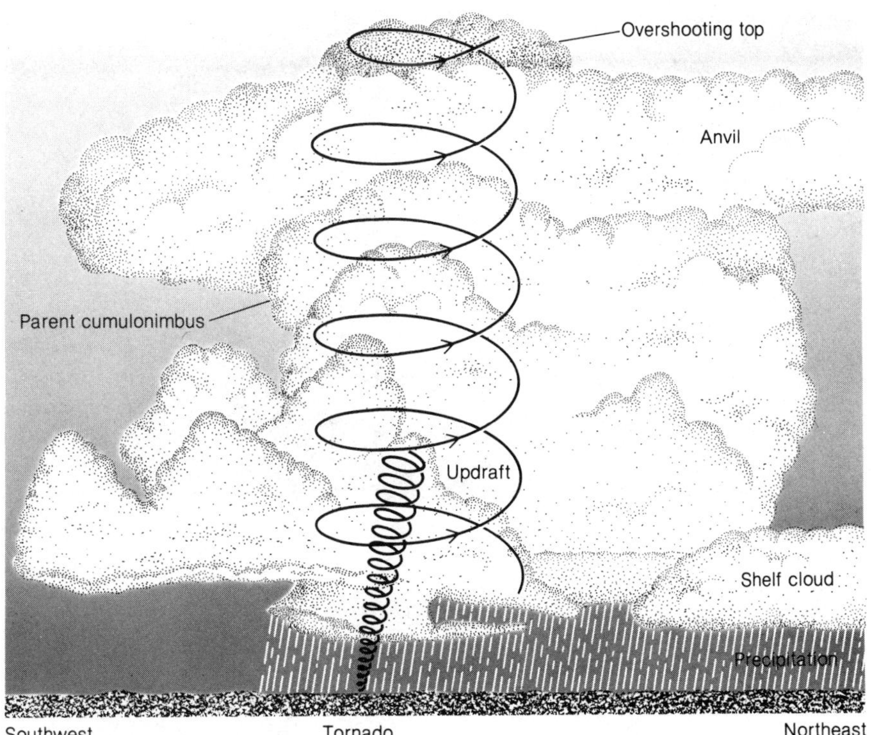

Waterspouts and dust devils

A **waterspout** is a tornadolike disturbance that occurs over the ocean or over a large inland lake (Figure 14.8). It is so named because it is composed of a whirling mass of water that appears to stream out of the base of the parent cumulonimbus cloud. A waterspout is usually considerably less energetic, smaller, and shorter lived than a tornado. The most intense waterspouts are probably tornadoes that formed over land and then moved over a body of water. In any event, waterspouts should be avoided by boaters.

A tornado does have some look-alikes in nature. Distant **virga** (rain or snow that vaporizes before reaching the ground) may be mistaken for a tornado because of its cylindrical or funnel shape (see Figure 8.2). The absence of any rotary motion, however, clearly distinguishes virga from a funnel cloud. In deserts or wherever exposed soil dries out, intense solar heating often gives rise to swirling masses of dust, called dust devils, which resemble tornadoes but form at ground level, are not associated with any clouds, and cause little if any damage (see Chapter 12).

FIGURE 14.8
Waterspouts like these off the Grand Bahama Islands are weak tornado-like disturbances that travel over or form over a large body of water. (NOAA photograph)

Hurricanes

Hurricanes are violent oceanic cyclones that originate in tropical latitudes, usually in the fall of the year. They produce torrential rains and winds of at least 120 km (74 mi) per hour. Wind gusts often top 250 km (155 mi) per hour, although they quickly diminish once the storm reaches land. As noted earlier, it is not unusual for hurricanes approaching coastal areas to spawn destructive tornadoes. Torrential rains may continue, even if the storm tracks well inland, and may be so heavy that flash flooding occurs.

More than heavy winds and rains, the most destructive aspect of hurricanes is the surge of ocean waters that accompanies the storm. Hurricanes cause ocean waters to inundate low-lying coastal areas rapidly, ravaging property and taking many lives. The situation is compounded when this **storm surge** coincides with high tide. In November 1970, for example, a storm surge hit the Bay of Bengal coast of Bangladesh, claiming more than 200,000 lives by drowning.

Just as scientists have developed an intensity scale for tornadoes, they have also developed one for hurricanes. Known as the Saffir/Simpson Hurricane Intensity Scale, after its designers H. S. Saffir and R. H. Simpson, it assigns ratings to hurricanes, from 1 to 5 corresponding to increasing intensity. As shown in Table 14.4, each intensity category specifies (1) range of central air pressure and wind speed and (2) potential for storm surge and damage. Of the 129 hurricanes that struck the U.S. Gulf and Atlantic coasts from 1900 to 1978, 53 (41 percent) are classified as "major," that is, they rated 3 or higher on the Saffir/Simpson scale. The 10 deadliest and 10 costliest U.S. hurricanes for the period 1900–1982 are identified in Tables 14.5 and 14.6, respectively.

Although much can be said about the disastrous aspects of hurricanes, their rains can be beneficial. For example, in the southeastern United States, tropical storms (hurricanes and their precursors) account for an average 10 to 15 percent of the June through October rainfall.

TABLE 14.4

Saffir/Simpson Hurricane Intensity Scale

SCALE NUMBER (CATEGORY)	CENTRAL PRESSURE		WIND SPEED		STORM SURGE		DAMAGE
	MB	IN.	MPH	KM/H	FT	M	
1	≥ 980	≥ 28.94	74–95	121–154	4–5	1–2	**Minimal**
2	965–979	28.50–28.91	96–110	155–178	6–8	2–3	**Moderate**
3	945–964	27.91–28.47	111–130	179–210	9–12	3–4	**Extensive**
4	920–944	27.17–27.88	131–155	211–250	13–18	4–6	**Extreme**
5	< 920	< 27.17	> 155	> 250	> 18	> 6	**Catastrophic**

TABLE 14.5

Ten Deadliest Hurricanes to Strike the United States in the Period 1900–1982

HURRICANE	YEAR	CATEGORY	DEATHS
1 **Texas** *(Galveston)*	**1900**	**4**	**6000**
2 **Florida** *(Lake Okeechobee)*	**1928**	**4**	**1836**
3 **Florida** *(Keys/S. Texas)*	**1919**	**4**	**600–900***
4 **New England**	**1938**	**3†**	**600**
5 **Florida** *(Keys)*	**1935**	**5**	**408**
6 **Audrey** *(Louisiana/Texas)*	**1957**	**4**	**390**
7 **Northeast U.S.**	**1944**	**3†**	**390‡**
8 **Louisiana** *(Grand Isle)*	**1909**	**4**	**350**
9 **Louisiana** *(New Orleans)*	**1915**	**4**	**275**
10 **Texas** *(Galveston)*	**1915**	**4**	**275**

Source: Hebert, P. J., and G. Taylor. "The Deadliest, Costliest, and Most Intense United States Hurricanes of This Century (and Other Frequently Requested Hurricane Facts)." NOAA Technical Memorandum, NWS, NHC 18, 1983

*Over 500 of these lost on ships at sea.

†Storm center moving more than 48 km (30 mi) per hour.

‡Some 344 of these lost on ships at sea.

TABLE 14.6

Ten Costliest Hurricanes to Strike the United States in the Period 1900–1982

HURRICANE	YEAR	CATEGORY	DAMAGE IN U.S. DOLLARS
1 **Frederic** *(Ala./Miss.)*	**1979**	**3**	**2,300,000,000**
2 **Agnes** *(Fla./Northeast U.S.)*	**1972**	**1**	**2,100,000,000**
3 **Camille** *(Mississippi/La.)*	**1969**	**5**	**1,420,700,000**
4 **Betsy** *(Florida/Louisiana)*	**1965**	**3**	**1,420,500,000**
5 **Diane** *(Northeast U.S.)*	**1955**	**1**	**831,700,000**
6 **Eloise** *(Northwest Florida)*	**1975**	**3**	**550,000,000***
7 **Carol** *(Northeast U.S.)*	**1954**	**3†**	**461,000,000**
8 **Celia** *(South Texas)*	**1970**	**3**	**453,000,000**
9 **Carla** *(Texas)*	**1961**	**4**	**408,000,000**
10 **Donna** *(Fla./Eastern U.S.)*	**1960**	**4**	**387,000,000**

Source: Hebert, P. J., and G. Taylor. "The Deadliest, Costliest, and Most Intense United States Hurricanes of This Century (and Other Frequently Requested Hurricane Facts)." NOAA Technical Memorandum, NWS, NHC 18, 1983

*Includes $60 million in Puerto Rico.

†Storm center moving more than 48 km (30 mi) per hour.

Characteristics of a hurricane

Perhaps the most convenient way to describe a hurricane is to contrast it with midlatitude, or "extratropical," cyclones, which we examined in Chapter 11. Hurricanes develop in a uniform mass of very warm, humid air, so they have no fronts or frontal weather. Air pressure is distributed symmetrically about the system center, and thus, as shown in Figure 14.9, isobars form a series of closely spaced concentric circles. Typically, the central pressure is considerably lower and the horizontal pressure gradient much stronger in a hurricane than in an extratropical cyclone. In addition, the tropical disturbance is a smaller system, averaging a third of the diameter of a midlatitude cyclone.

Structurally, a mature hurricane is a warm-core, low pressure system that weakens rapidly with altitude—especially above 3 km (9000 ft). In the upper troposphere, it is usual to find anticyclonic airflow above the hurricane at an altitude of 15 km. At the hurricane center is an area of almost cloudless skies, light winds, and gently subsiding air, called the **eye of the hurricane**, or eye of the storm (Figure 14.10). The eye is typically only 10 to 15 km (6 to 10 mi) across. Many people are lulled into a false sense of security when clearing skies and slackening winds follow a hurricane's initial blow. They may well be experiencing the hurricane's eye, and heavy rains and ferocious winds will soon resume, but from a different direction.

FIGURE 14.9

Surface pattern of isobars on the morning of 10 August 1980 as hurricane Allen tracked northwestward from the Gulf of Mexico and into extreme southeast Texas. The isobars (in millibars) form closely spaced concentric circles about the hurricane center. See Figure 14.11 for the corresponding satellite photograph. (Redrawn from NOAA weather map)

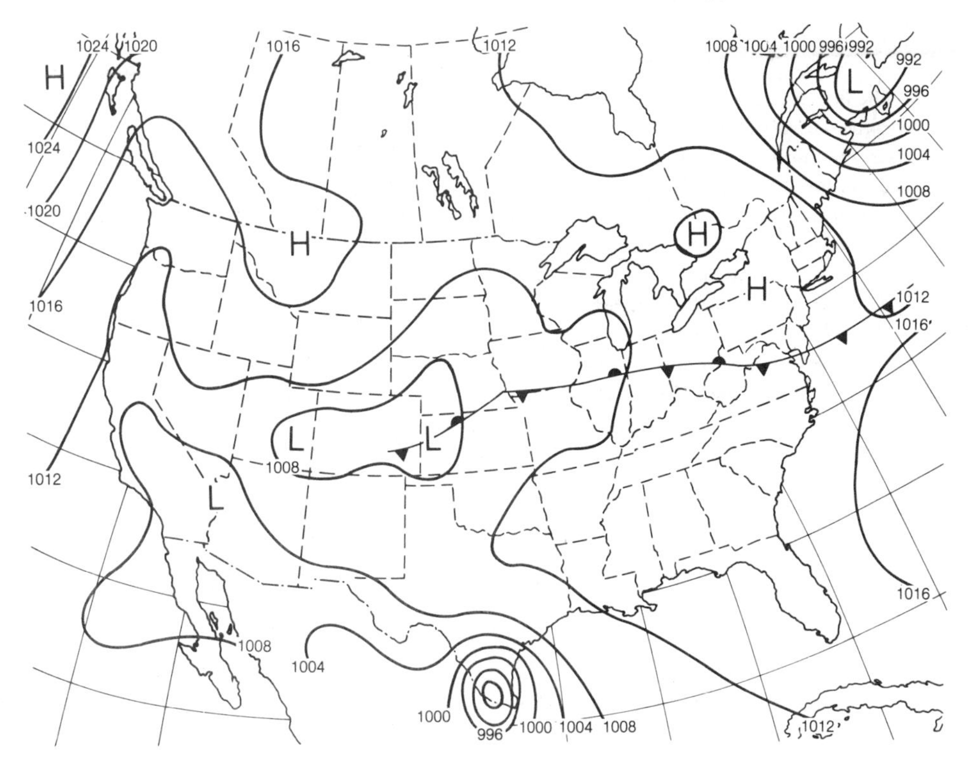

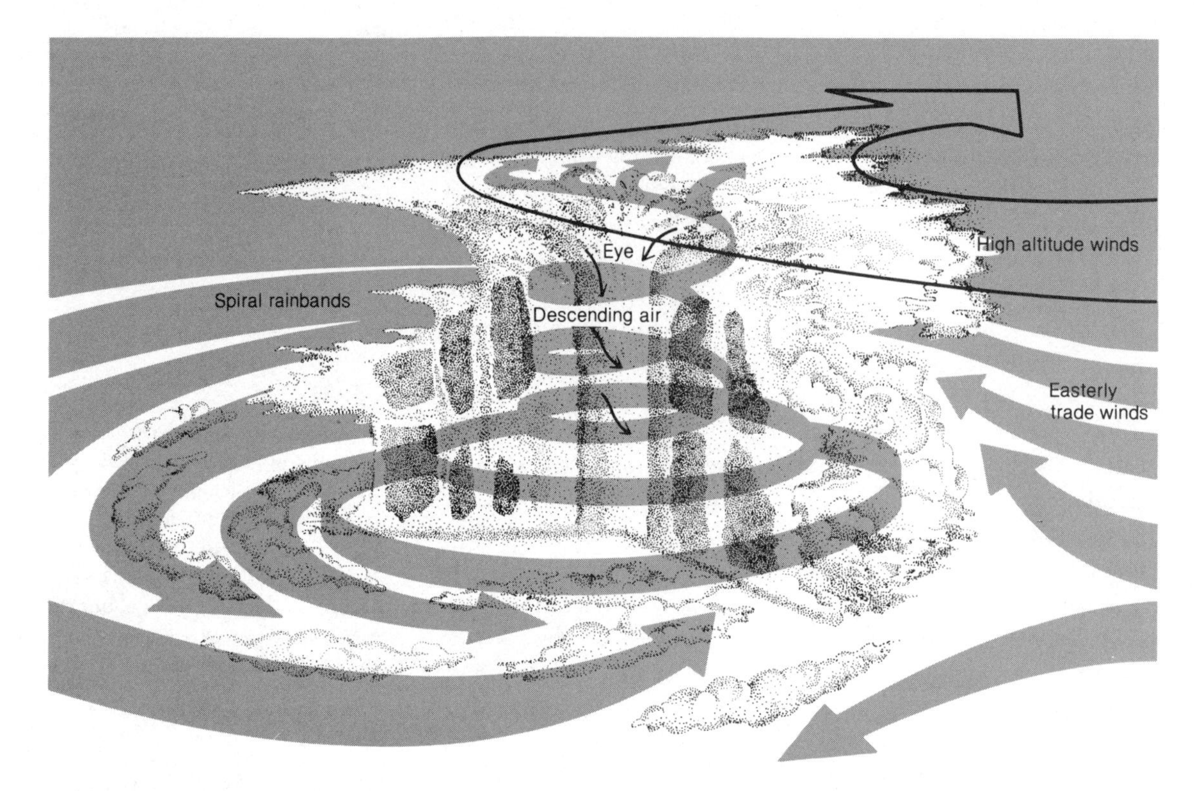

FIGURE 14.10

The internal structure of a hurricane as determined by radar and satellite monitoring. This artist's conception has a greatly exaggerated vertical dimension. Actual hurricanes are less than 15 km in altitude and have diameters of several hundred kilometers. (From NOAA, *Hurricane*. Washington, D.C.: Superintendent of Documents, 1977, p. 11.)

Bordering the eye of a mature hurricane is the **eye wall**, a circle of cumulonimbus clouds that produce heavy rains and the strongest winds. Spiral cloud bands accompanied by hurricane force winds and heavy convective showers curve outward from the eye wall. All this is surrounded by an outer region of high clouds (cirrus or cirrostratus) and cyclonic winds. The typical cloud pattern associated with a hurricane is shown in the satellite photograph in Figure 14.11.

The life cycle of a hurricane

Two conditions must be met before a hurricane can develop. One of these is very warm surface ocean water. In fact, it appears that 27 °C (81 °F) is the minimum surface water temperature for hurricane formation. Such exceptionally warm water sustains the hurricane circulation by the latent heat released when water evaporated from the ocean surface subsequently condenses within the storm. The warmer the water, the more readily the water will vaporize. Thus when a hurricane passes over land or over cold ocean water, it loses

FIGURE 14.11
A satellite photograph of the cloud pattern associated with hurricane Allen over southeast Texas on 10 August 1980. (NOAA photograph)

its energy source. As a result, the system weakens rapidly and winds diminish abruptly.

A second prerequisite for hurricane formation is a significant Coriolis effect. The influence of the earth's rotation must be strong enough to induce and sustain a cyclonic circulation. As noted in Chapter 9, the Coriolis effect weakens toward lower latitudes and becomes zero at the equator. The minimum latitude where the Coriolis effect is strong enough for hurricane formation is about 4 degrees.

The required combination of appropriate latitude and warm surface water temperatures is achieved only over certain portions of the world's oceans, identified in Figure 14.12. Major hurricane breeding grounds are (1) the western tropical North Pacific, where hurricanes are called **typhoons**,* (2) the Indian Ocean (including the Bay of Bengal and the Arabian Sea) and the tropical waters north of Australia, where hurricanes are termed simply cyclones, (3) the tropical North Atlantic west of the bulging west coast of Africa, and (4) the Pacific Ocean just southwest of Central America. The absence of hurricanes off either South American coast is noteworthy and is due to the relatively cold ocean water. Only hurricanes spawned over

*By convention, these storms over the Pacific Ocean are called "typhoons" when they occur west of 180 degrees longitude, and "hurricanes" when they occur east of 180 degrees longitude.

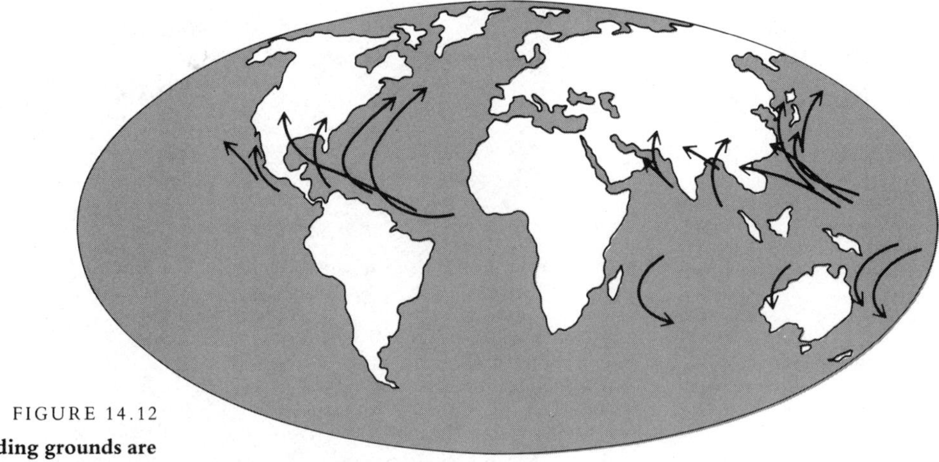

FIGURE 14.12

Hurricane breeding grounds are located in specific portions of the world's oceans. The arrows indicate average hurricane trajectories.

the North Atlantic pose a threat to coastal North America, although those that form west of Central America may strike Mexico or Hawaii and bring heavy rains to the desert Southwest.

The requirement of a relatively high surface water temperature makes hurricane occurrence distinctly seasonal. Because of the great thermal stability of ocean water, surface ocean water temperatures reach a seasonal maximum long after the time of peak solar radiation. Most Atlantic hurricanes consequently develop in late summer and in autumn, from August through October.

The first sign that a hurricane may be in the making is an area of organized thunderstorm activity over tropical seas. Chances are that this convective activity initially was triggered by a disturbance along the ITCZ, by a trough in the westerlies intruding into the tropics from the midlatitudes, or by a wave (or ripple) in the easterly trade winds. If conditions favorable to hurricane formation persist, a cyclonic circulation develops and the central air pressure begins to fall. A drop in air pressure signals the beginning of an energy cascade. Water vapor condenses within the storm, releasing latent heat. The heated air then rises, causing more condensation and the release of even more latent heat. Rising temperatures coupled with an anticyclonic outflow of air aloft cause a sharp drop in air pressure, which, in turn, induces convergence of air at the surface. The consequent uplift around the developing eye leads to more condensation and the release of more latent heat. Through this process, the **tropical depression** (low) intensifies, and winds strengthen. When wind speeds reach 60 km (40 mi) per hour, the system is classified as a **tropical storm**. If winds exceed 120 km (74 mi) per hour, the storm is officially designated a **hurricane**.

Hurricanes that threaten the eastern and southern coasts of the United States usually drift very slowly with the trade winds westward across the tropical North Atlantic and into the Caribbean. At this stage in the storm's trajectory, it is not unusual for the storm to move at a mere 10 to 20 km (6 to 12 mi) per hour. Once in the western Atlantic, however, the storm usually speeds up and begins to curve north and then northeastward, as it is caught in the midlatitude westerly airflow. Precisely where this trajectory change takes place determines whether the hurricane will enter the Gulf of Mexico, move up along the eastern seaboard, or curve back out to sea. By the time the storm reaches a latitude of 30 degrees N, it may begin to acquire extratropical characteristics as colder air is drawn into the system and as fronts develop. From then on, the storm follows a life cycle similar to that of any other midlatitude cyclone and ends by occluding over the North Atlantic.

Many hurricanes depart significantly from the track just described. The hurricane tracks shown in Figure 14.13, for example, are quite erratic. Sometimes a hurricane path makes a complete circle or reverses direction. In addition, some hurricanes maintain

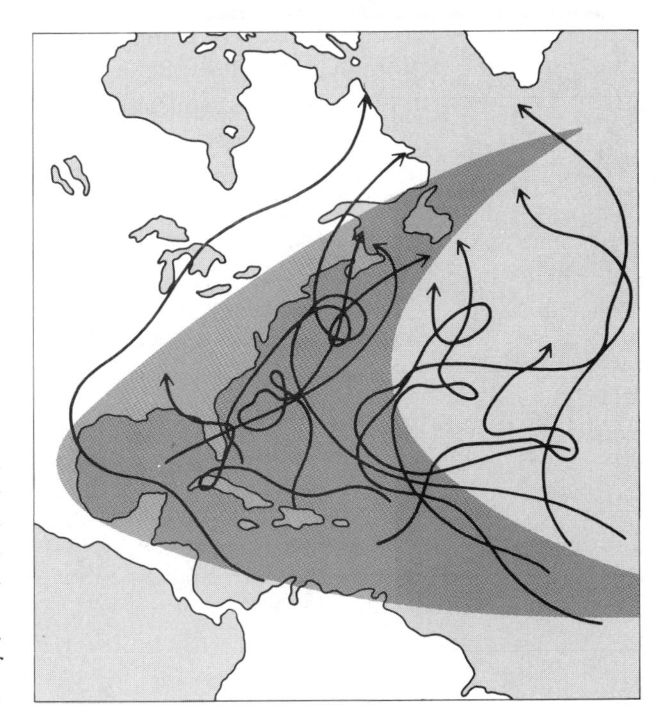

FIGURE 14.13

Hurricane trajectories are often erratic, as shown by these sample tracks. As indicated by the shaded area, however, hurricanes describe a general westerly drift and then a recurvature toward the northeast. (From NOAA, *Hurricane*. Washington, D.C.: Superintendent of Documents, 1977, p. 13.)

their tropical characteristics even after traveling far north along the Atlantic coast or well inland. The eye of hurricane Hazel, for example, was still discernible when the storm passed over Toronto in October 1954. New England, located more than 25 degrees of latitude north of the usual hurricane breeding ground, has been the target of full-blown hurricanes. Perhaps the most noteworthy of these was the disastrous hurricane of 21 September 1938 (intensity category 3). A storm surge ravaged the New England coast. Winds gusting over 200 km (120 mi) per hour severely damaged interior forests, and torrential rains triggered flash flooding by rivers and streams. Fatalities were estimated at 600.

Today, many atmospheric scientists are concerned about the hurricane threat to coastal areas of the southeast United States. Rapid population growth in coastal cities and resorts has necessitated evacuation plans for residents of low-lying areas prone to hurricane storm surges. The effectiveness of these evacuation plans was tested in the summer of 1985 when hurricane Elena (Plate 20) menaced the Gulf of Mexico coast. For 4 days, Elena followed an erratic path over the Gulf, first taking aim on southern Louisiana, then the Florida Panhandle, and later central Florida before reversing direction and finally coming ashore near Biloxi, Mississippi, on 2 September. Hundreds of thousands of people from Sarasota, Florida, to New Orleans were forced to leave beachfront communities and flee to inland shelters. Some returned home only to evacuate again as Elena changed course. Although property damage was considerable because of extensive flooding and winds that exceeded 160 km (100 mi) per hour, there were no fatalities when the hurricane finally struck the coastline. The evacuation of coastal residents likely saved many lives. For more on the hurricane threat to the Southeast, see the Special Topic.

Hurricane Agnes* provides an illustration of the life cycle of a particularly disastrous storm. It claimed 122 lives and caused more than $2.1 billion in property damage in the northeastern states.

Hurricane Agnes

Hurricane Agnes began as a weak tropical depression near Cozumel, Yucatan, on 14 June 1972 (Figure 14.14). The depression gradually intensified and moved eastward along the twentieth parallel. On 16 June, it curved northward and became a tropical storm, with wind speeds of 62 to 117 km (39 to 73 mi) per hour. By 17 June, the storm was just west of Cuba, and by 18 June, it was

*Hurricanes are now given both male and female names.

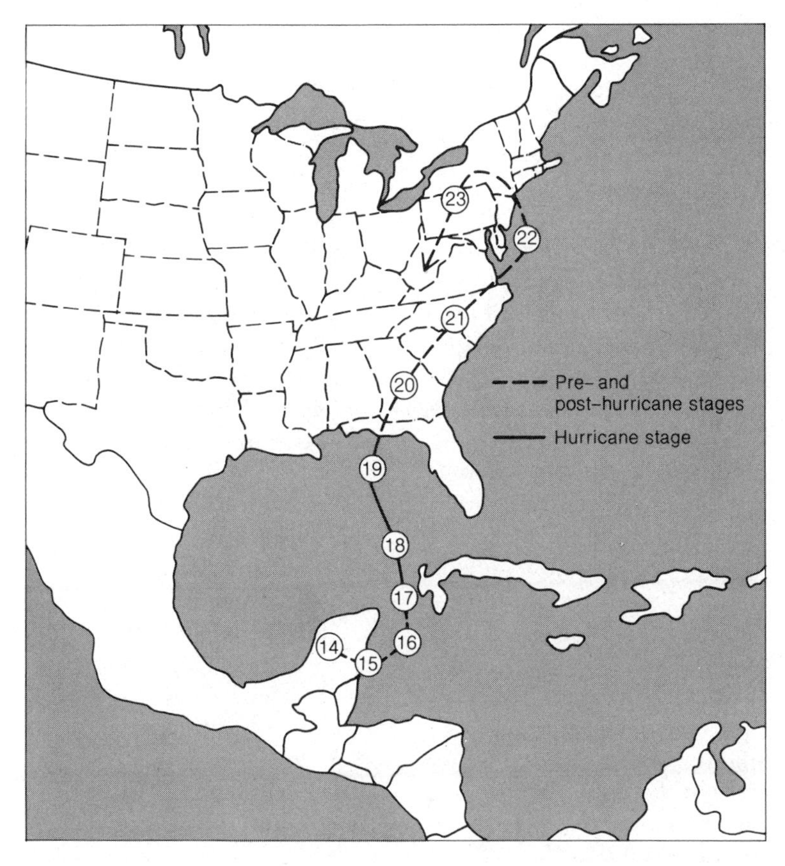

FIGURE 14.14
Trajectory of hurricane Agnes, 14–23 June 1972.

southwest of the Florida Keys. That afternoon, winds reached 136 km (85 mi) per hour, and Agnes was designated a full-fledged hurricane. The next day, Agnes shifted to the northeast and moved ashore, weakening to a tropical depression. Then on 18 and 19 June, Agnes spawned 15 tornadoes in Florida and 2 in Georgia.

By 20 June, as a result of Agnes's circulation, rain was falling from northern Florida to southern Virginia, and most of the eastern United States was covered by an unstable tropical air mass. On 21 June, the storm crossed into North Carolina, and the blanket of unstable air and heavy rains extended into Canada. That afternoon, Agnes strengthened to a tropical storm and crossed the Virginia coastline. On 22 June, the storm center was east of Baltimore, but later that day the storm curved toward the northwest and its center moved inland toward New York City. Finally, on 23 June, Agnes was absorbed in the center of a now dominant extratropical low over western Pennsylvania, and the storm drifted gradually toward the northeast and out over the Atlantic Ocean.

Although Agnes was a hurricane of intensity category 1 for only a few hours, it was the second most expensive hurricane ever to hit

SPECIAL TOPIC

Barrier islands: a fool's paradise?*

At the entrance to the U.S. National Hurricane Center [in Coral Gables, Florida] is posted a weather bulletin for the city of Galveston, Texas, dated September 8, 1900. It advised: "Rain, brisk to high northerly winds becoming variable."

That night, however, a mighty hurricane descended on Galveston, sending a wave of water 20 feet high and 50 miles wide across the city. At daybreak, Galveston was a crescent-shaped pile of rubble with 6,000 dead.

Hoping to prevent such a disaster from happening again, scientists at the hurricane center here are stationed within a glowing circle of computer screens 24 hours a day. On some screens, they can spot a cold front moving up the East Coast or a splotchy disturbance in the Caribbean. On another, a satellite spies on storm breeding grounds near the African coast. With more hurricanes striking the U.S. in September and October than in all other months combined, research planes are poised at a nearby airport to fly into the eye of any big storm.

But for all their scientific hardware, meteorologists at the hurricane center concede that predicting the direction of the big storms is difficult. For instance, when a hurricane is 72 hours offshore, the weather service has only a 10% chance of predicting accurately where it will finally come ashore—if it comes ashore at all. Even when a storm is just 24 hours off the coast, scientists say, their predictions are wrong more than half the time.

What's more, they say that the predictions are only slightly better today than they were in the 1950s, that they are no better than they were a decade ago and that they probably won't get any better in the foreseeable future.

Imperfect as it is, the forecasting system worked well in recent years. But while the ability to predict the course of hurricanes has plateaued, the population growth along coastal areas hasn't. From Brownsville, Texas, along more than 3,000 sandy miles to Eastport, Maine, the population of coastal counties has increased 34% since 1960 to some 40 million people. An even greater rate of growth has occurred on the 1.6 million acres of barrier islands that lie offshore, places where vacation homes and tourist towns sit shoulder to shoulder.

So herein lies the rub: With such enormous growth in the coastal lands, and no real improvement in hurricane forecasting, scientists have concluded that they can no longer predict the landfall of the big storms in time to guarantee that everyone can get out alive.

Hurricane scientists are obviously more concerned with a possible hurricane catastrophe than the public is. Along the nation's barrier islands, the summer homes and seaside towns extend across 300,000 acres today, compared with about 90,000 acres in 1950. Construction continues at 6,000 acres a year; development is four times as dense as on the mainland.

The public has a short memory for the darker side of the islands' history, Mr. [Neil] Frank [of the National Hurricane Center] says. He cites these examples:

At Westhampton Beach on Long Island in 1938, a hurricane destroyed all but 26 of 179 homes. Today there are more than 900 homes there.

On Long Beach Island south of Wilmington, N.C., in 1954,

*Excerpted from Calonius, L. Erik. "Storm Warnings: Hurricane Experts Say the State of Their Art Can't Avert a Disaster." *Wall Street Journal*, 14 October 1983, p. 1. Reprinted by permission of *The Wall Street Journal*,

Hurricane Hazel destroyed all but five of 357 homes. Now there are more than 1,000 homes on the island.

The barrier islands off Georgia and South Carolina were swept in 1893 by a 20-foot-high wave that killed 2,000 people. The islands now are the site of resorts that attract thousands of vacationers.

On most barrier islands, evacuation can be completed quickly. It is in the more populated areas that scientists see far greater risks. Consider, for instance, the Florida Keys, a 150-mile-long archipelago of coral rock rising at most five feet out of the sea. The home of 70,000 people, the keys are connected to the Florida mainland only by 42 narrow bridges, the longest of which spans seven miles of ocean like a piece of string. A Miami newspaper columnist noted of the long bridge:

"Already the smallest emergency puts it in a fearsome snarl. As a hurricane route, the prospects are scary."

The U.S. Army Corps of Engineers recently calculated that it would take 31 hours to evacuate the keys—48 hours counting the time needed to get the process under way. At 48 hours before landfall, a forecast that the storm would actually hit a stretch of the keys would have well under a 20% chance of being right, yet this would be the last time a decision could be made to evacuate the people with the assurance that everyone would get off the keys alive.

This is also assuming that there really *are* 48 hours left. Hurricanes are erratic storms, sometimes surprising the forecasters by arriving several hours early, sometimes looping back on themselves, and often changing intensity.

In the densely populated Miami area, where one hurricane can be expected to hit every seven years, a hurricane of much lesser intensity could send a 10-foot surge across Biscayne Bay, sweeping across the islands and into downtown Miami. With nearly half a million people in South Florida vulnerable to a big storm, evacuation time itself is estimated at 21 hours, with six to 12 hours beyond that required to get the process working.

But Florida isn't the only place vulnerable. Galveston, despite its 15-foot seawall, still requires about 36 hours to evacuate. Studies aren't available for most other regions, but the New Orleans area and the New Jersey shore are just two of many that probably need long evacuation times.

Until this year the hurricane center sent out a hurricane "watch" at 36 hours and hurricane "warnings" at 24 hours and 12 hours. But as evidence mounted that many areas needed 36 hours to clear out, forecasters decided to issue the odds that a storm would hit a particular city.

Under the new system, when a storm is estimated to be 72 hours from the coast, cities and towns in a 700-mile stretch get a one-in-10 chance that the eye of the storm will pass within 65 miles of any one of them. As the storm gets closer, the 700-mile zone is progressively narrowed—to 500 miles at 48 hours, 400 miles at 36 hours, 300 miles at 24 hours, and 100 miles at 12 hours. And the chances that the storm will hit any one city or town within the zone are progressively increased—to one in eight at 48 hours, one in six at 36 hours, one in three at 24 hours, and an almost even chance at 12 hours.

The system of probabilities raises almost as many questions as it answers. Should the decision maker in a town order an evacuation if there is only a 15% to 20% chance that the town will be hit? Will coastal residents evacuate knowing that the odds are good that they don't really have to?

Mr. Frank is aware that credibility is an issue, but he answers: "In the fairy tale, the boy cried wolf and there was no wolf. But here there is a wolf. It's just a question of whether he'll appear at your door or your neighbor's."

the United States (Table 14.6). One reason for the great damage was Agnes's unusually large size. Another reason for the extensive damage was that Agnes followed a period of unusually wet weather over the eastern United States. The storm's heavy rains triggered devastating floods from North Carolina to New York, with many record-breaking river crests (Figure 14.15). The damage total was almost twice that of Camille, a very violent 1969 hurricane that rated an intensity of category 5.

FIGURE 14.15
Waters of the Susquehanna River raged through downtown Wilkes-Barre, Pennsylvania, following heavy rainfall associated with hurricane Agnes in June 1972. (U.S. Coast Guard photograph)

Weather Radar

Weather **radar*** is a valuable tool for the detection and monitoring of severe weather systems. Thunderstorm cells are so small that they often are not sighted directly by the widely spaced network of weather-observing facilities. Weather radar, on the other hand, scans a wide area continuously and can locate small and isolated pockets of precipitation.

A conventional weather radar unit emits short pulses of microwaves with wavelengths of either 5 or 10 cm. Radar waves are scattered by precipitation but are not scattered by the very small droplets or ice crystals that compose clouds. Weather radar therefore reveals ("sees") rain or snow, but does not register the parent clouds. When precipitation occurs, a portion of the radar waves is

***_Ra_dio _d_etecting _a_nd _r_anging.

scattered back to a receiving unit, which displays the return signal, called a **radar echo**, as electrical pulses on a cathode ray tube similar to a television screen (Figure 14.16A). Because radar signals are sent out and received hundreds of times each second as the radar continuously scans a 360-degree circle (Figure 14.16B), the product is a map of the precipitation pattern surrounding the radar unit. The time interval between emission and reception of the radar signal is calibrated to give the distance to the precipitation, and the intensity of the echo is used as an index of rainfall intensity. Some radar units are equipped with electronic devices that show echo intensity on a color scale, so a patch of red indicates very heavy rain and, at the other end of the scale, green indicates very light rain (Plate 19).

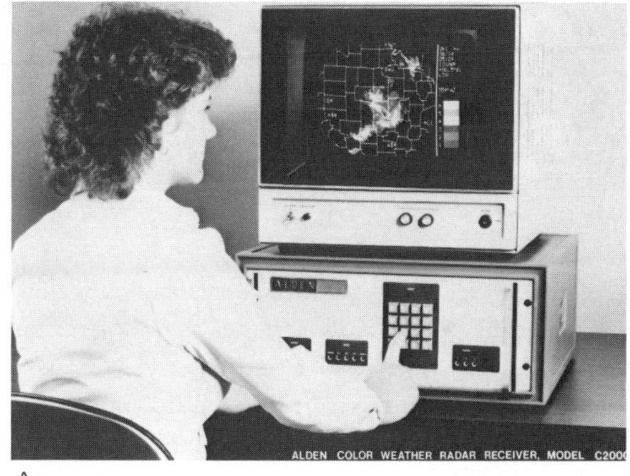

A

B

FIGURE 14.16

A weather radar receiving unit (A) and scanning dish (B). (Photographs courtesy of Alden Electronics and NCAR/NSF)

Weather radar monitors the development and dissipation of thunderstorm cells, the direction and speed of movement of those cells, and the spiral bands of rainfall associated with hurricanes (Figure 14.17). Radar cannot detect a tornado directly, but in some cases when a tornado is present, a hook-shaped echo appears on a radar screen (Figure 14.18). The **hook echo** apparently indicates rainfall being whirled about by the circulation within a severe thunderstorm.

To this point, the plan-position indicator (PPI) radar has been described. With this radar unit, a microwave beam sweeps in an almost horizontal circle. The observing circle is limited in size by the curvature of the earth and may have a radius of up to 400 km

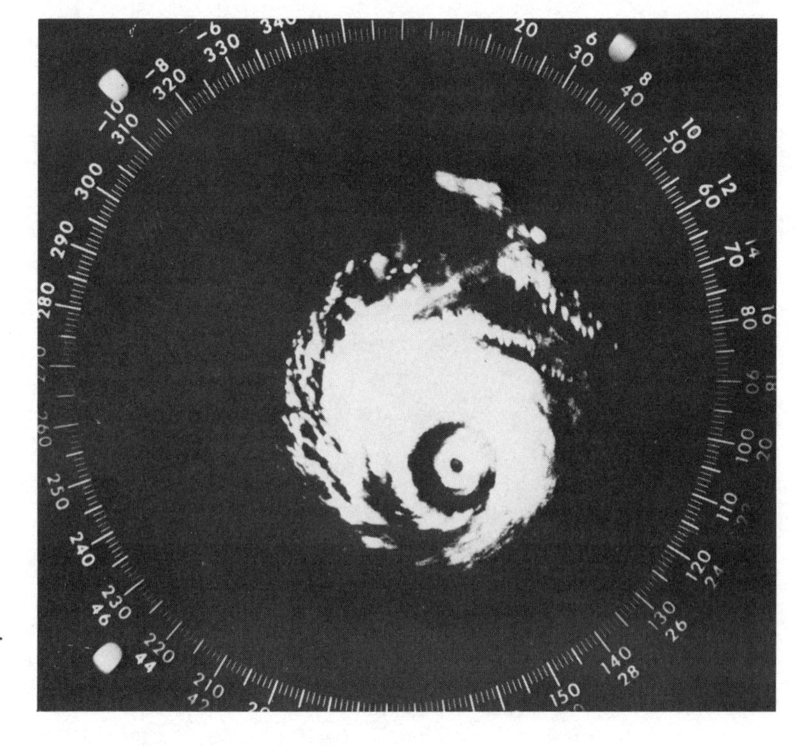

FIGURE 14.17

Radar display of spiral bands of heavy rainfall associated with a hurricane. (NOAA photograph)

(250 mi). A second type of radar display, the range-height indicator (RHI), scans up and down, rather than horizontally, and is used to determine the altitude of cloud tops. As discussed in Chapter 13, the altitude of cloud tops gives an indication of thunderstorm intensity. Conventional weather radar units can operate in either the PPI or the RHI mode.

The use of radar for purposes of weather analysis began shortly after World War II. The first weather radars were short-range surplus military units. Not until the mid-1950s, following major tornado and hurricane disasters, did the U.S. Congress allocate funds for the purchase of new long-range radar units designed specifically for meteorological applications. These weather radars were installed along the Atlantic and Gulf coasts for hurricane detection and in the midwestern United States for tornado and severe thunderstorm monitoring. In the ensuing years, aerial coverage by weather radar increased steadily, and by the late 1960s, radar had become a routine component of televised weathercasts.

Over the past decade, much research has focused on the development of Doppler weather radar. **Doppler radar** is a conventional

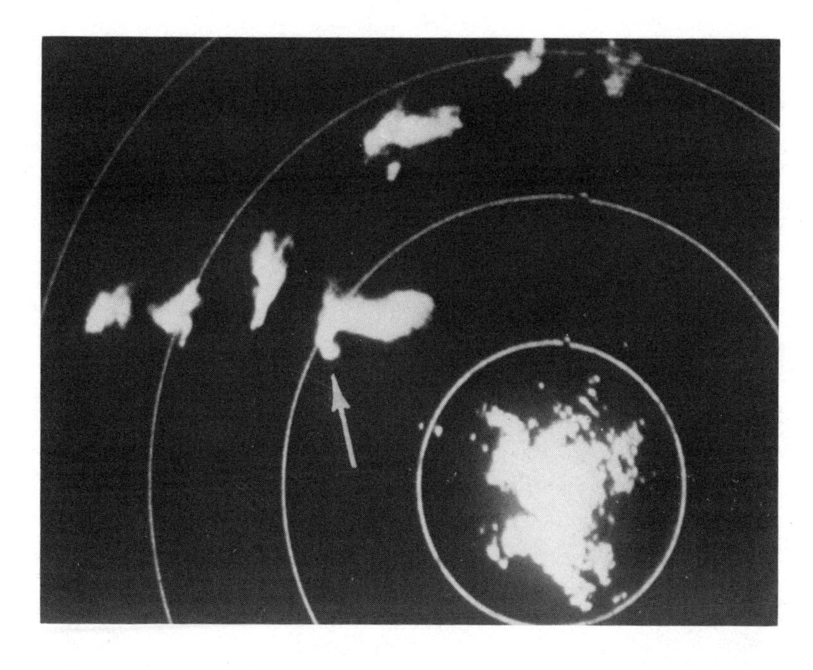

FIGURE 14.18

Conventional radar display of a hook echo, which may indicate tornadic circulation of air. (NOAA photograph)

radar that has the added capability of determining the detailed motion of the targeted precipitation toward or away from the radar unit. With this capability, it is possible to measure the air circulation pattern in a thunderstorm and the intensity of turbulence, and to detect microbursts and tornadic winds. Doppler radar works on the same principle as the police radar used to monitor traffic flow, or the "gun" used to measure the speed of a baseball pitch. As a raindrop moves away from or toward a radar unit, the frequency (or wavelength) of the radar signal shifts slightly between emission and return (echo). This frequency shift is calibrated in terms of the motion of the target precipitation. Multiple Doppler radar units viewing the same storm simultaneously create a three-dimensional pattern of air circulation.

Conclusions

This chapter completes our description of the genesis, life cycle, and characteristics of the various atmospheric circulation systems that affect midlatitude weather. Next, we consider weather forecasting.

SUMMARY STATEMENTS

- A tornado is a small mass of air that whirls rapidly about an almost vertical axis, and is made visible by clouds, dust, and debris sucked into the system.

- When a tornado strikes, damage is caused by very strong winds, an abrupt air pressure drop, and suction vortices. Most tornadoes occur during the spring in a corridor stretching from Texas northward into Missouri.

- Synoptic weather conditions favorable for the outbreak of tornadoes progress northward (with the sun) from the Gulf Coast in early spring to southern Canada by early summer.

- A tornado forms in the strong updraft of a severe thunderstorm by a process that is not yet well understood. The circulation in the tornado is apparently the consequence of an interaction between the updraft and the horizontal wind.

- A hurricane is a violent oceanic cyclone that originates in tropical latitudes where surface water temperatures are 27 °C (81 °F) or higher and where the Coriolis effect is significant.

- Hurricane damage is a consequence of strong winds, heavy rains, and, in some coastal areas, a storm surge.

- Once a hurricane makes landfall, it loses its warm-water energy source, and its circulation rapidly weakens. Heavy rains may continue, however, and can lead to flooding.

- Weather radar is a valuable tool for detecting the location and movement of areas of rainfall. Using Doppler radar, the detailed circulation within a severe thunderstorm may be determined.

KEY WORDS

tornado
funnel cloud
tornado alley
mesocyclone
waterspout
virga
hurricane
storm surge
eye of a hurricane
eye wall
typhoon
tropical depression
tropical storm
radar
radar echo
hook echo
Doppler radar

REVIEW QUESTIONS

1 What is a tornado?

2 Distinguish between a tornado and a funnel cloud.

3 Tornadoes usually move in what direction?

4 What is the principal force operating within a tornado?

5 What is the basis for the F-Scale of tornado intensity? Where do most U.S. tornadoes rank on this scale?

6 Why are tornadoes most likely to develop over relatively flat and dry terrain?

7 Where is tornado alley?

8 The locale of principal tornadic activity shifts northward during the spring. Why?

9 Why is our understanding of tornado genesis far from complete?

10 How do tornadoes compare in appearance and intensity with dust devils and waterspouts?

11 How can distant virga be distinguished from a funnel cloud?

12 What is a hurricane?

13 Describe the most destructive effect of a hurricane approaching a coastal area.

14 What is the basis for the Saffir/Simpson Hurricane Intensity Scale?

15 Describe the isobar pattern about an intense hurricane.

16 Describe the structure of a hurricane and the life cycle of the system.

17 While in tropical latitudes, a hurricane or other tropical storm drifts slowly toward the west. Why west?

18 Describe what happens when a hurricane reaches the middle latitudes.

19 Explain why hurricanes develop only in certain oceanic areas.

20 Distinguish among (a) a tropical depression, (b) a tropical storm, and (c) a hurricane.

POINTS TO PONDER

1 Can a tornado cause a building to explode? Explain your answer.

2 What is understood about the relationship between severe thunderstorms and tornadoes? What is a "mesocyclone"?

3 Describe synoptic weather conditions conducive to tornado development.

4 Why are tornadoes more frequent in spring than in fall?

5 Why do hurricane winds weaken as soon as the system moves over land?

6 Compare and contrast a hurricane with a mature, midlatitude, synoptic-scale cyclone.

7 Distinguish between conventional weather radar and Doppler radar. Why is radar particularly useful for monitoring thunderstorm cells?

PROJECTS

1 From the NOAA Climatic Summary for your area, determine the frequency of tornadoes in your region of the country. What local topographic and meteorological conditions make your locality more or less conducive to tornado development?

2 Find out if the circulation from a hurricane or tropical storm ever influences your locality.

SELECTED READINGS

Changnon, S. A., Jr. "User Beware: The Upward Trend in Tornado Frequencies." *Weatherwise* 35 (1982):64–69. *An analysis of the upward trend in Illinois tornado frequency.*

Galway, J. G. "Ten Famous Tornado Outbreaks." *Weatherwise* 34 (1981):100–109. *Great tornado outbreaks during the period 1870 to 1979.*

Hebert, P. J., and G. Taylor. "Everything You Always Wanted to Know About Hurricanes—Part I." *Weatherwise* 32 (1979):59–67. *Statistics on hurricanes in this century.*

Hebert, P. J., and G. Taylor. "Everything You Always Wanted to Know About Hurricanes—Part II." *Weatherwise* 32 (1979):100–107. *More statistics on hurricane climatology.*

Hughes, P. "The Great Galveston Hurricane." *Weatherwise* 32 (1979):148–156. *A description of probably the greatest natural disaster in U.S. history.*

Purett-Carrol, L. "First Measurements of Size and Velocity of a Violent Tornado." *Weatherwise* 35 (1982):127–130. *An example of the usefulness of Doppler radar.*

Snow, J. T. "The Tornado." *Scientific American* 250, no. 4 (1984):86–96. *An exceptionally well-illustrated review of tornado characteristics and genesis.*

Witten, D. "A Major Hurricane Disaster?" *Weatherwise* 33 (1980):159. *Identifies various factors that are heightening the potential for severe hurricane impact along the southeastern U.S. coast.*

Probable nor'east to sou'west winds, varying to the souhard and westard and eastard and points between;
high and low barometer, sweeping round from place to place;
probable areas of rain, snow, hail, and drought, succeeded or preceded by earthquakes with thunder and lightning.
MARK TWAIN
New England Weather

Today's televised weathercasts use sophisticated tools for observing and analyzing weather conditions, including satellites, radar, and computers. (Courtesy of The Weather Channel[SM])

FIFTEEN

Weather Forecasting

MOST PEOPLE can recall readily the occasions when an erroneous weather forecast upset their plans. It may have been an unexpected thundershower that brought an abrupt end to a softball game, or a raging blizzard that appeared instead of the anticipated clearing skies, or the promised springlike weekend that turned out to be anything but springlike. People seem to remember missed weather forecasts all too clearly and conveniently overlook the many times when the forecast was on target. Indeed, when viewed with the detached objectivity provided by statistical analysis, short-term weather forecasting is found to be surprisingly accurate. For example, the United States **National Weather Service** (NWS), an agency of the **National Oceanic and Atmospheric Administration** (NOAA), issues 24-hour forecasts that are correct nearly 85 percent of the time. The popular misconception that weather forecasting is seldom accurate probably stems from the simple fact that a missed forecast is noticed and remembered more readily than an accurate one because of the inconvenience experienced.

How are weather forecasts made? What are the limits of forecast accuracy? On the basis of what you have learned so far, how can you devise your own weather forecasts? These are some of the questions considered in this chapter.

World Meteorological Organization

Because the atmosphere is a continuous fluid that envelops the globe, weather observation, analysis, and forecasting require international cooperation. To this end, the **International Meteorological Organization** (IMO) was established in 1878. In 1947, the IMO changed its name to the **World Meteorological Organization** (WMO) and became an agency of the United Nations. Today, the WMO, headquartered in Geneva, Switzerland, coordinates the efforts of more than 145 member nations in a standardized global weather monitoring network called **World Weather Watch**.

At standard observation times, the state of the atmosphere is monitored daily by almost 4000 land stations, by more than 7000 ships at sea, by almost 1000 upper-air observing stations, and by reconnaissance aircraft and satellites. These data are transmitted to the three World Meteorological Centers at Washington, D.C.; Moscow, U.S.S.R; and Melbourne, Australia, where maps and charts are drawn up representing the present state of the atmosphere. From analyses of this information, generalized weather forecasts are prepared. Maps, charts, and forecasts are then sent to National Meteorological Centers (NMCs) in WMO member nations as well as to 26 Regional Meteorological Centers (RMCs). At NMCs and RMCs, weather information and forecasts are generated and interpreted for the respective areas, and are disseminated to local weather service offices and then to the public. The United States NMC is located at Camp Springs, Maryland; the Canadian NMC is in Toronto, Ontario.

From the above description we see that weather forecasting entails (1) acquisition of present weather data, (2) depiction of data on weather maps, (3) analysis of data and prediction, and (4) dissemination of weather information and forecasts to users.

Acquisition of Weather Data

In the three centuries since the invention of weather instruments, weather monitoring has undergone considerable refinement. Denser monitoring networks, more sophisticated instruments and communications systems, and better trained weather observers have produced an increasingly detailed, reliable, and representative record of weather and climate. Consider some of the milestones in weather monitoring that have occurred in North America.

Historical perspective

The first European explorers to set foot on North American soil were very interested in the weather and climate of the New World. French colonists were the first to spend the winter, in 1604–1605, and to establish a settlement on the Atlantic coast near the present border between Maine and New Brunswick. Samuel de Champlain, geographer for this expedition, took note of the sharp contrast in weather from that of his native France:

> Snow fell on the sixth of October. On the third of December we saw ice passing, which came from some frozen river. The cold was severe and more extreme than in France, and lasted much longer. I believe this is caused by the north-west winds, which pass over mountains continually covered with snow. This we had to a depth of three or four feet up to the end of the month of April; and I believe also that it lasts much longer than it would if the land were under cultivation.*

*From *The Works of Samuel de Champlain*. H. P. Biggar, ed. Toronto: The Champlain Society, 1922, pp. 302–303.

Almost half of the 79 French colonists did not survive that severe winter.

Some early weather observers used primitive weather instruments, while others made qualitative assessments of the weather—jotting down observations in journals or diaries. The first systematic weather observations in North America took place in 1644–1645 at Old Swedes Fort (now Wilmington, Delaware). The observer was John Campanius Holm, chaplain of the Swedish military expedition. Other temperature records were begun in Philadelphia in 1731; in Charleston, South Carolina, in 1738; and in Cambridge, Massachusetts, in 1753. Thomas Jefferson and James Madison are credited with making the first simultaneous weather observations in the United States, in 1777 and 1778.

In the late 1700s and early 1800s, weather recording was sparse and sporadic across North America. A notable exception is the New Haven, Connecticut, temperature record, which was begun in 1781 and continues today. Dr. James Tilton, Surgeon General of the U.S. Army, organized the first U.S. government-sponsored weather observation network in 1816. By 1839, army physicians were monitoring the weather daily at 13 midwestern forts. This network had expanded to 143 member stations by the late 1860s.

The coming of the telegraph in the mid-1800s spurred Joseph Henry, then Secretary of the new Smithsonian Institution, to form a network of 150 volunteer weather observers. Data were wired to the Smithsonian and displayed on maps at the institution in Washington, D.C. Henry's system later served as a model for the establishment of a national weather observation network.

The impetus for a national network was provided by the appalling loss of life and property in shipwrecks caused by surprise storms that swept the Great Lakes. Congress called on the Army to monitor weather conditions and to issue appropriate storm warnings. President Grant signed this resolution into law in early 1870. Personnel of the U.S. Army Signal Services (or Corps) subsequently monitored weather conditions at army facilities numbering 24 stations in 1870, and climbing in number to 284 stations by 1878. This was the birth of the United States National Weather Service.

Twenty years later, the weather program was transferred to a new Weather Bureau in the Department of Agriculture, with a special mandate to provide weather and climate guidance to farmers. With the rapid emergence of aviation after 1920, the Weather Bureau eventually came under the jurisdiction of the U.S. Commerce Department in 1940. As the complexity of the physical environment became better understood, the Weather Bureau was reorganized as the National Weather Service (NWS) and was placed under the supervision of the Environmental Science Services Administration (ESSA) in 1965. In 1971, the ESSA became the National Oceanic and Atmospheric Administration (NOAA).

In Canada, the first systematic weather observations were taken in 1768. Seventy-one years later, the Meteorological Service of Canada was founded and placed initially under military jurisdiction. In 1853, the service came under civilian control. The first national observation network was established in 1843–1844, and the first forecast service was begun in 1876. Cooperation with the U.S. Weather Bureau dates back to that year, and cooperation increased after World War II with the establishment of joint Arctic weather stations. Today, Canadian weather services are provided by the Atmospheric Environment Service (AES) in the Department of Fisheries and the Environment.

Surface weather observations

Across the United States, nearly 1000 stations routinely monitor surface weather. These stations may be operated by any of the following: (1) National Weather Service personnel, (2) the staff of other government agencies, including the Federal Aviation Administration (FAA), or (3) private citizens or businesses in cooperation with the NWS. At sea and in the Great Lakes, more than 2000 ships also voluntarily supply surface weather data. In addition, the NWS maintains a network of automated meteorological observation stations in locations where manned observations are not feasible.

Land-based weather observing stations are part of the **synoptic weather network** or the **basic weather network** or both. Member stations of the synoptic weather network gather weather information primarily to prepare weather maps and forecasts and to exchange data with other nations. These stations observe and report cloud type and sky cover, wind speed and direction, visibility, precipitation, air temperature, dew-point temperature, and air pressure. These reports are issued every 3 hours, beginning at midnight **Greenwich Mean Time** (GMT), which is the time at The Old Royal Observatory, Greenwich, England. (Greenwich is located at 0 degrees longitude, the prime meridian.) Greenwich Mean Time* is used by meteorologists all over the world.

*For reference, at 0600 GMT, it is midnight Central Standard Time (CST) in Chicago, and 10 P.M. Pacific Standard Time (PST) in Vancouver, British Columbia.

Member stations of the basic weather network provide weather observations hourly, mainly for aircraft but also to supplement weather forecasting. They measure the same variables as the synoptic weather network stations and in addition report cloud height and altimeter settings for aircraft. Pilots obtain weather briefings prior to and during flight and are advised on any weather conditions that may pose a hazard to flight. For a description of the chief aviation weather hazards, refer to the Special Topic, Aviation Weather Hazards.

Some weather stations are equipped with radar, which tracks the intensity and movement of severe thunderstorms, tornadoes, and hurricanes, as well as other areas of precipitation. The National Weather Service operates about 120 radar stations across the United States. Half of these operate continuously; the other half operate only when weather conditions warrant. The U.S. government currently has plans to replace these conventional radar units with a network of Doppler weather radars by 1988.

Besides the nearly 1000 land-based weather stations that provide information of potential use in weather forecasting, an additional 11,590 cooperative weather stations are scattered across the United States. These stations constitute the **National Substation Program**, and their principal function is to record daily precipitation totals and maximum and minimum temperatures for hydrologic, agricultural, and climatic purposes.

Upper-air weather observations

As noted in Chapter 1, meteorologists monitor the upper atmosphere with radiosondes. A **radiosonde** is a radio-equipped instrument package borne by a balloon. The instruments transmit to a ground station the vertical profiles of temperature, pressure, and relative humidity up to an altitude of about 30 km (19 mi). In addition, winds at various levels are computed by tracking the balloons with radar (rawinsonde observations). Worldwide readings are made twice each day, at 0000 GMT and 1200 GMT.

In the United States, there are 129 radiosonde stations, 25 of which use special high-altitude balloons that provide data from altitudes above 30 km. Meteorological rockets probe much higher altitudes (to 100 km), but the data obtained from these probes is used primarily for research. Some upper-air weather data are also obtained by aircraft, radar, and satellites. In some urban areas, low-level soundings (to 3000 m) monitor weather conditions to assess air pollution potential.

Canada's Atmospheric Environment Service operates similar networks of surface and upper-air weather observation stations. The Canadian service has 270 first-order weather stations and more than 3000 climatological stations nationwide.

Satellite observations

A variety of meteorological satellites are either currently in orbit or are planned for the future. Weather satellites have been launched by the United States, Russia, Japan, and the 11-nation European Space Agency. Most are in near-polar orbits at altitudes of 800 to 1000 km (500 to 650 mi), that is, orbits that take the satellites on longitudinal (meridional) trajectories over the poles. For example, the U.S. weather satellites NOAA-7 and NOAA-8 observe a strip 1800 km (1100 mi) wide from pole to pole in 102 minutes. The satellites then observe the next adjacent meridional strip. With 14 orbits each day, these satellites "see" a particular locality twice every 24-hour day, once at night and once during daylight.

Some weather satellites are in geosynchronous orbits at the equator, that is, they rotate with the earth, so they always scan the same region. From an altitude of about 35,800 km (22,240 mi), each **geosynchronous satellite** monitors almost one third of the earth's surface. The Western Hemisphere is monitored by two Geostationary Operational Environmental Satellites, GOES West and GOES East. GOES West, located at 135 degrees W longitude,* and GOES East, located at 75 degrees W, provide cloud cover pictures every 30 minutes both night and day.

GOES weather satellites can provide nighttime observations of cloud cover because they are equipped with infrared radiation (IR) sensors. These instruments measure infrared emissions from the earth's surface as well as from the top surface of clouds. The temperature dependence of the emitted radiation enables infrared sensors to distinguish warm surfaces from cool surfaces, so the sensors can distinguish clouds at low levels, which are relatively warm, from clouds at high levels, which are relatively cool. These measurements are then calibrated in terms of a gray scale on which the

*Longitude is measured in degrees east and west of the meridian that passes through Greenwich, England. The Greenwich meridian is called the "prime meridian" and is assigned a longitude of 0 degrees. Longitude reaches a maximum of 180 degrees E(ast) and 180 degrees W(est).

SPECIAL TOPIC

Aviation weather hazards

The Federal Aviation Administration estimates that 50 percent of all aircraft accidents are weather related. Three major aviation weather hazards are obstructions to visibility, turbulence, and icing. Chapter 13 described a fourth hazard, the microburst, which accompanies thunderstorms.

In the interest of aircraft safety, pilots obtain preflight and in-flight weather briefings tailored to their individual flight plans. These briefings report current and anticipated weather conditions including any potential weather hazards. The briefings are given in a special code, which is presented in Appendix III.

Visibility is the maximum horizontal distance at which prominent objects can be seen and identified. The *ceiling*, or altitude of the base of the lowest cloud layer, is important, because visibility can be restricted by fog or low stratus clouds. Other factors include polluted air, haze, precipitation, and blowing snow, sand, or dust. In general, visibility is good and ceilings are high in anticyclones (except for radiation fog); conversely, visibility is poor and ceilings are low in cyclones.

Depending on the degree of visibility, pilots fly under one of four flight rules listed in the table. When visibility is good, Visual Flight Rules (VFR) apply, and a pilot relies on vision to spot landmarks and other aircraft. When visibility is poor, Instrument Flight Rules (IFR) apply, and navigation must be aided by instruments.

Because low visibility is the chief reason for flight cancellation, airports should be located where local climate favors good visibility. The ideal site is a moderately elevated area, not a low area subject to cold air drainage and frequent radiation fogs. The location should be upwind of industrial areas to avoid the adverse effects of air pollution.

Turbulence is an irregular flow of air analogous to the white water in the rapids of a swiftly flowing stream. A series of eddies (swirls) of varying size are associated with the main current of air. As the average wind speed increases, the eddies generally become more energetic.

A shear in wind speed, that is, a change in wind speed over a relatively short distance, most often generates turbulence. The greater the wind shear, the more severe the turbulence. Convection currents, the jet stream, and fronts (especially cold fronts) all produce wind shear. In convection currents, for example, the wind

Flight Rule Criteria

CATEGORY	CEILING (FT)		VISIBILITY (MI)
Low Instrument Flight Rules (LIFR)	less than 500	and/or	less than 1
Instrument Flight Rules (IFR)	500 to less than 1000	and/or	1 to less than 3
Marginal Visual Flight Rules (MVFR)	1000 to 3000	and/or	3 to 5
Visual Flight Rules	more than 3000	and	more than 5

shear between upward and downward moving air produces turbulence commonly felt as bumpiness upon takeoff and landing. Cumulus clouds are visible evidence of convective turbulence: winds are upward where the sky is clouded and downward where the sky is clear. Convective turbulence usually poses no problem for aircraft unless the currents surge to high altitudes and trigger thunderstorms with great wind shear between updraft and downdraft. Because of this severe turbulence and the possibility of damaging hail, aircraft should never attempt to fly through a thunderstorm.

Mountain waves can also cause turbulence. Recall from Chapter 7 that a mountain range disturbs large-scale winds so that standing (stationary) waves develop to the lee of the range. Vigorous eddies form below mountain waves, and some turbulence may be encountered at altitudes above the mountain waves, perhaps up to the tropopause.

Often turbulence can be detected by merely observing clouds. In addition to cumuliform clouds, those with a wave pattern (stratocumulus, altocumulus, and cirrocumulus) are associated with turbulence. In other instances, clear air turbulence (CAT), defies detection except by sophisticated airborne instruments currently under development. Normally, however, CAT is confined to relatively thin layers of the atmosphere, and pilots readily escape turbulence by changing altitude.

The usual turbulence encountered by commercial airliners is so light as to be merely annoying. On rare occasions when an aircraft is flown into an area of severe turbulence, large and abrupt changes in altitude, attitude, or both occur. Passengers wisely keep seat belts buckled throughout a flight.

Aircraft flown through freezing rain or clouds composed of supercooled water droplets are prone to *icing*, the accumulation of ice on the leading edges of the aircraft. Supercooled raindrops or cloud droplets freeze on contact with the surface of the wings and fuselage. If the aircraft does not have deicing or anti-icing equipment, ice buildup adds weight and interferes with the aircraft's aerodynamics. Icing is most often a hazard for aircraft flown below about 6000 m (20,000 ft) and at air speeds of less than 400 knots. Above 6000 m, air temperatures are so low that clouds are composed exclusively of ice crystals, which do not accumulate on the aircraft; at air speeds above 400 knots, frictional heat usually inhibits ice accumulation.

low clouds appear dark gray and the high clouds appear white.

From satellite-derived cloud patterns and their displacement and infrared characteristics, meteorologists deduce vertical profiles of temperature, humidity, and wind, as well as storm locations and paths. Satellites also monitor the extent of snow cover and sea ice. Besides the satellites used for routine weather observation, some meteorological satellites, such as those in the Nimbus series, are designed mainly for research purposes.

The foremost advantage of weather observation satellites is that they provide a continuous picture of the state of the atmosphere. The networks of surface and upper-air observation stations detect weather conditions at discrete points only, some of which may be hundreds to thousands of kilometers apart. For example, satellites "see" subsynoptic weather systems, such as severe thunderstorms, that might escape detection by weather stations. Satellites have enabled forecasters to locate and track tropical storms over oceans where weather observation stations are few and far between.

Data Depiction on Weather Maps

Computers at the National Meteorological Centers use the special symbols presented and described in Appendix III to plot weather observations on synoptic and hemispheric weather maps. The weather for each observation station is depicted on a map by following a conventional **station model**. The station model in Figure 15.1 shows the symbols for surface weather conditions. By international agreement, the same symbols and station model are used throughout the world.

Weather systems are three dimensional, hence the need for both surface and upper-air weather maps. A very different approach is used for the two types of maps, however. Surface weather data are plotted on a constant-altitude (usually sea-level) surface, and upper-air weather data are plotted on constant-pressure surfaces.

Surface weather maps

It is standard practice for meteorologists to reduce surface air pressure readings to sea level (Chapter 5). This means that pressure readings are adjusted so the pressure is what it would be if the station were actually at sea level. This procedure is intended to

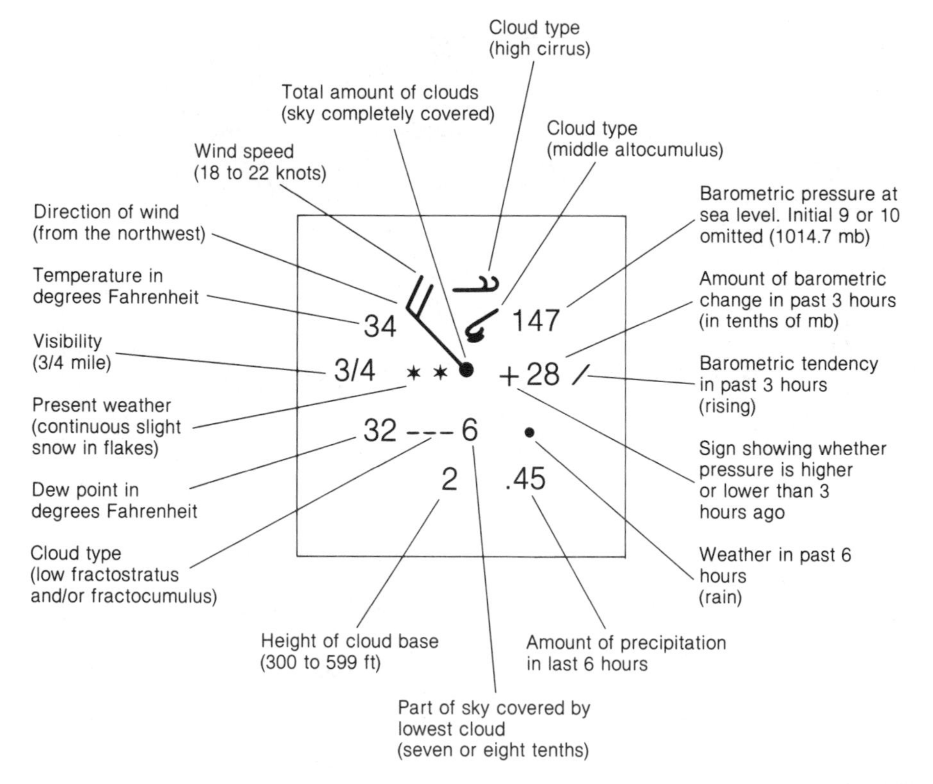

FIGURE 15.1

This conventional weather station model shows the symbols used to depict weather observations on surface weather maps.

remove the influence of topography on air pressure and to thus enable meteorologists to compare surface air pressure readings at weather stations that are situated at different elevations above sea level. The adjusted air pressure readings are plotted on surface weather maps. **Isobars**, lines of equal pressure, are then constructed by hand at 4-mb intervals to reveal such features as anticyclones and cyclones, troughs and ridges, and horizontal pressure gradients.

Figure 15.2 presents the surface weather maps compiled for 1–3 April 1982 along with the corresponding GOES East satellite photographs. The NOAA *Weekly Weather and Crop Bulletin* reports the following weather summary for each of the three days:

THURSDAY, 1 April 1982—A warm front reached from South Carolina to Arkansas and through central Texas. Warm, moist air from the Gulf of Mexico flowed into Texas and then veered northeastward into Arkansas. Light showers fell in Texas, but moderate to heavy showers and thunderstorms battered Arkansas and northern Mississippi. A complex frontal system through the West produced light to moderate rain and heavy snow in the mountains. The precipitation covered the area west of the Rockies and spread into the northern Plains.

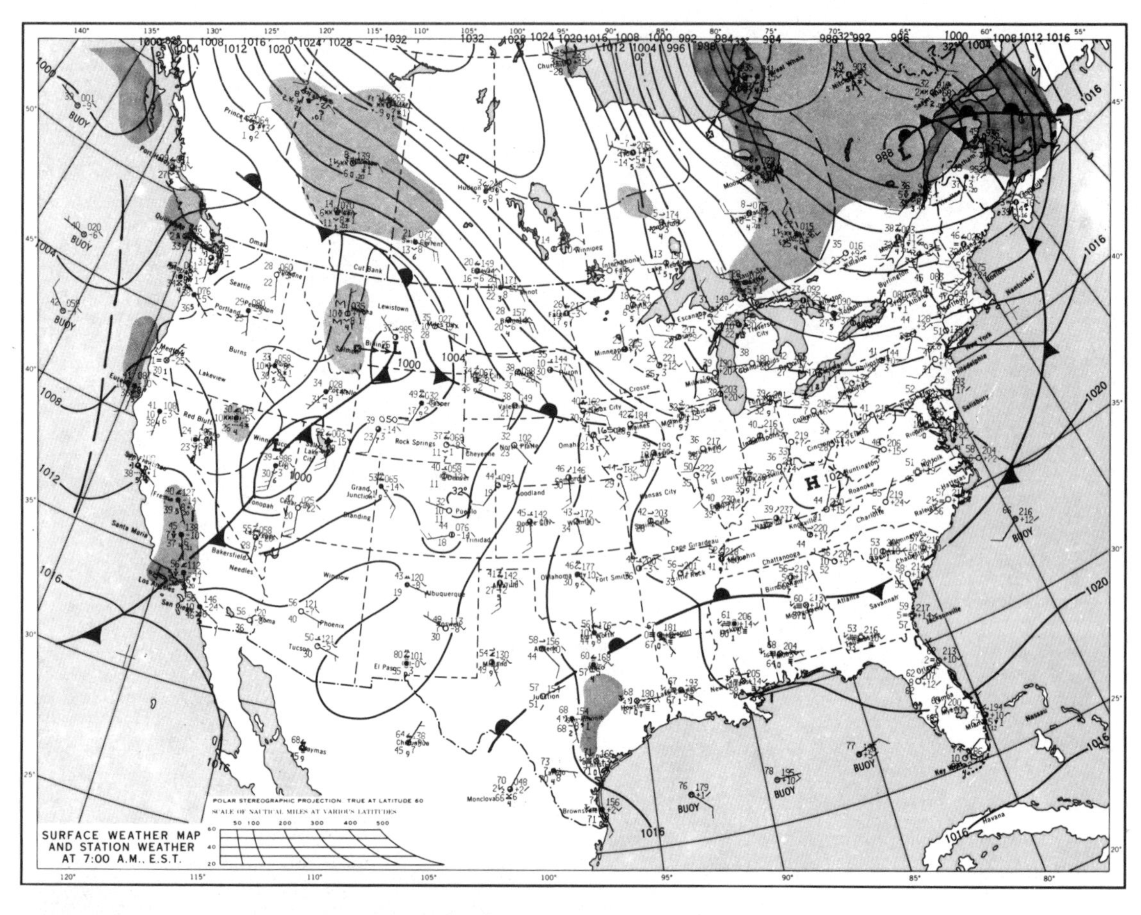

A

FIGURE 15.2

Surface weather maps (top) and GOES East satellite photographs (bottom) for (A) 1 April 1982, (B) 2 April 1982, and (C) 3 April 1982. (From NOAA)

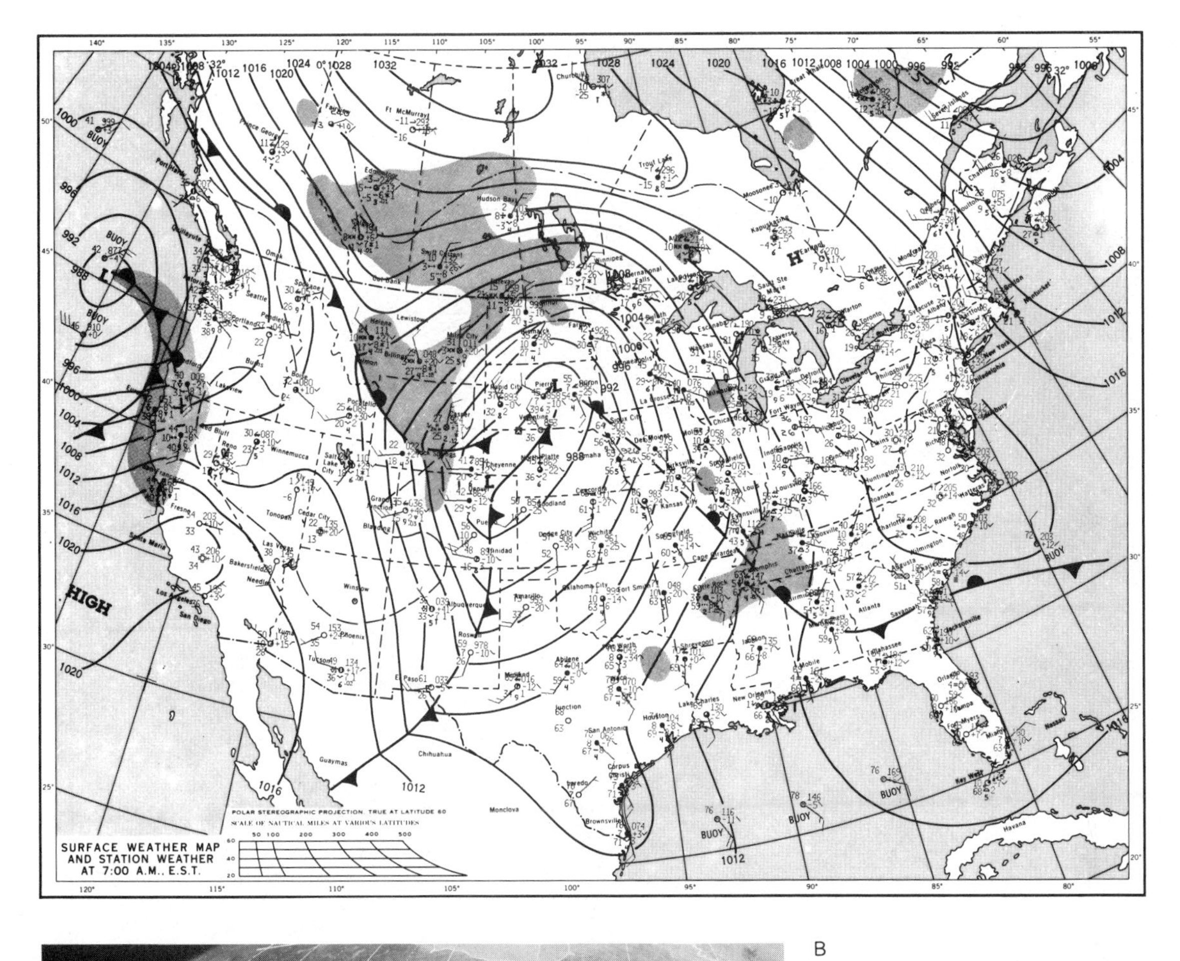

B

FIGURE 15.2 *(continued)*

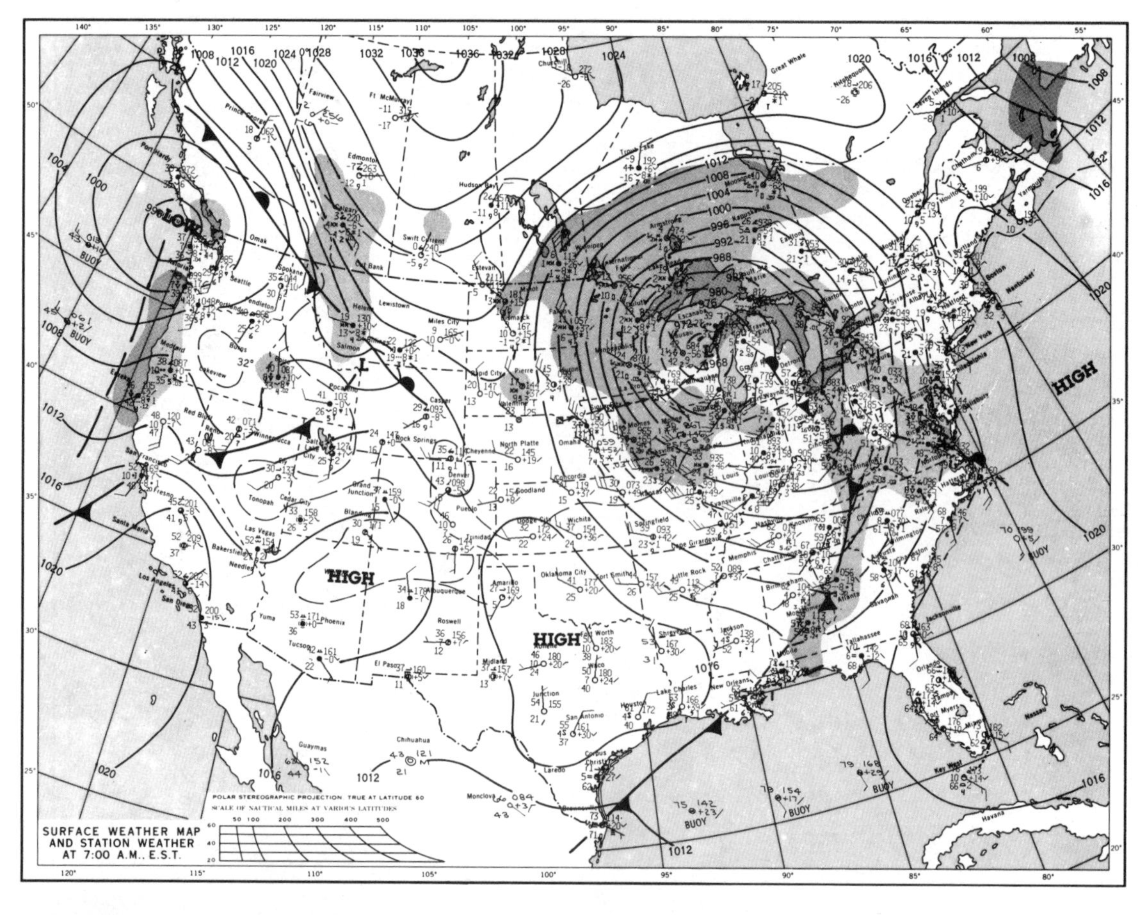

C

FIGURE 15.2 *(continued)*

FRIDAY, 2 April 1982—An intense low pressure system deepened in the central Plains. Severe weather moved ahead of the storm through the entire Mississippi Valley and spread slowly eastward to a line from the upper Ohio Valley through Alabama. Light showers and thunderstorms extended to the mid-Atlantic States by the end of the day. The low pressure center deepened to record intensity as it reached the Great Lakes area. Tornadoes, hail, high winds, and heavy rain lashed the area from northeastern Texas into Minnesota and Wisconsin. Rain, wind, and snow from another storm hit the Pacific Northwest.

SATURDAY, 3 April 1982—The intense low pressure system moved slowly into the central lakes area. The cold front, marking the line of severe weather, moved from the central Great Lakes, through the Appalachians to central New York State, and along the east coast to northern Florida. Thunderstorms were not as intense along the front as on Friday, but golf ball-sized hail and high winds were reported from New York to Georgia. Rainshowers, with snow in the mountains, spread into central California over the Plateau to the northern Plains.

The major weather event of these three days was the development of an intense cyclone that tracked into the Great Lakes region. On satellite photographs, the mature storm shows up as a giant comma-shaped area of cloudiness. On the surface weather maps, the storm is indicated by well-defined fronts and closely spaced isobars.

Synoptic surface weather maps are drawn every 3 hours for North America and at 6-hour intervals for the Northern Hemisphere. Special charts are also constructed that summarize a variety of weather elements including (1) maximum and minimum temperatures for 24-hour periods, (2) precipitation amounts for 6 hours and for 24 hours, and (3) observed snow cover. A radar summary chart is also issued many times daily.

Upper-air weather maps

Upper-air weather data acquired by radiosondes are plotted on constant-pressure surfaces. Applying the basic laws of atmospheric physics, meteorologists compute the actual altitudes corresponding to these measurements. For example, we can determine the altitude of a 500-mb surface, that is, the altitude at which the air pressure drops to about one half of the standard air pressure at sea level. By plotting altitudes of the 500-mb level as monitored simultaneously by all radiosonde stations across North America, meteorologists can then construct a map showing the "topography" of the 500-mb surface. They simply draw contours through localities where the 500-mb level is at the same altitude, a procedure that requires some interpolation between stations.

The altitude of a pressure surface (such as the 500-mb surface) varies from one place to another primarily because of air temperature differences. Air pressure drops with altitude more rapidly in cold air masses than in warm air masses. Because raising the temperature of air reduces its density, greater altitudes are required for warm air to exhibit the same drop in pressure as cold air. This means, for example, that the 500-mb level is at a low altitude where the air below is relatively cold and at a high altitude where the air below is relatively warm. The contours of the 500-mb height surface therefore show a gradual slope downward from the warm tropics to the cooler polar latitudes.

Upper-air observations indicate that horizontal winds parallel contour lines at the 500-mb level. Where contours are closely spaced (a steep contour gradient), winds are strong, and where contours are far apart (a weak contour gradient), winds are light. Why are winds associated with a contour gradient? A contour gradient develops where there is a horizontal temperature gradient, and where there is a horizontal temperature gradient, there is also a horizontal pressure gradient and wind. The 500-mb surface is so far above the friction layer that the synoptic-scale and global-scale winds are essentially geostrophic winds where contours are straight, and gradient winds where contours are curved (Chapter 9). The large-scale wind at the 500-mb level is thus the product of interactions among a horizontal contour gradient, the Coriolis effect, and centripetal force.

On upper-air weather maps, contours exhibit both cyclonic (counterclockwise) and anticyclonic (clockwise) curvature. These are the ridges and troughs described in Chapter 10. Contours may also define a series of concentric circles, indicating a cutoff or blocking circulation. At the center of a ridge, the air column is relatively warm and contour heights are high, so we label the ridge with an "H." An upper-air ridge may be linked to a warm-core anticyclone at the surface. In contrast, at the center of a trough, the air column is relatively cold and contour heights are low, so the trough is labeled with an "L." An upper-air trough may be linked to a cold-core cyclone at the surface.

From the preceding discussion, it follows that warm-core cyclones (thermal lows) and cold-core anticyclones (polar and arctic highs) do not appear on 500-mb weather maps. These pressure systems are simply too shallow to influence the air circulation pattern at the 500-mb level.

The 500-mb maps for 1–3 April 1982 (Figure 15.3) illustrate upper-air analyses. Solid lines are altitude contours of the 500-mb surface labeled in feet above sea level; the contour interval is 200 ft. Dashed lines are isotherms labeled in degrees Celsius (°C). Flags indicate wind direction and speed at the 500-mb level. Winds paralleling contour lines describe cyclonic flow in troughs and anticyclonic flow in ridges. Note that the Great Lakes storm in Figure 15.2C is linked to a cold trough aloft in Figure 15.3C.

Winds flowing across isotherms produce cold or warm air advection. **Cold air advection** takes place when winds blow from colder localities toward warmer localities. **Warm air advection** occurs when winds blow in the opposite direction. Warm air advection causes the 500-mb surface to rise, and cold air advection causes the 500-mb surface to lower. Cold air advection at 500 mb thus deepens troughs and weakens ridges, while warm air advection strengthens ridges and weakens troughs.

The National Meteorological Centers (NMCs) issue 500-mb maps twice each day based on upper-air observations at 0000 GMT and 1200 GMT. One set of maps covers North America and another the Northern Hemisphere. Although we have focused our discussion of upper-air weather maps on the 500-mb level, similar analyses are routinely constructed twice daily for the 850-mb, 700-mb, 300-mb, 250-mb, 200-mb, and 100-mb levels. The 500-mb weather chart is particularly useful, however, because winds at that level closely approximate the upper-air steering winds and the trough and ridge patterns responsible for the development and displacement of surface weather systems. The 300-mb level analysis is also quite important, because that level is near the strongest winds of the midlatitude jet stream.

Weather Prediction

Meteorologists at National Meteorological Centers analyze satellite observations and weather data plotted on surface and upper-air weather maps. From these analyses, weather forecasts are then prepared. Forecasting the weather is extremely challenging, primarily because of the many variables involved and the vast quantity of weather data. For these reasons, weather forecasts are generated by computerized numerical models.

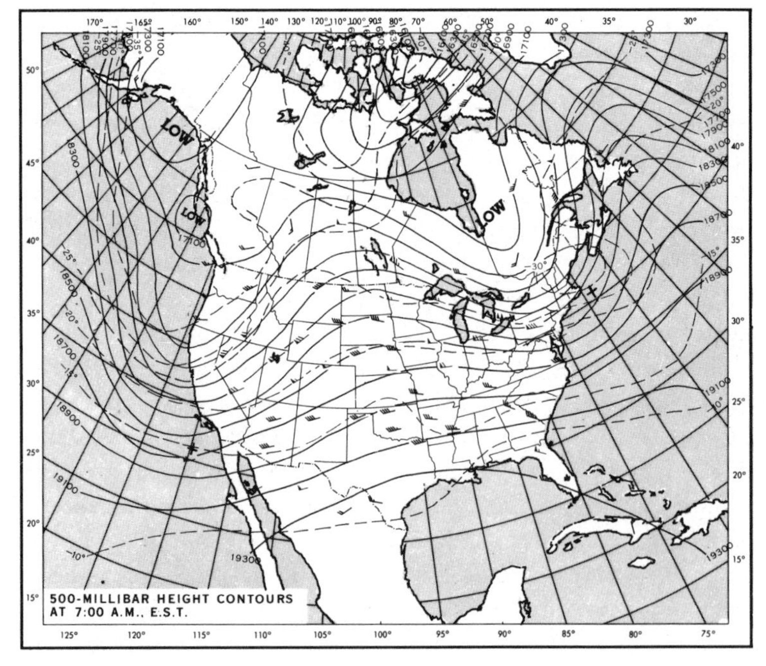

A

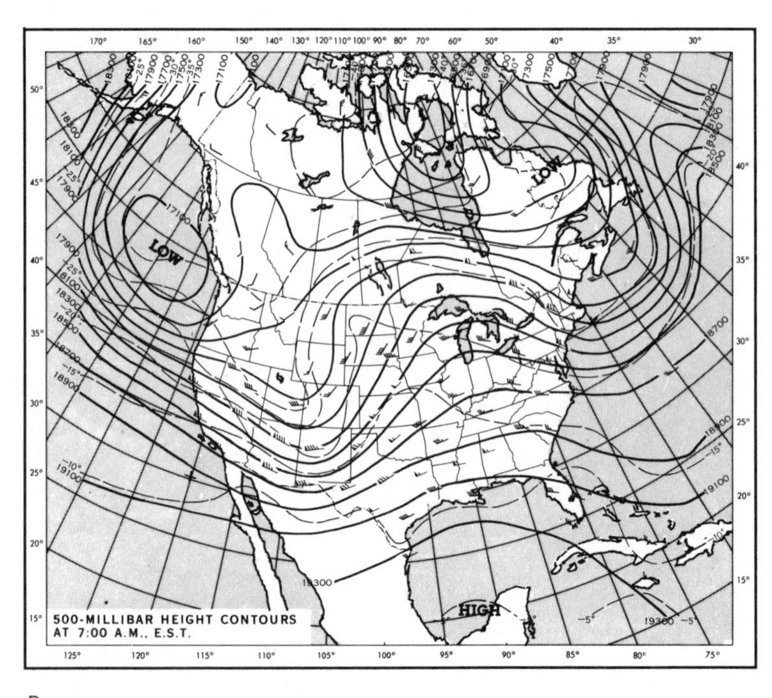

B

FIGURE 15.3

The 500-mb analyses for (A) 1 April 1982, (B) 2 April 1982, and (C) 3 April 1982. (From NOAA)

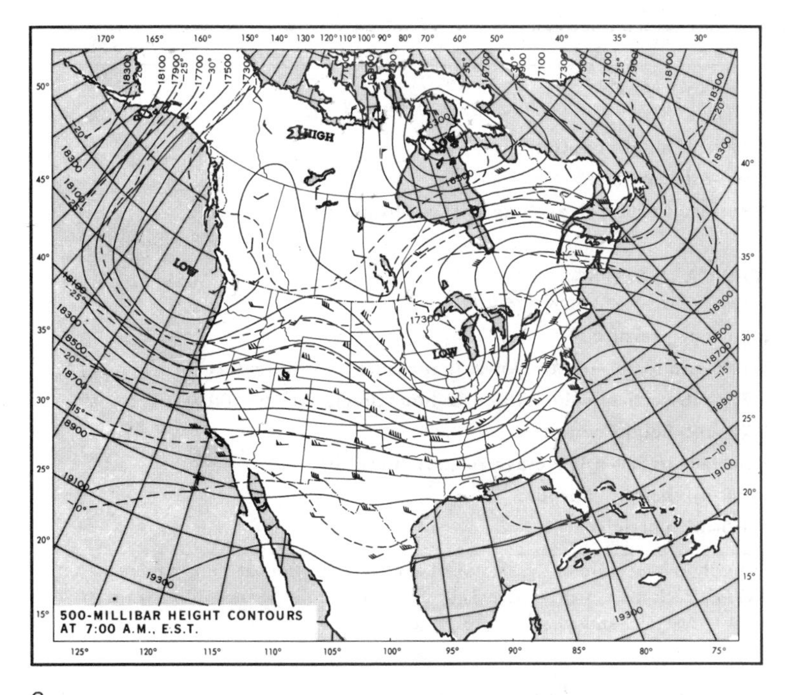

C

Numerical weather forecasting

Since 1955, computers have generated weather forecasts routinely from surface and upper-air weather observations. The computers are programmed with numerical models of the atmosphere—models that use mathematical equations relating winds, temperature, pressure, and water vapor concentrations. Starting with the immediate weather data, numerical models predict pressure surface heights at some future time, say 10 minutes hence. Using these predicted conditions as a new starting point, another forecast is then computed for, say, the subsequent 10 minutes. The computer repeats this process again and again at very high speeds until a weather map is generated for the next 12, 24, 36, and 48 hours. Tens of millions of computations must be performed each second, hence the need for a computer. NMC computers apply three different numerical models to specific portions of the atmosphere. One of these, the Limited Area Fine Mesh (LFM) model, generates the analysis and forecast maps listed in Table 15.1 twice daily.

Special forecast centers

Not all weather forecasts are prepared by the National Meteorological Center. Responsibility for forecasting tropical storms

TABLE 15.1

Weather Guidance and Forecast Information Generated by NMC Computers Using the LFM Numerical Model of the Atmosphere

ANALYSIS MAPS
(based on observations at 0000 GMT and 1200 GMT

1 500-mb height contours

2 500-mb vorticity*

3 Surface to 500-mb average relative humidity

FORECAST MAPS
(for periods of 12, 24, 36, and 48 hours)

1 Surface pressure

2 1000-mb to 500-mb thickness (a measure of temperature)

3 700-mb vertical velocity and 12-hour precipitation totals

4 700-mb height contours

5 Surface to 500-mb average relative humidity

6 500-mb height contours

7 500-mb vorticity*

* *Vorticity* is a measure of the rotational tendency of a fluid at a given point.
NOTE: Each map, issued twice daily, covers the United States, Canada, most of Mexico, and adjacent oceanic areas.

and hurricanes is divided among three centers: The National Hurricane Center in Miami, Florida; the Eastern Pacific Hurricane Center in San Francisco, California; and the Central Pacific Hurricane Center in Honolulu, Hawaii. Local and regional weather service offices transmit information from these centers to the public as advisories, warnings, or statements.

In recent years, the National Hurricane Center in Miami has been including a probability forecast as part of its public advisory statements. The probability is defined as the percent chance that the center of a hurricane or tropical storm will pass within 105 km (65 mi) of any of 45 designated Gulf and East Coast communities from Brownsville, Texas, to Eastport, Maine. The first probability forecast is usually issued 72 hours before the storm's anticipated landfall. At that time, by convention, the probability is set no higher than 10 percent for any community. Probabilities increase to 13 to 18 percent 48 hours in advance of landfall, 20 to 25 percent at 36 hours, 35 to 45 percent at 24 hours, and up to 60 percent at 12 hours before the storm's predicted landfall.

The National Severe Storms Forecast Center (NSSFC), located in Kansas City, Missouri, also monitors atmospheric conditions for the potential development of severe local storms, and issues watches for severe thunderstorms and tornadoes. Actual severe weather warnings are made by local or regional weather service offices.

Forecast accuracy

Just how accurate are today's computerized weather forecasts? This question can be answered by comparing the accuracy of modern weather forecasting with predictions based on persistence (forecasting no change in present weather) or with predictions based on climatology (forecasts derived from past weather records). Table 15.2 rates the accuracy of computerized weather forecasting according to whether or not it exceeds the accuracy of persistence or climate forecasting. Results show that the accuracy of all types of weather forecasting deteriorates rapidly for periods longer than 48 hours and that accuracy is minimal beyond 10 days.

Nevertheless, the accuracy of short-term weather forecasting has shown slow but steady improvement over the past 25 years, and prospects for continued progress are good. This trend is due largely to the maturation of computerized weather prediction that makes use of more realistic numerical models of the atmosphere, larger and faster computers, and improved observational tools, including radar and satellites. Establishment of denser observation networks involving satellite monitoring, especially over the oceans, and the development of more precise atmospheric models are expected to further upgrade the reliability of short-term computerized weather forecasts.

Computerized weather forecasts are not likely, however, to replace human weather forecasters. This is because computerized forecasts are only as accurate as the input data and predictive equations (numerical model) allow them to be. There is nothing magical about a computer. Incomplete weather observation networks and imprecise numerical models will limit the accuracy of com-

TABLE 15.2

Accuracy of Weather Forecasting Compared With Accuracy of Persistence and Climatological Forecasting*

FORECAST PERIOD	FORECAST ACCURACY
To 12 hours	**Considerable**
12–48 hours	**Considerable**
2–5 days	**Moderate for temperature** **Slight for precipitation**
6–10 days	**Low for average temperature** **Slight for precipitation amounts**
10–30 days	**Minimal**
More than 30 days	**Minimal**

Source: "Policy Statement of the American Meteorological Society on Weather Forecasting." *Bulletin of the American Meteorological Society* 60 (1979):1453–1454

*In the middle latitudes of the Northern Hemisphere.

puterized weather forecasts. Current observation points are too widely spaced to detect all mesoscale weather systems, and the predictive equations constituting numerical models are based on many assumptions and first approximations. In addition, special local or regional conditions may call for the significant modification of the computerized overall weather forecast. Meteorologists must therefore analyze and interpret computerized forecasts and adapt those forecasts as necessary to regional and local circumstances.

Let us now consider an example of how a knowledge of regional conditions can be used to improve a weather forecast. Suppose that an intense early winter storm tracks northeastward through the Midwest, across central Illinois and into eastern lower Michigan, as shown in Figure 15.4. Muskegon, Michigan, is on the cold, snowy side of the storm's path, so residents experience strong northeast winds accompanied by heavy, blowing snow. After the storm passes, winds shift to the north and northwest, and cold air advection begins. Ordinarily we would expect clearing skies, but regional conditions dictate a different result.

Strong winds from the northwest advect cold, dry air across the relatively warm waters of Lake Michigan, giving rise to the infamous **lake effect snows**. Water evaporates from the lake and

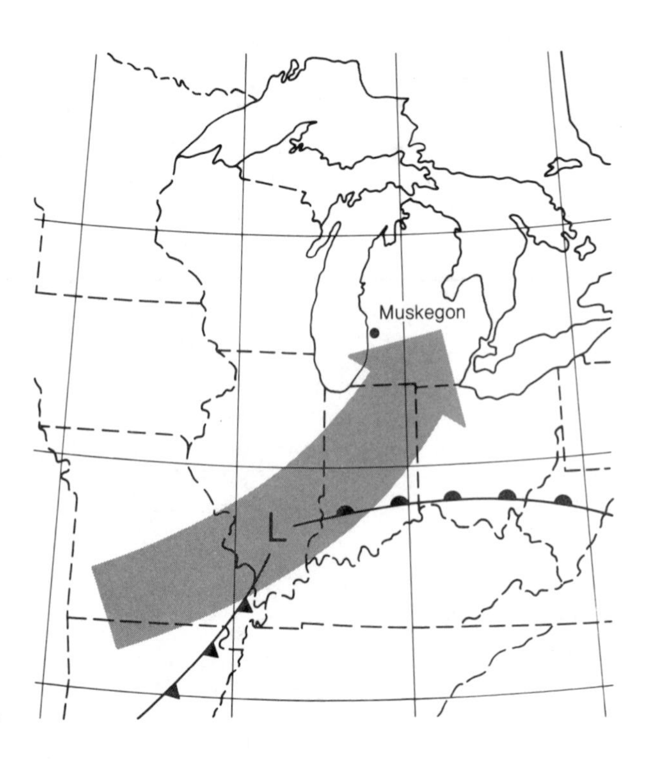

FIGURE 15.4

An early winter storm tracks northeastward through the Midwest and causes heavy snowfall at Muskegon, Michigan.

soon saturates the lowest layer of the advecting cold air mass. The cold, humid air converges as it approaches the shoreline, causing uplift, cloud development, and locally heavy snowfalls. Because of this local effect, northwest winds may bring more snow to Muskegon (long after the storm has passed) than northeast winds did (when the storm was nearby).

Long-range forecasting

As noted earlier, the reliability of detailed weather forecasts for periods longer than 10 days is minimal. Nevertheless, the Long Range Prediction Branch of the Climate Analysis Center in Camp Springs, Maryland, prepares 30-day (monthly) and 90-day (seasonal) generalized weather "outlooks" that identify areas of expected positive and negative **anomalies** (departures from long-term averages) in temperature and precipitation. An example is shown in Figure 15.5. The greatest success so far has been with winter temperature outlooks, which have been accurate about 65 percent of the time.

Forecasting the prevailing circulation pattern at the 700-mb level is the first step in forecasting monthly temperature and precipitation anomalies at the surface. Basically, the present circulation pattern is extrapolated into the future, although an effort is also made, based on historical data, to identify features of the present pattern that are most likely to persist. The prevailing westerly flow at the 700-mb level permits identification of areas of strong warm and cold air advection as well as of principal storm tracks. From these data, surface temperature and precipitation anomalies are derived.

A somewhat different approach is taken for seasonal forecasts. Forecasters rely more on long-term trends and recurring events, and attempt to isolate persistent features from prior months and seasons. In some localities, the November temperature anomaly is a good predictor of the sign (positive or negative) of the temperature anomaly for the following three months. For example, if the average November temperature at Green Bay, Wisconsin, is above the long-term average for the month, then chances for a warmer than normal winter (December through February) are better than 70 percent.

A promising area of research in long-range weather forecasting concerns atmospheric teleconnections. A **teleconnection** is a linkage between weather changes occurring in widely separated regions of the globe—often many thousands of kilometers apart. An example is the seesawing of surface air pressure between Darwin, Aus-

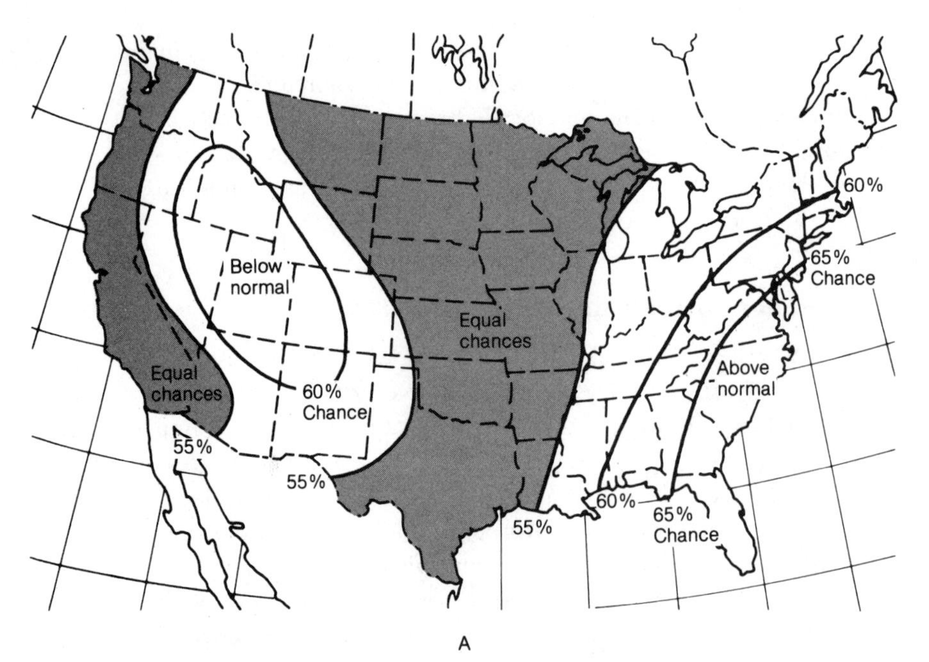

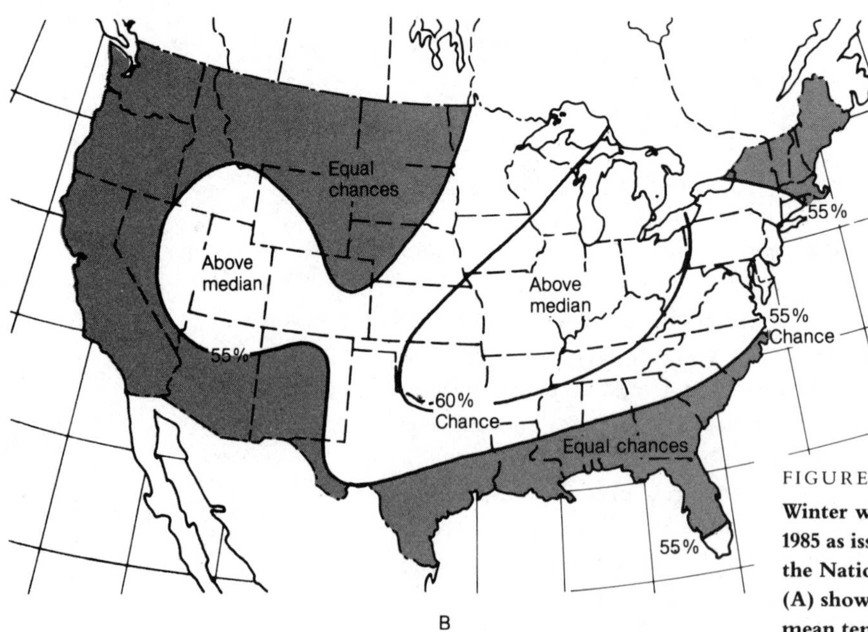

FIGURE 15.5

Winter weather outlook for 1984–1985 as issued in late fall 1984 by the National Weather Service. (A) shows the probability of the mean temperature being above or below normal. (B) shows the probability of the total precipitation being above or below the median. Winter encompasses the 90-day period of December, January, and February. (From NOAA)

tralia, and Tahiti in the central South Pacific. In the 1920s, Sir Gilbert Walker demonstrated that low pressure at Darwin tends to be accompanied by high pressure at Tahiti and vice versa. These opposing swings of air pressure actually occur between the eastern and western tropical Pacific. Known as the **southern oscillation**, these pressure changes are linked to anomalies in sea-surface temperatures in the equatorial Pacific and are accompanied by large-scale anomalies in weather.

Jerome Namias and his colleagues at the Scripps Oceanographic Institution have been studying a teleconnection that may aid in seasonal weather forecasting for North America. They found a relationship between anomalies in sea-surface temperatures over the Pacific and the westerly circulation pattern that prevails over North America during the following season. That the ocean somehow influences atmospheric circulation follows from the discussion of global radiation in Chapters 3 and 4. After all, the ocean is in contact with the atmosphere over about 72 percent of the earth's surface, and the ocean is the principal sink for solar radiation.

Among the teleconnections noted thus far, anomalies in sea-surface temperatures over the eastern equatorial Pacific appear to have the most important influence on midlatitude weather. Even this teleconnection, however, is not very reliable for forecasting purposes, because so many other factors interact to shape North American weather patterns. Although these teleconnections are an interesting avenue of research for long-range seasonal weather forecasting, a better understanding is needed of the causes of ocean temperature anomalies and of the physical basis of atmospheric teleconnections.

Single-station forecasting

Short-term weather forecasts based on weather observations at one location, known as **single-station forecasts**, may be derived from the principles of weather behavior discussed in the previous chapters of this book. Because such forecasts are based on rules applied at only one location, they tend to be quite generalized and tentative, and forecasting complications may crop up as local conditions are modified by changes elsewhere. Table 15.3 is a sample list of rules of thumb applicable to the midlatitudes. You may wish to add other rules of your own. Sometimes, but not usually, weather proverbs are useful in forecasting. This is the subject of the Special Topic, Weather Proverbs: Fact or Fiction.

SPECIAL TOPIC

Weather proverbs: fact or fiction?

Over the centuries, human beings have developed weather proverbs to serve as guides for predicting weather. These proverbs are based on casual observations, usually of a weather event or a change in animal behavior, that is associated with a subsequent weather event. In recent decades, these assumed relationships have been subjected to closer scientific scrutiny. A few have proved to be substantially correct, but most have turned out to be only picturesque myths with little or no scientific basis.

The Zuni tribe of New Mexico has a weather proverb that often proves to be true:

When the sun is in his house,
it will rain soon.

The phrase "in his house" refers to the sun's being encircled by a halo. Recall from Chapter 8 that a halo is formed when the sun's rays are refracted by the ice crystals that compose high, thin cirrus clouds. Recall also from Chapter 11 that cirrus clouds are the first clouds to appear in advance of an approaching storm. The appearance of a halo around the sun thus indicates the approach of a storm, but a halo does not ensure that precipitation will follow. Because cirrus clouds may be as much as 1000 km ahead of the storm center, there is a chance that the storm may change direction or die out before it reaches your area.

Another reliable proverb uses rainbows to predict the day's weather:

A rainbow in the morning
Is the sailor's warning.
A rainbow at night
Is the sailor's delight.

The basis for this proverb's validity is described in Chapter 8 and is linked to (1) the fact that in midlatitudes, weather systems usually progress from west to east and (2) an observer of a rainbow must face the rainshower with the sun at his or her back.

Most weather proverbs, however, have no validity. For example, many cultures have proverbs that deal with the ability of animals to forecast weather changes. Probably the most famous of these is the belief that the groundhog can determine on the second day of February whether winter is about to end. If the sun is shining and the groundhog can see its shadow, winter will supposedly last another 6 weeks. If it is cloudy, winter will soon end. Interestingly, the proverb originated in medieval Europe, and Groundhog Day was known as Candlemas Day. At that time, no definite period was assigned for winter to stay. Then the proverb read:

If Candlemas Day be fair and bright
Winter will have another fight;
But if Candlemas Day brings clouds and rain
Winter is gone and won't come again.

Several observations show this proverb to be false. Many parts of the United States and Canada are snow covered on the second of February, and it is unlikely that the groundhog even makes an appearance. What if the day dawns clear and then turns cloudy? Would the weather for the next 6 weeks depend on whether or not the groundhog was an early riser? Ridiculous, isn't it? Moreover, no evidence exists that the presence or absence of sunshine on one particular day can determine the weather for weeks to come. Weather is too variable and too complex to support such a simple cause and effect relationship.

Other weather proverbs deal with the ability of animals such as squirrels, deer, rabbits, and caterpillars to predict the severity of the upcoming winter. For example: "When squirrels lay in a large store of nuts, expect a cold winter."

There is, however, no evidence of any linkage between the two. Squirrels are more likely to store more nuts if more nuts are available! The squirrels are simply responding to a good acorn growing season, rather than exhibiting an innate ability to predict the future.

Another familiar, but false, weather proverb concerns lightning: "Lightning never strikes twice in the same place."

Recall that the process of cloud-ground lightning was described in Chapter 13. This discharge usually takes place along the shortest distance between the ground and the cloud, so the lightning often involves a tree, an electrical pole, or a tower or other tall object. As long as this object is not destroyed by the first lightning strike, there is no physical reason why lightning cannot strike the object again.

This concluding verse, attributed to E. V. Lucas (1863–1938), does illustrate one universal truth about weather—no matter what the weather, not everybody will be happy.

The Duke of Rutland urged the folk to pray
For rain; the rain came down the following day,
The pious marvelled, the skeptics murmured fluke,
The farmers late with hay, said "Damn the Duke."

TABLE 15.3

Rules to Aid Single-Station Weather Forecasting

1. **Nighttime temperatures will be lower if the sky is clear than if the sky is cloud covered.**
2. **Clear skies, light winds, and a fresh snow cover favor extreme nocturnal radiational cooling and very low air temperatures by dawn.**
3. **Falling air pressure may indicate the approach of stormy weather, while rising air pressure suggests that fair weather is in the offing.**
4. **The appearance of cirrus, cirrostratus, and altostratus clouds, in that order, indicates overrunning ahead of a warm front and the possibility of precipitation.**
5. **A counterclockwise wind shift from northeast to north to northwest (called *backing*) is usually accompanied by clearing skies and cold air advection.**
6. **A clockwise shift from east to southeast to south (called *veering*) is usually accompanied by clearing skies and warm air advection.**
7. **A wind shift from northwest to west to southwest is usually accompanied by warm air advection.**
8. **If radiation fog lifts by late morning, a fair afternoon is likely.**
9. **With west or northwest winds, a steady or rising barometer, and scattered cumulus clouds, fair weather is likely to persist.**
10. **Vertical development of cumulus clouds by midmorning may indicate afternoon showers or thunderstorms.**

Observations derived from studying records of past weather events may aid single-station weather forecasting. Records of past weather exhibit a **fair-weather bias**, that is, fair-weather days outnumber stormy days almost everywhere. If we boldly predict that all days will be fair, we will probably be correct more than half of the time. The only merit of this exercise would be to establish a baseline for evaluating the accuracy of more sophisticated weather forecasting techniques. (We would expect traditional forecasting methods to score higher than forecasting based solely on a fair-weather bias.)

Another characteristic of past weather records is **persistence**, or the tendency for weather episodes to persist for some period of time. For example, if the weather has been cold and stormy for several days, the weather may well continue that way for many more days. The same weather records also show, however, that an episode of one weather type typically gives way to another weather type very abruptly, usually in a day or less. Weather forecasts based on persistence alone are therefore prone to serious error.

A third approach is to base a weather forecast for a particular day directly on the record of the weather that occurred on that same day in years past. Suppose, for example, that in your town it has rained on August 7 only 12 times in the past 100 years. On this basis, you predict that the probability of rain next August 7 is only 12 percent, and you confidently plan a picnic. The problem with this approach is that, statistics apart, there is no guarantee that it will not rain next August 7.

Private forecasting

In our description of weather forecasting, we have focused mainly on the role of government agencies. In addition, numerous private weather forecasters and forecast services analyze weather maps and other materials supplied by the National Meteorological Center and tailor them for the private sector's special needs. For example, a private weather forecaster retained by an appliance store chain might warn store executives of a pending heat wave so that stores might be stocked with an adequate supply of fans and air conditioners. Or a private forecaster might advise an electric power company of expected winter temperatures so the energy supplier could better anticipate public fuel demands. In this way, private forecasters supplement the efforts of government weather forecasting.

Communication and Dissemination

Weather maps, charts, forecasts, and outlooks issued by the U.S. National Meteorological Center are transmitted to regional Weather Service Forecast Offices (WSFOs). Three times daily, these offices issue weather forecasts covering the individual states for the next 48 hours. Once each day, these same offices issue forecasts for the following 5 days. The states are subdivided into climatically homogeneous zones, each the size of several counties, and weather forecasts are prepared for each zone. In the event of severe weather, the WSFOs prepare and disseminate the necessary warnings. In addition, local Weather Service Offices (WSOs) issue local weather forecasts based on zone forecasts prepared by the regional WSFOs.

When hazardous weather appears either possible or probable, the National Weather Service issues **weather watches** and **weather warnings** covering the affected geographical area for a specified time period. In general, a *watch* is indicated when hazardous weather is considered possible on the basis of current or anticipated atmospheric conditions. People in the designated area need not interrupt their normal activities except to remain alert for threatening weather and to keep the television or radio on for further advisories. A *warning* is issued when hazardous weather is likely or is occurring somewhere in the region. People are then advised to take all necessary safety precautions.

Watches and warnings are issued for tornadoes, severe thunderstorms, hurricanes, floods, and winter storms. A tornado warning is issued only after a tornado has actually been sighted or appears on radar as a hook echo. The warning bulletin specifies the location of the tornado, its anticipated path, and the time at which the tornado is expected to traverse the warning area. Winter storm warnings may specify heavy snow, blizzard conditions, or an ice storm. A **blizzard warning** means that falling or blowing snow will be accompanied by winds in excess of 55 km (35 mi) per hour. Ice storm warnings mean that potentially dangerous accumulations of freezing rain or sleet are expected on the ground and other exposed surfaces.

Because weather is highly changeable, weather observations and such guides as weather maps, forecasts, watches, and warnings must be communicated as rapidly as possible both nationally and internationally. For this reason, the World Meteorological Organization and its member nations maintain elaborate communications networks consisting of a variety of systems. Weather information is relayed by teletypewriter, radio, and facsimile systems like the one shown in Figure 15.6, which reproduce maps, charts, and satellite photographs.

FIGURE 15.6

A conventional weather facsimile machine used to relay and reproduce weather maps, satellite photographs, and other weather guidance materials. (NOAA photograph)

Recent conversion to the **Automation of Field Operations and Service** (AFOS) system has made the U.S. weather communications network more efficient and reliable. Each AFOS system has two main components: (1) a minicomputer that collects weather data and transmits them to the forecaster and other AFOS stations, and (2) a console of television-type screens for displaying data and maps, and a keyboard for data and message input (Figure 15.7). The AFOS system has replaced the slower and less dependable teletypewriters and facsimile machines at all Weather Service Forecast Offices and at most Weather Service Offices. While facsimile transmissions took 5 to 10 minutes and teletypewriters processed about 100 words per minute, the new AFOS system can reproduce data and maps in only 15 seconds and can transmit 3000 words per minute. In addition, the AFOS system is less prone than the old system to service interruptions during severe weather, when the system is most needed. All AFOS systems tie into and are coordinated through a master computer in Maryland.

The public receives regular weather reports on the radio, on commercial television and cable-TV weather channels, and in the newspapers. In many U.S. localities, weather information is also

FIGURE 15.7
The console unit of the new National Weather Service weather data communications system, called AFOS for Automation of Field Operations and Service. (NOAA photograph)

available via recorded telephone announcement systems and the NOAA weather radio, which is a low-power VHF-FM radio transmitter that broadcasts continuous weather information over a 65-km (40-mi) radius. A taped message is repeated every 4 to 6 minutes and is revised routinely every 2 to 3 hours. Should hazardous weather threaten, watches, warnings, and advisories are issued.

Conclusions

Weather forecasting is a complex and challenging science that depends on the efficient interplay of weather observation, data analysis by computers and meteorologists, and rapid communications. With these systems, meteorologists have achieved a very respectable level of accuracy for short-term weather forecasting. Further improvement is expected with denser surface and upper-air observation networks, and with the development of more precise numerical models of the atmosphere. If these advances are to be realized, however, continued international cooperation is essential, for the atmosphere is a continuous fluid that knows no political boundaries.

What, then, are the prospects for climate forecasting? So far, accuracy in long-range forecasting has been minimal, and the prospects for improvement in the near future are not promising. The reasons for this will become evident in Chapters 17 and 18, in which we discuss principles of climate, climatic trends, and climatic variability. Before then, however, let us turn our attention to the topic of air pollution meteorology.

SUMMARY STATEMENTS

- The fluid nature of the atmosphere requires international cooperation in the gathering and interpretation of surface and upper-air weather data. The World Meteorological Organization coordinates an international effort of weather observation, analysis, and forecasting.
- Surface weather is monitored at land stations and by automated weather stations and ships at sea. Upper-air weather data are profiled by rawinsonde measurements, and additional weather observations are supplied by radar, aircraft, and satellites.
- Upper-air weather data are plotted on constant-pressure surfaces, and surface weather data are plotted on a constant-altitude (sea-level) surface.
- Because many variables are involved in the atmospheric system and because huge quantities of weather observations are generated, weather data are analyzed and forecasts are prepared by high speed electronic computers using numerical models of the atmosphere.
- Although the accuracy of short-term weather forecasting has improved steadily over the past 25 years, forecasting skill still deteriorates rapidly for periods longer than 48 hours and is minimal beyond 10 days.
- One-station weather forecasting may be based on principles of weather behavior, fair-weather bias, climatic records, or persistence of weather episodes.
- Weather information and forecasts are communicated to users via teletypewriters, radio, facsimile systems, television, and newspapers.

KEY WORDS

National Weather Service (NWS)
National Oceanic and Atmospheric Administration (NOAA)
International Meteorological Organization (IMO)
World Meteorological Organization (WMO)
World Weather Watch
synoptic weather network
basic weather network
Greenwich Mean Time (GMT)
National Substation Program
radiosonde
geosynchronous satellite
station model
isobars
cold air advection
warm air advection
lake effect snows
anomalies
teleconnection
southern oscillation
single-station forecasts
fair-weather bias
persistence
weather watches
weather warnings
blizzard warning
Automation of Field Operations and Service (AFOS)

REVIEW QUESTIONS

1 Why does weather forecasting require international cooperation?

2 Describe the major steps involved in the preparation of weather forecasts.

3 What events provided the impetus for the establishment of a national weather observational network?

4 Distinguish between the activities of the synoptic weather network and the activities of the basic weather network.

5 What is Greenwich Mean Time (GMT)?

6 What time is it at your locality at 0100 GMT?

7 What is the chief purpose of the National Substation Program?

8 Distinguish between a satellite in polar orbit and a satellite in geosynchronous orbit.

9 List the advantages of satellites in weather observation.

10 What is the basic difference between surface weather maps and upper-air weather maps?

11 What is meant by a numerical model of the atmosphere?

12 Why does the reliability of numerical weather prediction deteriorate rapidly for forecast periods beyond 48 hours?

13 Why are computerized weather forecasts not likely to replace human weather forecasters?

14 Describe the basis for long-range (30-, 60-, 90-day) weather forecasts.

15 What is a teleconnection?

16 Describe the linkage between sea-surface temperature anomalies and atmospheric circulation.

17 Speculate on how long-term weather records could be used to formulate seasonal weather forecasts.

18 An episode of one weather type usually gives way to another weather type very abruptly—in a day or less. Explain.

19 What is meant by a fair-weather bias?

20 Distinguish between weather watches and weather warnings.

POINTS TO PONDER

1 What is the advantage of simultaneous weather observations?

2 Explain how weather satellites are able to provide nighttime observations of cloud cover.

3 For a sea-level location, determine the fraction of the atmosphere that lies below (a) the 100-mb level, (b) the 700-mb level, and (c) the 850-mb level.

4 Explain why warm air advection causes the 500-mb surface to rise and why cold air advection causes the 500-mb surface to lower.

5 Explain why cold-core anticyclones and warm-core cyclones do not appear on 500-mb maps.

6 Why are winds associated with contour gradients on upper-air weather maps?

7 Describe how computers are used to forecast the weather.

PROJECTS

1 Design an experiment in which you test the validity of (a) persistence weather forecasting and (b) weather forecasting based on the climatic record.

2 Speculate on why there is a "fair-weather bias" and test the notion that there really is a "fair-weather bias" for your locality.

SELECTED READINGS

FAA and NOAA. *Aviation Weather Services.* Washington, D.C.: U.S. Government Printing Office, 1979. 123 pp. *Describes the weather maps and forecast charts available for pilots.*

Gilman, D. "Predicting the Weather for the Long Term." *Weatherwise* 36 (1983):290–297. *Includes a discussion on how long-range weather outlooks are prepared.*

Glahn, H. R. "Yes, Precipitation Forecasts Have Improved." *Bulletin of the American Meteorological Society* 66 (1985):820–830. *Analysis of precipitation data showing an improvement in precipitation probability forecasts for the period 1967–1982.*

Hughes, P. "American Weather Services." *Weatherwise* 33 (1980):100–111. *Principal people and events in the history of the National Weather Service.*

Kawamoto, T. M. "Via U.S. Mail—Early Weather Forecasts." *Weatherwise* 34 (1981):110–115. *Describes dissemination of weather information in the 1890s by mail.*

NOAA. *Operations of the National Weather Service.* Washington, D.C.: U.S. Government Printing Office, 1981. 251 pp. *Describes in detail observational activities, communications, and forecast dissemination activities of the National Weather Service.*

Smith, D. L. "Eighty-Five Percent and Holding—A Limit to Forecast Accuracy?" *Bulletin of the American Meteorological Society* 60 (1979): 788–790. *An evaluation of the future of weather forecast accuracy.*

Thaler, J. S. "West Point—152 Years of Weather Records." *Weatherwise* 32 (1979):112–115. *History of one of the longest weather records in the nation.*

Thomas, C. G. "Volunteers Gather Valuable Weather Information." *Weatherwise* 32 (1979):200–201. *Brief report on the more than 11,650 volunteer weather observers across the United States.*

Walsh, J. E., and D. Allen. "Testing the *Farmer's Almanac.*" *Weatherwise* 34 (1981):212–215. *An evaluation of long-range weather forecasts published in the* Old Farmer's Almanac.

This goodly frame, the earth, seems to me a sterile promontory; this most excellent canopy the air, look you, this brave o'erhanging firmament, this majestical roof fretted with golden fire—why, it appeareth no other thing to me than a foul and pestilent congregation of vapours.

WILLIAM SHAKESPEARE
Hamlet

The quality of the air we breathe is strongly influenced by the speed of the wind and the stability of the air. (Photograph by Mike Brisson)

SIXTEEN

Air Pollution Meteorology

AIR POLLUTION is not new. Indeed, it is at least as old as civilization itself. The first air pollution episode probably occurred when early humans tried to make a fire in a poorly ventilated cave. Reference to polluted air appears as early as Genesis (19:28): "Abraham beheld the smoke of the country go up as the smoke of a furnace." About 400 B.C., Hippocrates noted the pollution of city air. And in 1170 A.D., Maimonides, referring to Rome, wrote, "The relation between city air and country air may be compared to the relation between grossly contaminated, filthy air, and its clear, lucid counterpart."

The Industrial Revolution was the single greatest contributor to air pollution as a chronic problem in Europe and North America. As industrial innovations spread from one nation to another, so, too, did air pollution. In the United States, in post-Civil War days, cities swelled with new industries and with new immigrants to work in those industries. By the turn of the century, the urban environment was becoming increasingly fouled by the fumes of foundries and steel mills. In those days, a city took pride in smokestacks like those shown in Figure 16.1; they signaled a prosperous

FIGURE 16.1

Industrial smokestacks in 1906 Pittsburgh meant a prosperous economy. (Carnegie Library of Pittsburgh)

economy. Efforts to regulate air quality were meager, and little was known about the health effects of polluted air. In an attempt to placate a wheezing and coughing populace, some physicians even argued that polluted air had medicinal value.

Recent concern over polluted air does not stem from disenchantment with the fruits of industrialism. Belching smokestacks still reassure many people that the economy is healthy, but others are troubled by the possibility that polluted air may be adversely affecting their health, agricultural productivity, and the weather. In this chapter, we examine two questions related to these concerns: (1) How do weather conditions influence air pollution levels? (2) What is the impact of air pollution on weather?

Many gases and aerosols that can be air pollutants are actually normal constituents of the atmosphere (see Chapter 1). These substances become pollutants only when their concentrations threaten the well-being of living things or disrupt physical or biological processes. In the Special Topic, Principal Air Pollutants, we survey the major air pollutants—noting their natural cycling within the environment, the contributions to pollution made by human activity, and some health effects of air pollution.

Air Pollution Episodes

On the morning of 26 October 1948, a fog blanket reeking of pungent sulfur dioxide fumes spread over the town of Donora in Pennsylvania's Monongahela Valley. Before the fog lifted 5 days later, almost half of the area's 14,000 inhabitants had fallen ill, and 20 had died. This killer fog resulted from the combination of mountainous topography and stable weather conditions that trapped and concentrated deadly effluents from the community's steel mill, zinc smelter, and sulfuric acid plant.

Air pollutants are especially dangerous when atmospheric conditions favor their concentration. Once pollutants are emitted into the atmosphere, their concentrations usually begin to decline. The rate of concentration decrease, or dilution, is determined in part by the extent to which pollutants mix with cleaner air: the more thorough the mixing, the more rapid the dilution. When conditions in the atmosphere favor rapid dilution, the impact of pollutants is usually minor. On other occasions—termed **air pollution episodes**—conditions in the atmosphere minimize dilution, and the impact of air pollution can then be severe, particularly on human health. The

SPECIAL TOPIC

Principal air pollutants

In this Special Topic we summarize the sources, cycling, and impacts of the major air pollutants.

Oxides of Carbon

The burning of carbon-laden fossil fuels, such as coal and oil, releases carbon dioxide (CO_2) into the atmosphere. Much of this carbon dioxide is absorbed by ocean water, and some is taken up by vegetation through photosynthesis. Today the concentration of carbon dioxide in the atmosphere is about 340 parts per million (ppm) and is rising at a rate of about 18 ppm per decade. The increased burning of coal and oil has been primarily responsible for this rise, which, as we will see in Chapter 18, may affect the global climate.

By far the most important natural source of atmospheric carbon monoxide (CO) is the combination of oxygen with methane (CH_4), a product of the anaerobic decay* of vegetation. Carbon monoxide is removed from the atmosphere by the activity of certain soil microorganisms. The net result is a harmless Northern Hemisphere average concentration of less than 0.15 ppm.

The principal source of carbon monoxide derived from human activity is motor vehicle exhaust. Breathing CO causes drowsiness, slows reflexes, and impairs judgment; at high concentrations death ensues. Because the gas is odorless and tasteless, carbon monoxide defies detection by human senses and constitutes a serious health hazard, especially when the concentration is high, as it can be in highway tunnels and underground parking garages.

Hydrocarbons

In urban areas, the principal source of reactive hydrocarbons, which cause air pollution, is the incomplete combustion of gasoline by motor vehicles. This source is in fact responsible for hundreds of different hydrocarbons. Because gasoline is very volatile, some hydrocarbons (perhaps as much as 15 percent of the total in some cities) enter the air during gasoline delivery and refueling operations at service stations.

The general term *hydrocarbons* encompasses a wide variety of chemical compounds that contain only hydrogen and carbon. Of the hydrocarbons that occur naturally in the atmosphere, methane (CH_4) is present in highest concentrations (1 to 1.5 ppm). Even at relatively high concentrations, methane is nonreactive (that is, it does not chemically interact with other substances) and causes no detrimental health effects. Occurring naturally in lower concentrations (less than 0.1 ppm), but much more reactive, are the terpenes, a variety of volatile (readily vaporized) hydrocarbons emitted by vegetation. These compounds are responsible for the aromas of pine, sandalwood, and eucalyptus trees. The terpenes also form particles that scatter sunlight and produce the bluish haze that is often seen hanging over forests such as the Great Smokies.

Although our understanding of the natural cycling of hydrocarbons in the atmosphere is as yet minimal, we do know that the typically low con-

*Decay in the absence of oxygen.

centrations of most hydrocarbons found in city air appear to pose no environmental threat. The most serious health threats arise from the products of hydrocarbons reacting with other pollutants in the presence of sunlight. These photochemical reaction products include formaldehyde, ketones, and PAN (peroxyacetylnitrate)—substances that irritate the eyes and damage the respiratory system. Some hydrocarbons, such as benzene (C_6H_6) (a component of many consumer goods, including rubber cement) and benzoapyrene (a product of fossil fuel and tobacco combustion) are also carcinogenic.

Oxides of Nitrogen

The action of soil bacteria is responsible for most of the nitric oxide (NO) that is produced naturally and released to the atmosphere. Within the atmosphere, NO combines readily with oxygen to form nitrogen dioxide (NO_2). Together, these two oxides of nitrogen are usually referred to as NO_x.

Although human activities contribute only about 10 percent of the atmosphere's total load of NO_x, our contributions tend to be much more concentrated than the natural atmospheric average. Based on the mode of formation, scientists distinguish between thermal NO_x and fuel NO_x. Thermal NO_x forms when high combustion temperatures like those inside an internal combustion engine cause nitrogen (N_2) and oxygen (O_2) in the air to combine. Fuel NO_x forms when nitrogen contained in a fuel such as coal is oxidized, that is, the N_2 in the fuel combines with O_2 in the air. For both modes of formation, NO is generated initially, and then when the products are vented and cooled, a portion of the NO is converted to NO_2. About half of fuel NO_x comes from stationary sources (power plants, primarily) and the other half comes from motor vehicle emissions.

Nitrogen dioxide (NO_2) is a much more serious air pollutant than its precursor NO. The toxicity of NO_2 is about four times that of NO. Nitrogen dioxide at high levels is believed to contribute to heart, lung, liver, and kidney damage, and is linked to the incidence of bronchitis and pneumonia. Moreover, because nitrogen dioxide occurs as a brownish haze, it reduces visibility. When NO_2 combines with water vapor, nitric acid (HNO_3), a corrosive substance, is formed. The oxides of nitrogen are also major precursors of smog.

Compounds of Sulfur

Many human activities also release sulfur into the atmosphere, contributing perhaps one third of the amount of sulfur that is emitted into the air by natural sources. Most of the sulfur dioxide we contribute comes from the burning of fuels that contain sulfur as an impurity. These fuels, such as coal and oil, are burned in electric generating plants (69 percent) and industrial boilers (8 percent). In addition, the smelting of sulfur-bearing ores—lead, zinc, and copper sulfides—is a source of sulfur dioxide (8 percent), as are petroleum refining (5 percent), motor vehicles (5 percent), and residential and commercial space heating (5 percent).

Sulfur also enters the atmosphere naturally as sulfur dioxide (SO_2) from volcanic eruptions, as sulfate particles from sea spray, and as hydrogen sulfide (H_2S) produced when organic matter decays anaerobically. These sulfur compounds are washed from the air by precipitation and are taken up by soil, vegetation, and surface waters.

(continued)

Principal air pollutants

(continued)

In the atmosphere, sulfur dioxide is converted to sulfur trioxide (SO_3) and sulfate particles (SO_4). Sulfate particles restrict visibility and, in the presence of water, form sulfuric acid (H_2SO_4), a highly corrosive substance that also lowers visibility. Both SO_2 and SO_3 irritate respiratory passages and can aggravate asthma, emphysema, and bronchitis. Sulfate particles and sulfuric acid droplets are thought to increase vulnerability to respiratory infection.

Certain industrial activities, including paper and pulp processing, emit hydrogen sulfide (H_2S) as well as a family of organic sulfur-containing gases called mercaptans. Even in extremely small concentrations, these compounds are foul smelling. Hydrogen sulfide tarnishes silverware and copper facings, and discolors lead-based paints.

Smog

When vehicular traffic is congested—during morning and evening rush hours, for example—*photochemical smog* is likely to form. The oxides of nitrogen and the hydrocarbons in auto exhaust react in the presence of sunlight to produce a noxious, hazy mixture of suspended particles and gases. While this reaction is most common in urban areas, winds occasionally transport auto exhaust into suburban and rural areas, where the sun's rays trigger smog development. Smog can therefore be a serious problem downwind from, as well as within, a large metropolitan area.

The constituents of photochemical smog include PAN (peroxyacetylnitrate) and ozone (O_3). PAN damages vegetation and stings the eyes. While levels of ozone at the earth's surface average only about 0.02 ppm, during very smoggy periods ozone concentrations may exceed 0.5 ppm. At these relatively high concentrations, ozone irritates the eyes and the mucous membranes of the nose and throat. It also degrades rubber and fabrics, and damages some crops.

Suspended Particulates

To this point our survey of the major air pollutants has focused primarily on gases. Now we turn our attention to the millions of metric tons of tiny solid particles and liquid droplets, such as acid mist, that are suspended in the atmosphere. Collectively these particles and droplets are termed suspended particulates, or *aerosols*. Sea-salt spray, soil erosion, volcanic activity, and various industrial emissions account for about one half of the atmosphere's total aerosol load. The other half is largely the consequence of the atmospheric reactions among various gases.

Perhaps the most common particulates are dust and soot. Most dust is produced when wind erodes soil; this erosion is often accelerated by agricultural activity. Soot—tiny solid particles of carbon—is emitted during the incomplete combustion of fossil fuels and refuse. Particulates may be composed of a wide variety of materials, depending on the specific types of mining, milling, or manufacturing carried on in a given area. In urban-industrial air, particulates usually include a diverse array of trace metals such as lead, nickel, iron, zinc, copper, magnesium, and cadmium. These particulates pose a significant health hazard, because their typically small size allows them to be inhaled readily. In addition, air may contain asbestos fibers, pesticides, and fertilizer dust.

Air normally also contains fungal spores and pollen. Disturbance of the land by farming and construction promotes the abundant growth of ragweed and other weeds with pollen that evokes allergic reactions, such as hayfever, in roughly 1 out of every 20 people.

two weather conditions that most influence the rate of dilution are wind speed and air stability.

Wind speed

We know intuitively that air is likely to mix more vigorously on a windy day than on a calm day. As a general rule, if the wind speed doubles, the concentration of air pollutants is cut in half (see Figure 16.2). Certain weather patterns favor light winds and thus inhibit the dispersal of contaminants. Within an anticyclone, for example, horizontal pressure gradients are weak, and winds near the center are very light or calm, so pollutants do not disperse readily.

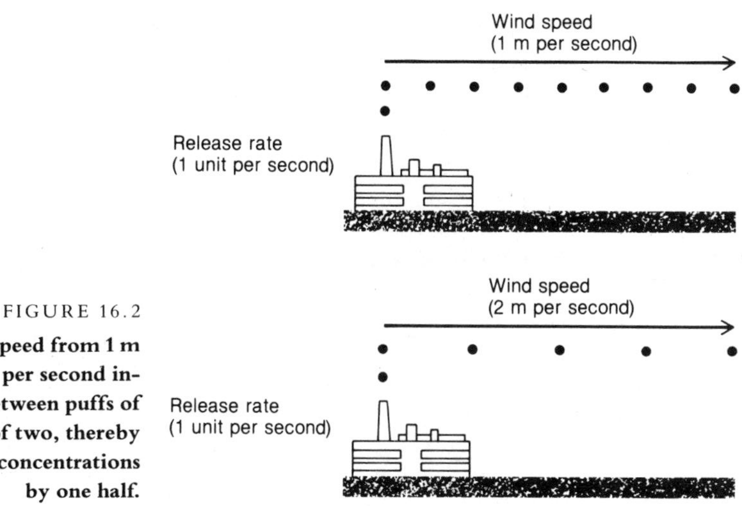

FIGURE 16.2

Doubling the wind speed from 1 m per second to 2 m per second increases the spacing between puffs of smoke by a factor of two, thereby reducing pollution concentrations by one half.

Wind speed is influenced not only by horizontal pressure gradients but also by friction. In a city, winds are slowed by the rough surface created by the canyonlike topography of tall buildings and narrow streets. In fact, average wind speeds may be 25 percent slower in a city than in the surrounding countryside. When light regional winds (less than 15 km, or 10 mi, per hour) prevail, the contrast between city and country is even more pronounced, amounting to a wind-speed reduction of up to 30 percent. The dilution of air pollutants by wind is therefore particularly impeded in urban localities—the very places where most of the contaminants are generated.

Air stability

As we saw in Chapter 6, stability affects vertical motion within the atmosphere. Convection and turbulence are enhanced when the air is unstable, and are inhibited when the air is stable. The stability of air thus influences the rate at which polluted air mixes with clean air. A parcel of polluted air emitted into unstable air undergoes more mixing than polluted air emitted into stable air. Stable air inhibits the upward transport of air pollutants, and a layer of stable air aloft may act as a lid over the lower troposphere to trap air pollutants. The continual emission of contaminants into stable air results in the accumulation and concentration of pollutants.

Mixing depth is the vertical distance between the ground and the altitude to which convection (i.e., mixing) extends. When mixing depths are great, for example, many kilometers, the relative abundance of clean air allows pollutants to mix and dilute rapidly. When mixing depths are shallow, air pollutants are restricted to a smaller volume of air, and concentrations may approach hazardous levels. When air is stable, convection is suppressed and mixing depths are low. When air is unstable, convection is enhanced and mixing depths increase. Because solar heating triggers convection, mixing depths tend to be greater in the afternoon than in the morning, greater during the day than at night, and greater in summer than in winter.

We can sometimes estimate the stability of air layers by observing the behavior of a plume of smoke belching from a smokestack. If the smoke is entering an unstable air layer, the plume undulates, as in Figure 16.3. In general, this plume behavior indicates that pol-

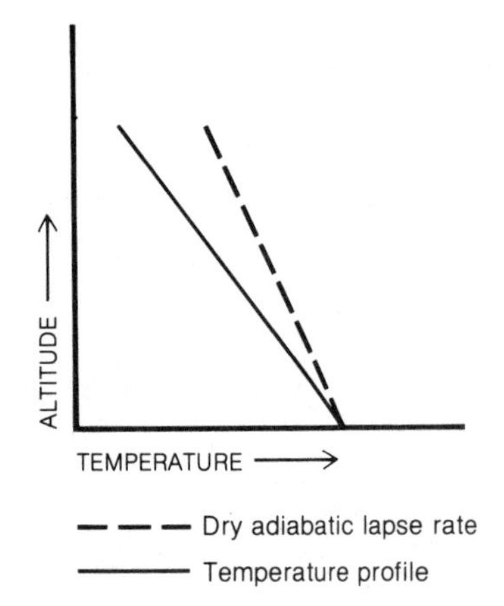

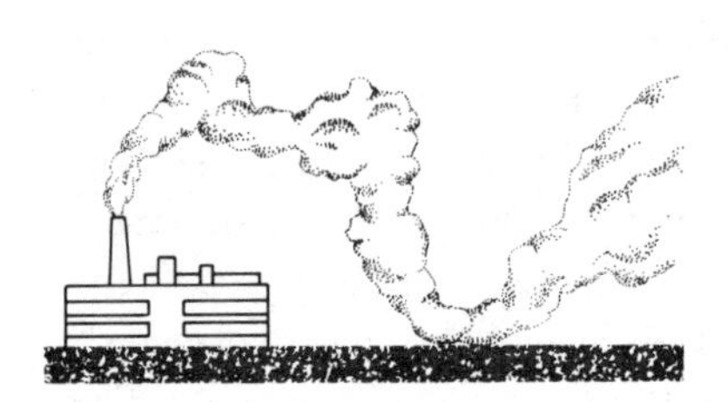

FIGURE 16.3

With unstable atmospheric conditions, the looping of smoke plumes favors dilution of air pollutants.

luted air is mixing readily with the surrounding cleaner air, thereby facilitating dilution. The net effect is improved air quality (except where the plume loops to the ground). On the other hand, a smoke plume that flattens and spreads slowly downwind, as in Figure 16.4, indicates very stable conditions and minimal dilution.

In summary, air stability influences the rate at which polluted air and clean air mix. If air layers are stable, dilution is inhibited, but if air layers are unstable, dilution is enhanced.

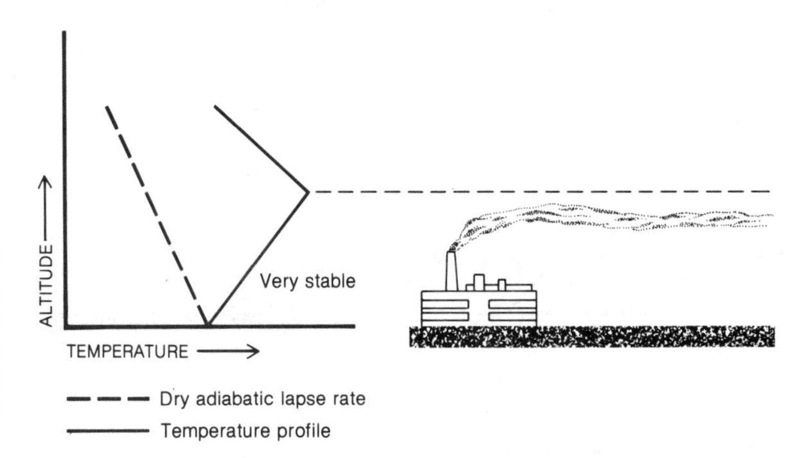

FIGURE 16.4
A temperature inversion within a surface air layer indicates very stable conditions. A smoke plume forms a thin ribbon extending downwind from the stack and dilution of pollutants is minimal.

Temperature inversions

An air pollution episode is most likely when a persistent **temperature inversion** develops. Recall from Chapter 6 that in an air layer with a temperature inversion, air temperatures increase with altitude. Warm, light air thus overlies cooler, denser air. This is an extremely stable stratification that strongly inhibits convective mixing and dilution. A temperature inversion can form by (1) subsidence of air, (2) extreme radiational cooling, or (3) advection of air masses. The resulting inversion may occur aloft or at the surface.

A **subsidence temperature inversion** forms a lid over a wide area, often encompassing several states at one time. It develops during a period of fair weather when the hemispheric weather pattern causes a warm anticyclone to stall. A high is characterized by descending and compressionally warmed air currents that spread toward the earth's surface. The warm air is prevented from reaching the surface by the **mixing layer** in which air is thoroughly mixed by convection (Figure 16.5). Air temperatures within the mixing layer decrease with altitude, but air just above the mixing layer, having been warmed by adiabatic compression, is significantly warmer than air at the top of the mixing layer. An elevated temperature inversion thus separates the mixing layer from the compressionally warmed air above. Under these conditions, pollutants are distributed throughout the mixing layer up to the altitude of the temperature inversion. This situation is sometimes referred to as **fumigation**.

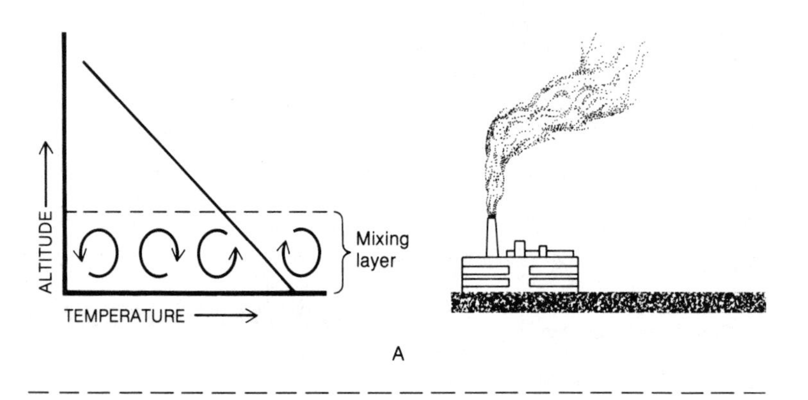

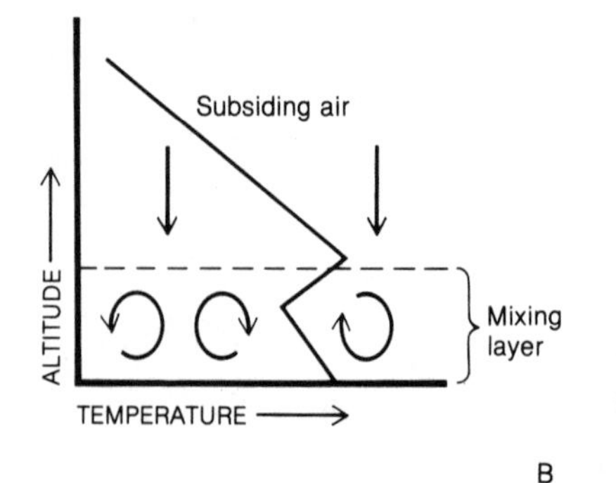

FIGURE 16.5

An elevated temperature inversion develops through subsidence of air. A sounding prior to subsidence (A) is compared with a sounding during subsidence (B). The elevated temperature inversion acts as a lid over the lower atmosphere and may trap air pollution.

Radiation temperature inversions are perhaps more common and are often more localized than subsidence temperature inversions. At night, under clear skies, the earth's warmth is lost rapidly to space through infrared radiation, called **radiational cooling**. Surface air layers are then chilled by contact with the cooler ground. Because the air at the surface is coldest, a surface temperature inversion develops. Smoke emitted into such an air layer forms a thin ribbon that drifts slowly downwind. After sunrise, as solar radiation is absorbed by the ground and heat is radiated and conducted to the overlying air, the inversion gradually disappears and a normal temperature lapse rate is restored. In winter, however, when snow covers the ground and the sun's rays are weak, a radiation temperature inversion can persist for several days or even for weeks at a time, and may severely inhibit the dispersal of air pollutants.

Advecting air masses can also give rise to temperature inversions. This is sometimes the case at the base of the Rocky Mountains. As shown in Figure 16.6, a westerly airflow is compressionally warmed as it is drawn down the leeward slope of the mountain range. At the foot of the range, however, northerly surface winds advect cold air. Stratified air and an elevated temperature inversion are the consequences. Although temperature inversions also characterize warm and cold fronts, these inversions have little adverse effect on air quality because the fronts are moving and the accompanying precipitation washes pollutants from the air.

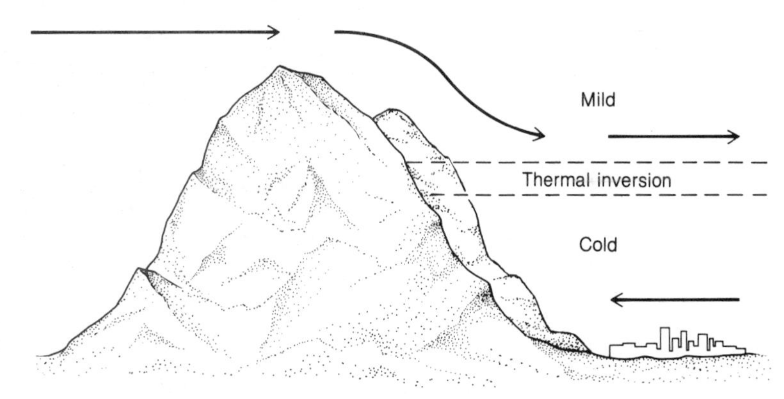

FIGURE 16.6

An elevated temperature inversion develops on the leeward side of the Rocky Mountains. Compressionally warmed air overlies cold air advected by northerly winds at the surface. This situation creates a high air pollution potential for the city of Denver.

Air pollution potential

Weather conditions that favor air stagnation and air pollution episodes occur with varying frequency in different places and at different times of the year. Areas having particularly high air pollution potentials include southern and coastal California, portions of the Rocky Mountain states, and the mountainous portions of the mid-Atlantic states. In general, in much of the West, air quality is lowest in winter, while in the East, air quality is lowest in autumn. In southern California, air pollution potential is highest in summer. These seasonal changes in air quality are due to normal seasonal shifts in circulation patterns.

Many regions with high air pollution potential are also locales of great topographic relief, since hills and mountain ranges can block horizontal winds that would disperse polluted air. In addition, radiation temperature inversions that form in lowlands, such as river valleys, are often strengthened by an accumulation of cold, dense air that drains downward from nearby highlands. As illustrated in Figure 16.7, the result is a persistent stratification of mild, light air over cool, dense air.

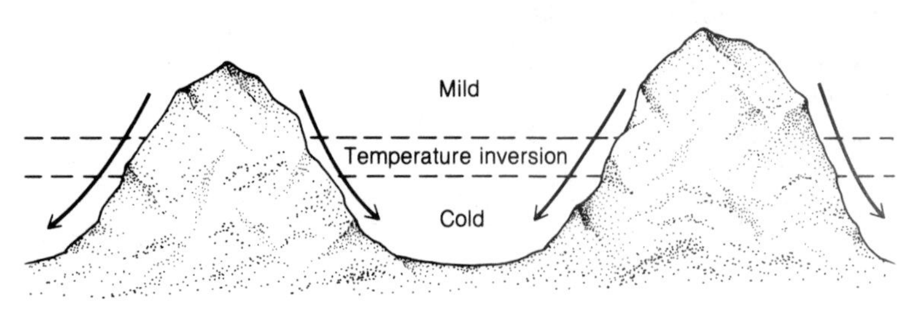

FIGURE 16.7
Under the influence of gravity, cold air slides downhill and the temperature inversion in valleys is strengthened.

Los Angeles is particularly susceptible to air pollution episodes because of its topographic setting, a high concentration of pollutant sources (more than 8 million cars), and frequent periods of air stability. Figure 16.8 shows the air circulation and topographic features influencing air quality in that city. Weather in the Los Angeles area, like the weather throughout much of California, is strongly influenced by the eastern edge of the semipermanent Pacific anticyclone. This high is responsible not only for California's famous fair weather, but also for descending air currents aloft that generate a subsidence temperature inversion at about 700 m (2300 ft) over Los Angeles. In consequence, low mixing depths occur on about two thirds of the days of the year. The exceptionally high incidence of extremely stable atmospheric conditions is further aggravated by topographic barriers. Los Angeles is situated on a plain that opens to the Pacific and is rimmed on three sides by mountains. Because of this topography, cool breezes sweeping inland from the ocean are unable to flush pollutants out of the city. The mountains and an elevated temperature inversion encase the city in its own fumes, and within this crucible a complex photochemistry takes place to produce smog. (See the Special Topic, Principal Air Pollutants.)

Because air pollutants can be a serious threat to human health, weather reports in many cities regularly include air quality bulletins. These advisories commonly use the Pollutant Standards In-

SPECIAL TOPIC

Pollutant Standards Index

Polluted air can have serious and even fatal effects on the human body. An index of air quality was devised by the U. S. Environmental Protection Agency (EPA) to indicate the relative dangerousness of the air. Elderly persons, runners, those with heart or lung disease, as well as the general population, may consult this index to judge the safety of the air on any given day.

The EPA established uniform air quality standards for six air pollutants (Table 1). These standards apply to the ambient (outdoor) air and are of two types, primary and secondary. A *primary* air quality standard is the maximum exposure level of an air pollutant that can be tolerated by human beings without ill effects. A *secondary* air quality standard is the maximum allowable concentration of an air pollutant when considering its potential harmful impact on vegetation, visibility, personal comfort, and climate.

TABLE 1

National Ambient Air Quality Standards for Criteria Pollutants

POLLUTANT	AVERAGING TIME*	PRIMARY STANDARD	SECONDARY STANDARD
Total suspended particulates (micrograms per cubic meter)†	1 year	75	60
	24 hours	260‡	150
Sulfur oxides (ppm)	1 year	0.03	—
	24 hours	0.14‡	—
	3 hours	—	0.5
Carbon monoxide (ppm)	8 hours	9‡	9
	1 hour	35‡	35
Nitrogen dioxide (ppm)	1 year	0.053	0.053
Ozone (ppm)	1 hour	0.12‡	0.12‡
Lead (micrograms per cubic meter)	3 months	1.5	—

Source: U.S. Environmental Protection Agency

*Averaging time is the period over which concentrations are measured and averaged.

†A microgram is one millionth of a gram.

‡Concentration not to be exceeded more than once (on separate days) per year.

The *Pollutant Standards Index* (PSI) is computed for a particular location using the air pollutant that exhibits the highest concentration when compared with its primary air quality standard. If the concentration is at the primary standard, the PSI is assigned a value of 100; if it is twice the standard, the PSI value is 200, and so on. As shown in Table 2, PSI values are divided into six ranges that correspond to the risk to human health. PSI values higher than 100 are considered unhealthy, particularly for elderly persons and those with heart or respiratory ailments. A PSI value of 300 or higher is considered hazardous, and at such a time everyone should avoid outdoor activity.

(continued)

Pollutant Standards Index

(continued)

When PSI values exceed 100, what is the potential impact on human health? Although many parts of the human body are eventual targets, the initial attack of pollutants occurs primarily through the respiratory system. Inhaled air follows a long pathway before finally reaching the air sacs of the lungs. Within the sacs, oxygen is removed from the air by red blood cells, which then carry it to all parts of the body. If the blood is deprived of its oxygen supply, the person succumbs to *asphyxiation*, or oxygen starvation. Because pollutants are inhaled into the respiratory tract along with air, they follow the same route to the air sacs. In sufficiently high concentrations, some air pollutants are asphyxiating agents. Other pollutants irritate the tissues of the respiratory tract itself.

Carbon monoxide, a constituent of motor vehicle exhaust, is the most common asphyxiating air pollutant. If enough reaches the air sacs, it displaces the oxygen on the hemoglobin molecules (the transporters of oxygen in red blood cells). As increasing concentrations of carbon monoxide are inhaled, the quantity of life-sustaining oxygen decreases. Initially, the victim experiences dizziness, headache, and impaired perception; higher concentrations result in nausea, visual problems, and, ultimately, death.

Gases that act mainly as respiratory tract irritants include ozone, sulfur dioxide, and nitrogen dioxide. Each causes somewhat different reactions, but the effects generally include persistent cough and heavy secretion of mucus. The latter clogs smaller pathways leading to the lungs' air sacs, thereby decreasing the amount of oxygen available to the blood.

Particulates that reach the lungs may cause illness and even death. The most severe problems result from lengthy exposure to relatively high concentration levels, as in such occupations as mining, metal grinding, and manufacturing of abrasives. Resultant lung diseases, which are usually named for the type of particulates involved, include *black lung* (from coal dust), *silicosis* (from quartz dust generated during mining), *asbestosis* (from asbestos fibers), and *brown lung* or *byssinosis* (from cotton dust). Depending on the type of particulate and the concentrations inhaled, the lungs sustain irritation, allergic reactions, or scarring of tissue, which becomes a potential site for tumor development. Victims typically experience coughing and shortness of breath and, in the long run, may have pneumonia, chronic bronchitis, emphysema, or lung cancer.

Air quality in many urban areas is improving. Between 1974 and 1980, the average number of unhealthy air quality days declined in the 23 metropolitan areas where *Pollutant Standards Index (PSI)* data are available. A particularly dramatic improvement was reported for Chicago, where the number of unhealthy days (PSI of at least 100) fell 80 percent, from 240 in 1974 to 48 in 1980. During the same period, New York City had a substantial decline of 51 percent.

Not all cities have been so fortunate, however. Los Angeles experienced only slight improvement during this time, and Houston's air quality actually deteriorated. Also, the trend toward improvement does not hold for the number of most hazardous days, when the PSI is 300 or above.

TABLE 2

Pollutant Standards Index (PSI) of Air Quality

POLLUTANT STANDARDS INDEX VALUE STATEMENT	DESCRIPTION	EPISODE LEVEL	GENERAL HEALTH EFFECTS AND CAUTIONS
0–50	Good		
51–100	Moderate		
101–200	Unhealthy	Primary standard	Symptoms are slightly aggravated in susceptible persons. Symptoms of irritation occur in healthy population.
			Persons with heart or respiratory ailments should reduce physical exertion and outdoor activity.
201–300	Very unhealthy	Alert	Symptoms are significantly aggravated and exercise tolerance is decreased in persons with heart or lung disease. Symptoms are widespread in healthy population.
			Elderly persons and those with heart or lung disease should stay indoors and reduce physical activity.
301–400	Hazardous	Warning	Premature onset of some diseases is noted, in addition to significant incidence of symptoms and decreased exercise tolerance in healthy persons.
			Elderly persons and those with heart or lung disease should stay indoors and avoid physical exertion. General population should avoid outdoor activity.
401 or more	Hazardous	Emergency	Premature death occurs in ill and elderly persons. Healthy persons experience adverse symptoms affecting normal activity.
			All persons should minimize physical exertion, avoid traffic, and remain indoors, keeping windows and doors closed.

Source: Wisconsin Department of Natural Resources, Bureau of Air Quality Management

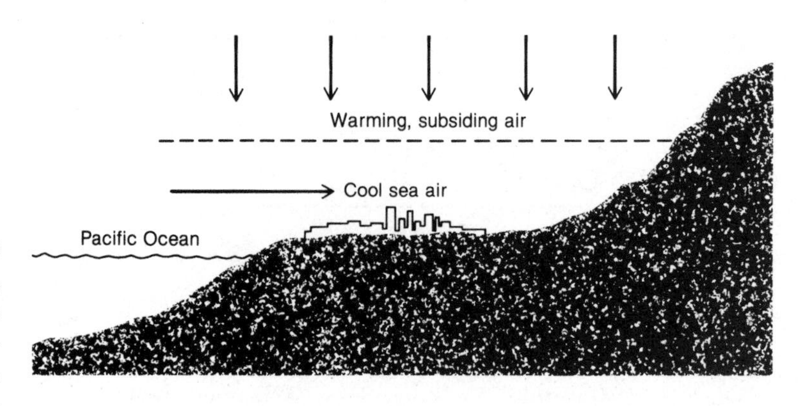

FIGURE 16.8

Atmospheric circulation patterns and topographic features give Los Angeles an unusually high air pollution potential. (From Moran, J. M., M. D. Morgan, and J. H. Wiersma. *Introduction to Environmental Science*. San Francisco: W. H. Freeman, © 1980, p. 313.)

dex (PSI) as the measure of air quality. In the Special Topic, Pollutant Standards Index, we describe how the PSI is determined and some effects of air pollutants on human health.

Natural cleansing processes

Conditions favoring the accumulation and concentration of pollutants in the air are countered to some extent by natural removal mechanisms. Some particulates are removed from the air when they strike and adhere to buildings and other structures, a process called **impaction**. Aerosols are also subject to **gravitational settling**, which is most effective for aerosols with radii greater than a tenth of a micrometer. The heavier and larger aerosols will settle more rapidly than the smaller aerosols. For this reason, larger aerosols tend to settle nearer their source, whereas small aerosols may be carried for many kilometers and to great altitudes before finally settling to the ground. The combined processes of impaction and gravitational settling are sometimes referred to as **dry deposition**.

The most effective natural pollution removal mechanism is **scavenging** by rain and snow. In fact, in localities with moderate precipitation, as much as 90 percent of the suspended aerosols are removed by scavenging. Gaseous pollutants are somewhat less susceptible to scavenging than are aerosols, but they do dissolve to some extent in raindrops and in cloud droplets. Although air pollutant scavenging enhances air quality, it degrades the quality of the precipitation—sometimes to the point of polluting surface water so much that aquatic life is threatened. We discuss this effect later in the chapter.

Air Pollution Impact on Weather

We have seen how weather conditions influence air pollution potential. Let us now examine the impact of air pollution on weather. (In Chapter 18 we discuss how air pollution may be affecting climate.)

Urban weather and climate

Certain air pollutants usually found in urban air, including a variety of dust particles and acid droplets, can influence the development of clouds and the precipitation within and downwind from a city. These pollutants, many of which are hygroscopic, serve as nuclei for cloud droplets, and thus accelerate condensation. The tendency for cloudiness and precipitation to occur more frequently in and near urban-industrial areas than in more rural areas is further enhanced by the rising warm air over the urban heat island (Chapter 12), which helps lift the air to saturation.

The influence of urban air pollution on condensation and precipitation is illustrated by the typical climatic contrasts between urban and rural regions. Winter fogs occur about twice as frequently in cities as in the surrounding countryside. Downwind from cities, rainfall may be enhanced by 5 to 10 percent. The greater contrasts tend to be exhibited on weekdays, when urban-industrial activity is at its peak, suggesting that increased precipitation is at least partially attributable to urban-industrial air pollutants.

Data from the Metropolitan Meteorological Experiment, **METROMEX**, indicate significantly greater precipitation enhancement downwind of St. Louis. METROMEX scientists analyzed weather observations during a 5-year intensive field study (1971–1975) and concluded that downwind of St. Louis the summer rainfall was up to 30 percent greater than upwind of the city. This rainfall anomaly was attributed to the combined effect of urban contributions of heat and "giant" cloud condensation nuclei.

Because precipitation, fog, and cloudiness in urban areas often have adverse effects on both surface and air transportation, any artificial increase in these conditions is potentially troublesome. Reduced visibility, for example, slows surface traffic, curtails air travel, and contributes to auto accidents. In recent years, some improvement in local urban visibilities has been reported and is apparently the consequence of stricter air quality standards.

Jet plane traffic is modifying the cloud cover, especially along heavily traveled air corridors between major cities. The visible jet contrails etching the sky, like those shown in Figure 16.9, are feathery cirrus clouds traceable to the water vapor and condensation nuclei produced by jet engines as combustion products. Increased cloudiness in turn reduces sunshine penetration and may enhance local precipitation by serving as a source of ice-crystal nuclei.

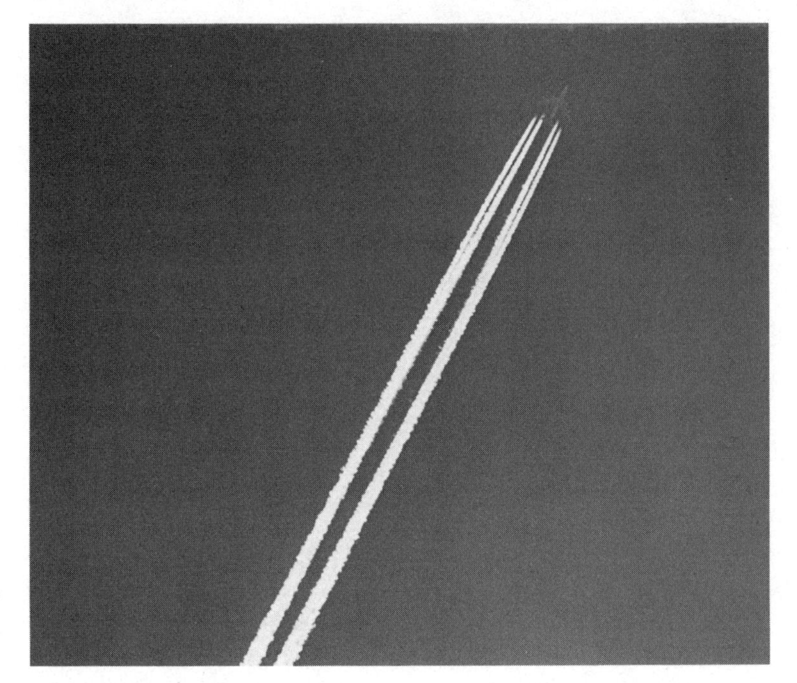

FIGURE 16.9
Contrails of jet aircraft. (Photograph by Mike Brisson)

Acid rains

As we saw in Chapter 6, the atmospheric subcycle of the hydrologic cycle purifies water. As rain and snow fall from the clouds to the ground, they wash pollutants from the air and in the process take up pollutants. Rainfall is normally slightly acidic because it dissolves some atmospheric carbon dioxide, producing weak carbonic acid (H_2CO_3). Where the air is polluted with oxides of sulfur and oxides of nitrogen, however, rainfall produces relatively strong sulfuric acid (H_2SO_4) and nitric acid (HNO_3). Precipitation that falls through such contaminated air may become 200 times more acidic than normal. Rainfall that is more acidic than normal is called **acid rain**. In addition, dry deposition delivers acidic materials to the earth's surface.

The range of acidity and alkalinity, called the **pH scale**, is shown in Figure 16.10, which compares the normal acidity of rainwater with the pH values of some other familiar substances. Note that the pH scale is logarithmic, that is, each unit increment corresponds to a tenfold change in acidity. A drop on the pH scale from the normal rain value of 5.6 to a value of 3.6 thus indicates a hundredfold increase (10 × 10) in acidity.

Where acid rains fall on soils or rock that cannot neutralize the acidity, regional surface waters become more acidic. Localities in North America where the lakes are susceptible to acid precipitation are shown in Figure 16.11. Acids in lake or river water disrupt the reproductive cycles of fish, and acid rains leach heavy metals from the soil—washing them into lakes and streams where they may harm fish, aquatic plants, and microorganisms. The fish populations in many lakes and streams in Norway, Sweden, eastern Canada, and the northeastern United States have declined or been eliminated because acid rains have so increased the acidity of aquatic habitats. Recent studies suggest that acid rains may also be involved in the decline and dying back of conifer forests in West Germany, upstate New York, and northern New England. Another costly impact of acid rains is the accelerated weathering of building materials, especially limestone, marble, and concrete. Metals, too, corrode faster than normal when exposed to acidic moisture.

Gene E. Likens of Cornell University and his associates reported an increase in rainfall acidity over the eastern United States between 1955 and 1973. Their findings were later confirmed and updated by measurements made by the National Atmospheric Deposition Program in the United States and by the Canadian Network for Sampling Precipitation (Figure 16.12). In the summer of 1983, the National Research Council of the National Academy of Sciences issued a long-awaited report that attributed 90 to 95 percent of acid rains in the Northeast to industrial sources and motor vehicle exhaust. Coal burning electric power plants in the Midwest may be the chief emitters of acid rain precursors.

Winds aloft can transport oxides of sulfur and nitrogen for thousands of kilometers from the tall stack sources, so acid rain is becoming a global problem. In fact, acid rains have been reported from such isolated localities as the Hawaiian Islands and the central Indian Ocean. Long-range transport of acid rain precursors has even strained the traditionally amiable relationship between the United States and Canada.

Canadian scientists are concerned about southerly winds transporting pollutants from the United States into Canada. These pol-

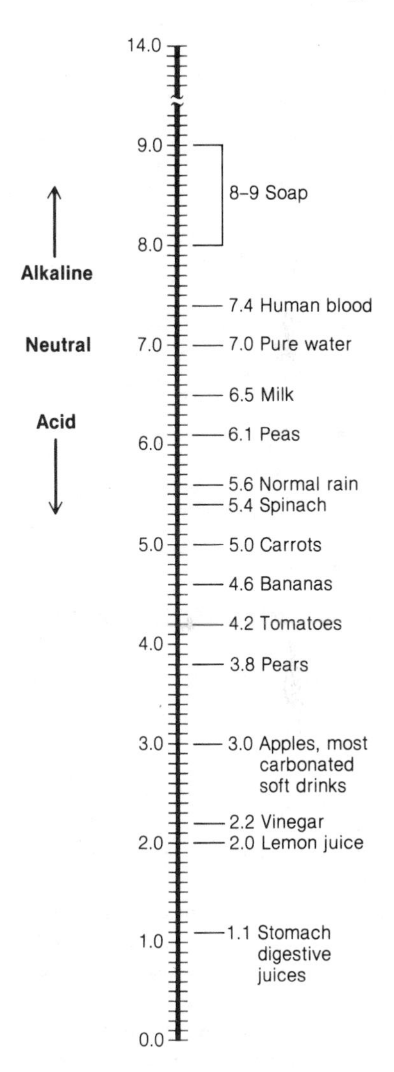

FIGURE 16.10

The scale of acidity and alkalinity, the pH scale, and the pH of some common substances.

FIGURE 16.11

The surface waters in many areas of North America are particularly susceptible to acidification. These areas lack natural buffers (neutralizers) in the soil or bedrock. (From Galloway, J. N., and E. B. Cowling. "The Effects of Precipitation on Aquatic and Terrestrial Ecosystems: A Proposed Precipitation Chemistry Network." ***Journal of the Air Pollution Control Association*** **28 (1978):233.)**

lutants give rise to the acid rains that threaten Canada's primary industries, lumber and fisheries. Perhaps half of the acidic rainfall in Canada originates with industrial emissions from Ohio Valley industries and power plants. Canadian government officials are pressing their Washington counterparts to enact stricter controls on polluting industries and power plants in order to ease the problem, which some Canadians feel will become severe by the end of this decade.

Threats to the Ozone Shield

In general, the residents of urban-industrial areas run the greatest risk of adverse health effects due to polluted air, but some pollutants threaten the well-being of all living things everywhere because they

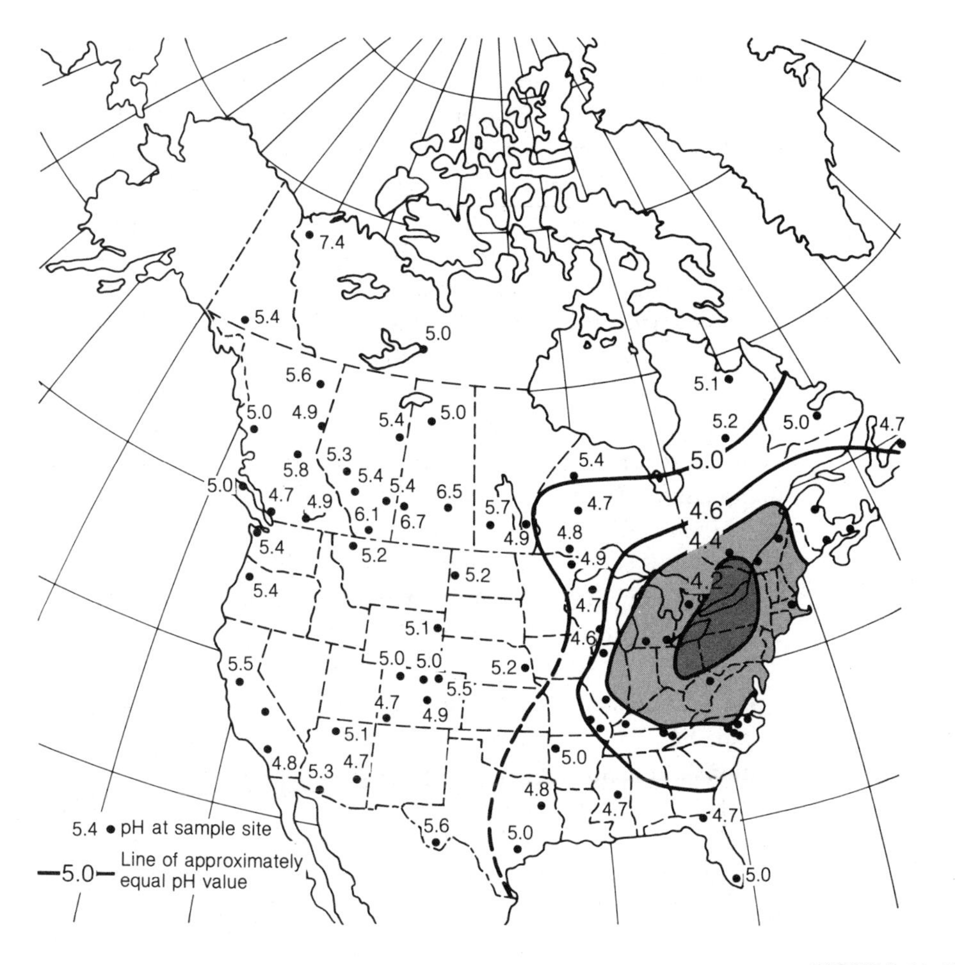

FIGURE 16.12

The acidity of precipitation over North America. Figures are average annual pH of rain and snow for 1982. (Data supplied by the National Atmospheric Deposition Program in the United States and the Canadian Network for Sampling of Precipitation.)

threaten the delicate ozone layer that shields organisms from dangerous ultraviolet radiation.

Ozone (O_3) is ironically a pollutant in the lower atmosphere, but its presence in minute concentrations (less than 12 parts per million) in the stratosphere is essential for the continuation of life on earth. Ozone forms primarily at altitudes between 15 and 35 km (9 and 22 mi) when oxygen absorbs solar ultraviolet radiation (UV). Ozone itself also absorbs UV. These absorption processes shield organisms from exposure to potentially lethal intensities of UV radiation, hence the term **ozone shield**.

In the longer wavelength region of the ultraviolet radiation band, called UV-B, atmospheric absorption is only partial and is very sensitive to changes in ozone concentration. UV-B radiation is responsible for sunburn, and causes or contributes to skin cancer.

Scientists have distinguished three forms of skin cancer: basal cell, the more serious squamous cell, and the most virulent, called malignant melanoma. Malignant melanoma afflicts about 15,000 U.S. residents each year and eventually kills about one third of its victims. A real possibility exists that disruption of the atmospheric absorption processes may deplete stratospheric ozone, and thereby increase the intensity of UV-B radiation reaching the ground, thus raising the incidence of skin cancer.

Many scientists consider the most serious threat to the ozone shield to be from chlorofluorocarbons (CFCs), commonly called Freons. Most of us have used several CFCs, knowingly or unknowingly, in our daily lives. One, a propellant in common household aerosol sprays such as deodorants and hairsprays, was banned in the United States in late 1978. The second is still widely used as a coolant in refrigerators and air conditioners. Both of these substances are relatively inert (nonreactive) in the troposphere, where they have been accumulating for many years. When CFCs enter the stratosphere, however, they break down chemically and release chlorine, which reacts with and destroys ozone.

Reports on the potential health effects of ozone depletion are disquieting. In 1982 the National Research Council (NRC) of the National Academy of Sciences projected that continued emissions of CFCs at the 1977 rate could eventually reduce stratospheric ozone concentrations by 5 to 9 percent. For every 1 percent drop in stratospheric ozone, the amount of ultraviolet radiation reaching the earth's surface increases by as much as 2 percent. The NRC estimated that every 2 percent increase in ultraviolet radiation would in turn raise the incidence of basal cell skin cancer by 2 to 5 percent, and the incidence of squamous cell skin cancer by 4 to 10 percent. The number of U.S. residents contracting these diseases is currently 400,000 to 500,000 per year. Because of the complexity of factors involved in the development of malignant melanoma, the NRC was unable to project the effect of ozone depletion on the incidence of this form of skin cancer.

In early 1984, after careful evaluation of the available evidence, the NRC retreated from its earlier estimates, and predicted that CFCs would reduce ozone levels by only 2 to 4 percent by the year 2100. Their report also stated that there was "no discernible" change in stratospheric ozone levels between 1970 and 1980, and that atmospheric ozone levels could actually increase as a consequence of photochemical reactions involving motor vehicle exhaust.

More research and monitoring are clearly needed to resolve questions surrounding threats to the ozone shield. Although the potential health effects are our prime concern, the effects of more intense ultraviolet radiation on vegetation and climate also need to be addressed.

Conclusions

Weather influences and is influenced by air pollutants. Circulation of the atmosphere affects the rate of dilution of pollutants, and some pollutants affect the quality and quantity of precipitation. In Chapter 18 we consider the possible impact of air pollution on climate, but next, let us turn to the nature of climate and climatic behavior.

SUMMARY STATEMENTS

- Most air pollutants are cycled naturally within the atmosphere, but concentrations may reach levels that threaten human health or disrupt physical and biological processes.
- Strong winds and unstable air enhance the rate of dilution of air pollutants, while weak winds and stable air suppress dilution.
- A temperature inversion consists of an extremely stable stratification of light, mild air over heavier, cooler air. An inversion may develop at the earth's surface or aloft through subsidence of air, extreme radiational cooling, or air mass advection. Temperature inversions greatly inhibit the dispersal of air pollutants.
- Many regions with high air pollution potential are also locales of great topographic relief, because hills and mountain ranges can block horizontal winds that disperse polluted air.
- Conditions that favor the accumulation and concentration of pollutants in the atmosphere are countered to some extent by natural cleansing processes including gravitational settling and washout by rain and snow.
- Urban air pollutants affect precipitation, cloudiness, and fog development within and downwind from large metropolitan areas.
- In localities where the air is polluted by oxides of sulfur or nitrogen, precipitation becomes strongly acidic. Acidic precipitation threatens aquatic life and corrodes structures.
- The formation of ozone in the stratosphere protects life on earth by filtering out harmful intensities of solar ultraviolet radiation. This ozone shield is threatened by chlorofluorocarbons accumulating in the atmosphere.

KEY WORDS

air pollution episode
mixing depth
temperature inversion
subsidence temperature inversion
mixing layer
fumigation
radiational temperature inversion
radiational cooling
impaction
gravitational settling
dry deposition
scavenging
METROMEX
acid rain
pH scale
ozone shield

REVIEW QUESTIONS

1 Under what conditions are components of the atmosphere considered to be pollutants?

2 As a general rule, how does wind speed affect the concentration of air pollutants?

3 Why is the dilution of air pollutants particularly impeded in urban areas?

4 Define mixing depth. How does the mixing depth vary with (a) time of day and (b) season?

5 How does the stability of air influence the rate of dilution of air pollutants?

6 Describe the behavior of smoke plumes in (a) stable air and (b) unstable air.

7 Why do subsidence temperature inversions generally cover a larger area than radiation temperature inversions?

8 Why are radiation temperature inversions often short lived?

9 In areas of great topographic relief, cold-air drainage strengthens radiation temperature inversions. Explain this statement.

10 What factors contribute to the relatively high air pollution potential of Los Angeles?

11 Compare the natural atmospheric cleansing processes for the troposphere and the stratosphere.

12 What is the most effective natural air pollution removal mechanism?

13 How do air pollutants influence urban weather?

14 Describe the potential climatic influence of jet plane contrails.

15 Why is rainfall normally slightly acidic?

16 What air pollutants are acid rain precursors?

17 What is the ozone shield?

18 What is thought to be the most serious threat to the ozone shield?

19 Speculate on the major sources and types of air pollutants in nonindustrialized nations.

20 How do gaseous air pollutants influence human health? Provide specific examples.

POINTS TO PONDER

1 Midlatitude cyclones are beneficial to air quality. Explain this statement.

2 How do taller smokestacks contribute to the acid rain problem?

3 Comment on the notion that at least some air pollution is an inevitable consequence of our way of life.

4 Why is a temperature inversion a case of "extreme" atmospheric stability?

5 What types of weather patterns favor (a) low air quality and (b) high air quality?

6 Describe how a temperature inversion develops by (a) subsidence of air, (b) extreme radiational cooling, and (c) air mass advection.

7 Many of the air pollutants in urban air are hygroscopic. What does this imply about the development of clouds downwind from a city?

8 It is possible for temperature inversions to form simultaneously at the surface and aloft in the atmosphere. Speculate on the sequence of weather events that would give rise to this situation.

PROJECTS

1 Evaluate the air pollution potential of your community in terms of (a) frequency of stable atmospheric conditions, (b) topographic influences, and (c) locations of major pollutant sources.

2 The types of pollutants that foul the air of a particular locality depend on the specific kinds of industrial, domestic, and agricultural activities that take place there. Prepare a list of pollution sources and types of air pollutants for your community. Is information available on the amounts of air pollutants emitted by each source?

3 Maintain a daily log describing the appearance of local industrial smoke plumes. Determine whether the shape and behavior of the plumes indicate unstable or stable conditions. If a temperature inversion is evident, does it usually form at the surface or aloft? Taking daily photographs of plumes may aid your analysis.

4 Has your community ever experienced a severe air pollution episode? If so, did an increased incidence of respiratory illnesses accompany the incident? You may wish to refer to past issues of local newspapers or consult with your local public health agency.

5 Collect a sample of snow, melt it, and filter the meltwater. Examine the residue on the filter paper under a microscope. Describe what you see and speculate on its origins. Compare the appearance of samples taken from different locations in your community.

SELECTED READINGS

Changnon, S. A., Jr. "More on the LaPorte Anomaly: A Review." *Bulletin of the American Meteorological Society* 61 (1980):702–711. *Concludes that precipitation anomalies at LaPorte, Indiana, from the late 1930s to the 1960s were linked to urban-industrial air pollution.*

The Conservation Foundation. *State of the Environment 1982.* Washington, D.C.: The Conservation Foundation, 1982. 439 pp. *Includes a report on recent trends in air quality.*

Heidorn, K. C. "A Chronology of Important Events in the History of Air Pollution Meteorology to 1970." *Bulletin of the American Meteorological Society* 59 (1978):1589–1597. *Includes a useful bibliography.*

Likens, G. E., et al. "Acid Rain." *Scientific American* 241, no. 4 (1979):43–51. *Discusses the trends and causes of acid precipitation in North America and western Europe.*

National Research Council. *Causes and Effects of Stratospheric Ozone Reduction: An Update.* Washington, D.C.: National Academy Press, 1982. 339 pp. *A summary report on what is currently understood about the causes and implications of stratospheric ozone depletion.*

Office of Technological Assessment (U.S. Congress). "Balancing the Risks: What Do We Know." *Weatherwise* 37 (1984):240–249. *Excerpt of a major study of public policy questions involved in the regulation of acid rain precursors.*

Postel, S. "Air Pollution, Acid Rain, and the Future of Forests." *Worldwatch Paper*, no. 58 (1984): 1–54. *Reviews possible link between acid deposition and sick and dying trees in Europe and North America.*

Wisniewski, J., and J. D. Kinsman. "An Overview of Acid Rain Monitoring Activities in North America." *Bulletin of the American Meteorological Society* 63 (1982):598–618. *Includes a summary of precipitation chemistry monitoring in the United States, Canada, and Mexico.*

Antiphanes said merrily, that in a certain city the cold was so intense that words were congealed as soon as spoken, but that after some time they thawed and became audible; so that the words spoken in winter were articulated next summer.

PLUTARCH
Of Man's Progress in Virtue

The earth is a mosaic of different climate types. The variety of the earth's climates is best viewed in mountainous regions where in several thousand meters of altitude, the climate exhibits the same variations as in thousands of kilometers of latitude. (Visuals Unlimited photograph by P. Armstrong)

SEVENTEEN

World Climates

Now that we have examined the governing principles of weather and the major characteristics of atmospheric circulation, let us turn to climate. *Climate* is defined as the weather of some locality averaged over some time period. This definition of climate includes any extremes in weather behavior that may have occurred during the specified time period. The extremes must be considered along with other weather conditions of the same period to produce a total climate picture.

In this chapter, we show how climate is described and classified. In the next chapter, we review the climatic record and the possible causes of climatic variability.

Describing Climate

Climate is usually described in terms of normals, means, and extremes of a variety of weather elements. Weather extremes for the world and for North America are listed in Appendix IV. Climatic summaries are available in tabular form for each of the states and provinces as well as for major cities. A sample climatic summary is shown in Table 17.1 and climatic data for selected North American cities are presented in Appendix V. A narrative description of local or regional climate usually accompanies these data. In addition, a vast variety of climatic maps have been constructed for individual continents, countries, and regions. The NOAA publication *Climates of the United States*, for example, contains 62 climatic maps summarizing a variety of weather elements.

In the United States, the National Weather Service is responsible for gathering the basic weather data used in climatological summaries. Data are processed, entered into archives, and made available to the public at the National Climate Data Center (NCDC) in Asheville, North Carolina. The NCDC also disseminates climatic information and prepares analyses for specialized uses. In Canada, comparable services are provided by the Atmospheric Environment Service, headquartered at Downsview, Ontario.

The climatic norm

A potentially misleading term used in climatic summaries is "normal." It does not indicate just average values, because the **climatic norm** includes the total variation in the climatic record. For example, the occurrence of an exceptionally cold winter may not be "abnormal" since it may fall well within the expected climatic range of variability of winter temperature. Even the cold summer of 1816, described in the Special Topic as "1816, the Year Without a Summer?," may have been a normal, albeit extreme, event.

By international convention, climatic norms are computed from averages and extremes of weather elements compiled over a 30-year period. Current climatic summaries are based on weather records from 1951 to 1980. The average July rainfall, for example, is the simple average of the total rainfall during each of 30 consecutive Julys from 1951 through 1980. Selection of a 30-year period was arbitrary, however, and may be inappropriate for many applications. Climate can change significantly in periods shorter than 30 years. For other purposes, a 30-year period provides a myopic view of the climatic record. Compared with the long-term climatic record, the current 1951–1980 "norm," for example, was an unusually mild period.

Climatic anomalies

Climatologists often compare the weather of a specific week, month, or year with the past climatic record. Such comparisons carried out over wide geographical areas show that departures from long-term climatic averages, called **anomalies**, never occur with the same sign or magnitude everywhere.

As an illustration, consider Figure 17.1, which shows the temperature anomaly pattern for the winter of 1980–1981 (December through February) across the United States. Note that average temperatures were above the long-term averages (positive anomaly) in the western two thirds of the nation and below the long-term averages (negative anomaly) in the eastern third of the nation. Furthermore, the magnitude of the anomaly, positive or negative, varied from one place to another.

The geographic nonuniformity of climatic anomalies is linked to the prevailing westerly pattern. Average temperatures, for example, mirror the air mass advection that is controlled by the dominating westerly flow pattern. During the winter of 1980–1981, the prevailing westerly pattern favored more than the usual cold air advection into the East and anomalous warm air advection into the West. The airflow aloft (the westerlies) determines the location of weather extremes such as drought or very cold temperatures. As a consequence, a single weather extreme never occurs over an area as large

TABLE 17.1

Summary of Climatic Elements Recorded at the National Weather Service Station at General Mitchell Field, Milwaukee, Wisconsin, in 1976

	TEMPERATURE °F							NORMAL DEGREE DAYS BASE 65°F		PRECIPITATION IN INCHES										
	NORMAL			EXTREMES						WATER EQUIVALENT							SNOW, ICE PELLETS			
MONTH	DAILY MAXIMUM	DAILY MINIMUM	MONTHLY	RECORD HIGHEST	YEAR	RECORD LOWEST	YEAR	HEATING	COOLING	NORMAL	MAXIMUM MONTHLY	YEAR	MINIMUM MONTHLY	YEAR	MAXIMUM IN 24 HRS	YEAR	MAXIMUM MONTHLY	YEAR	MAXIMUM IN 24 HRS	YEAR
(a)				36		36					36		36		36		36		36	
Jan.	27.3	11.4	19.4	62	1944	−24	1963	1414	0	1.63	4.04	1960	0.31	1961	1.71	1960	28.4	1943	12.8	1962
Feb.	30.3	14.6	22.5	65	1976	−19	1951	1190	0	1.13	3.10	1974	0.05	1969	1.67	1960	42.0	1974	16.7	1960
Mar.	39.4	23.4	31.4	81	1945	−10	1962	1042	0	2.24	6.93	1976	0.31	1968	2.57	1960	26.7	1965	11.2	1961
Apr.	54.6	34.7	44.7	85	1962	13	1971	609	0	2.76	7.31	1973	0.81	1942	3.11	1976	15.8	1973	11.6	1973
May	65.0	43.3	54.2	92	1975	21	1966	348	13	2.88	5.27	1945	0.90	1971	2.06	1948	0.4	1960	0.4	1960
June	75.3	53.6	64.5	99	1953	33	1945	90	75	3.58	8.28	1954	0.85	1965	3.13	1950	0.0		0.0	
July	80.4	59.3	69.9	101	1955	40	1965	15	167	3.41	7.66	1964	0.95	1946	4.35	1959	0.0		0.0	
Aug.	79.7	58.7	69.2	100	1955	44	1965	36	166	2.68	7.07	1960	0.46	1948	4.05	1953	0.0		0.0	
Sept.	71.5	50.7	61.1	98	1953	28	1974	140	23	3.02	9.87	1941	0.30	1956	5.28	1941	T	1960	T	1960
Oct.	61.4	40.6	51.0	89	1963	21	1976	440	6	1.98	6.42	1959	0.15	1956	2.60	1959	4.0	1976	4.0	1976
Nov.	44.4	28.5	36.5	77	1950	−5	1950	855	0	2.01	3.37	1958	0.62	1949	2.18	1943	9.6	1951	6.3	1954
Dec.	31.5	16.8	24.2	63	1970	−15	1963	1265	0	1.75	4.34	1971	0.29	1976	1.93	1942	26.5	1951	11.1	1959
Yr.	55.1	36.3	45.7	101	Jul. 1955	−24	Jan. 1963	7444	450	29.07	9.87	Sep. 1941	0.05	Feb. 1969	5.28	Sep. 1941	42.0	Feb. 1974	16.7	Feb. 1960

SOURCE: NOAA

NOTE: Means and extremes above are from existing and comparable exposures. Annual extremes have been exceeded at other sites in the locality as follows: Highest temperature 105 in July 1934; lowest temperature −25 in January 1875; maximum monthly precipitation 10.03 in June 1917; maximum precipitation in 24 hours 5.76 June 1917; maximum monthly snowfall 52.6 in January 1918; maximum snowfall in 24 hours 20.3 in February 1924.

(a) Length of record, years, through the current year unless otherwise noted, based on January data.

(b) 70° and above at Alaskan stations.

* Less than one half.

T Trace.

NORMALS: Based on record for the 1941-1970 period.

DATE OF AN EXTREME: The most recent in cases of multiple occurrence.

PREVAILING WIND DIRECTION: Record through 1963.

WIND DIRECTION: Numerals indicate tens of degrees clockwise from true north. 00 indicates calm.

FASTEST MILE WIND: Speed is fastest observed 1-minute value when the direction is in tens of degrees.

RELATIVE HUMIDITY PCT. HOUR 00 (Local time)	HOUR 06	HOUR 12	HOUR 18	WIND: MEAN SPEED (MPH)	WIND: PREVAILING DIRECTION	WIND: FASTEST MILE: SPEED (MPH)	WIND: FASTEST MILE: DIRECTION	WIND: FASTEST MILE: YEAR	% OF POSSIBLE SUNSHINE	MEAN SKY COVER, TENTHS, SUNRISE TO SUNSET	MEAN NUMBER OF DAYS: SUNRISE TO SUNSET: CLEAR	PARTLY CLOUDY	CLOUDY	PRECIPITATION .01 INCH OR MORE	SNOW, ICE PELLETS 1.0 INCH OR MORE	THUNDERSTORMS	HEAVY FOG, VISIBILITY 1/4 MILE OR LESS	TEMPERATURES °F: MAX. 90° AND ABOVE (b)	MAX. 32° AND BELOW	MIN. 32° AND BELOW	MIN. 0° AND BELOW	AVERAGE STATION PRESSURE (mb) ELEV. 693 FEET MEAN SEA LEVEL
16	**16**	**16**	**16**	**36**	**14**	**36**	**36**		**36**	**36**	**36**	**36**	**36**	**36**	**36**	**36**	**36**	**16**	**16**	**16**	**16**	**4**
74	**75**	**68**	**71**	**12.9**	**WNW**	**62**	**W**	**1950**	**44**	**6.8**	**7**	**6**	**18**	**11**	**3**	*	**2**	**0**	**20**	**30**	**9**	**991.6**
74	**75**	**67**	**70**	**12.8**	**WNW**	**58**	**NE**	**1960**	**47**	**6.7**	**7**	**5**	**16**	**9**	**2**	*	**2**	**0**	**16**	**27**	**4**	**991.1**
77	**79**	**66**	**70**	**13.3**	**WNW**	**73**	**SW**	**1954**	**51**	**6.9**	**6**	**8**	**17**	**12**	**3**	**2**	**3**	**0**	**7**	**24**	*	**989.0**
76	**79**	**62**	**65**	**13.2**	**NNE**	**66**	**SW**	**1947**	**54**	**6.5**	**7**	**8**	**15**	**12**	*	**4**	**3**	**0**	*	**12**	**0**	**990.5**
76	**79**	**61**	**62**	**12.0**	**NNE**	**72**	**SW**	**1950**	**59**	**6.3**	**7**	**10**	**14**	**12**	**0**	**5**	**3**	*	**0**	**2**	**0**	**988.0**
79	**82**	**62**	**63**	**10.5**	**NNE**	**62**	**W**	**1971**	**64**	**5.9**	**8**	**10**	**12**	**11**	**0**	**7**	**2**	**2**	**0**	**0**	**0**	**988.2**
80	**82**	**60**	**63**	**9.7**	**SW**	**59**	**W**	**1952**	**71**	**5.2**	**11**	**11**	**9**	**9**	**0**	**6**	**1**	**4**	**0**	**0**	**0**	**990.3**
84	**87**	**61**	**67**	**9.6**	**SW**	**50**	**W**	**1974**	**67**	**5.1**	**11**	**11**	**9**	**9**	**0**	**5**	**2**	**3**	**0**	**0**	**0**	**992.5**
83	**87**	**63**	**71**	**10.7**	**SSW**	**62**	**S**	**1941**	**60**	**5.5**	**10**	**9**	**11**	**9**	**0**	**4**	**1**	**1**	**0**	*	**0**	**992.9**
77	**82**	**63**	**69**	**11.5**	**SSW**	**60**	**S**	**1949**	**56**	**5.6**	**10**	**9**	**12**	**8**	*	**2**	**2**	**0**	**0**	**5**	**0**	**993.4**
78	**80**	**67**	**73**	**12.8**	**WNW**	**72**	**W**	**1955**	**41**	**7.1**	**6**	**6**	**18**	**10**	**1**	**1**	**2**	**0**	**3**	**19**	*	**990.9**
78	**80**	**73**	**75**	**12.5**	**WNW**	**62**	**SW**	**1948**	**38**	**7.2**	**6**	**6**	**19**	**11**	**3**	*	**2**	**0**	**16**	**28**	**3**	**991.7**
78	**81**	**64**	**68**	**11.8**	**WNW**	**73**	**SW**	**Mar. 1954**	**56**	**6.2**	**96**	**99**	**170**	**122**	**13**	**36**	**26**	**9**	**62**	**146**	**16**	**990.8**

SPECIAL TOPIC

1816, the year without a summer?

The summer of 1816 is famous for its anomalous cold. In fact, 1816 is sometimes described as "the year without a summer." Unseasonable snowfall and freezes damaged many crops in the then agrarian northeastern United States and adjoining portions of Canada. More than 90 percent of the corn crop, the prime food staple, was lost. Some superficial accounts of this summer give the impression that unusual cold persisted throughout the summer and afflicted the entire civilized world. Such was not the case.

In the Northeast, the summer of 1816 was punctuated by several outbreaks of unusually cold weather. Killing frosts occurred in northern and interior southern New England as well as in Quebec in early June and July and again in late August. The June cold snap was accompanied by moderate to heavy snowfalls in the highlands. No sooner had farmers replanted their frozen crops than a killing freeze would strike again. The miserably short and disastrous growing season came to an end with a general hard freeze on 27 September.

These brief but unusually cold episodes interrupted longer spells of seasonably warm weather. Mean June and July temperatures were 1.6 °C (3 °F) to 3.3 °C (6 °F) below average. Individually, these monthly anomalies fall within the expected range of climatic variability. In fact, some previous and subsequent Junes, Julys, and Augusts have been at least as cold as the summer months of 1816, if not colder. What is most notable about the summer of 1816 is the persistence of a strong negative temperature anomaly through all three summer months. Nevertheless, the summer in the Northeast as a whole fell within the range of expected climatic extremes.

What about the weather elsewhere during the summer of 1816? We know that it was probably not the same everywhere as in the Northeast, for climatic anomalies are usually geographically nonuniform in both magnitude and sign. Unfortunately, little weather information is available from the central and western United States and Canada because few settlements existed there at the time. France, Germany, and Great Britain did experience an unusually cold summer, but this is not surprising. The meridional Rossby long-wave pattern that would advect unseasonably cold air into the Northeast would also favor western Europe with the same anomalous weather. The two localities are typically one wavelength apart. In contrast, data suggest that east central and eastern Europe experienced a warmer than usual summer.

The purpose of this discussion is not to diminish the climatic significance of the summer of 1816 or the hardships that people suffered then. It is, rather, to show that weather extremes typically fall within the expected range of climatic variability. We have more to say about the summer of 1816 in Chapter 18, where the controversy surrounding the impact of volcanic eruptions on climate is discussed.

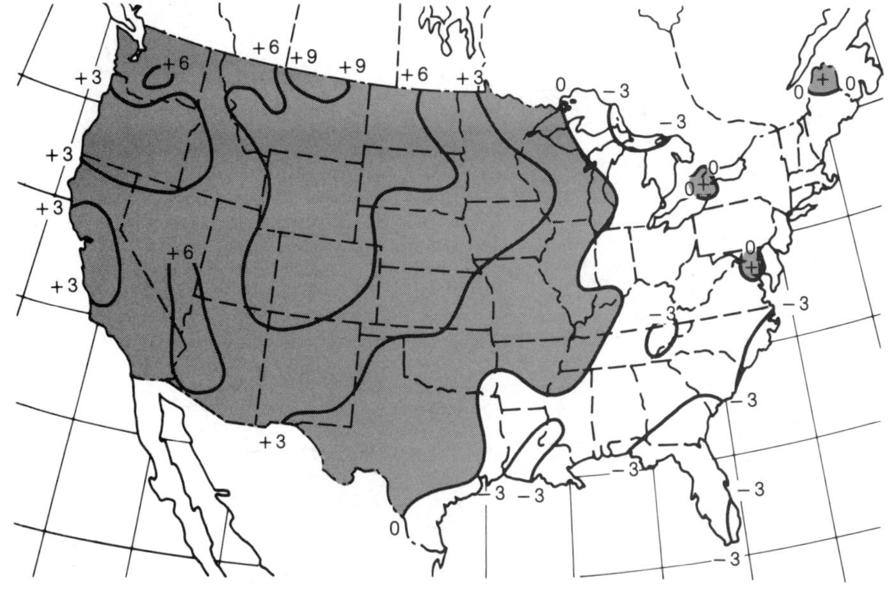

Mean winter temperature above the long-term average

Mean winter temperature below the long-term average

FIGURE 17.1

Temperature anomaly pattern across the United States for the winter of 1980–1981 (December through February). Numbers are departures from the long-term average in °F. (NOAA data)

as the United States or Canada, that is, severe cold or drought never grips the entire nation at the same time.

Rainfall typically forms considerably more complex anomaly mosaics than does temperature (Figure 17.2). This is due to the greater spatial variability of rainfall arising from the variability of storm tracks and the almost random distribution of convective showers. In a spring month in the midlatitudes, for instance, adjoining counties will quite possibly experience opposite rainfall anomalies (one having above average rainfall and the other below average rainfall) because of the sporadic nature of air mass thunderstorms.

From an agricultural perspective, the geographic nonuniformity of climatic anomalies may be advantageous in that some compensation is implied. Poor growing conditions and consequent low yields in one area may be compensated for to some extent by better growing conditions and increased yields elsewhere. This is known as **agroclimatic compensation** and is discussed further in the Special Topic, Agroclimatic Compensation: The Benefits and Limitations.

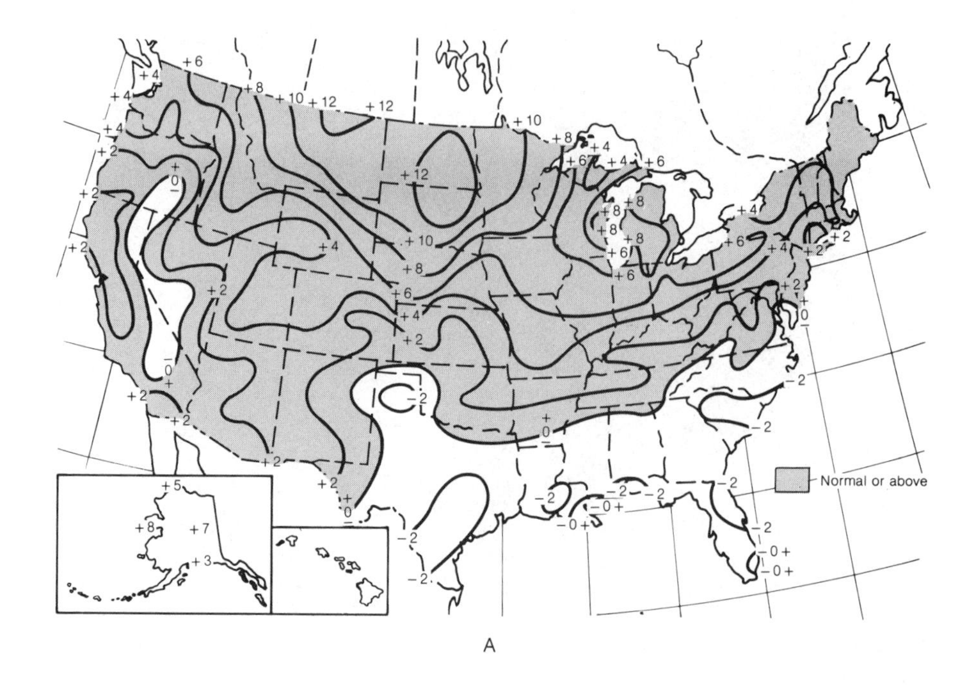

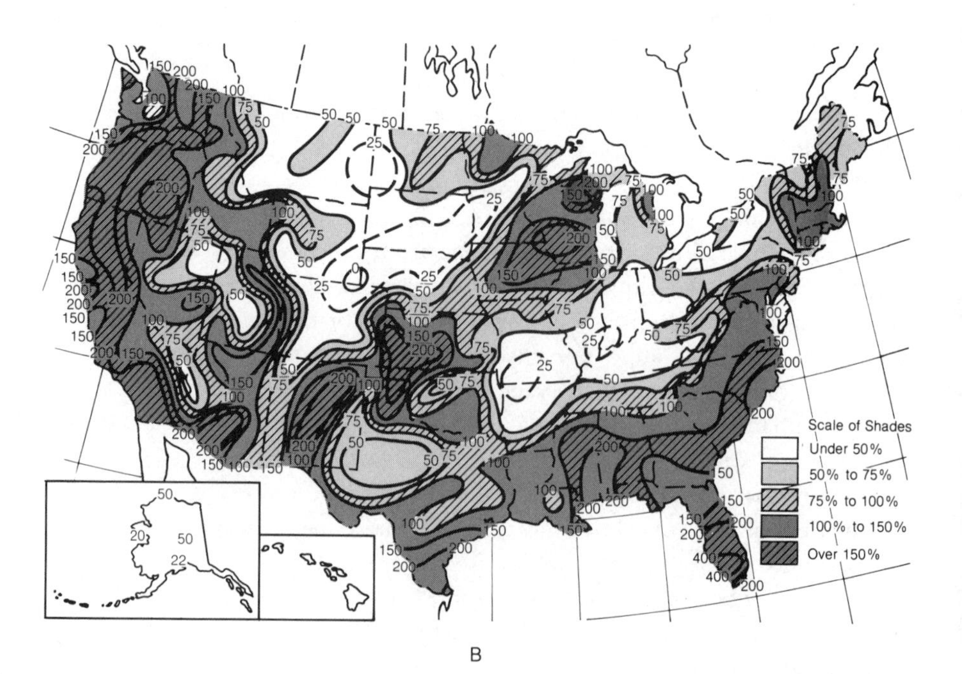

FIGURE 17.2

A precipitation anomaly pattern is usually much more complex than a temperature anomaly pattern. In (A) the temperature anomalies across the United States for February 1983 are departures from the 30-year averages and are expressed in °F. In (B) the precipitation anomalies are expressed as percentages of long-term averages. (NOAA data)

Geographic nonuniformity also characterizes large-scale trends in climate. A trend in the average temperature of the Northern Hemisphere is therefore not necessarily representative of all localities within the Northern Hemisphere. For the same period, some locations will experience cooling trends while others experience warming trends, regardless of the general direction of the hemispheric temperature trend. Not only is it misleading to assume that the direction of large-scale climatic trends is true for all localities, it is also erroneous to assume that the magnitude of climatic trends is the same everywhere. In fact, a small change in the average hemispheric temperature typically translates into a much larger change in certain areas and to little or no change in other areas.

Air Mass Climatology

Although the climate of a locality is traditionally described by normals, means, and extremes of various weather elements, an alternative approach, known as **air mass climatology**, has some interesting implications. In this approach, the frequency with which various types of air masses develop over a locality or are advected into a locality is used to describe the climate of that locality. For example, during January a northern U.S. city on average might have cold, dry air 60 percent of the time; mild, humid air 30 percent of the time; and mild, dry air 10 percent of the time.

Earlier we identified the major source regions for North American air masses. Air masses actually originate in the clockwise divergent flow that characterizes surface winds in anticyclones. As air streams away from an anticyclone, the temperature and humidity of the air change to some extent, depending on the nature of the surface over which it travels. In this way, a single anticyclone can be the source of many different types of air mass depending on the modification that takes place. Recall, for example, that winds spiraling outward from the subtropical anticyclones are dry on the eastern flank and humid on the western flank. Air masses to the east of subtropical highs are therefore dry, while those to the west are humid.

Wayne M. Wendland of the Illinois Institute of Natural Resources and Reid A. Bryson of the University of Wisconsin recently identified the anticyclonic sources for Northern Hemisphere air masses. They did this by determining the average surface streamline pattern across the hemisphere for each month of the year. The average

streamlines represent the average paths of air moving horizontally. Figure 17.3 is an example of the Wendland-Bryson analysis. Anticyclonic flow is indicated by streamlines that turn into a clockwise and outward spiral. Wendland and Bryson identified 19 anticyclonic source regions. Some were over the oceans, some were over the continents, and a few were in the Southern Hemisphere. Five

---

SPECIAL TOPIC

Agroclimatic compensation: the benefits and limitations

Because of the geographic nonuniformity of weather, crop yields in any one season are likely to vary from one place to another. As a consequence, lower crop yields in one area may be compensated for by higher crop yields in another area. For example, in the spring of 1982, corn planting was delayed for nearly a month in Iowa and Kansas by wet fields. As a consequence, per hectare yields in these states declined by 5 and 9 percent, respectively, from the previous year's levels. Meanwhile, eastern corn belt states such as Ohio and Indiana experienced quite favorable weather throughout the growing season and their per hectare yields jumped 22 and 18 percent, respectively, from 1981 levels. Interestingly, these yields resulted in a record high average yield per hectare for the entire corn belt.

The impact of a climatic anomaly on crop production will depend on its time of occurrence. For example, by the time severe heat and drought struck the Midwest and Plains states in 1983, much of the wheat crop had already matured. While the yields per hectare of the later maturing corn and soybeans declined by over 28 percent and 19 percent, respectively, from the previous year's, the 1983 wheat harvest broke a record for yield per hectare.

The impact of climate can also vary by continent. Although the United States experienced very favorable growing conditions for corn production during May to September of 1982, the Republic of South Africa received less than half of its normal summer rainfall (December 1981 through March 1982). As a consequence, the United States enjoyed a record corn harvest, while the Republic of South Africa suffered

sources persisted through the entire year, and three persisted for about 11 months. The others were prominent from 1 to 9 months.

Where streamlines from different anticyclones meet, a zone of confluence forms. A **confluence zone** is a boundary between air masses and is equivalent to an average frontal position. Confluence zones are indicated by dashed lines in Figure 17.3. As a rule, these

---

from a nearly 30 percent reduction in corn yield. Yields declined over much of Africa and South America that year while yields increased in North America and Europe. For 1982, the gains more than compensated for the declines as the worldwide average yield per hectare increased by 3 percent.

Agroclimatic compensation has its limitations, however. More favorable weather conditions in a region where low soil fertility already limits crop yields would probably not compensate for a yield decline caused by unfavorable weather in a region with highly fertile soils. For example, many of the soils in Illinois and Iowa developed from loess—a dusty, wind-borne material that was deposited during the Ice Age. The deeper the deposit of loess, the more fertile the soil. Even with fertilizer applications, corn yields on the shallower loess soils are often only half of what is achieved on the deeper loess soils. Improved corn-growing weather in areas with thin loess soils would probably not compensate for the decline in yields resulting from less favorable weather in areas with deep loess soils.

Another limitation of agroclimatic compensation is that intensive cultivation of most crops remains generally confined to specific regions. In the United States, nearly half of the vegetables for fresh market and food processing are grown in California. In March and April of 1983, cold winds and high rainfall delayed the planting and harvesting of vegetables, thereby reducing significantly the nation's supply of these vital foodstuffs at that time. Although the return of favorable weather spurred vegetable production, the inclement weather contributed to a 4 percent annual production decline coupled with a nearly 6.5 percent increase in market value, a rate twice that of the inflation rate.

For a nation as large as the United States with a variety of climatic regimes that support a variety of crops, agroclimatic compensation provides some flexibility in feeding our citizens, but it may mean switching to more abundant foodstuffs and food may also be somewhat more expensive. For smaller nations, agroclimatic compensation may not be possible within their borders, and severe climatic anomalies will likely require the importation of foodstuffs, often at high prices.

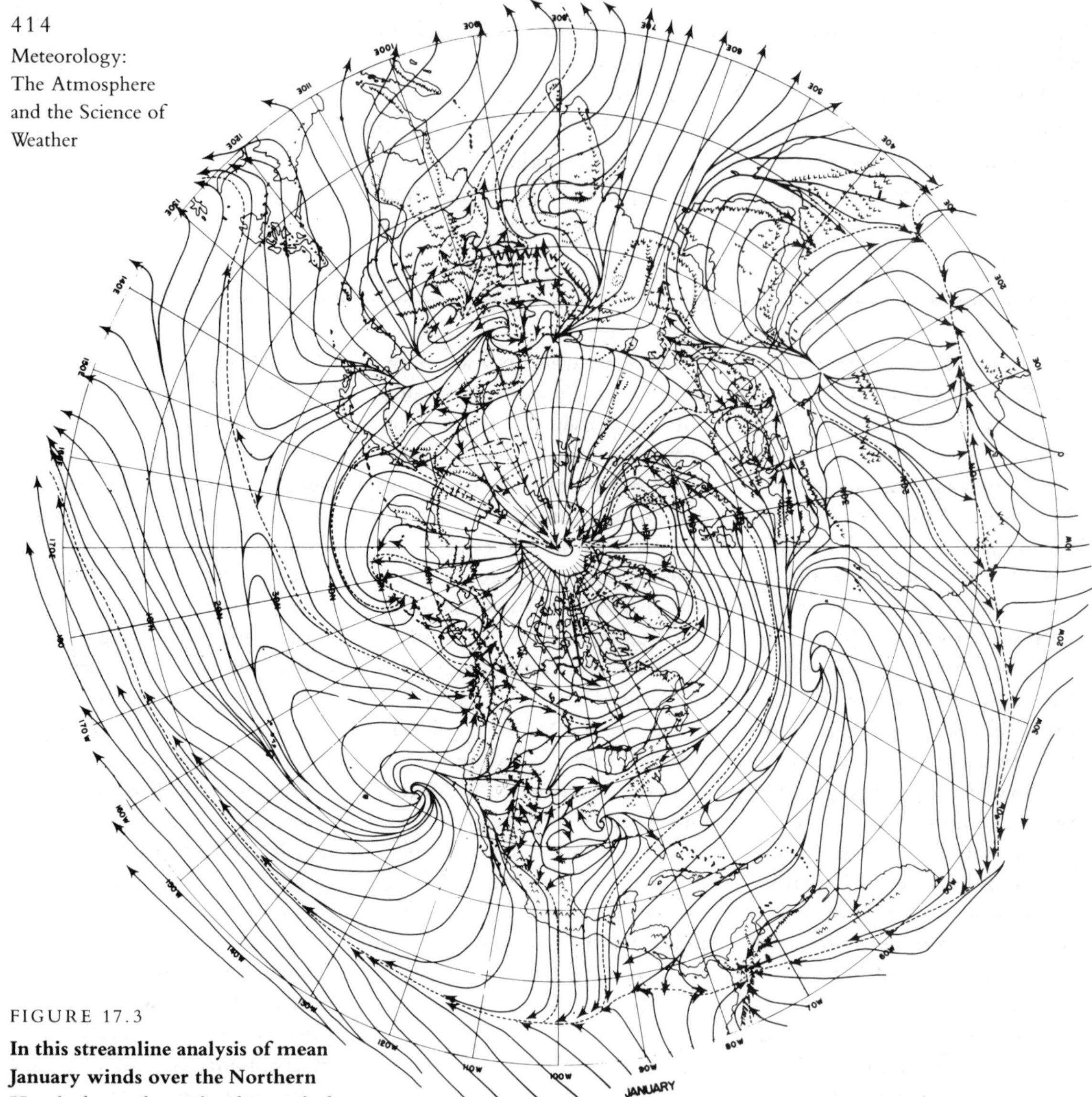

FIGURE 17.3

In this streamline analysis of mean January winds over the Northern Hemisphere, the anticyclone spirals are air mass source regions. Dashed lines mark confluence zones that are climatic air mass boundaries. (From Wendland, W. M., and R. A. Bryson. "Northern Hemisphere Airstream Regions." ***Monthly Weather Review*** **109 (1981):257. American Meteorological Society publication. Graphic by Raymond Steventon, Center for Climatic Research, University of Wisconsin—Madison.)**

zones separate distinctly different types of climate. Climate is quite uniform within air streams, but changes significantly across confluence zones.

Using air mass frequency to describe climate appears to be a valid approach in light of the apparent air mass effect on certain vegetational communities. For example, Bryson has demonstrated a close correspondence between the region dominated by cold, dry arctic air and the location of the coniferous boreal forest of Canada. The southern boundary of the boreal forest nearly coincides with the average position of the leading edge of arctic air (the arctic front) during winter, and the northern border of the forest corresponds closely to the average position of the leading edge of arctic air during summer (Figure 17.4).

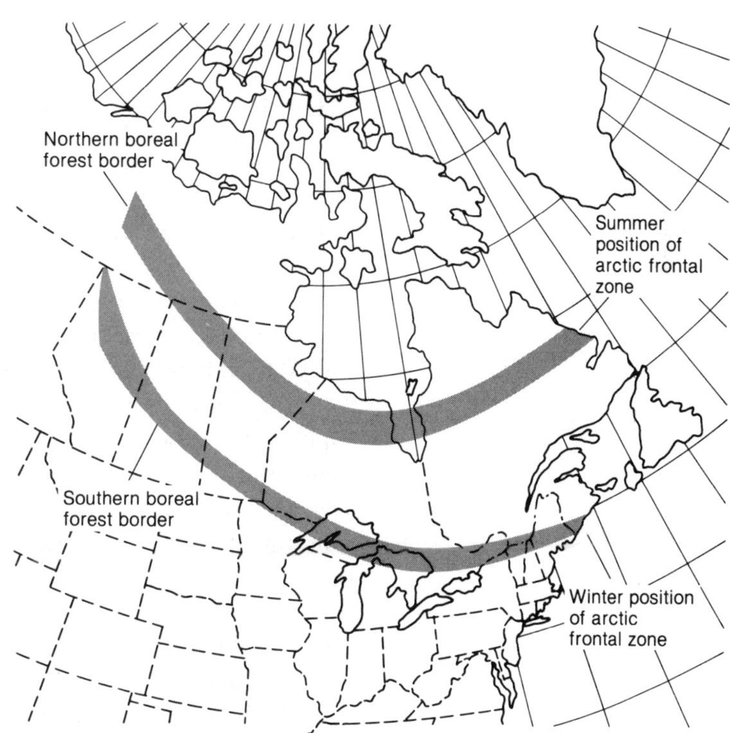

FIGURE 17.4
The northern and southern borders of the boreal forest of Canada correspond to the average position of the leading edge of arctic air in summer and in winter. (From Bryson, R. A. "Air Masses, Streamlines and the Boreal Forest." *Geographical Bulletin* 8 (1966):266. Reproduced by permission of the Minister of Supply and Services Canada.)

Climate Controls

Many controls work together to shape climate. These include solar radiation, which changes in a periodic and predictable manner (Chapter 3), and permanent features of the earth's surface, such as topography and the distribution of land and water. Other climatic controls are considerably less regular and include the combined influence of the various weather systems. This variability is especially characteristic of synoptic- and subsynoptic-scale weather systems. The global-scale circulation systems, such as the prevailing wind belts and the subtropical anticyclones, exert a somewhat more systematic impact on climate.

In previous chapters, we have seen many illustrations of how these climatic controls work. In Chapter 2, we examined the influence of the ocean water's great thermal stability on downwind air temperatures. In Chapter 4, we learned how air temperature responds to the regular variations of solar energy input. In Chapter 6, we saw how mountain ranges can influence precipitation patterns. How these and other controls interact to shape the climates of the continents will become clearer as we classify and describe the earth's major climate groups.

Climate Classification

In response to the interaction of many controls, the world's climates form a complex mosaic. Climatologists have attempted to simplify and organize the myriad of climate types by devising classification schemes that group together climates having common characteristics. Classification schemes typically group climates according to (1) the meteorological basis of climate or (2) the environmental effects of climate. The first is a genetic climate classification, and the second is an empirical climate classification. One of the most popular classification systems, designed by Wladimir Köppen, combines the two approaches.

Recognizing that vegetation indigenous to a region is a natural indicator of regional climate, Köppen delineated climatic boundaries throughout the world based on the limits of vegetational communities and the monthly means of temperature and precipitation. Köppen's climate classification scheme uses letters to symbolize five main groupings of world climates: (A) tropical rainy; (B) dry; (C) midlatitude rainy, mild winter; (D) midlatitude rainy, cold winter; and (E) polar. Each of these climates is further subdivided into principal climate types, designated by the addition of another letter symbol to form a pair. Even further qualification of the climate may be specified by adding a third letter symbol. With use of a combination of only two or three letters, the general climate characteristics of a region can thus be described.

Since its introduction in 1901, Köppen's climate classification has undergone numerous and substantial revisions by Köppen himself and by other climatologists. One of the more recent and useful of the Köppen-based climate classifications is presented by G. T. Trewartha and L. H. Horn in their fifth edition of *An Introduction to Climate* (1980). Trewartha and Horn identify six main climate groups (Table 17.2); five are temperature based and one is precipitation based. The worldwide distribution of these climate groups is shown in Figure 17.5.

TABLE 17.2

Climate Classification by G. T. Trewartha and L. H. Horn

CLIMATE GROUPS	CLIMATE TYPES	PRECIPITATION
A Tropical humid	**Ar,** tropical wet	Not over 2 dry months
	Aw, tropical wet-and-dry	High-sun wet (zenithal rains), low-sun dry
C Subtropical	**Cs,** subtropical dry summer	Summer drought, winter rain
	Cf, subtropical humid	Rain in all seasons
D Temperate	**Do,** oceanic	Rain in all seasons
	Dc, continental	Rain in all seasons, accent on summer; winter snow cover
E Boreal	**E,** boreal	Meager precipitation throughout year
F Polar	**Ft,** tundra	Meager precipitation throughout year
	Fi, ice cap	Meager precipitation throughout year
B Dry	**BS,** semiarid (steppe)	
	BSh (hot), tropical-subtropical	Short moist season
	BSk (cold), temperate-boreal	Meager rainfall, most in summer
	BW, arid (desert)	
	BWh (hot), tropical-subtropical	Constantly dry
	BWk (cold), temperate-boreal	Constantly dry
H Highland	**H,** variable	Variable

Source: Trewartha, G. T., and L. H. Horn. *An Introduction to Climate*. 5th ed. New York: McGraw-Hill, 1980, p. 227

Tropical humid climates (A)

Tropical humid climates constitute a discontinuous belt straddling the equator and extending poleward to near the Tropic of Cancer in the Northern Hemisphere and the Tropic of Capricorn in the Southern Hemisphere. Temperatures are high year-round, with the coolest monthly mean temperature no lower than 18 °C (64 °F). There is no frost. Mean monthly temperatures show little variability through the year. The temperature contrast between the warmest and coolest month is typically less than 10 °C (18 °F). In fact, the diurnal temperature range exceeds the annual temperature range. This monotonous air temperature regime is the consequence of consistently intense insolation and nearly uniform day length year-round.

While tropical humid climate types are not readily distinguishable on the basis of temperature, there are important differences in precipitation regimes. Tropical humid climates are therefore subdivided into tropical wet climates (Ar) and tropical wet-and-dry climates (Aw). Both climate types feature abundant annual rainfall—more than 100 cm (40 in.)—but important differences in seasonality occur.

In tropical wet climates, the yearly rainfall of 175 to 250 cm (70 to 100 in.) supports the world's most luxuriant vegetation (Figure

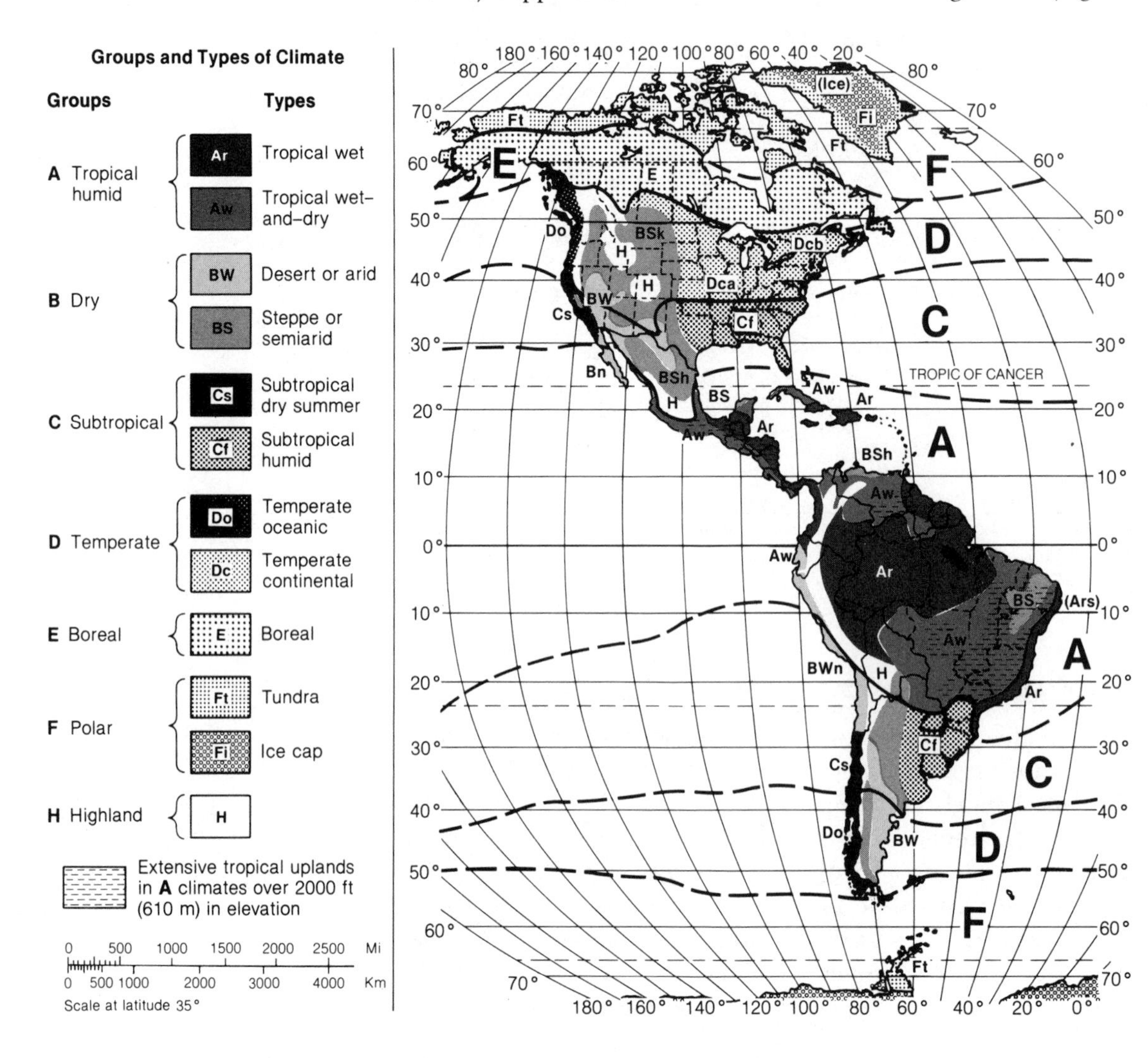

17.6). Tropical rain forests occupy the Amazon basin of Brazil, the Congo basin of Africa, and the islands of Micronesia. For the most part, rainfall is distributed uniformly throughout the year, although in some areas there is a brief (1 to 2 months) relatively dry season. Rainfall occurs as heavy downpours in the frequent thunderstorms triggered by local convection and by surges of the ITCZ. Convective rainfall, controlled by insolation, typically peaks in midafternoon, the warmest time of day. Because water vapor concentrations are very high, even the slightest cooling during the early morning hours results in dew and fog, which gives the region a sultry, steamy appearance.

FIGURE 17.5

Distribution of major climate groups across the globe. (Modified from Trewartha, G. T., and L. H. Horn. *An Introduction to Climate*. 5th ed. New York: McGraw-Hill, 1980. Used with permission.)

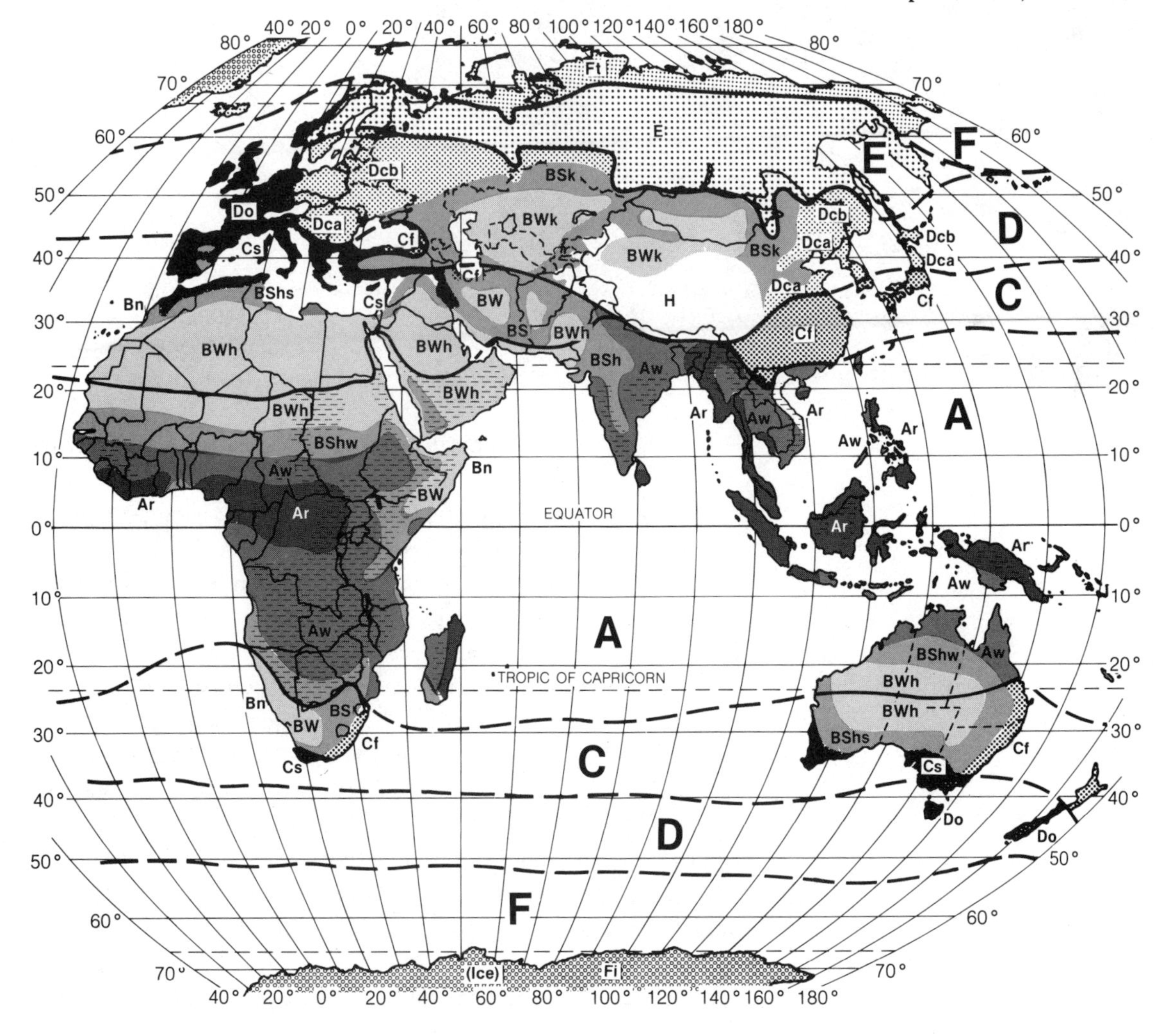

FIGURE 17.6

The luxuriant tropical rain forest of the Amazon Basin. (FAO photograph by Peyton Johnson)

For the most part, tropical wet-and-dry climates (Aw) border tropical wet climates and are transitional poleward to the subtropical dry climates. Aw climates support the savanna—tropical grasslands with scattered deciduous trees (Figure 17.7). Summers are wet and winters are dry, with the dry season lengthening poleward. This marked seasonality of rainfall is linked to shifts of the ITCZ and to subtropical anticyclones, which follow the seasonal excursions of the sun. In summer, surges of the ITCZ trigger convective rainfall, while in winter the weather is dominated by the dry eastern flank of the subtropical anticyclones. Much of the region of Aw climates is characterized by monsoon circulations (Chapter 12).

Annual mean temperatures in Aw climates are only slightly lower and the annual temperature range is only slightly greater than in the tropical wet climates (Ar). Diurnal temperature variations are noticeably larger, however. In summer, cloudy skies and high humidities suppress the diurnal temperature range, but in winter, fair

FIGURE 17.7
The grassland savanna of Tanzania. (Visuals Unlimited photograph by E. C. Williams)

skies enhance the diurnal temperature range. Cloudy, rainy summers also mean that the year's highest temperatures typically occur toward the close of the dry season in late spring.

Dry climates (B)

A dry climate characterizes those regions where annual potential evaporation exceeds annual precipitation. Because evaporation is proportional to air temperature, specifying some rainfall amount as the criterion for dry climates is not possible. Rainfall is not only limited in B climates, it is also highly variable and unreliable. In fact, as a general rule, the lower the mean annual rainfall is, the greater its variability.

The world's dry climates encompass a larger land area than any other single climate grouping. Perhaps 30 percent of the earth's land surface—stretching from the tropics into midlatitudes—is characterized by a moisture deficit of varying degree. These are the climates of the world's deserts and steppes where vegetation is sparse and equipped with special adaptations that permit survival under conditions of severe moisture stress (Figure 17.8). Dryness is the consequence of either subtropical anticyclones or the rain-shadow effect of high mountain barriers. Mean annual temperatures are latitude dependent, as is the variation of mean monthly temperatures through the year.

On the basis of degree of dryness, we distinguish two climate types: steppe or semiarid (BS) and arid or desert (BW). The steppe

FIGURE 17.8

The sparsely vegetated desert of the southwestern United States. (Photograph by John D. Cunningham)

or semiarid climates are transitional to more humid climates and usually border desert or arid climates. We further distinguish warm, dry climates of tropical latitudes (BSh and BWh) from cold, dry climates of midlatitudes (BSk and BWk).

Descending stable air on the eastern flanks of the subtropical anticyclones gives rise to the tropical dry climates (BSh and BWh). These huge semipermanent pressure cells dominate the weather year-round near the tropics of Cancer and Capricorn. The result is a swath of dry climates from northwest Africa eastward to northwest India and the deserts of the southwestern United States and northern Mexico, coastal Chile and Peru, southwest Africa, and much of interior Australia.

Although persistent and abundant sunshine is the general rule in dry tropical climates, some important exceptions occur. In coastal deserts bordered by cold ocean waters, a shallow layer of cool maritime air drifts inland. The desert air thus features high relative humidity, persistent low stratus clouds and fog, and considerable dew formation. Examples are the Atacama Desert of Peru and Chile, the Namib Desert of southwest Africa, and portions of the coastal Sonoran Desert of Baja California, and the coastal Sahara Desert of northwest Africa. These anomalous desert climates are designated Bn or BWn.

Cold, dry climates of midlatitude (BWk and BSk) lie in the rain shadow of great mountain barriers. They occur primarily in the Northern Hemisphere and are found to the lee of the coastal ranges, the Sierra Nevada and Cascade ranges in North America, and the

Himalayan chain in Asia. Because these dry climates are at higher latitudes than their tropical counterparts, mean annual temperatures are lower and the seasonal temperature contrast is greater. High pressure systems dominate winter, resulting in cold, dry conditions, while summers are hot and dry. The meager precipitation is principally the product of scattered convective showers in summer.

Subtropical climates (C)

Subtropical climates are situated just poleward of the tropics of Cancer and Capricorn and are dominated by seasonal shifts of the subtropical anticyclones. There are two basic climate types: subtropical dry summer (i.e., Mediterranean) climates (Cs), and subtropical humid climates (Cf).

Mediterranean climates occur on the west side of continents between about 30-degree and 45-degree latitudes. In North America, mountain ranges limit this climate type to a narrow coastal strip of California. Elsewhere, Cs climates rim the Mediterranean Sea and occur in extreme southwestern and southeastern Australia. Summers are dry because at that time of year Cs regions are under the stable subsiding air on the eastern flanks of the subtropical highs. The shift toward the equator of the highs in winter allows oceanic cyclones to migrate inland, bringing moderate rainfall. Annual precipitation totals are in the range of 40 to 80 cm (16 to 32 in.).

While Mediterranean climates exhibit a pronounced seasonality in precipitation (dry summers, wet winters), the temperature regime is quite variable. In coastal areas, cool onshore breezes prevail, lowering mean annual temperatures and reducing seasonal temperature contrasts. Well inland, however, away from the ocean's moderating influence, summers are considerably warmer, resulting in higher mean annual temperatures and greater seasonal temperature differences than in coastal Cs localities. An illustration of thermal contrasts within Cs regions is provided by the climatic records of coastal San Francisco and inland Sacramento, California (Figure 17.9).

Subtropical humid climates (Cf) occur on the eastern sides of continents between about 25-degree and 40-degree latitudes. Cf climates are situated primarily in the southeastern United States, southeastern South America, eastern China, and southern Japan and on the southeastern coasts of South Africa and Australia. These climates feature abundant precipitation (76 to 165 cm, or 30 to 65 in.), which is distributed throughout the year. In summer, Cf re-

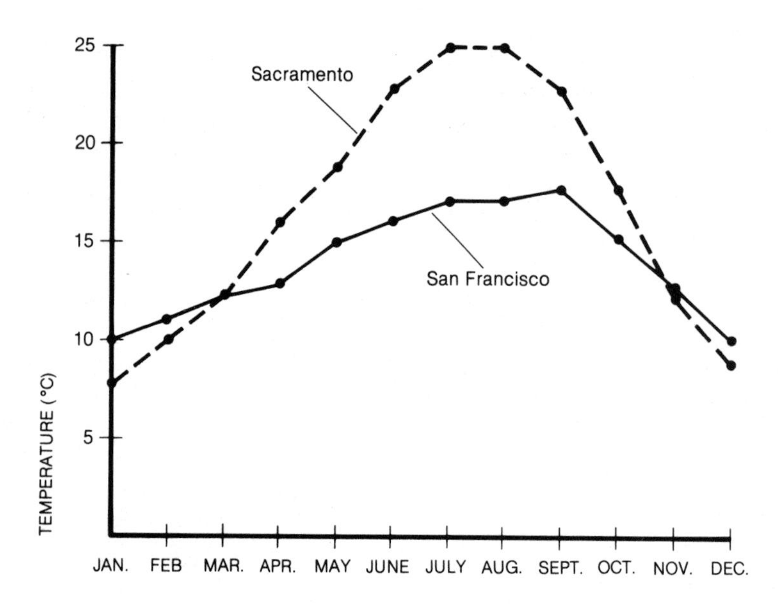

FIGURE 17.9

Note the contrast in the annual temperature cycles between Sacramento, California, and more maritime San Francisco. (NOAA data)

gions are dominated by the flow of sultry maritime tropical air on the western flanks of the subtropical anticyclones. Summers are consequently hot and humid with frequent thunderstorms, and as noted in Chapter 14, tropical storms contribute significant rainfall to some North American and Asian Cf regions. In winter, as subtropical highs shift toward the equator, Cf regions come under the influence of migrating midlatitude cyclones and anticyclones, and frontal weather.

A noteworthy exception to the nonseasonality of precipitation in Cf regions is the interior of southern China where winters are relatively dry. This is because prevailing winter winds are from the cold and dry north and northwest. This anomalous Cf climate is designated Cw.

In Cf localities, summers are hot and winters are mild. Mean temperatures of the warmest month are typically in the range of 24 to 27 °C (75 to 80 °F). For the coolest month, average temperatures are typically in the 4 to 13 °C (40 to 55 °F) range. Freezes and snowfall are infrequent.

Temperate climates (D)

The primary distinction among temperate climates is between temperate oceanic (Do) and temperate continental (Dc). As the names imply, the first features a strong maritime influence while the second is highly continental. These are midlatitude climates.

Temperate oceanic climates are found on the windward side of continents, mainly poleward of 40 degrees latitude. These climates occur along a narrow coastal strip in the Pacific Northwest from northern California northwestward into Alaska, along the southwestern coastal plain of South America, throughout most of western Europe, and in small portions of Australia and New Zealand. In these regions, a strong maritime influence prevails year-round. Temperatures are consequently relatively mild for the latitude and seasonal and diurnal temperature contrasts are reduced. Cold waves and heat waves are rare, and the growing season is long (up to 210 freeze-free days).

The precipitation regime also reflects the maritime influence in Do climates. Maritime polar (mP) air masses dominate to produce persistent episodes of low clouds and light-to-moderate rainfall with rainfall amounts dependent on topography. Droughts are infrequent. The mP air is stable, so convective showers are unusual.

Temperate continental climates (Dc) occur only in the Northern Hemisphere from about 40 degrees to 50 degrees N. These Dc climates are inland and on the leeward side of continents: the northeastern third of the United States, Eurasia, and extreme eastern Asia. A marked continentality increases inland with maximum temperature contrasts between the coldest and warmest month as great as 25 to 35 °C (45 to 63 °F). Dc climates are divided into the southerly Dca climates featuring cool winters and warm to hot summers, and the northerly Dcb climates with cold winters and mild summers. The frost-free season varies in length from 7 months in the south to 3 months in the north.

The weather in Dc regions is very changeable and dynamic. This is because these areas are swept by cyclones and anticyclones and by surges of contrasting air masses. Polar front cyclones dominate winter—bringing episodes of light-to-moderate frontal precipitation. These storms alternate with surges of dry polar and arctic air masses. In summer, cyclones are weak and infrequent because the principal storm track shifts northward (into Canada, for example). Summer rainfall is mostly convective and sometimes very heavy in severe thunderstorms. Although precipitation is distributed fairly uniformly throughout the year, in most places there is a summer maximum.

In the northern portions of Dc climates, winter snowfall becomes an important factor. The amount of snow and the persistence of snow cover increase northward. Because of its high albedo for solar radiation and its excellent emission of infrared, a snow cover chills and stabilizes the overlying air. For these reasons, a snow cover is self-preserving.

Boreal climate (E)

The boreal climate occurs only in the Northern Hemisphere as an east-west band between 50 or 55 degrees N and 65 degrees N latitude. It is a region of extreme continentality and very low mean annual temperatures. Summers are short and cool, and winters are long and bitterly cold. Both continental polar (cP) and arctic (A) air masses originate here, and this area is the site of an extensive coniferous forest. Midsummer freezes are possible, so the growing season is precariously short.

Weak cyclonic activity occurs throughout the year and yields meager precipitation (typically less than 50 cm, or 20 in.). Convective activity is rare. A summer precipitation maximum is due to the winter dominance of cold, dry air masses. Snow cover is persistent in winter.

Polar climates (F)

Polar climates occur poleward of the 66 degree 30 minute latitude circles, the Arctic and Antarctic circles. These boundaries correspond roughly to localities where the mean temperature for the warmest month is 10 °C (50 °F). These limits also approximate the tree line, the poleward limit of tree growth. Poleward is tundra (Figure 17.10), and the Greenland and Antarctic ice sheets. A distinction is made between tundra (Ft) and ice cap (Fi) climates, with the dividing criterion being 0 °C (32 °F) for the mean temperature of the warmest month. Vegetation is sparse in Ft regions and nonexistent in Fi areas.

Polar (F) climates are characterized by extreme cold and slight precipitation, which is mostly in the form of snow (less than 25 cm, or 10 in., melted). Although summers are cold, the winters are so extremely cold that F climates feature a marked seasonal temperature range. Mean annual temperatures are the lowest in the world.

Highland climates (H)

Highland climates encompass a wide variety of climate types that characterize mountainous terrain. Altitude, latitude, and exposure

FIGURE 17.10
The tundra of northern Canada. (Photograph by Steve McCutcheon)

are among the factors that shape a complex mosaic of climate types. Climate-ecological zones are telescoped in mountainous areas, that is, in climbing several thousand meters of altitude, we encounter the same biotic-climatic zones that we would experience in traveling several thousand kilometers of latitude. As a general rule, every 300 m (1000 ft) ascended corresponds roughly to a northward advance of 500 km (310 mi).

Conclusions

This chapter described and classified climate with a focus on present climates. Climate, however, changes with time. How and why climate changes is the subject of the next chapter.

SUMMARY STATEMENTS

- Climate is defined as weather conditions averaged over some time period plus extremes in weather observed during the same period. Climate is usually specified for a particular location and for some time interval.
- Climatic controls include the various weather systems operating at all spatial and temporal scales along with insolation, land-water distribution, and topography.
- Anomalies in climate, departures from long-term averages, are geographically nonuniform in both sign (positive or negative) and magnitude.
- The climatic norm encompasses both means and extremes of weather elements and is variable through time.
- An alternate way of describing the climate of a locality is by frequency of occurrence of the various types of air masses that develop over or are advected into that locality.
- The globe is a mosaic of many different types of climate that may be grouped and classified by meteorological causes, environmental effects, or both.

KEY WORDS

climatic norm
anomalies
agroclimatic compensation
air mass climatology
streamlines
confluence zone

REVIEW QUESTIONS

1 Define climate.

2 What is meant by the climatic norm?

3 By international convention, what period of time is used as the basis for computation of climatic norms?

4 What is a climatic anomaly?

5 How is air mass frequency used to describe the climate of a locality?

6 How can a single anticyclone be a source of many different air mass types?

7 What is the purpose of a streamline analysis?

8 How does arctic air mass frequency correspond to the location of Canada's boreal forest?

9 List the major controls of climate.

10 What are the bases for climate classifications?

11 How does Köppen's classification scheme group global climates?

12 In tropical humid climates how does the diurnal temperature range compare with the annual temperature range?

13 On the continents, which climate grouping covers the largest area?

14 How does the variability (and hence, reliability) of rainfall change as yearly rainfall totals decrease?

15 How is a desert climate defined? What are the principal controls of desert climates?

16 How do seasonal shifts of the subtropical anticyclones induce seasonal changes in rainfall in subtropical latitudes?

17 Explain the seasonality of rainfall in (a) India and (b) coastal California.

18 Contrast temperate oceanic climates with temperate continental climates.

19 How does a seasonal snow cover influence climate?

20 Why is convective activity rare in boreal climates and polar climates?

POINTS TO PONDER

1 Explain why a 30-year period provides a myopic view of the climatic record.

2 Why are climatic anomalies geographically nonuniform?

3 Explain why a precipitation anomaly map is considerably more complex than a temperature anomaly map.

4 What is the significance of the geographic nonuniformity of climatic anomalies for (a) agricultural productivity and (b) home heating and cooling demands?

5 Speculate on how long-term changes in climatic controls might influence climate.

6 Climatic controls are really not independent of one another. Provide some examples of linkages.

PROJECT

1 In what climate grouping is your locality? How closely does the climate of your area match the general description given in this chapter?

SELECTED READINGS

Bryson, R. A., and F. K. Hare, eds. *World Survey of Climatology* 11, *Climates of North America.* New York: Elsevier Scientific, 1974. 420 pp. *Includes review articles on climates of Canada, the United States, and Mexico.*

Hughes, P. "1816—The Year Without a Summer." *Weatherwise* 32 (1979):108–111. *Reviews the events of the relatively cold summer and speculates on possible causes.*

Oliver, J. E., and J. J. Hidore. *Climatology.* Columbus, Ohio: C. E. Merrill, 1984. 381 pp. *An introductory text on the basics of climatology.*

Trewartha, G. T., and L. H. Horn. *An Introduction to Climate.* 5th ed. New York: McGraw-Hill, 1980. 416 pp. *A basic textbook on the meteorological basis of climate plus descriptions of primary climate belts.*

Wendland, W. M., and R. A. Bryson. "Northern Hemisphere Airstream Regions." *Monthly Weather Review* 109 (1981):255–270. *Determination of air mass source regions on the basis of streamline analysis.*

Now from the smooth deep
ocean-stream the sun
Began to climb the heavens,
and with new rays
Smote the surrounding fields.
HOMER
Iliad

The climatic future, like the climatic past, will be shaped by many interacting factors. The major influences on future climates, though, will most likely be variations in solar radiation and the great thermal inertia of the oceans. (Photograph by Arjen Verkaik)

EIGHTEEN

Climatic Record and Climatic Variability

CLIMATE is inherently variable. As we saw in the previous chapter, climate differs from one place to another. Climate also varies with time. In this closing chapter, we describe what is understood about past variations in climate, the factors that may be responsible for climatic variability, and the prospects for the climatic future.

The Climatic Past

As we go back in time, we must describe climate in increasingly broad generalities, and longer and longer gaps in data appear. Here, we summarize what is currently understood about the earth's climatic record.

Geologic time

As we journey back through the millions and millions of years that constitute **geologic time**, the climatic record becomes extremely fragmented and unreliable. From various lines of evidence, including fossils and sedimentary rock layers, some generalized sketches of global climate emerge (Table 18.1). When dealing with time frames of hundreds of millions of years, however, complications arise because of geologic processes such as mountain building and continental drift. Today we consider topography and land-water distribution to be fixed climatic controls, but from the perspective of geologic time, they are variable. Mountain ranges have risen and eroded away; seas have invaded and withdrawn from the land, continually altering the shapes of continents; and land masses have drifted slowly across the face of the globe.

Continental drift, an idea first proposed in 1912 but not widely accepted until the 1960s, might explain such seemingly anomalous finds as 200-million-year-old glacial deposits in India and in the Sahara Desert, fossils of tropical plants in Greenland, and fossil coral reefs in Wisconsin. These materials were most likely formed when the continents were situated at different latitudes from their present locations. The continents are carried along as parts of gigantic, rigid plates that drift slowly (about 2 to 10 cm per year) over the globe. Geologic evidence suggests that approximately 200 million years ago there was just one continent, called Pangaea, which subsequently split into smaller land masses. Its constituent land masses, the continents we now know, drifted slowly apart, even-

TABLE 18.1

Summary of Earth's Climatic History Over the Past 350 Million Years

YEARS BEFORE PRESENT	CLIMATIC EVENT
350–250 million	**Pangaean ice age**
250–50 million	**Relatively mild episode**
50 million	**Onset of cooling**
15–10 million	**Glaciation of Antarctica**
2 million	**Beginning of major glacial epoch in North America**
1 million	**Major interglacial episode**
25,000	**Last major glacial advance**
18,000	**Last glacial maximum** **Onset of warming**
10,000	**Glacier retreats from northern United States**
7000–5000	**Peak mild episode**

tually reaching their present positions shown in Figure 18.1. What preceded Pangaea is the subject of much speculation.

The past million years

As we focus on the climatic record of the past million years, we do not have to be concerned with the effects of continental drift and mountain building. For all practical purposes, mountain ranges and continents were essentially as they are today. Because climate varies over a very wide range of time scales, it is, however, useful to view the climatic record of the past million years in the perspective provided by increasingly narrow time frames. Such an approach helps resolve the complex oscillations of climate into somewhat simpler fluctuations.

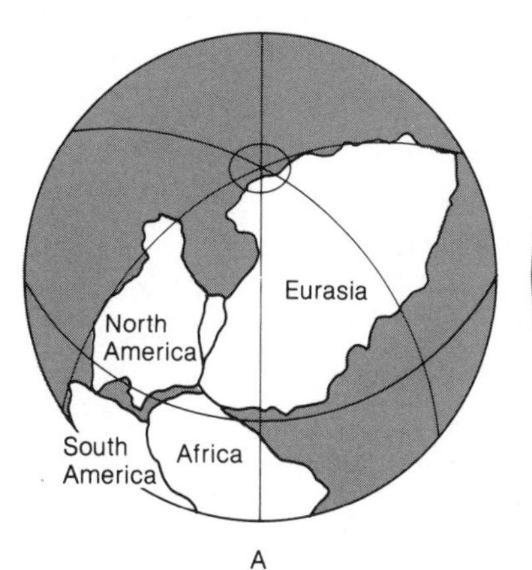

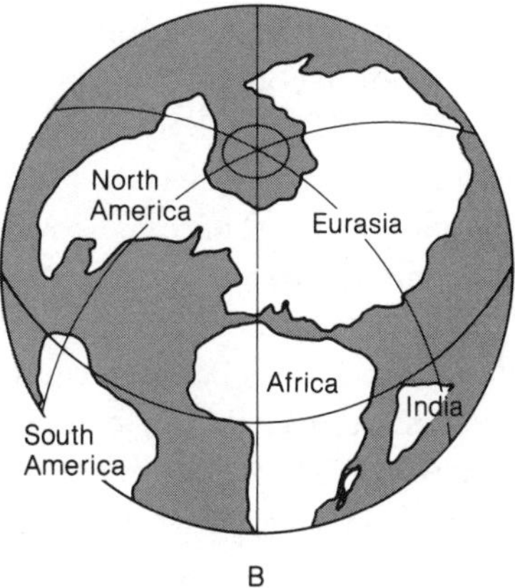

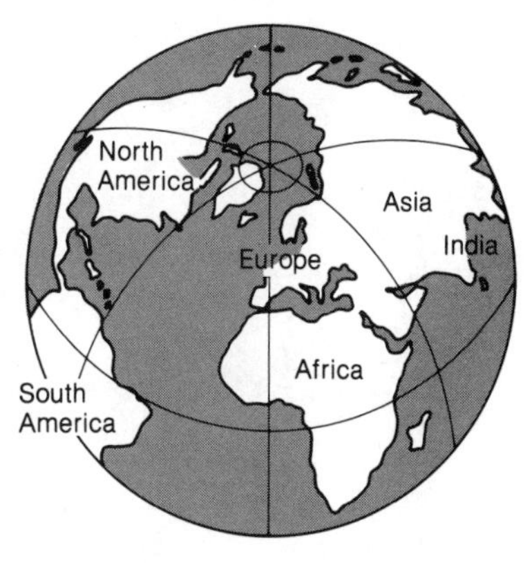

FIGURE 18.1

About 200 million years ago, today's continents formed a single supercontinent. (A) The supercontinent broke up and the fragments began to drift apart. Approximately 65 million years ago, the configuration of the drifting plates probably resembled (B). Today, the continents are arranged as in (C), but the plates continue to drift. (From Wyllie, P. J. *The Way The Earth Works*. New York: John Wiley, 1976, p. 2)

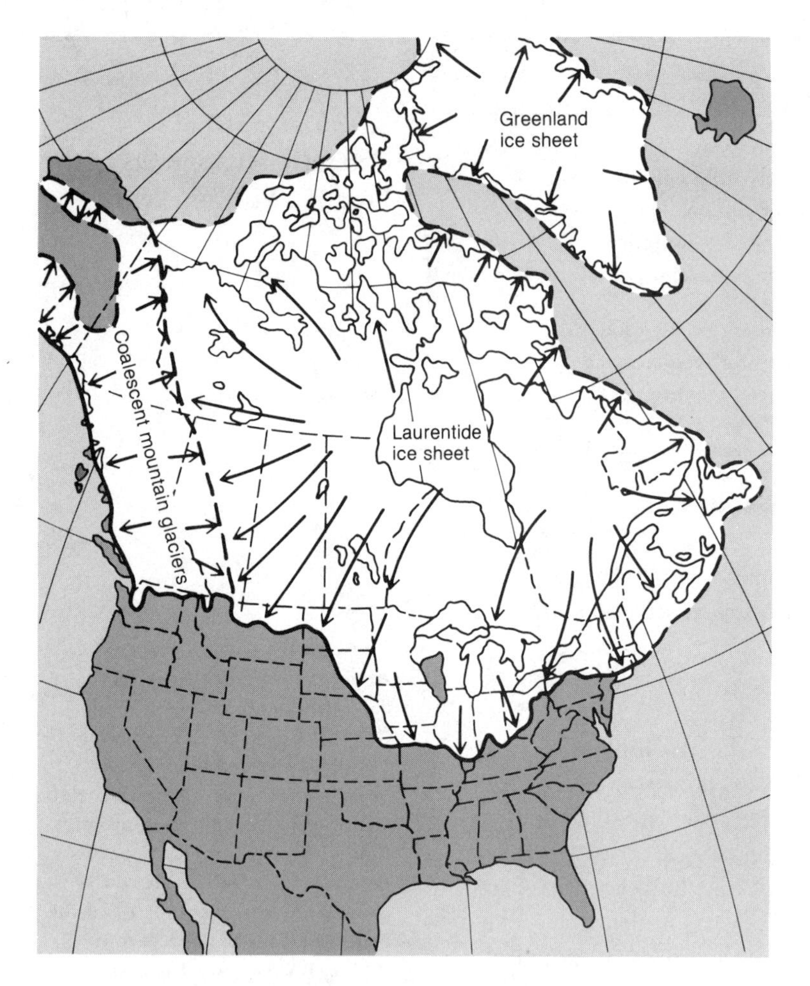

FIGURE 18.2

The extent of glaciation over North America 18,000 years ago, the time of the last glacial maximum.

When compared with the climatic conditions that have prevailed over most of geologic time, the climate of the last million years has been anomalous in favoring the development of huge glacial ice sheets. In fact, for most of the earth's history the average global temperature may have been 10 °C (18 °F) warmer than the last million years. The trend toward colder conditions began perhaps 100 million years ago and culminated in the Ice Age. Deep-sea sediment cores indicate that during the Ice Age the climate shifted numerous times between conditions conducive to glacier growth, a **glacial climate**, and conditions conducive to glacier decay, an **interglacial climate**. During major glacial climatic episodes, the Laurentide ice sheet developed over central Canada and spread westward almost to the mountains, eastward to the ocean, and southward over the northern tier states of the United States (Figure 18.2). At the same time, a much smaller ice sheet formed over northwestern Europe. Because a vast quantity of water was locked up in these ice sheets, sea level fell by perhaps 130 m (400 ft), exposing portions of the continental shelf, including a land bridge linking Siberia and North America. The Laurentide and European ice sheets thinned and retreated, and may even have disappeared entirely during the relatively mild interglacial episodes, which typically lasted about 10,000 years. Throughout this period, however, glacial ice cover persisted over most of Greenland and Antarctica, as it still does today.

Resolution of the past climatic record improves somewhat when we shift focus to the last 100,000 years. The temperature curve in Figure 18.3 was derived from a combination of midlatitude sea-surface temperature indicators, fossil pollen data, and reconstructed sea-level fluctuations. The last major glacial climatic episode and

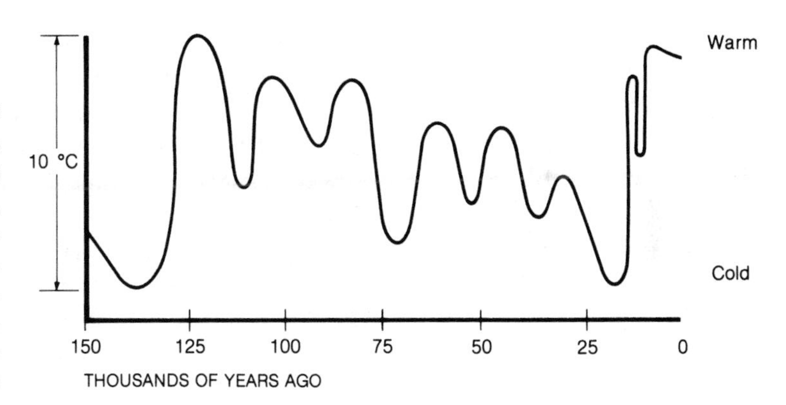

FIGURE 18.3
A generalized curve of midlatitude air temperature over the past 150,000 years is based on pollen records, sea-level fluctuations, and reconstructed sea-surface temperatures. (From Gates, W. L., and Y. Mintz. *Understanding Climatic Change.* Washington, D.C.: National Academy of Sciences, 1975, p. 130.)

the present interglacial period are visible. In the perspective of the past 25,000 years, even more climatic detail appears. The temperature curve in Figure 18.4 is based primarily on reconstructed glacial fluctuations and pollen studies. It shows the last glacial maximum at 18,000 years ago.

At the peak of the last major glacial ice advance, 18,000 years ago, global temperatures averaged 4 to 6 °C (7.2 to 10.8 °F) colder than at present. Because climatic trends are nonuniform geographically, the cooling was much smaller in some latitude belts, and much greater in others. A variety of evidence indicates that temperature fluctuations between glacial and interglacial climatic episodes typically amounted to about 2 °C (3.6 °F) in the tropics, 3 to 4 °C (5.4 to 7.2 °F) at midlatitudes, and as much as 10 °C (18 °F) at high latitudes.

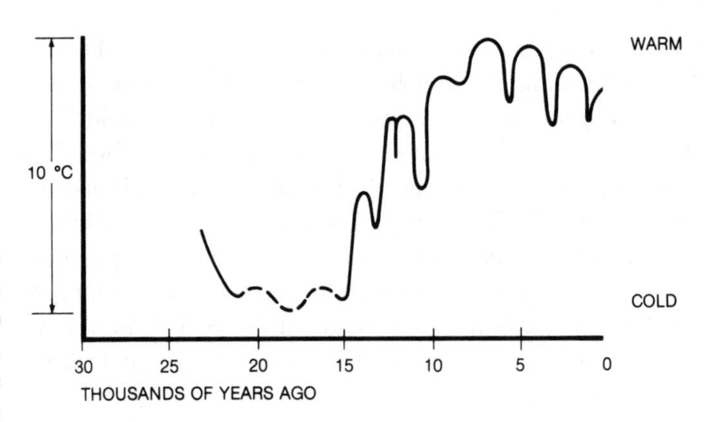

FIGURE 18.4

A generalized curve of midlatitude air temperature over the past 22,000 years is based on pollen records, tree line changes, and glacial ice fluctuations. (From Gates, W. L., and Y. Mintz. *Understanding Climatic Change*. Washington, D.C.: National Academy of Sciences, 1975, p. 130.)

The last glacial maximum was followed by a warming trend that triggered oscillatory glacial retreat. This postglacial warming trend culminated in the so-called **altithermal episode** 5000 to 7000 years ago, a period when global temperatures were warmer than at present.

The climatic record of the past 1000 years, shown in Figure 18.5, is based on a winter severity index developed by the British meteorologist and historian Hubert H. Lamb. By examining documentary records of old farming communities in eastern Europe, Lamb derived a generalized temperature curve. The most notable feature of this record is the relatively mild conditions of the Middle Ages, which were followed by sharp cooling into the period from about 1400 to 1850—a period that has come to be known as the **Little Ice Age**. Other independent lines of evidence have confirmed that this was indeed a relatively cool period throughout the world, with

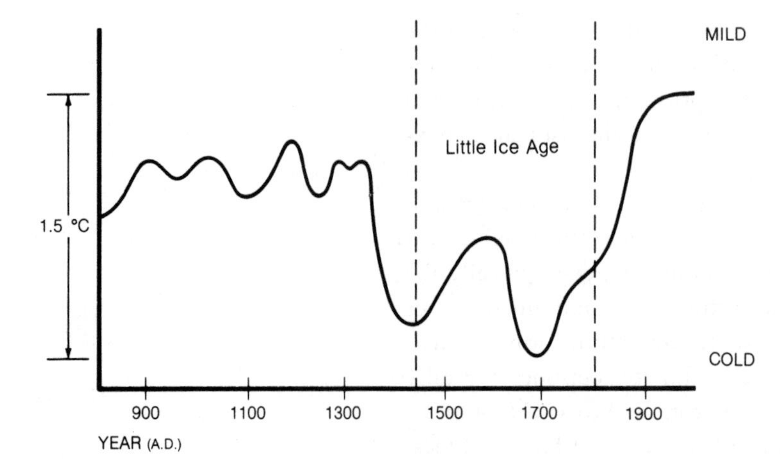

FIGURE 18.5

Winter severity for eastern Europe based on an index by H. H. Lamb. (Adapted from Lamb, H. H. "Climatic Fluctuations." In *World Survey of Climatology*, vol. 2, *General Climatology*, H. Flohn, ed. New York: Elsevier, 1969, p. 236.)

mean annual global temperatures perhaps 1 to 2 °C (1.8 to 3.6 °F) lower than at present. Sea ice expanded toward the equator, mountain glaciers advanced, and growing seasons shortened—bringing much hardship to many people.

With the invention of weather instruments and the establishment of weather observational networks throughout the world, the climatic record becomes much more detailed for the past 100 years. As shown in Figure 18.6, mean annual Northern Hemisphere temperatures rose from the 1890s through the late 1930s and have subsequently declined into the late 1970s. The total temperature fluctuation amounts to only about ±0.6 °C (±1.1 °F) about the century average. Remember, however, that this is a hemispheric mean, so this trend was amplified or reversed or both in specific localities. For example, Figure 18.7 shows that winter trends (December through February) of mean temperature for essentially the same period varied considerably from west to east across the United States.

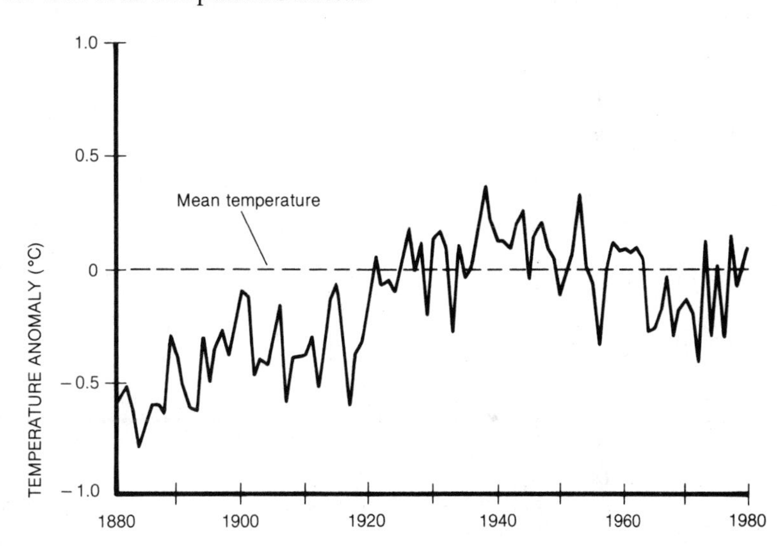

FIGURE 18.6
Variation in mean annual temperature of the Northern Hemisphere, 1881–1980, is expressed as departures from the 100-year mean temperature (in °C). Note the warming trend to about 1938 and the subsequent cooling trend. (From Jones, P. D., T. M. L. Wigley, and P. M. Kelly. "Variations in Surface Air Temperatures: Part 1. Northern Hemisphere, 1881–1980." ***Monthly Weather Review*** **110 (1982):67. American Meteorological Society publication.)**

We gain a somewhat different perspective on temperature variations over the past 100 years when we consider the global rather than the hemispheric scale. Recent research by J. E. Hansen and his colleagues at NASA indicates that the post-1940 cooling trend, although pronounced in the Northern Hemisphere in the area north of the tropics, was minor elsewhere. Mean global temperatures actually increased about 0.4 °C (0.7 °F) from 1880 to 1980, and the global mean temperature in 1980 was almost as high as it was in 1940.

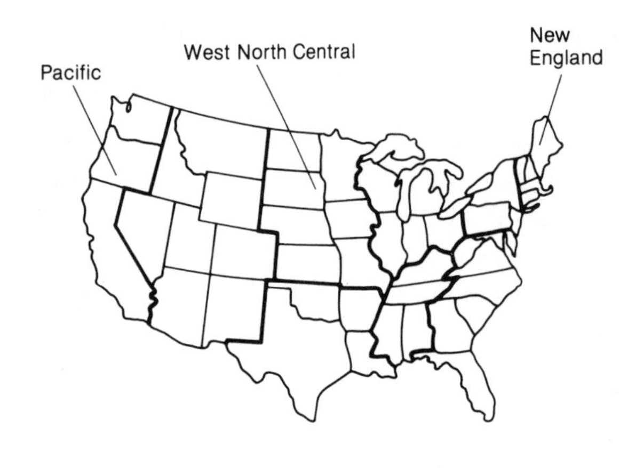

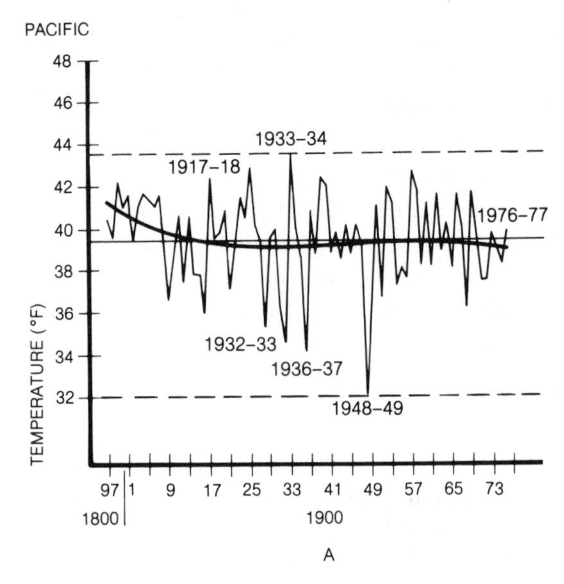

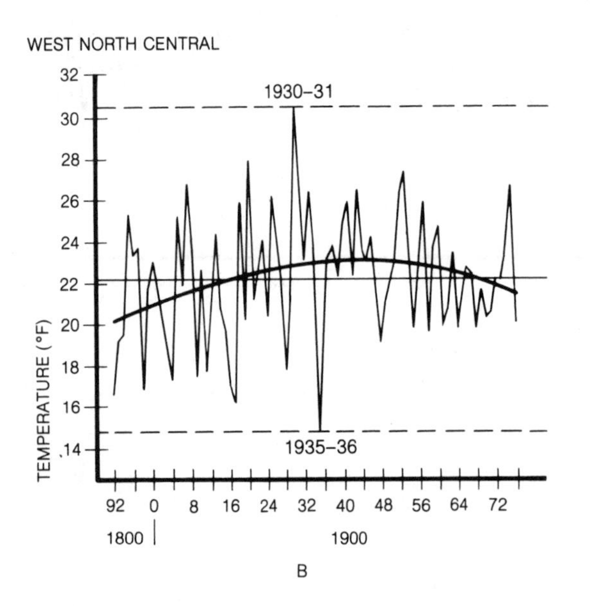

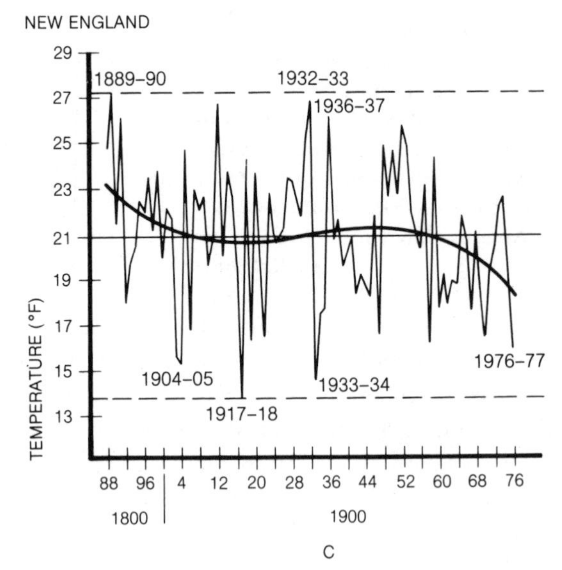

FIGURE 18.7

The geographic differences in climatic trends are illustrated by differences in trends of mean winter (December through February) temperatures in three regions of the United States: (A) Pacific, (B) West North Central, and (C) New England. Data are from the late 1800s to 1977. (From Diaz, H. F., and R. G. Quayle. "The 1976–77 Winter in the Contiguous United States in Comparison With Past Records." ***Monthly Weather Review*** **106 (1978):1402–1405. American Meteorological Society publication.)**

Critics sometimes question the validity of hemispheric and global temperature records—arguing that many possibilities for error exist. They cite (1) improved sophistication and reliability of weather instruments through the period, (2) changes in location and exposure of instruments at most long-term weather stations, and (3) huge gaps in monitoring networks, especially over the oceans, as possible sources of error. By careful statistical treatment of available data, however, large-scale temperature trends have been confirmed with a reasonable degree of certainty.

Lessons of the Climatic Record

What does the climatic record indicate about climatic behavior? After all, that is a major reason for journeying into the climatic past. First and foremost, climate varies over a wide range of time scales, from years to millennia. Other conclusions may be drawn as well.

Climate and society

There is little doubt that climate and climatic variations have played a role in the history of society since humans emerged during the Ice Age. In some cases, climate was probably the key factor in the course of history.

In their book, *Climates of Hunger*, R. A. Bryson and T. J. Murray describe several cases in the distant past when climatic change severely affected societies. They argue convincingly, for example, that prolonged drought contributed to the decline and fall of the Harappan civilization of the Indus valley region of northwestern India about 1700 B.C., the Mycenaean civilization of Greece in 1200 B.C., and the Mill Creek culture of northwest Iowa about 1200 A.D. Other scientists propose that a succession of severe droughts forced the Pueblo Indians of Mesa Verde in southwestern Colorado to abandon their homes around 1300 A.D. In these and other studies of the impact of climatic change on society, the message is clear: The people most vulnerable to climatic change are those living in areas where the climate is just marginal for survival. These are typically climatic zones where barely enough rain falls or where the growing season is barely long enough for crops to grow. Even a small change in these critical parameters in the wrong direction can spell disaster. This is apparently what happened to the early Greenland settlements with the onset of the cooling trend that heralded the Little Ice Age (discussed in the Special Topic).

SPECIAL TOPIC

Climatic change and the Greenland tragedy

In the late ninth century, a lengthy episode of unusually mild conditions began in the Northern Hemisphere. This warming enabled Viking explorers to probe the far northerly reaches of the Atlantic Ocean. Severe cold and extensive drift ice had previously been insurmountable obstacles to European navigators. By 930 A.D., the Vikings established the first permanent settlement in Iceland, some 970 km (600 mi) west of Norway and just south of the Arctic Circle.

Among Iceland's early inhabitants was Eric the Red, a troublesome individual whose exploits eventually caused him to be banished from Iceland in 982 A.D. He sailed west and discovered a new land, which he named Greenland. Some historians think he called it that to entice others to follow him. Then, as now, much of this new land was buried under a massive ice sheet. In the late tenth century, however, the climate there was so mild that some sheltered areas probably were indeed quite green.

In such a coastal place, on Greenland's southwest shore, Eric founded the first of three colonies (Figure 1). Although far from prosperous, the settlement developed as a successful agrarian society. Initially, the colony fared well, and its population climbed to roughly 3000. By the fourteenth century, however, the climate was deteriorating rapidly. Drift ice again expanded in the North Atlantic, hampering and eventually halting navigation between Iceland and Greenland. Greenland was thus cut off from the rest of the world. What happened to the Norse settlers in succeeding years can only be inferred from their graves and the ruins of their homes. There were no survivors.

In 1921, an expedition from Denmark examined the remains of the Greenland settlements. The expedition reported that the colony had lasted for 500 years and had suffered a slow, painful annihilation. Grazing land was buried under advancing lobes of glacial ice and most farmland was rendered useless by permafrost. Near the end, the descendants

of a once robust and hardy people were ravaged by famine. They became crippled, dwarflike, and diseased. There is also some indication that the malnourished colonists were attacked by pirates when they turned to the sea for food.

The Greenland tragedy may be the only historical example of the extinction of a European society in North America. What lesson does it teach contemporary society? The lesson is that climate changes, and that it can change rapidly—sometimes with serious, even disastrous consequences. Nowhere are people more vulnerable to climatic shifts than in regions where the climate is just marginal for their survival. These regions include areas where barely enough rain falls to sustain crops and livestock, and places where temperatures are so cold and the growing season so short that only a few hardy crops can be cultivated successfully. In such regions, even a slight deterioration in climate makes agriculture impossible. If the inhabitants have no alternate food source, and if they cannot migrate to more hospitable lands, their fate may be similar to that of the early Greenland population.

FIGURE 1

Location of ancient Norse settlements on coastal Greenland. (From Bryson, R. A., and T. J. Murray. Climates of Hunger. *Madison, Wis.: The University of Wisconsin Press, 1977, p. 48.)*

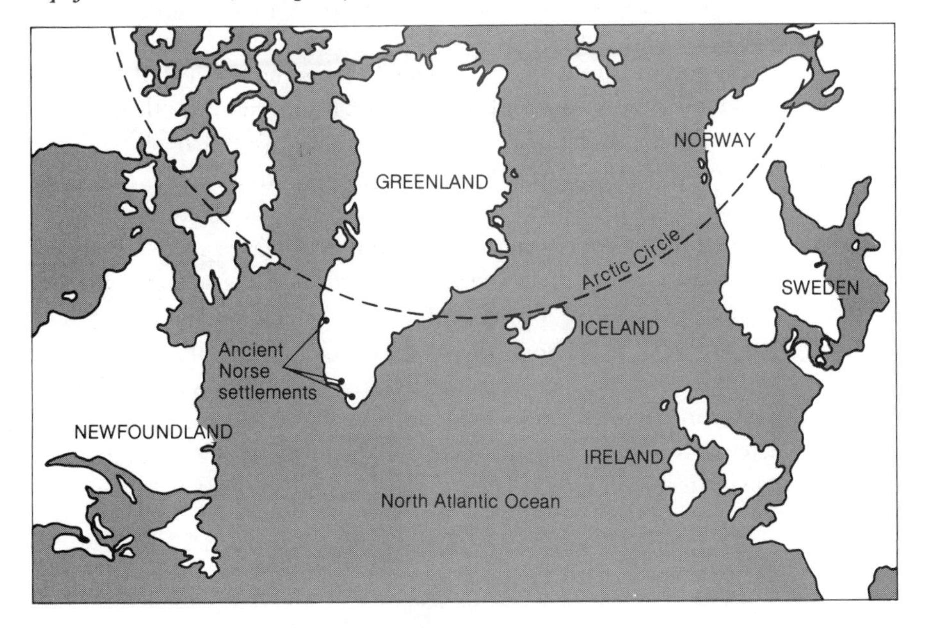

Interestingly, the meanderings of climate that seriously affected the course of civilizations in the past fall within the range of climatic anomalies that appear in the historical climatic record. Even the very cold summer of 1816 falls within the range of expected climatic variability—albeit at an extreme end. Theoretically, then, the atmospheric system could shift to any weather pattern at any time. In effect, what has happened weatherwise in the past can happen again.

This also lends credence to the notion that past climatic episodes differ from present climatic episodes in frequency of occurrence of those episodes rather than in type of episode. For example, an exceptionally cold winter is the consequence of a usual cold weather circulation pattern that occurs with unusually high frequency or persistence. A climatic shift to drier conditions results from more frequent dry weather patterns. The individual weather patterns are not themselves unusual, but the higher frequency at which they occur may be anomalous.

How rapidly do significant climatic variations take place? Are climatic shifts abrupt or gradual? The conclusion depends on our time frame of reference. A climatic shift that takes 100 years is abrupt in the context of the long-term climatic record, which stretches back a million years, but the same shift is gradual when viewed in the limited human perspective of historical climate. With this rate criterion in mind, the climatic record indicates that significant climatic variations tend to occur abruptly rather than gradually. Now, as in the past, a significant variation in climate can occur so rapidly that humankind may be unable to adjust to the shift.

We must also be careful to distinguish between a short-lived meandering from the climatic norm and a semipermanent climatic change. The Great Plains drought in the 1930s, the three severe midwestern winters in the late 1970s, and the tragic Sahelian drought from the late 1960s through the mid-1980s were all temporary climatic fluctuations. In contrast, the climatic cooling of the late thirteenth century that heralded the Little Ice Age constituted a longer-term climatic change.

Rhythms in the record

The recorded and reconstructed climatic record has long been the target of painstaking analyses for possible trends or regular rhythms. One formidable challenge in this search is to separate signal from noise. Some climatic elements are so variable (the noise) through time that detection of any cycles or trends (the signal) requires close scrutiny and meticulous dissection of the climatic record. In recent years, use of high-speed electronic computers programmed with sophisticated statistical routines has greatly facilitated the search for climatic rhythms and trends. The motivation behind all this effort is obvious: identification of any statistically real periodicities or trends in the climatic record would be a powerful tool in both weather and climate forecasting. What is the result of such investigations?

So far, few cycles have been identified in the climatic record that are significant in a rigorous statistical sense, and none of the cycles has had much practical value for weather or climate forecasting. Cycles established as statistically significant are (1) the familiar annual and diurnal radiation-temperature cycles and (2) a less familiar quasibiennial (almost every 2 years) cycle in various midlatitude climatic elements. The first means merely that winters tend to be cooler than summers and nights tend to be cooler than days. Examples of the second type of cycle include an approximate 2-year fluctuation in midwestern rainfall and a 25.5-month oscillation in a lengthy temperature record (1659 to present) from central England. A 20- to 22-year recurrence of drought on the western High Plains, and the major glacial-interglacial climatic fluctuations are actually only quasiperiodic (not quite regular) and need to be thoroughly understood before they can be used to reliably forecast future weather and climate. Trends may be visible in the climatic record, but unless a trend is demonstrated to be part of a statistically significant cycle, there is no guarantee that the trend will not end abruptly or reverse direction at any time.

Explaining Climatic Variability

There is no simple explanation for why climate varies. The complex spectrum of climatic variability is a response to the interactions of many processes both internal and external to the earth-atmosphere system. These processes are shown schematically in Figure 18.8. Internal processes are indicated by open arrows, and external processes are represented by solid arrows.

There are perhaps as many hypotheses about the causes of climatic variations as there are scientists who seriously investigate the question. Many of these ideas evolved out of efforts to explain the great Ice Age, but some stemmed from attempts to explain shorter-

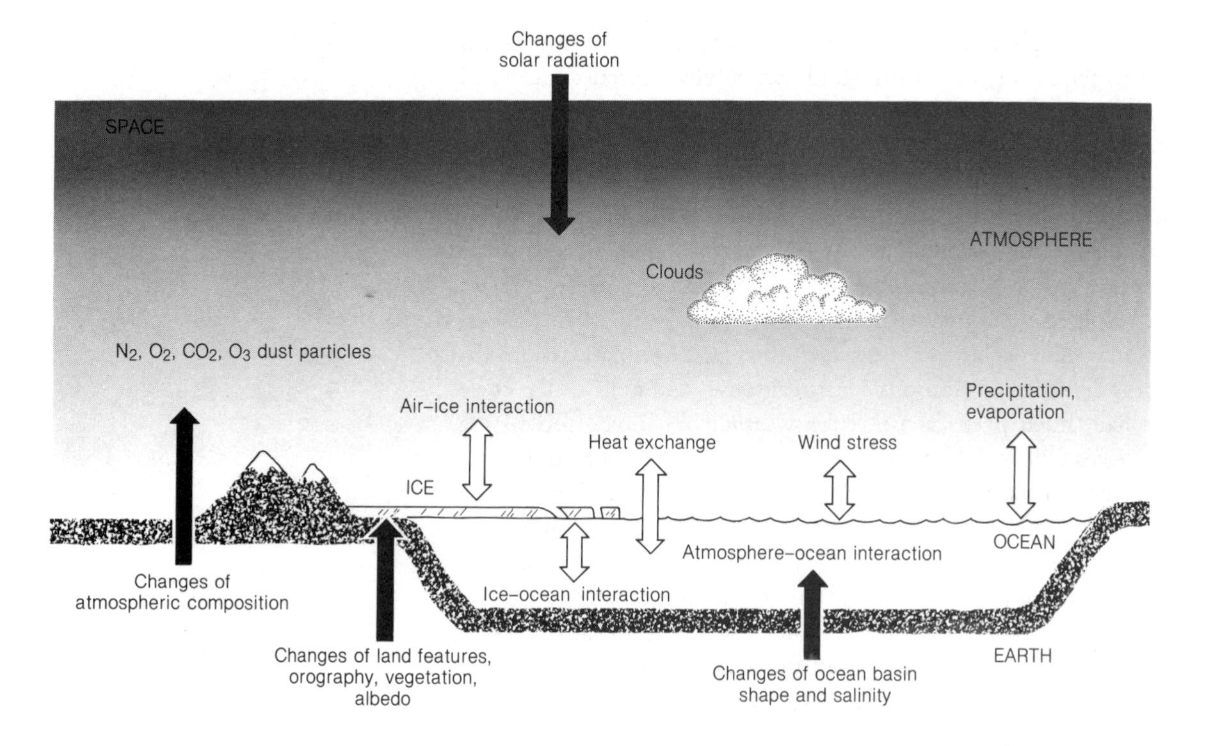

FIGURE 18.8

Climatic variability is influenced by many processes both internal (white arrows) and external (black arrows) to the earth-atmosphere-ocean-ice system. (From Gates, W. L., and Y. Mintz. *Understanding Climatic Change*. Washington, D.C.: National Academy of Sciences, 1975, p. 14.)

term fluctuations of climate. One way to organize the many hypotheses on the causes of climatic variability is to match a possible cause (or forcing) with a specific climatic oscillation on the basis of the period of that oscillation. This approach is displayed schematically in Figure 18.9. For example, mountain building and continental drift (plate tectonics) might explain climatic changes over periods of hundreds of millions of years. Systematic changes in the earth's orbit about the sun may account for climatic shifts of the order of 1000 to 100,000 years. Volcanic eruptions may induce climatic fluctuations lasting a few months or years. Note, however, that matching the possible cause and effect is no guarantee of a real physical relationship. We could well be dealing with mere coincidence.

Hypotheses about climatic variability can also be organized in the context of the global radiation balance. Energy that goes in must ultimately equal energy that goes out. The net input of solar radiation must thus be balanced by an output of infrared radiation from the earth-atmosphere system. Anything that alters this state of radiative equilibrium could change the earth's climate. This input-output energy balance might be altered by fluctuations in (1) the solar constant, (2) the planetary albedo, or (3) the gas and aerosol

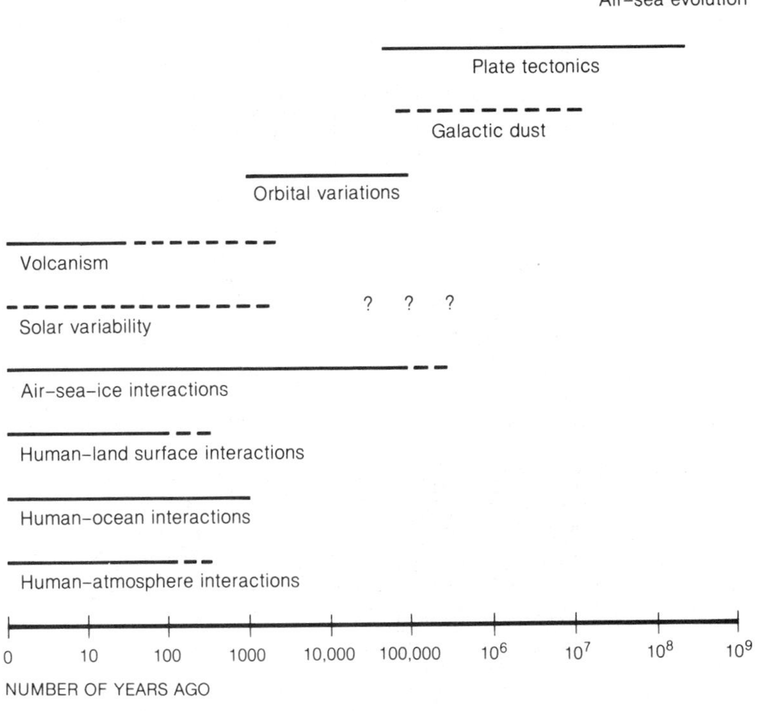

FIGURE 18.9

One way to speculate on the possible cause of a climatic oscillation is to match the period of that oscillation (on baseline) with the period of the appropriate forcing phenomena. Lines are dashed where there is great uncertainty. (After the National Research Council. *Geological Perspectives on Climatic Change*. Washington, D.C.: National Academy of Sciences, 1978, p. 20.)

composition of the atmosphere. This approach appears in mathematical form in the Mathematical Note at the end of the chapter.

Climate and Solar Variability

Changes in the sun's energy output or variations in the earth's orbit about the sun could alter the solar constant or the distribution of solar radiation over the earth's surface. Any of these changes could in turn alter climate.

The solar constant

Is the solar constant* really constant? As demonstrated in the Mathematical Note, even a 1 percent change in the solar constant could significantly alter the radiative equilibrium temperature of the earth-atmosphere system. In fact, a 1 percent fluctuation in the solar constant could account for the total variation of the mean hemi-

*In Chapter 3, earth's solar constant was defined as the flux of solar radiation on a surface perpendicular to the solar beam at the top of the atmosphere when the earth is at its average distance from the sun.

spheric temperature over the past century. Until recently, we have unfortunately been unable to monitor adequately the variations in the solar constant. Older ground-based instruments were simply unable to detect possible changes in the solar constant, and modern satellite-borne instruments have not been operating long enough to provide reliable records. This situation began to change on 14 February 1980 when NASA's Solar Maximum Mission (SMM) satellite (Figure 18.10) was launched into circular orbit 550 km (340 mi) above the earth's surface. Sensitive instruments aboard the satellite can detect changes as small as 0.10 percent in the total solar radiative output, called **solar irradiance**.

FIGURE 18.10

An artist's concept of NASA's Solar Maximum Mission spacecraft, designed to provide the most comprehensive observations of solar activity. The spacecraft is in a 550-km (340-mi) high orbit around the earth. It was launched from Cape Kennedy, Florida, on February 14, 1980. (Courtesy of NASA)

Analysis of measurements by the SMM satellite indicates a decline in solar irradiance from February 1980 to August 1981, amounting to 0.10 percent. Almost immediately this finding spurred speculation linking the weakening of solar output to the harsh winter of 1981–1982. Superimposed on this 18-month downward trend in solar irradiance were short-term variations lasting from days to weeks. Slight increases were apparently due to the appearance of **faculae**, small, bright hot spots on the sun, and slight decreases were due to sunspots. Faculae and sunspots alter the local photospheric radiation emission.

The day-to-day variation of solar irradiance as measured by the SMM satellite is too small to have a detectable influence on earth temperature. In fact, the thermal effect of this variation is lost in the usual diurnal temperature cycle. The unfortunate failure of instruments aboard the SMM satellite after only about 18 months of data collection brought a temporary halt to analysis of possible longer term trends in solar irradiance. As of this writing, however, the SMM instruments are back in operation having been repaired during a recent space shuttle mission.

Another promising avenue of inquiry regarding solar irradiance variations focuses on the relationship between solar energy output and the sun's radius. For reasons that are beyond the scope of this book, the sun, a hot gaseous sphere, expands or contracts slightly as solar irradiance increases or decreases. Changes in solar radius are so slight, however, that until recently scientists were unable to resolve variations in solar radius accurately enough to infer possible climatic effects. This situation should change, however, with the scheduled 1986 launch of the Solar Disk Sextant satellite, which is capable of precise measurement of the sun's radius.

Sunspots

Both popular and technical literature contain much speculation on the possible link between the earth's weather and climate and sunspot activity. **Sunspots** are relatively large (typically thousands of kilometers in diameter) dark blotches that appear on the face of the sun. As shown in Figure 18.11, a sunspot consists of a dark central area, called an **umbra**, which is ringed by an outer, lighter area, termed a **penumbra**. The umbra radiates at about 4000 K, the penumbra at 5400 K, and the surrounding surface of the sun, the **photosphere**, at 5800 K.

As early as 28 B.C., Chinese astronomers observed sunspots with the unaided eye by viewing the sun's reflection in a quiet pond.

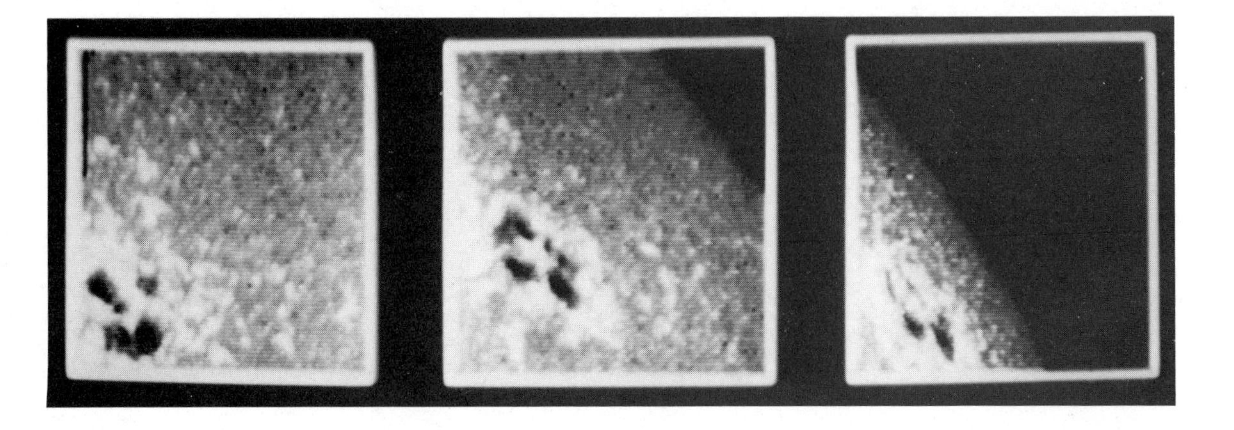

FIGURE 18.11
This sequence of images, taken by monitoring instruments aboard NASA's Solar Maximum Mission spacecraft, shows an active solar region containing four large sunspots, which are moving toward the edge of the sun. Images in these photographs represent an area of the sun's surface approximately 360,000 km (224,000 mi) wide. (Courtesy of NASA)

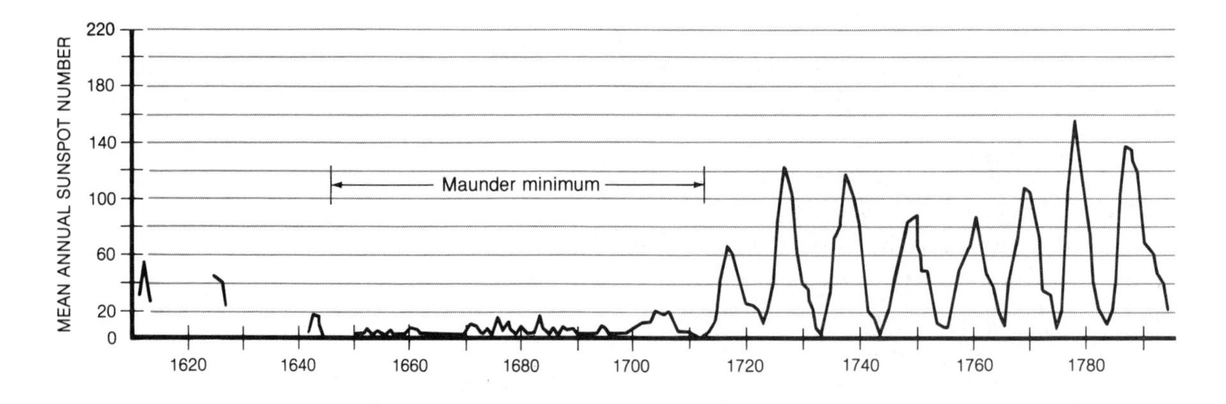

FIGURE 18.12

On this graph of mean annual sunspot number, note the Maunder minimum from 1645 to 1715. Note also that the time between sunspot maxima is not regular, but varies between 8 and 13 years. (From Eddy, J. A. "The Case of the Missing Sunspots." *Scientific American* 236, no. 5 (1977):82–83. Copyright © 1977 by Scientific American, Inc. All rights reserved.)

Galileo is credited with being the first to study sunspots telescopically in 1611, and thereafter sunspots became objects of considerable scientific curiosity. Speculation on a possible sunspot-climate link was spurred by Heinrich Schwabe's discovery in 1843 of the regularity of sunspot activity. As shown in Figure 18.12, the number of sunspots varies systematically with an average 11-year shift between maximum and minimum sunspot number and an approximate 22-year oscillation ("double" sunspot cycle) in a strong magnetic field that is associated with sunspots. Note, however, that the sunspot record is not precisely periodic and that some variations of 8 to 13 years or more occur between maxima or minima.

The exact relationship, if any, between sunspots and climate is not known, although there are some possible correspondences. In 1893, E. Walter Maunder of the Old Royal Observatory at Greenwich, England, while searching old records, discovered that sunspot activity was greatly reduced in the 70 years between 1645 and 1715. In fact, the total number of sunspots observed during this so-called **Maunder minimum** was fewer than is typical in a single year today. Strangely, Maunder's finding was largely ignored by the scientific community until the 1970s, when Maunder's work was reinvestigated and confirmed by John A. Eddy, a solar astronomer at the Center for Astrophysics at the Harvard College Observatory.

Eddy pointed out that the Maunder minimum and an earlier period of reduced sunspot activity, called the **Spörer minimum** (1410–1540), happened to coincide with two relatively cold pulses of the Little Ice Age in western Europe. However, skeptics dismiss the climatic significance of this correspondence, arguing that anomalous cold did not prevail throughout the sunspot minima and was not global in extent. The climatologist Helmut Landsberg of the

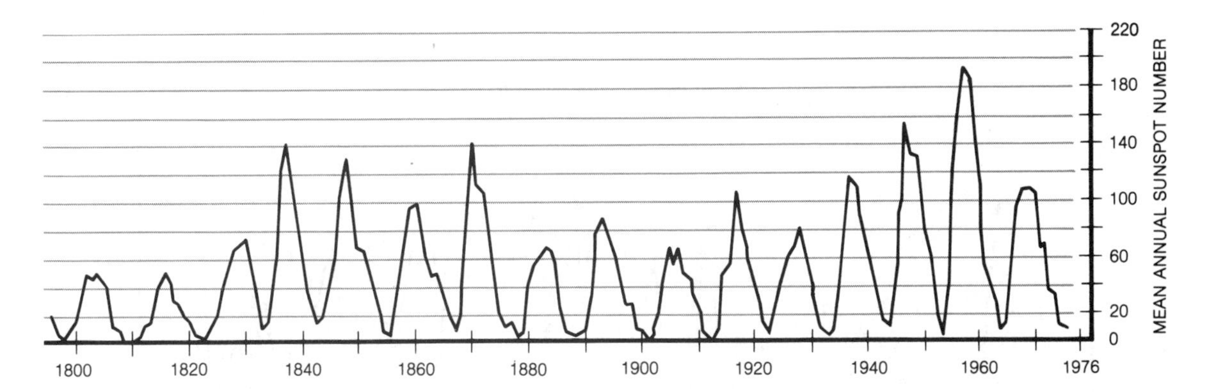

University of Maryland notes that globally the coldest episode of the Little Ice Age occurred 100 years after the Maunder minimum.

There is also a proposed match between the 22-year "double" sunspot cycle and the frequency of drought on the western High Plains. By analyzing tree growth-ring records from 40 sites in the western United States, Charles Stockton of the University of Arizona's Laboratory of Tree-Ring Research was able to extend the region's drought chronology back to the seventeenth century. He discovered that drought recurred every 20 to 22 years.

Are these correlations actual cause-effect relationships or are they merely coincidences? So far, investigators have not come up with an adequate explanation of the mechanism by which sunspot activity might influence climate, but the structure of sunspots is an interesting avenue of inquiry. In 1979, Douglas V. Hoyt, a solar researcher for NOAA, reported that the ratio of the area of the sunspot umbra (*U*) to the area of its penumbra (*P*) may be proportional to the brightness of the sun, or the solar constant. As shown in Figure 18.13, the *U* to *P* ratio was initially high and then low from 1874 to 1970. This trend closely parallels the variation in the mean annual Northern Hemisphere temperature. It is therefore possible that the brightness of the sun is influenced by sunspot structure, and if that is the case, then some possible linkages exist between solar and climatic variability. The current scientific consensus is not, however, definitive. A 1982 report by the National Research Council* summarizes a century of research on this issue by stating ". . . in our view, none of these endeavors, nor the combined weight of all of them, has proved sufficient to establish unequivocal connections between solar variability and meteorological response." The report

*National Research Council. *Solar Variability, Weather, and Climate*. Washington, D.C.: National Academy Press, 1982, p. 5.

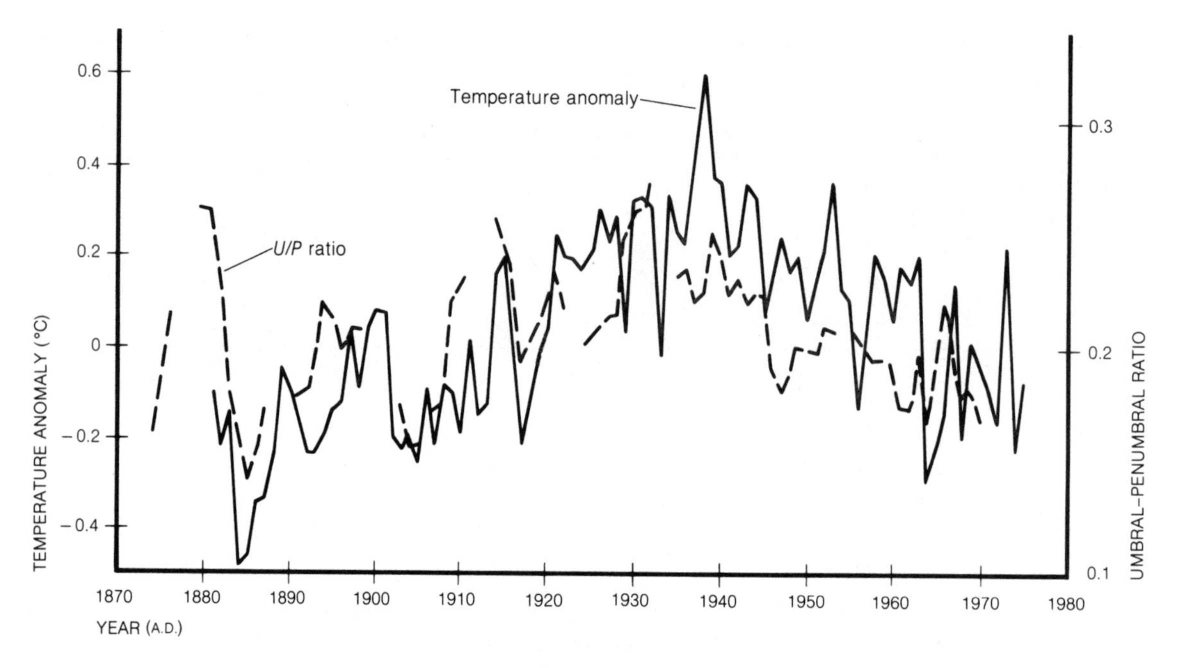

FIGURE 18.13

Over the past hundred years there has been a close parallel between the variation in mean annual temperature of the Northern Hemisphere (solid line) and the ratio of sunspot umbra (*U*) area to penumbra (*P*) area (dotted line). This suggests a possible cause-effect relationship. (From Hoyt, D. V. "Climatic Change and Solar Variability." ***Weatherwise*** **33, no. 2 (1980):69.)**

goes on to urge a shift in research emphasis from a search for past correlations to a search toward better understanding of the physical relationship between the sun and the earth-atmosphere-ocean system.

Astronomical changes

The regular oscillations of glacial and interglacial climates as revealed by analyses of deep-sea sediment cores may be related to systematic changes in the geometry of the earth's orbit about the sun. These changes, first identified in the middle of the nineteenth century, were extensively investigated in the 1920s and 1930s by the Serbian mathematician Milutin Milankovitch. Milankovitch studied cyclical variations in three elements of earth-sun geometry: (1) times of perihelion and aphelion, which alter the earth-sun distance; (2) tilt of the earth's axis of rotation (varying between 22.0 and 24.5 degrees), which affects seasonal contrasts; and (3) eccentricity (departure from a circle) of the earth's orbit. Although these variations do not change the solar constant appreciably, they do alter significantly the latitudinal and seasonal distribution of solar radiation received by the earth. Milankovitch proposed that glacial climatic episodes were initiated at times when the earth-sun geometry favored minimum summer insolation at high northern latitudes. During these times, a portion of the winter snows falling in Canada would presumably survive the summer, and a repetition of many such years would initiate continental glaciation.

Milankovitch hand-calculated latitudinal variations in insolation for 600,000 years prior to the year 1800. More recently, with some refinement in methodology and use of high-speed electronic computers, these calculations were repeated and extended over a longer time period. This research demonstrated that systematic changes in earth-sun geometry would induce climatic cycles of about 22,000 years as a result of perihelion-aphelion changes and cycles of about 41,000 years due to variations in the earth's tilt. Evidence for climatic cycles of similar periodicity has been discovered in deep-sea sediment cores, strongly suggesting that astronomical variations are responsible for the regular fluctuations of the earth's glacial ice cover. The Milankovitch theory does not, however, clearly account for the dominant 100,000-year glacial-interglacial climate cycle that appears in deep-sea sediment cores, although this cycle may stem from periodic changes in eccentricity of the earth's orbit.

Although the correspondence between astronomical changes and the deep-sea sediment climate record is impressive, the precise physical linkage between glacial-interglacial climates and periodic change in earth-sun geometry is not yet understood. Furthermore, the Milankovitch cycles do not account for the lengthy periods of climatic quiescence in the millions of years prior to major glacial episodes.

Volcanoes and Climatic Variability

The spectacular eruption of Mount St. Helens on 18 May 1980 (Figure 18.14) spurred much speculation on the possible climatic impact of volcanic eruptions. An estimated 5 billion metric tons of volcanic ash and gases were blasted 10 km (6 mi) into the stratosphere.

The notion that volcanoes somehow influence weather and climate has been around for more than two centuries. Benjamin Franklin proposed that the eruption of the Laki volcano in Iceland in the summer of 1783 was responsible for the severe winter of 1783–1784. The unusually cool summer of 1816, described in the Chapter 17 Special Topic, followed the violent eruption of Tambora, an Indonesian volcano, in the spring of 1815. Several relatively cold years also occurred after Krakatoa, also in Indonesia, blew its top in 1883. Is the relationship between volcanic eruptions and atmospheric cooling real or merely coincidental?

FIGURE 18.14

Although spectacular, the ejected smoke and ash from the eruption of Mount St. Helens on 18 May 1980 had an apparently negligible impact on hemispheric temperatures. (Photograph courtesy of the U.S. Geological Survey, EROS Data Center)

Until recently, climatologists were concerned primarily with the potential climatic impact of the relatively fine ash particles thrown high into the stratosphere during violent volcanic eruptions. Theoretically, volcanic ash should raise the planetary reflectivity (albedo), and a higher albedo means less solar radiation input and lower air temperatures. Today, however, most climatologists agree that ash particles, being relatively large, quickly settle out of the stratosphere to the ground with little or no long-term effect on the radiation balance. The large-scale climatic impact of violent volcanic eruptions now appears to depend more on the volume of sulfurous gases ejected. Once in the stratosphere, sulfurous gases convert to tiny droplets of sulfuric acid. The extremely small size of the acid droplets (less than 1 micrometer in diameter), coupled with the stratosphere's extreme stability and absence of precipitation, means that the sulfuric acid droplets remain suspended in the stratosphere for periods ranging from many months to several years before finally settling to the earth's surface. While in the stratosphere, the veil of sulfuric acid droplets interacts with solar radiation, so that a portion of the radiation is absorbed, causing stratospheric warming and reducing the amount of radiation available to the troposphere. Furthermore, some solar radiation is scattered by the sulfuric acid droplets back to space, also contributing to cooling at the earth's surface. Over the years, a succession of volcanic eruptions has produced a permanent sulfur layer in the lower stratosphere, and occasionally sulfurous emissions from particularly violent eruptions will further enhance the sulfur layer.

Apparently only those volcanic eruptions rich in sulfurous gases and sufficiently explosive to send ejecta well into the stratosphere can disturb hemispheric or global climate. Even then, surface hemispheric mean temperatures are not likely to be lowered by more than 1 °C (1.8 °F). A volcanic eruption in this category was the 1963 eruption of Agung in Bali, which, according to one estimate, lowered mean global air temperatures by about 0.2 °C (0.4 °F) for a year or two. Although the 1980 eruption of Mount St. Helens produced about as much ash as the Agung explosion, the Mount St. Helens ejecta was low in sulfurous gases and had no detectable influence on large-scale climate.

Although the Mount St. Helens eruption did not influence hemispheric or global climate, some short-term localized effects on surface temperatures did occur immediately downwind of the volcano over eastern Washington, Idaho, and western Montana. Over these areas, the ash plume, clearly visible in the satellite photograph on Plate 18, was sufficiently thick to alter the local radiation balance for 12 to 24 hours. Clifford Mass and Alan Robock of the University of Maryland report that on the day of the eruption, blockage of the sun by volcanic ash lowered temperatures over eastern Washington by up to 8 °C (14.4 °F). That night, over Idaho and western Montana, low-level ash impeded infrared cooling, thereby elevating temperatures by up to 8 °C.

The violent eruption of the Mexican volcano El Chichón on 4 April 1982 appears to have had considerably greater potential than the Mount St. Helens eruption for large-scale climatic impact. In fact, the eruption of El Chichón may turn out to be a more significant perturber of climate than any volcanic eruption of the past two centuries. Although many volcanic eruptions have been more violent and have spewed out much more ejecta, the El Chichón ejecta was unusually rich in sulfurous gases. Conversion of these gases to sulfuric acid droplets and the gradual spreading of the aerosol veil through the stratosphere of the Northern Hemisphere may have influenced climate on a large scale.

Several computerized numerical models of the atmosphere have predicted the effect of the El Chichón veil on surface temperatures. There is general agreement on a potential hemispheric cooling of about 0.5 °C (0.9 °F). Precisely when this cooling took place is, however, a point of disagreement. On the basis of their study of past Northern Hemisphere volcanic eruptions, researchers at Great Britain's University of East Anglia concluded that rapid cooling should have occurred within a few months of the eruption and been followed by a slow temperature recovery over the ensuing two years. Other estimates involve a longer time frame for peak cooling and place peak cooling in late 1983 or 1984 (NASA's Ames Research Center), winter of 1984–1985 (University of Maryland), or mid-1984 (NASA's Goddard Space Flight Center). Even at this point, years after the eruption, we cannot pinpoint the peak cooling period because distinguishing the contribution of the volcanic eruption from other influences on climate is very difficult.

Air Pollution and Climatic Variability

Among the many contemporary environmental issues is the possibility that certain human activities may disrupt the earth-atmosphere radiation balance and thereby contribute to variations in climate. From our study of the fundamentals of weather, we can deduce that human activity may influence climate in at least three ways: (1) by altering radiative and thermal properties of the earth's surface (changing the albedo, for example), (2) by venting waste heat into the atmosphere, and (3) by changing the concentrations of certain key gaseous or aerosol components of the atmosphere. Although many questions remain unanswered, the first two activities seem to be important only on a local scale, that is, in large metropolitan areas. We examined these influences in Chapter 16. In this chapter, we consider the third potential influence on climate: the impact of air pollution on climatic variations at the hemispheric or global scale. We focus specifically on the climatic implications of elevated levels of atmospheric carbon dioxide and aerosols.

The carbon dioxide question

In recent years, atmospheric scientists have expressed concern about the possible climatic ramifications of the steadily rising concentrations of carbon dioxide in the atmosphere. Higher concentrations of carbon dioxide are generally thought to intensify the "greenhouse effect"—that is, to increase absorption and reradiation of infrared radiation and consequently to warm the lower atmosphere. This idea was first proposed nearly 100 years ago by two scientists working independently: Thomas C. Chamberlin, an American geologist, and Svante Arrhenius, a Swedish chemist.

The increased concentration of atmospheric carbon dioxide is probably due to the following: (1) the burning of fossil fuels (coal and oil, particularly) and, to a much lesser extent, the burning of wood, and (2) the clearing of tropical forests, which reduces the photosynthetic removal of carbon dioxide from the atmosphere. Although a rising carbon dioxide level was first reported in 1939 by G. S. Callendar, a British engineer, the increase apparently began during the Industrial Revolution. Callendar suggested that a rising carbon dioxide concentration was responsible for the global warming trend of the previous 60 years.

Systematic measurements of atmospheric carbon dioxide levels were begun in 1958 at the Mauna Loa Observatory in Hawaii and at the South Pole station of the U.S. Antarctic Program. The Mauna Loa record is shown in Figure 18.15. At 3400 m (10,200 ft) above sea

FIGURE 18.15

Upward trend in atmospheric carbon dioxide levels measured at Mauna Loa Observatory, Hawaii. Horizontal lines represent average annual values. Open circles are estimated values. (Data from C. D. Keeling, et al., Scripps Institution of Oceanography, as reported in MacCracken, M. C., and H. Moses. "The First Detection of Carbon Dioxide Effects: Workshop Summary." ***Bulletin of the American Meterorological Society*** **63 (1982):1165.**

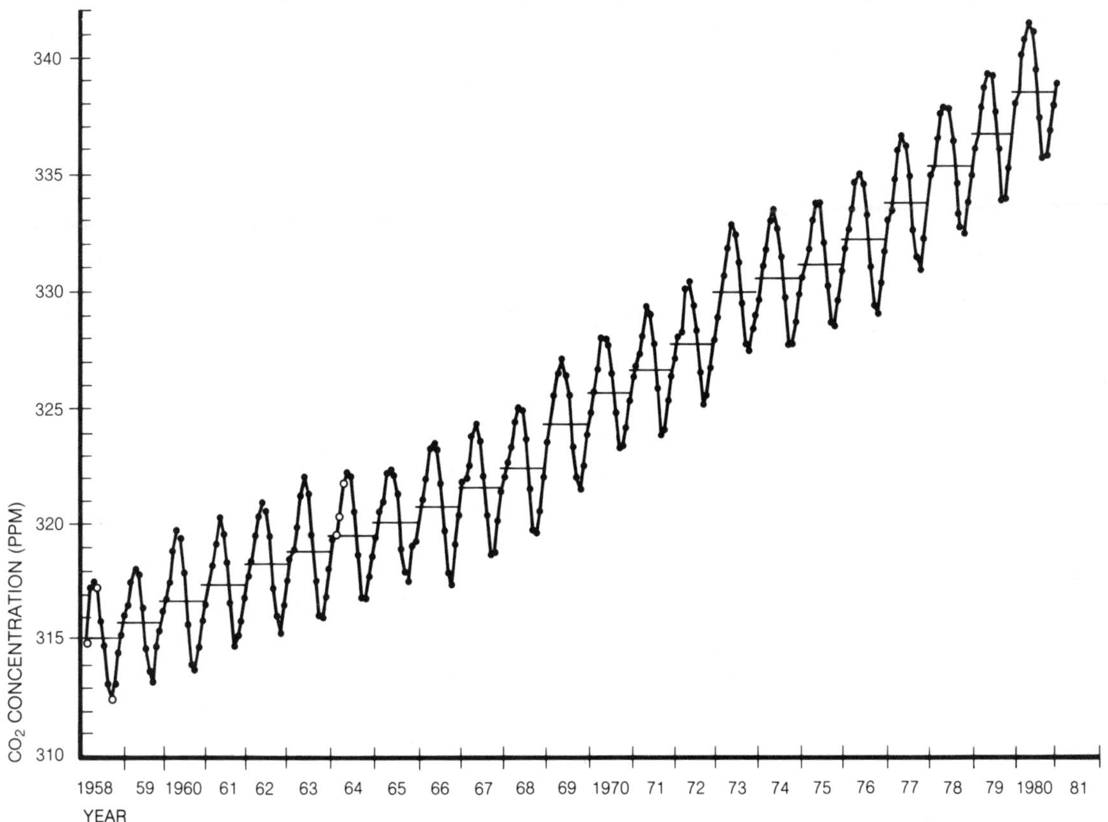

level, the Mauna Loa Observatory is far enough removed from major industrial sources of air pollution that carbon dioxide levels monitored there are considered representative of at least the Northern Hemisphere. The South Pole station is also free from local contamination and its record parallels closely that at Mauna Loa.

Both records show an annual carbon dioxide cycle due to seasonal changes in vegetation; carbon dioxide levels fall during the growing season because of photosynthetic removal. This annual cycle is superimposed on an upward trend that is nearly exponential. In 1958, the rate of carbon dioxide increase was about 0.7 ppm (parts per million) per year. Twenty-three years later the yearly growth rate was 1.8 ppm. This translates into an 8 percent increase in total global carbon dioxide in the atmosphere. Sketchy data suggest that over the past 100 years, atmospheric CO_2 concentration may have climbed by 17 to 26 percent.

What might this exponential rise in atmospheric carbon dioxide mean for the climatic future? Researchers have tried to answer this question by using numerical models of the earth-atmosphere system and high-speed electronic computers. In such experiments, the typical procedure is to test the sensitivity of the model by changing one of the variables—in this case, carbon dioxide concentration. Initially, the model is run to a state of equilibrium using current earth-atmosphere conditions. The model's atmospheric carbon dioxide concentration is then raised, and the numerical model is run to a new equilibrium state. The difference between the initial and final equilibrium states is assumed to be the consequence of elevated carbon dioxide concentration.

In a recent review of these modeling efforts, the National Research Council suggested that a doubling of atmospheric CO_2 levels would elevate the global mean surface temperature by 3 °C ±1.5 °C (5.4 °F ±2.7 °F). Polar warming would be several times this magnitude while the global stratosphere cooled. Because projected warming is greater in the lower troposphere and is less aloft, the troposphere would be less stable than it is today (steeper thermal lapse rates). This could translate into deeper convection and perhaps more intense thunderstorms.

What would be the societal impact of CO_2-induced warming? If CO_2 levels actually double by the middle of the next century, as some studies predict, the consequent global climatic change could be greater than any experienced in the 10,000-year history of civilization. Virtually every sector of society would be affected to some degree, with agriculture likely to be the most seriously disrupted.

Traditional farming practices would have to change. In many regions, warmer weather would also mean drier weather, so drought might be more frequent. More lands would then require irrigation, and subtropical deserts would expand.

Perhaps the most ominous consequence of CO_2-induced warming would be elevated sea level. Amplification of a global warming trend in polar latitudes has prompted much speculation on the fate of the Greenland and Antarctic ice sheets. Some studies suggest that a doubling of global carbon dioxide levels would trigger polar warming sufficient to melt enough ice to raise the ocean to a level that would inundate currently populated coastal areas. However, before residents of Boston and New Orleans quit the city and head for high ground, a note of caution is warranted. Because the atmosphere is a complex and highly interactive system, other processes may compensate for any CO_2-induced warming. For instance, the ocean has a high thermal inertia, which for a time may slow or delay a CO_2-induced warming. Warmer conditions in high latitudes may also mean more snowfall and eventually more, not less, glacial ice!

Some researchers hasten to point out that some benefits may come from CO_2-induced warming. Warmer winters would reduce fuel demand for space heating in middle and high latitudes. F. Kenneth Hare of the University of Toronto reports that Canada could realize a 12 to 18 percent drop in space heating costs should CO_2 levels double. Warmer winters would also lengthen the navigation season on lakes, rivers, and harbors where ice cover is a problem.

Future trends in atmospheric carbon dioxide level hinge on the rate of growth of our energy demand and on the portion of that demand that is met by fossil fuels. However, even conservative estimates of our carbon dioxide output point to climatically significant increases over the next century. In the summer of 1981, researchers at NASA's Goddard Space Flight Center reported on the responses of numerical models of the atmosphere to elevated carbon dioxide levels specified by various energy-use scenarios. They found that a slow growth in energy demand (about a third of the current rate) with equal reliance on both fossil and nonfossil fuels will elevate global mean temperature by about 2.5 °C (4.5 °F) during the twenty-first century.

It is also possible that CO_2-induced warming will be enhanced by rising levels of certain trace atmospheric gases, the concentrations of which are typically expressed in parts per billion (ppb).

These gases include methane (CH_4) at 1600 ppb, nitrous oxide (N_2O) at 300 ppb, and chlorofluorocarbons (CFCs) at less than 1 ppb. The reason for the methane increase is not known, but increases in the other two gases are probably linked to air pollution. The climatic importance of these trace gases lies in their strong absorption of the outgoing infrared radiation in the atmospheric windows. (Recall from Chapter 3 that within atmospheric windows infrared absorption by atmospheric gases is minimal.) In fact, for trace gases absorption is directly proportional to concentration, so doubling the concentration of a gas doubles the absorption. The combined climatic impact of rising levels of these gases could equal the impact of an increase in CO_2 level.

The dust question

In the opinion of some atmospheric scientists, increased dustiness, called **turbidity**, of the atmosphere has the opposite effect of elevated carbon dioxide levels. Instead of warming the earth, an increase in turbidity adds to the reflectivity of the earth-atmosphere system, reducing the amount of solar radiation that reaches the earth's surface and thus cooling the lower atmosphere. Other atmospheric scientists disagree, arguing that increased turbidity resulting from human activity, such as industry and agriculture, promotes warming at the earth's surface. They contend that these aerosols tend to scatter solar radiation back toward the earth's surface, where it is absorbed, and that the larger dust particles absorb and reemit infrared radiation to enhance the "greenhouse effect."

Resolution of this disagreement hinges on a better understanding of the net effect of aerosols on the radiation balance. We know that aerosols absorb and reemit infrared radiation, and absorb and scatter solar radiation. However, the percentage of insolation that is scattered downward toward the earth's surface versus the percentage that is scattered back into space varies with the optical properties of the aerosol. In addition, the albedo of the earth's surface underlying an aerosol layer may play a role in determining a net heating or cooling effect.

The impact of aerosols on climate is thus a complex and unresolved problem. While stratospheric aerosols of volcanic origin probably trigger cooling at the earth's surface, the net impact of aerosols produced by human activity—most of which do not reach the stratosphere—is uncertain.

Earth's Surface and Climatic Variability

In Chapter 3, we saw that the earth's surface, which is mostly water, is the prime absorber of solar radiation. Any change in the characteristics of the earth's land and water surface or in the relative distribution of land and sea may affect the earth's radiation balance and hence the climate.

On land, variations in regional snow cover may trigger important climatic fluctuations. This is because an extensive heavy snow cover has a refrigerating effect on the atmosphere (see Chapter 4). Fresh-fallen snow typically reflects 80 percent or more of the incident solar radiation, thereby substantially reducing the amount of solar heating and lowering the daily maximum temperature. Snow is also an excellent emitter of infrared radiation, so heat is quickly radiated off into space at night—especially on nights when skies are clear. Because of this radiative feedback, a snow cover tends to be self-sustaining. This effect may be further enhanced because storms tend to track along the periphery of a regional snow cover, where horizontal temperature gradients are steep. This places the snow-covered area on the cold, snowy side of migrating midlatitude cyclones—again perpetuating the chill. Radiative feedback triggered by an unusually extensive snow or ice cover clearly favors persistence of an anomalously cold climate.

Although large-scale changes in surface characteristics of the continents may affect climate, changes in ocean characteristics and circulation may be much more important. This follows from our discussion of radiative transfer within the earth-atmosphere system (Chapter 3). Because ocean waters cover nearly three quarters of the global surface and exhibit a very low albedo for solar radiation, the ocean is the principal absorber of insolation. Anything that alters this strong absorption, such as fluctuations in sea ice cover, is likely to affect radiative equilibrium and climate. Because ocean currents contribute to poleward heat transport, changes in current patterns could also trigger major shifts in climate. As noted in Chapter 15, a connection appears to exist between anomalies in sea-surface temperatures and atmospheric long-wave circulation patterns. Shifts in sea-surface temperature anomalies may therefore alter the prevailing circulation patterns and climate. The atmosphere-ocean system involves a two-way interaction, however, and determining which dominates is difficult: the ocean's influence on the atmosphere or the atmosphere's influence on the ocean. Atmospheric impacts on the ocean are relatively short term, however, while the ocean's impacts on the atmosphere are relatively long term due to the great thermal stability of the ocean.

Factor Interaction

Now that we have identified and elaborated on many of the potential causes of climatic variability, a few cautionary words are in order. First, this is not an exhaustive list of all of the possible controls of climatic variability. For example, subtle astronomical shifts may contribute to short-term climatic fluctuations. We have also treated the various climatic controls as if each acted independently of the others. This is clearly not the case. The earth-atmosphere is a highly interactive system in which many factors are linked in complex cause-effect chains, as illustrated in Figure 18.16.

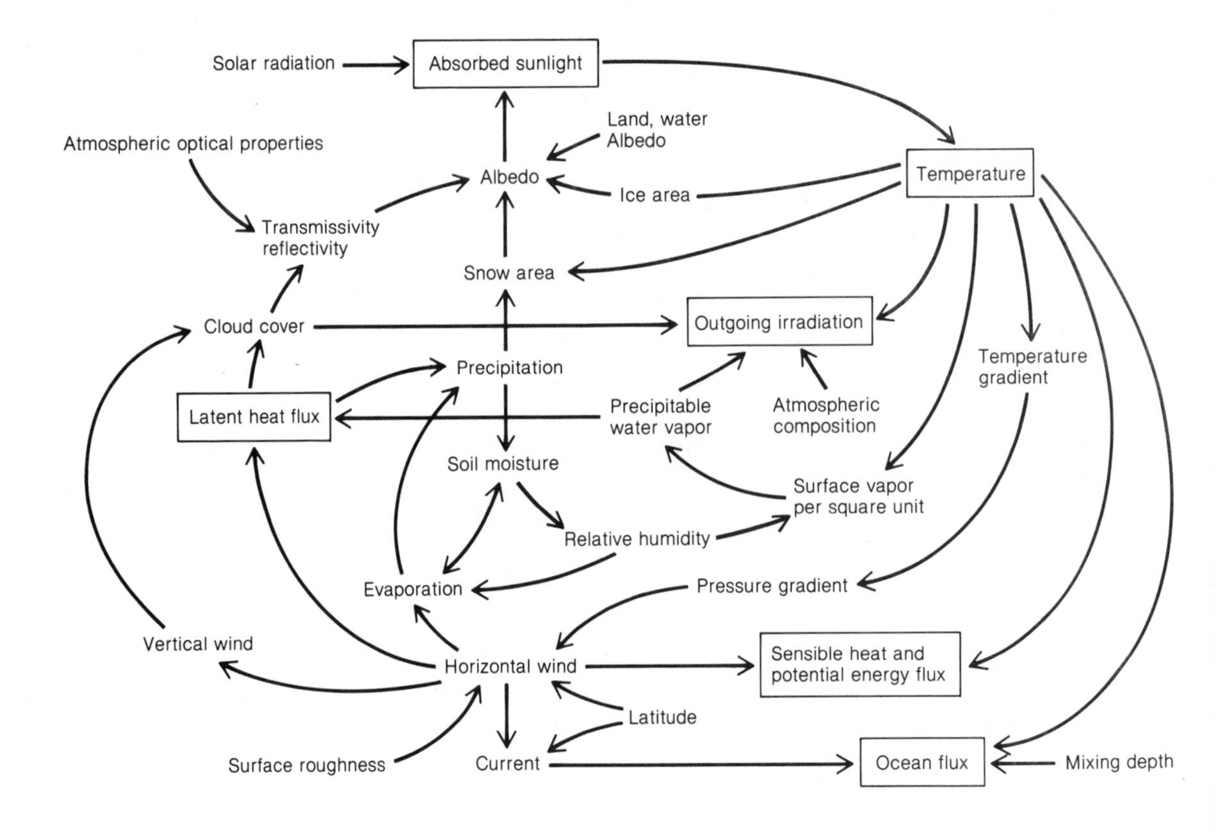

FIGURE 18.16

Climatic variability is influenced by the complex interaction of many processes operating the earth-atmosphere system. (From Kellogg, W. W., and S. H. Schneider. "Climate Stabilization: For Better or for Worse?" *Science* 186 (1974):1164. Copyright 1974 by the AAAS.)

Internal and external forces probably work together to induce climatic shifts.

An illustration of how several factors might interact to shape large-scale variations in climate is provided by James E. Hansen and his colleagues at NASA's Goddard Institute for Space Studies. Using a numerical model of the atmosphere, they predicted how rising carbon dioxide levels would affect mean global temperature from 1880 to 1980. In Figure 18.17A, the predicted temperature is compared with the actual variation in global temperature for the same period. The fit between the predicted and actual temperature variation improves substantially when the influence of volcanic aerosols on temperature is added (Figure 18.17B). The similarity between the two curves becomes even closer when the influence of solar irradiance is entered into the model (Figure 18.17C). Even with these three climatic controls considered together, however, the two curves do not match—suggesting that other influences are affecting global climate variations.

FIGURE 18.17

Using a numerical model of the atmosphere, J. E. Hansen and his colleagues at NASA predicted the response of mean global temperature to (A) rising carbon dioxide level alone, (B) the combined effects of increasing levels of carbon dioxide and volcanic aerosols, and (C) the effects of these factors plus the variability of solar output. In each frame, the predicted temperature (dashed line) is compared with the mean global temperature for the same 1880–1980 period (solid line). (From Revelle, R. "Carbon Dioxide and World Climate." ***Scientific American*** **247, no. 2 (1982):41.**

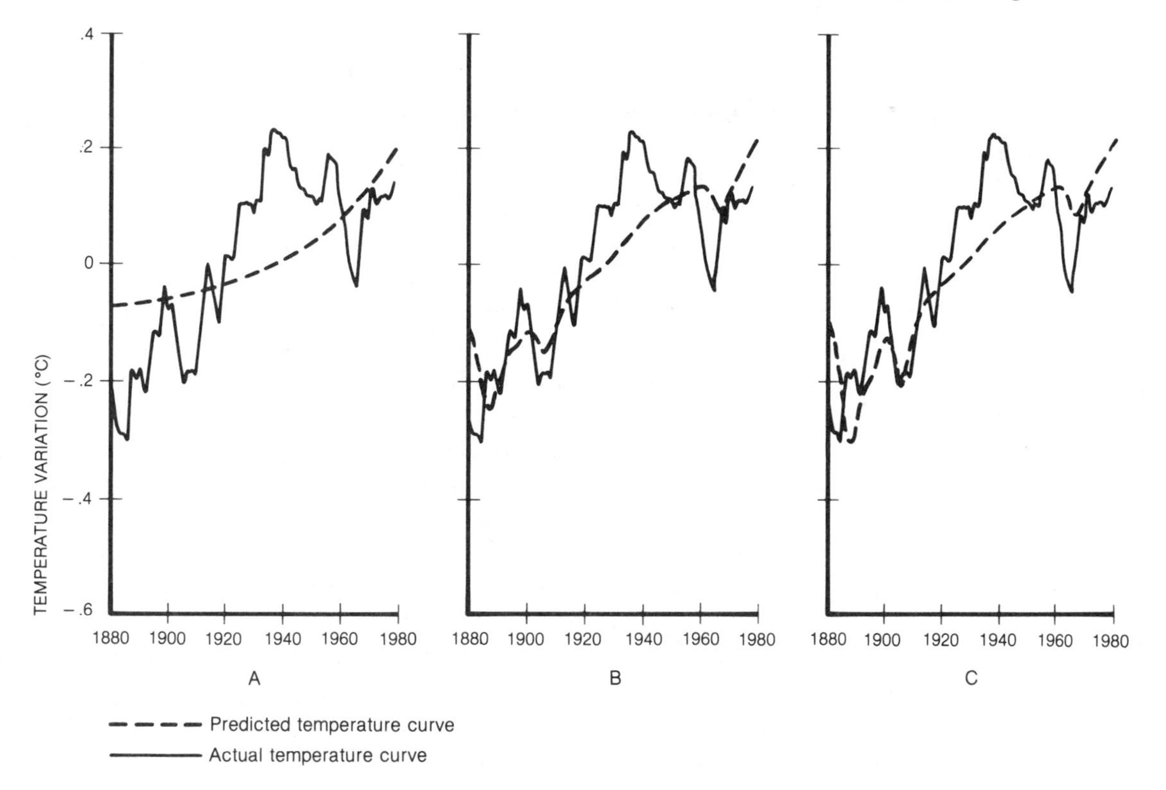

Factor interactions involve feedback loops that may at one extreme amplify (positive feedback) and at the other extreme weaken (negative feedback) climatic fluctuations. Consider an illustration. If rising carbon dioxide levels accentuate the "greenhouse effect," tropospheric temperatures will rise, and the pack ice cover in polar regions will be reduced. Less pack ice, in turn, would lower the albedo of polar seas, further warming the troposphere (positive feedback). On the other hand, increased solar radiation reaching the earth's surface increases evaporation, which results in thicker and more persistent cloud cover, which, in turn, decreases radiation and thereby cools the earth's surface (negative feedback).

In part because of factor interaction, the isolation of simple cause-effect relationships is difficult. Even when a near-perfect match in periodicities occurs between some internal or external forcing factor (such as a volcanic eruption) and a climatic response, there is no guarantee that the relationship is actually one of cause-effect. In the absence of a demonstrated physical linkage, the relationship may well be coincidental.

The Climatic Future

What does the climatic future hold for us? Theoretically, we could take two different approaches in attempting to answer this question. The first approach is dynamic modeling of the atmosphere. We could run a predictive numerical model of the atmosphere as far into the future as we desired, but numerical models of the atmosphere suffer from gaps in monitoring networks and from errors of approximation. Dynamic modeling is therefore reliable only for short-term weather forecasting.

A second approach to climatic forecasting is to analyze the various factors that may have contributed to past climatic fluctuations and to extrapolate their impact into the future. This is an empirical method. Atmospheric scientists have long probed climatic records in search of cycles that might be extrapolated into the future. As noted earlier, however, none of the statistically significant oscillations that appear in the climatic record has much practical value for climate forecasting. Even if we isolate a statistically real cycle of some climatic element, it is likely that the climatic future, like the climatic past, will be shaped by many interacting factors both external and internal to the earth-atmosphere-ocean system.

The possibility that we are headed for another ice age has received considerable attention in recent years. This has prompted publication of numerous popular and semitechnical books and articles espousing the doomsayer notion that northern cities will soon be overtaken and buried by the relentless advance of enormous lobes of glacial ice. Equal billing has been given to the opposing but no less catastrophic scenario in which rising concentrations of carbon dioxide cause global warming that melts polar ice caps and triggers massive flooding of coastal cities. Although the prospect of a huge ice sheet engulfing Montreal or of ocean waters inundating peninsular Florida is exciting material for a novel or screenplay, there is no substantial scientific basis for postulating that either climatic calamity is in the offing—at least not for many hundreds to thousands of years to come.

The climatic future will evidently remain a mystery until we have more accurate numerical models of the atmosphere, denser weather observation networks, and a better understanding of how climatic controls interact. At present, we are confident only that our climate will vary (as it has in the past), but precisely how it will vary is simply not known.

Conclusions

Many factors interact to cause climate to vary on many time scales. Although we can isolate specific climatic controls that are internal and external to the earth-atmosphere-ocean system, our picture of how these controls interact is far from complete. This state of the art severely limits our ability to forecast the climatic future. We should therefore be wary of simplistic scenarios of the climatic future, for not enough is understood about the causes of climatic variations.

Continued research on climate is needed. It is reasonable to assume that variations in climate are governed by physical laws, that is, variations in climate are not arbitrary, random events. Once we comprehend the laws governing climatic variability, we can then develop some degree of predictability. Meanwhile we must continue monitoring trends in climate, especially in view of the strong dependence of our food supply and energy demands on climate. In view of the uncertainties, some experts argue that we should plan for the worst possible future climate scenarios. What is your opinion?

MATHEMATICAL NOTE

Radiative equilibrium and climatic change

We can express the condition of radiative equilibrium within the earth-atmosphere system in a mathematical expression that equates energy input with energy output. Then, by perturbing the variables in the equation, we can simulate climatic change.

Radiational energy incident on the earth-atmosphere system is given by the solar constant, S. A portion of this energy is reflected or scattered back into space, the amount depending on the planetary albedo, α. The energy actually available to drive the atmosphere is therefore

$$S(1 - \alpha)$$

The earth intercepts this energy as a disc having an area of πR^2, where R is the radius of the earth. The net energy input is then given by:

$$(\pi R^2)S(1 - \alpha)$$

Energy is emitted by the earth-atmosphere system in the form of infrared radiation. If we assume the earth-atmosphere system to be a perfect radiator, then we can describe the energy output by the *Stefan-Boltzmann law* (see the Mathematical Note at the end of Chapter 3). This law states that energy output is proportional to the radiating temperature, T (K), raised to the fourth power, T^4. In this relationship, the constant of proportionality is σ, the Stefan-Boltzmann constant. However, the earth-atmosphere system is not a perfect radiator,* so we must introduce a correction factor to account for its actual radiating properties. The correction factor, ϵ, is called the *effective emissivity*. The energy output of the earth-atmosphere system is therefore

$$\epsilon\sigma T^4$$

This energy is emitted from the entire surface area ($4\pi R^2$) of the spherical earth. The total energy output is thus given by:

$$(4\pi R^2)\ \epsilon\sigma T^4$$

At radiative equilibrium, what comes in must equal what goes out, so

$$(\pi R^2)S(1 - \alpha) = (4\pi R^2)\epsilon\sigma T^4$$

This simplifies to:

$$S(1 - \alpha) = 4\epsilon\sigma T^4$$

Solving for T, the temperature of the earth-atmosphere system at radiative equilibrium, we have:

$$T = [S(1 - \alpha)/4\epsilon\sigma]^{1/4}$$

The radiative equilibrium temperature thus depends on (1) the solar constant, (2) the planetary albedo, and (3) the effective emissivity. A change in any one or any combination of these variables will change the value of T and hence the earth's climate. Consider some illustrations.

First, keep α and ϵ constant. By simple mathematical manipulation, we can show that a 1 percent change in the solar constant translates into a 0.6 °C (1.1 °F) change in radiative equilibrium temperature. This does not appear to be a major temperature change, but recall that climatic trends are geographically nonuniform in magnitude (and direction). In some localities, therefore, this temperature change will be amplified considerably. If we keep S and ϵ constant and increase the planetary albedo from its present value of 30 percent up to 35 percent (perhaps by increasing the ocean's ice cover), the radiative equilibrium temperature will drop by about 4.5 °C (8.1 °F).

As a third illustration, we could hold S and α constant and vary the effective emissivity. As noted earlier, the effective emissivity depends on the radiative properties of the earth-atmosphere system, so any change in the chemistry of the earth-atmosphere system could alter ϵ. Such chemical changes might include alterations in levels of carbon dioxide or aerosols.

*Recall that the Stefan-Boltzmann law applies to a black body, a perfect radiator.

SUMMARY STATEMENTS

▪ Because of the rapid deterioration of climatic detail as we go back in time, the most reasonable view of ancient climates is in terms of average climates prevailing over long periods of time.

▪ Continental drift and mountain building complicate our investigation of the earth's climatic record over the hundreds of millions of years that constitute geologic time.

▪ The climatic record becomes more detailed and reliable as we approach the present, especially since the beginning of instrument-based records.

▪ The people most vulnerable to climatic fluctuations are those living in areas where the climate is marginal for survival.

▪ The rate of climatic shift depends on our time frame of reference, but climate can change so rapidly that humankind may be unable to adjust.

▪ Few statistically significant cycles have been identified in the climatic record, and none of these has much practical value for weather or climate forecasting.

▪ Climatic variability may be explained by forces both internal and external to the earth-atmosphere-ocean system that affect the global radiation balance.

▪ One approach to explaining climatic variability is to match an appropriate cause with a specific climatic oscillation on the basis of the period of that oscillation. Such a match is, however, no guarantee of a real physical relationship.

▪ Measurements by instruments aboard the SMM satellite indicate that the solar constant is not constant.

▪ The exact linkage between sunspots and climate variability is not known, although correspondences may exist between prolonged sunspot minima and relatively cool periods, and between the "double" sunspot cycle and drought on the western High Plains.

▪ Regular long-term changes in earth-sun geometry may explain large-scale glacial-interglacial climatic shifts as recorded in deep-sea sediment cores.

▪ The potential impact of violent volcanic eruptions on large-scale climate is negligible unless the eruptive cloud is relatively rich in sulfurous gases. In the stratosphere, these gases are converted to sulfuric acid droplets that increase the planetary albedo.

▪ Fossil fuel combustion is raising the atmospheric carbon dioxide levels. Uncompensated, this trend will likely intensify the "greenhouse effect."

▪ The potential climatic impact of elevated dust loads in the troposphere is not known.

▪ Changes in the radiative characteristics of the earth's land and water surface or in the relative distribution of land and sea may influence the global radiation balance and hence the climate.

▪ Factor interactions involve feedback loops that might at one extreme amplify and at the other extreme weaken climatic oscillations.

▪ With regard to the climatic future, atmospheric scientists are confident only that the climate will vary (as it has in the past), but precisely how it will vary is simply not known.

KEY WORDS

geologic time
continental drift
glacial climate
interglacial climate
altithermal episode
Little Ice Age
solar irradiance
faculae
sunspots
umbra
penumbra
photosphere
Maunder minimum
Spörer minimum
turbidity

REVIEW QUESTIONS

1 How do we know that climate is inherently variable?

2 What happens to the reliability of the climatic record as we go back in time? Explain your response.

3 How might continental drift influence the climate of some locality over millions of years?

4 What is the difference between a glacial climate and an interglacial climate?

5 Characterize the altithermal episode and the Little Ice Age.

6 What considerations have led some scientists to question the validity of global and hemispheric temperature trends compiled from measurements made over the past 100 years?

7 Provide some examples of the societal impact of climatic variability.

8 The people most vulnerable to climatic change are those living in areas where the climate is just marginal for survival. What is meant by this statement? Give some illustrations.

9 Are there cycles in the climatic record that are useful for forecasting the climatic future?

10 What is meant by the globe's radiative equilibrium?

11 What is a sunspot? Describe its structure.

12 Is the sunspot record really periodic in a rigorous statistical sense? Explain why or why not.

13 What is the Maunder minimum and what might have been its influence on climate?

14 How does the Milankovitch theory seek to explain the major climatic shifts of the Ice Age?

15 How might violent volcanic eruptions influence global climate?

16 Identify those climatic forcing phenomena that might contribute to short-term variations in climate, that is, variations over years and decades.

17 Evaluate humankind's contribution thus far to climatic variability. How might this change in the future?

18 What human activities are contributing to the upward trend in atmospheric carbon dioxide concentration?

19 How is the ocean involved in the controversy now surrounding the potential climatic impacts of rising carbon dioxide levels?

20 Explain why aerosols injected into the stratosphere may have a more prolonged impact on climate than aerosols that are confined to the troposphere.

POINTS TO PONDER

1 A high statistical correlation between two factors does not necessarily indicate a cause-effect relationship. Give an example of how this statement applies to the search for the causes of climatic variability.

2 If the Milankovitch theory is correct, then the next ice age will peak about 23,000 years from now. Does this imply that each successive year from now on will get progressively colder? Explain your response.

3 How does the geographic nonuniformity of climatic trends complicate the search for causes of climatic variability?

4 In general terms, describe how sensitivity to climatic change varies geographically. Speculate on why polar regions are most sensitive to large-scale climatic trends.

5 In attempting to explain climatic variability, we cannot focus simply on changes that occur in the atmosphere alone. We must also consider changes in the oceans and in the earth's land surface characteristics. Explain.

6 How might ocean currents play an important role in climatic variability?

7 Are climatic shifts abrupt or gradual?

8 Is the solar constant really constant?

9 Speculate on the nature of the climatic future.

PROJECTS

1 Is there any indication that recent climatic trends have had any impact on your region's economy?

2 Speculate on the possible economic impact in your area of future climatic (a) cooling and (b) warming.

SELECTED READINGS

Brooks, C. E. P. *Climate Through the Ages.* New York: Dover Publications, 1970. 395 pp. *A classic treatment of the earth's climatic history, first published in 1926 and revised in 1949.*

Bryson, R. A., and T. J. Murray. *Climates of Hunger.* Madison, Wis.: The University of Wisconsin Press, 1977. 171 pp. *How past climatic variations have affected people living in agriculturally marginal regions.*

Harrison, M. R. "The Media and Public Perceptions of Climatic Change." *Bulletin of the American Meteorological Society* 63 (1982): 730–738. *A survey of 422 residents of southern Ontario indicated that two thirds believed that Canadian climate is changing.*

Le Roy Ladurie, E. *Times of Feast, Times of Famine: A History of Climate Since the Year 1000.* Garden City, N.J., and New York: Doubleday, 1971. 426 pp. *A fascinating climatic reconstruction by a French economic historian based primarily on documentary sources.*

Mitchell, J. M. "El Chichón: Weather-Maker of the Century?" *Weatherwise* 35 (1982):252–259. *Describes the eruption of El Chichón and possible climatic consequences.*

National Research Council, CO_2/Climate Review Panel. *Carbon Dioxide and Climate: A Second Assessment.* Washington, D.C.: National Academy Press, 1982. 72 pp. *A summary report concluding that a doubling of atmospheric carbon dioxide could raise global mean temperature by 3 °C ±1.5 °C.*

Radok, U. "The Antarctic Ice." *Scientific American* 253, no. 2 (1985):98–105. *Includes a discussion of the reconstruction of Antarctica's climatic history.*

Rampino, M. R., and S. Self. "The Atmospheric Effects of El Chichón." *Scientific American* 250, no. 1 (1984):48–57. *Reviews the atmospheric effects of this sulfur-rich 1982 eruption.*

Revelle, R. "Carbon Dioxide and World Climate." *Scientific American* 247, no. 2 (1982):35–43. *An excellent overview of the carbon dioxide problem and its potential implications.*

Rotberg, R. I., and T. K. Rabb. *Climate and History.* Princeton, N.J.: Princeton University Press, 1981. 280 pp. *A collection of informative articles primarily on the methodology of climatic reconstruction.*

Sofia, S., P. Demarque, and A. Endal. "From Solar Dynamo to Terrestrial Climate." *American Scientist* 73, no. 4 (1985):326–333. *Describes how variations in solar irradiance are monitored.*

Winstanley, D. "Africa in Drought: A Change in Climate?" *Weatherwise* 38, no. 2 (1985):74–81. *A discussion of recurring droughts in Africa and the possible linkage to long-term climatic shifts.*

APPENDIX I:

Conversion Factors

In an effort to facilitate technical communication, the scientific community is adopting worldwide the standard International System of Units (SI) of weights and measures. SI units are preferred metric units for various physical quantities. For example, the *newton* is the SI unit of force, and the *pascal* is the SI unit of pressure. Canada has almost completed its transition to SI units, but the United States is only gradually adopting the new system. English units and a variety of non-SI metric units therefore still appear in U.S. meteorological literature and data compilations.

Selection of units in this textbook reflects a stage that is transitional to total SI usage. Physical quantities are presented in metric units with the English equivalent in parentheses. However, the metric unit is not always the standard SI unit and was chosen because of customary usage. For example, we use *calorie* as the metric unit of heat, whereas the corresponding SI unit is the *joule*.

The following table shows how to convert between metric units and English units.

	MULTIPLY	BY	TO OBTAIN
LENGTH	**inches**	**2.54**	**centimeters**
	centimeters	**0.3937**	**inches**
	feet	**0.3048**	**meters**
	meters	**3.281**	**feet**
	statute miles	**1.6093**	**kilometers**
	kilometers	**0.6214**	**statute miles**
	kilometers	**0.54**	**nautical miles**
	nautical miles	**1.852**	**kilometers**
	kilometers	**3281**	**feet**
	feet	**0.0003048**	**kilometers**
WEIGHTS AND MASS	**ounces**	**28.350**	**grams**
	grams	**0.0353**	**ounces**
	pounds	**0.4536**	**kilograms**
	kilograms	**2.205**	**pounds**
	tons	**0.9072**	**metric tons**
	metric tons	**1.102**	**tons**
LIQUID MEASURE	**fluid ounces**	**0.0296**	**liters**
	gallons	**3.7854**	**liters**
	liters	**0.2642**	**gallons**
	liters	**33.8140**	**fluid ounces**
AREA	**acres**	**0.4047**	**hectares**
	square yards	**0.8361**	**square meters**
	square miles	**2.590**	**square kilometers**
	hectares	**0.010**	**square kilometers**
	hectares	**2.471**	**acres**

	MULTIPLY	BY	TO OBTAIN
PRESSURE	**pounds force**	**4.448**	**newtons**
	newtons	**0.225**	**pounds**
	millimeters mercury	**133.32**	**pascals (newtons per square meter)**
	pounds per square inch	**6.895**	**kilopascals (1000 pascals)**
	pascals	**0.0075**	**millimeters mercury at 0 °C**
	kilopascals	**0.1450**	**pounds per square inch**
	bars	**1000**	**millibars**
	bars	**100000**	**pascals**
	bars	**0.9869**	**atmospheres**
ENERGY	**joules**	**0.2389**	**calories**
	kilocalories	**1000**	**calories**
	joules	**1.0**	**watt-seconds**
	kilojoules	**1000**	**joules**
	calories	**0.00397**	**Btus**
	Btus	**252.0**	**calories**
POWER	**joules per second**	**1.0**	**watts**
	kilowatts	**1000**	**watts**
	megawatts	**1000**	**kilowatts**
	kilocalories per minute	**69.93**	**watts**
	watts	**0.0013**	**horsepower**
	kilowatts	**5.7**	**Btus per minute**
	Btus per minute	**0.235**	**horsepower**
TEMPERATURE	**degrees Fahrenheit plus 459.67**	**0.5555**	**kelvins**
	degrees Celsius plus 273.15	**1.0**	**kelvins**
	kelvins	**1.80**	**degrees Fahrenheit minus 459.67**
	kelvins	**1.0**	**degrees Celsius minus 273.15**
	degrees Fahrenheit minus 32	**0.5555**	**degrees Celsius**
	degrees Celsius	**1.80**	**degrees Fahrenheit plus 32**

In discussions in this book, it is important to understand the distinction between *energy* and *power*, as well as the corresponding metric and English units.

Energy in its various forms (for example, heat, chemical energy, radiant energy) represents the ability to perform work. Energy is therefore expressed in units of *work*, which is defined as a force applied over some distance. In the metric system, units of energy are the joule and the erg. By definition, 1 *joule* is the energy required when a unit force, called a *newton*,* acts through a distance of 1 meter. Alternately, 1 *erg* is the energy required when a unit force, called a *dyne*,† acts through a distance of 1 centimeter. One joule is the equivalent of 10^7 ergs, and 1 calorie of heat (defined in Chapter 2) equals 4.186 joules.

Power is the rate at which energy is used, that is, released or converted. One *watt* is defined as 1 joule per second, and 1 *kilowatt* is 1000 joules per second. One *megawatt* equals 1000 *kilowatts*.

When solar radiation is absorbed by something (air, water, or land, for example), radiant energy is converted to heat. We can describe the amount of heat in terms of calories or joules. We may also be concerned about the rate at which solar radiation travels through some cross-sectional area, or the rate at which it is absorbed by some surface. The flux of energy is then described in terms of watts per square meter, calories per square centimeter per minute, joules per square meter per minute, or langleys per minute. One *langley* is defined as 1 calorie per square centimeter.

*One *newton* is the force that accelerates a 1-kilogram mass by 1 meter per second per second.

†One *dyne* is the force that accelerates a 1-gram mass by 1 centimeter per second per second.

APPENDIX II:

Psychrometric Tables

Recall from Chapter 6 that a psychrometer is a standard instrument for measuring how close the air is to saturation. It consists of two thermometers mounted side-by-side: a dry-bulb thermometer and a wet-bulb thermometer. The dry bulb gives the actual air temperature while the wet bulb gives the temperature produced by evaporative cooling. The difference between the dry-bulb and wet-bulb readings is the wet-bulb depression.

The relative humidity and the dew point can be obtained from the dry-bulb temperature and the wet-bulb depression. Use Table A to obtain the relative humidity: find the dry-bulb temperature in the left-hand column and then read across to the relative humidity that corresponds to the wet-bulb depression. Follow the same procedure to determine the dew point in Table B.

TABLE A **Relative Humidity (Percent)**

DRY-BULB TEMP. (°C)	WET-BULB DEPRESSION (°C)														
	0.5	**1.0**	**1.5**	**2.0**	**2.5**	**3.0**	**3.5**	**4.0**	**4.5**	**5.0**	**7.5**	**10.0**	**12.5**	**15.0**	**17.5**
−10.0	85	69	54	39	24	10	—	—	—	—	—	—	—	—	—
− 7.5	87	73	60	48	35	22	10	—	—	—	—	—	—	—	—
− 5.0	88	77	66	54	43	32	21	11	0	—	—	—	—	—	—
− 2.5	90	80	70	60	50	41	31	22	12	3	—	—	—	—	—
0.0	91	82	73	65	56	47	39	31	23	15	—	—	—	—	—
2.5	92	84	76	68	61	53	46	38	31	24	—	—	—	—	—
5.0	93	86	78	71	65	58	51	45	38	32	1	—	—	—	—
7.5	93	87	80	74	68	62	56	50	44	38	11	—	—	—	—
10.0	94	88	82	76	71	65	60	54	49	44	19	—	—	—	—
12.5	94	89	84	78	73	68	63	58	53	48	25	4	—	—	—
15.0	95	90	85	80	75	70	66	61	57	52	31	12	—	—	—
17.5	95	90	86	81	77	72	68	64	60	55	36	18	2	—	—
20.0	95	91	87	82	78	74	70	66	62	58	40	24	8	—	—
22.5	96	92	87	83	80	76	72	68	64	61	44	28	14	1	—
25.0	96	92	88	84	81	77	73	70	66	63	47	32	19	7	—
27.5	96	92	89	85	82	78	75	71	68	65	50	36	23	12	1
30.0	96	93	89	86	82	79	76	73	70	67	52	39	27	16	6
32.5	97	93	90	86	83	80	77	74	71	68	54	42	30	20	11
35.0	97	93	90	87	84	81	78	75	72	69	56	44	33	23	14
37.5	97	94	91	87	85	82	79	76	73	70	58	46	36	26	18
40.0	97	94	91	88	85	82	79	77	74	72	59	48	38	29	21

TABLE B **Dew-Point Temperature (°C)**

DRY-BULB TEMP. (°C)	WET-BULB DEPRESSION (°C)														
	0.5	**1.0**	**1.5**	**2.0**	**2.5**	**3.0**	**3.5**	**4.0**	**4.5**	**5.0**	**7.5**	**10.0**	**12.5**	**15.0**	**17.5**
−10.0	−12.1	−14.5	−17.5	−21.3	−26.6	−36.3	—	—	—	—	—	—	—	—	—
− 7.5	− 9.3	−11.4	−13.8	−16.7	−20.4	−25.5	−34.4	—	—	—	—	—	—	—	—
− 5.0	− 6.6	− 8.4	−10.4	−12.8	−15.6	−19.0	−23.7	−31.3	−78.6	—	—	—	—	—	—
− 2.5	− 3.9	− 5.5	− 7.3	− 9.2	−11.4	−14.1	−17.3	−21.5	−27.7	−41.3	—	—	—	—	—
0.0	− 1.3	− 2.7	− 4.2	− 5.9	− 7.7	− 9.8	−12.3	−15.2	−18.9	−23.9	—	—	—	—	—
2.5	1.3	0.1	− 1.3	− 2.7	− 4.3	− 6.1	− 8.0	−10.3	−12.9	−16.1	—	—	—	—	—
5.0	3.9	2.8	1.6	0.3	− 1.1	− 2.6	− 4.2	− 6.1	− 8.1	−10.4	−47.7	—	—	—	—
7.5	6.5	5.5	4.4	3.2	2.0	0.7	− 0.8	− 2.3	− 4.0	− 5.8	−21.6	—	—	—	—
10.0	9.1	8.1	7.1	6.0	4.9	3.8	2.5	1.2	− 0.2	− 1.8	−12.8	—	—	—	—
12.5	11.6	10.7	9.8	8.8	7.8	6.7	5.6	4.5	3.2	1.9	− 6.8	−28.2	—	—	—
15.0	14.2	13.3	12.5	11.6	10.6	9.6	8.6	7.6	6.5	5.3	− 1.9	−14.5	—	—	—
17.5	16.7	15.9	15.1	14.3	13.4	12.5	11.5	10.6	9.6	8.5	2.3	− 7.0	−35.1	—	—
20.0	19.3	18.5	17.7	16.9	16.1	15.3	14.4	13.5	12.6	11.6	6.1	− 1.4	−14.9	—	—
22.5	21.8	21.1	20.3	19.6	18.8	18.0	17.2	16.3	15.5	14.6	9.6	3.2	− 6.3	−37.5	—
25.0	24.3	23.6	22.9	22.2	21.4	20.7	19.9	19.1	18.3	17.5	12.9	7.3	− 0.2	−13.7	—
27.5	26.8	26.2	25.5	24.8	24.1	23.3	22.6	21.9	21.1	20.3	16.1	11.1	4.7	− 4.7	−31.7
30.0	29.4	28.7	28.0	27.4	26.7	26.0	25.3	24.6	23.8	23.1	19.1	14.5	9.0	1.6	−11.1
32.5	31.9	31.2	30.6	29.9	29.3	28.6	27.9	27.2	26.5	25.8	22.1	17.8	12.8	6.6	− 2.4
35.0	34.4	33.8	33.1	32.5	31.9	31.2	30.6	29.9	29.2	28.5	24.9	21.0	16.4	11.0	3.9
37.5	36.9	36.3	35.7	35.1	34.4	33.8	33.2	32.5	31.9	31.2	27.7	24.0	19.8	14.9	8.9
40.0	39.4	38.8	38.2	37.6	37.0	36.4	35.8	35.1	34.5	33.9	30.5	26.9	23.0	18.5	13.3

APPENDIX III:

Weather Map Symbols

By international agreement, a standard set of symbols is plotted on weather maps to represent weather conditions. This standardization facilitates the international exchange of weather information. Presented here is an abridged listing of weather symbols.

Air Pressure Tendency

Rising, then falling; same as or higher than 3 hours ago

Barometric pressure now higher than 3 hours ago

Rising, then steady; or rising, then rising more slowly

Rising steadily, or unsteadily

Falling or steady, then rising; or rising, then rising more rapidly

Steady; same as 3 hours ago

Falling, then rising; same as or lower than 3 hours ago

Barometric pressure now lower than 3 hours ago

Falling, then steady; or falling, then falling more slowly

Falling steadily, or unsteadily

Steady or rising, then falling; or falling, then falling more rapidly

Cloud Abbreviations

St—Stratus

Fra—Fractus

Sc—Stratocumulus

Cu—Cumulus

Cb—Cumulonimbus

Ac—Altocumulus

Ns—Nimbostratus

As—Altostratus

Ci—Cirrus

Cs—Cirrostratus

Cc—Cirrocumulus

Cloud Types

Cu of fair weather, little vertical development and seemingly flattened

Cu of considerable development, generally towering, with or without other Cu or Sc bases all at same level

Cb with tops lacking clear-cut outlines, but distinctly not cirriform or anvil shaped; with or without Cu, Sc, or St

Sc formed by spreading out of Cu; Cu often present also

Sc not formed by spreading out of Cu

St or StFra, but no StFra of bad weather

StFra and/or CuFra of bad weather (scud)

Cu and Sc (not formed by spreading out of Cu) with bases at different levels

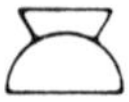

Cb having a clearly fibrous (cirriform) top, often anvil shaped, with or without Cu, Sc, St, or scud

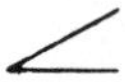

Thin As (most of cloud layer semitransparent)

Thick As, greater part sufficiently dense to hide sun (or moon), or Ns

Thin Ac, mostly semitransparent; cloud elements not changing much and at a single level

Thin Ac in patches; cloud elements continually changing and/or occurring at more than one level

Thin Ac in bands or in a layer gradually spreading over sky and usually thickening as a whole

Ac formed by the spreading out of Cu or Cb

Double-layered Ac, or a thick layer of Ac, not increasing; or Ac with As and/or Ns

Ac in the form of Cu-shaped tufts or Ac with turrets

Ac of a chaotic sky, usually at different levels; patches of dense Ci usually present also

Filaments of Ci, or "mares tails," scattered and not increasing

Dense Ci in patches or twisted sheaves, usually not increasing, sometimes like remains of Cb; or towers or tufts

Dense Ci, often anvil shaped, derived from or associated with Cb

Ci, often hook shaped, gradually spreading over the sky and usually thickening as a whole

Ci and Cs, often in converging bands, or Cs alone; generally overspreading and growing denser; the continuous layer not reaching 45° altitude

Ci and Cs, often in converging bands, or Cs alone; generally overspreading and growing denser; the continuous layer exceeding 45° altitude

Veil of Cs covering the entire sky

Cs not increasing and not covering entire sky

Cc alone or Cc with some Ci or Cs, but the Cc being the main cirriform cloud

Cloud Cover

No clouds

One-tenth or less

Two-tenths or three-tenths

Four-tenths

Five-tenths

Six-tenths

Seven-tenths or eight-tenths

Nine-tenths or overcast with openings

Completely overcast (ten-tenths)

Sky obscured

Wind Speed

Knots	*Miles per hour*	*Kilometers per hour*
Calm	Calm	Calm
1–2	1–2	1–3
3–7	3–8	4–13
8–12	9–14	14–19
13–17	15–20	20–32
18–22	21–25	33–40
23–27	26–31	41–50
28–32	32–37	51–60
33–37	38–43	61–69
38–42	44–49	70–79
43–47	50–54	80–87
48–52	55–60	88–96
53–57	61–66	97–106
58–62	67–71	107–114
63–67	72–77	115–124
68–72	78–83	125–134
73–77	84–89	135–143
103–107	119–123	192–198

Fronts

Fronts are shown on surface weather maps by the symbols below. (Arrows—not shown on maps—indicate direction of motion of front.)

Cold front (surface)

Warm front (surface)

Occluded front (surface)

Stationary front (surface)

Warm front (aloft)

Cold front (aloft)

Cloud development NOT observed or NOT observable during past hour	Clouds generally dissolving or becoming less developed during past hour	State of sky on the whole unchanged during past hour	Clouds generally forming or developing during past hour	Visibility reduced by smoke
Light fog (mist)	Patches of shallow fog at station, NOT deeper than 6 feet on land	More or less continuous shallow fog at station, NOT deeper than 6 feet on land	Lightning visible, no thunder heard	Precipitation within sight, but NOT reaching the ground
Drizzle (NOT freezing) or snow grains (NOT falling as showers) during past hour, but NOT at time of observation	Rain (NOT freezing and NOT falling as showers) during past hour, but NOT at time of observation	Snow (NOT falling as showers) during past hour, but NOT at time of observation	Rain and snow or ice pellets (NOT falling as showers) during past hour, but NOT at time of observation	Freezing drizzle or freezing rain (NOT falling as showers) during past hour, but NOT at time of observation
Slight or moderate dust storm or sandstorm, has decreased during past hour	Slight or moderate dust storm or sandstorm, no appreciable change during past hour	Slight or moderate dust storm or sandstorm has begun or increased during past hour	Severe dust storm or sandstorm, has decreased during past hour	Severe dust storm or sandstorm, no appreciable change during past hour
Fog or ice fog at distance at time of observation, but NOT at station during past hour	Fog or ice fog in patches	Fog or ice fog, sky discernible, has become thinner during past hour	Fog or ice fog, sky NOT discernible, has become thinner during past hour	Fog or ice fog, sky discernible, no appreciable change during past hour
Intermittent drizzle (NOT freezing), slight at time of observation	Continuous drizzle (NOT freezing), slight at time of observation	Intermittent drizzle (NOT freezing), moderate at time of observation	Continuous drizzle (NOT freezing), moderate at time of observation	Intermittent drizzle (NOT freezing), heavy at time of observation
Intermittent rain (NOT freezing), slight at time of observation	Continuous rain (NOT freezing), slight at time of observation	Intermittent rain (NOT freezing), moderate at time of observation	Continuous rain (NOT freezing), moderate at time of observation	Intermittent rain (NOT freezing), heavy at time of observation
Intermittent fall of snowflakes, slight at time of observation	Continuous fall of snowflakes, slight at time of observation	Intermittent fall of snowflakes, moderate at time of observation	Continuous fall of snowflakes, moderate at time of observation	Intermittent fall of snowflakes, heavy at time of observation
Slight rain shower(s)	Moderate or heavy rain shower(s)	Violent rain shower(s)	Slight shower(s) of rain and snow mixed	Moderate or heavy shower(s) of rain and snow mixed
 Moderate or heavy shower(s) of hail, with or without rain, or rain and snow mixed, not associated with thunder	 Slight rain at time of observation; thunderstorm during past hour, but NOT at time of observation	 Moderate or heavy rain at time of observation; thunderstorm during past hour, but NOT at time of observation	 Slight snow, or rain and snow mixed, or hail at time of observation; thunderstorm during past hour, but NOT at time of observation	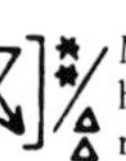 Moderate or heavy snow, or rain and snow mixed, or hail at time of observation; thunderstorm during past hour, but NOT at time of observation

Haze

Widespread dust in suspension in the air, NOT raised by wind, at time of observation

Dust or sand raised by wind at time of observation

Well-developed dust whirl(s) within past hour

Dust storm or sandstorm within sight of or at station during past hour

Precipitation within sight, reaching the ground but distant from station

Precipitation within sight, reaching the ground, near to but NOT at station

Thunderstorm, but no precipitation at the station

Squall(s) within sight during past hour or at time of observation

Funnel cloud(s) within sight of station at time of observation

Showers of rain during past hour, but NOT at time of observation

Showers of snow, or of rain and snow, during past hour, but NOT at time of observation

Showers of hail, or of hail and rain, during past hour, but NOT at time of observation

Fog during past hour, but NOT at time of observation

Thunderstorm (with or without precipitation) during past hour, but NOT at time of observation

Severe dust storm or sandstorm has begun or increased during past hour

Slight or moderate drifting snow, generally low (less than 6 ft)

Heavy drifting snow, generally low

Slight or moderate blowing snow, generally high (more than 6 ft)

Heavy blowing snow, generally high

Fog or ice fog, sky NOT discernible, no appreciable change during past hour

Fog or ice fog, sky discernible, has begun or become thicker during past hour

Fog or ice fog, sky NOT discernible, has begun or become thicker during past hour

Fog depositing rime, sky discernible

Fog depositing rime, sky NOT discernible

Continuous drizzle (NOT freezing), heavy at time of observation

Slight freezing drizzle

Moderate or heavy freezing drizzle

Drizzle and rain, slight

Drizzle and rain, moderate or heavy

Continuous rain (NOT freezing), heavy at time of observation

Slight freezing rain

Moderate or heavy freezing rain

Rain or drizzle and snow, slight

Rain or drizzle and snow, moderate or heavy

Continuous fall of snowflakes, heavy at time of observation

Ice prisms (with or without fog)

Snow grains (with or without fog)

Isolated starlike snow crystals (with or without fog)

Ice pellets or snow pellets

Slight snow shower(s)

Moderate or heavy snow shower(s)

Slight shower(s) of snow pellets, or ice pellets with or without rain, or rain and snow mixed

Moderate or heavy shower(s) of snow pellets, or ice pellets, or ice pellets with or without rain or rain and snow mixed

Slight shower(s) of hail, with or without rain or rain and snow mixed, not associated with thunder

Slight or moderate thunderstorm without hail, but with rain and/or snow at time of observation

Slight or moderate thunderstorm, with hail at time of observation

Heavy thunderstorm, without hail, but with rain and/or snow at time of observation

Thunderstorm combined with dust storm or sandstorm at time of observation

Heavy thunderstorm with hail at time of observation

KEY TO AVIATION WEATHER OBSERVATIONS

LOCATION IDENTIFIER TYPE AND TIME OF REPORT *	SKY AND CEILING	VISIBILITY WEATHER AND OBSTRUCTION TO VISION	SEA-LEVEL PRESSURE	TEMPERATURE AND DEW POINT	WIND	ALTIMETER SETTING	REMARKS AND CODED DATA
MCI SA Ø758	**15 SCT M25 OVC**	**1R-F**	**132**	**/58/56**	**/18Ø7**	**/993/**	**RØ1VR2ØV4Ø**

SKY AND CEILING

Sky cover contractions are for each layer in ascending order. Figures preceding contractions are base heights in hundreds of feet above station elevation. Sky cover contractions used are:

CLR = Clear: Less than Ø.1 sky cover.
SCT = Scattered: Ø.1 to Ø.5 sky cover.
BKN = Broken: Ø.6 to Ø.9 sky cover.
OVC = Overcast: More than Ø.9 sky cover.

— = Thin (When prefixed to SCT, BKN, OVC).
—X = Partly obscured: Ø.9 or less of sky hidden by precipitation or obstruction to vision (bases at surface).
X = Obscured: 1.Ø sky hidden by precipitation or obstruction to vision (bases at surface).

A letter preceding the height of a base identifies a ceiling layer and indicates how ceiling height was determined. Thus:

E = Estimated
M = Measured
W = Vertical visibility into obscured sky

V = Immediately following the height of a base indicates a variable ceiling.

VISIBILITY

Reported in statute miles and fractions (V = Variable)

WEATHER AND OBSTRUCTION TO VISION SYMBOLS

A	Hail	IC	Ice crystals	S	Snow
BD	Blowing dust	IF	Ice-fog	SG	Snow grams
BN	Blowing sand	IP	Ice pellets	SP	Snow pellets
BS	Blowing snow	IPW	Ice pellet showers	SW	Snow showers
D	Dust	K	Smoke	T	Thunderstorms
F	Fog	L	Drizzle	T+	Severe thunderstorm
GF	Ground fog	R	Rain	ZL	Freezing drizzle
H	Haze	RW	Rain showers	ZR	Freezing rain

Precipitation intensities are indicated thus: - Light; (no sign) Moderate; + Heavy

WIND Direction in tens of degrees from true north, speed in knots. ØØØØ indicates calm. G indicates gusty. Q indicates Squalls. Peak wind speed in the past 1Ø minutes follows G or Q when gusts or squalls are reported. The contraction WSHFT, followed by GMT time group in remarks, indicates windshift and its time of occurrence. (Knots x 1.15 = statute mi/hr).

EXAMPLES: 3627 = wind from 36Ø Degrees at 27 knots;
3627G4Ø = wind from 36Ø Degrees at 27 knots, peak speed in gusts 4Ø knots

ALTIMETER SETTING

The first figure of the actual altimeter setting is always omitted from the report.

RUNWAY VISUAL RANGE (RVR)

RVR is reported from some stations. For planning purposes, the value range during 1Ø minutes prior to observations and based on runway light setting 5 are reported in hundreds of feet. Runway identification precedes RVR report.

PILOT REPORTS (PIREPs)

When available, PIREPs in fixed-format may be appended to weather observations. PIREPs are designated by UA or UUA for urgent PIREPs.

DECODED REPORT

Kansas City International: Record observation completed at Ø758 GMT 15ØØ feet scattered clouds, measured ceiling 25ØØ feet overcast, visibility 1 mile, light rain, fog, sea-level pressure 1Ø13.2 millibars, temperature 58°F, dewpoint 56°F, wind from 18Ø°, at 7 knots, altimeter setting 29.93 inches. Runway Ø1, visual range 2ØØØ feet lowest 4ØØØ feet highest in the past 1Ø minutes.

* TYPE OF REPORT

SA = a scheduled record observation
SP = an unscheduled special observation indicating a significant change in one or more elements
RS = a scheduled record observation that also qualifies as a special observation

The designator for all three types of observations (SA, SP, RS) is followed by a 24 hour-clock-time-group in Greenwich Mean Time (GMT or Z).

U.S. DEPARTMENT OF COMMERCE—NATIONAL OCEANIC AND ATMOSPHERIC ADMINISTRATION—NATIONAL WEATHER SERVICE

KEY TO AVIATION WEATHER FORECASTS

TERMINAL FORECASTS contain information for specific airports on expected ceiling, cloud heights, cloud amounts, visibility, weather and obstructions to vision, and surface wind. They are issued 3 times/day, amended as needed, and are valid for up to 24 hours. The last six hours of each forecast period are covered by a categorical statement indicating whether VFR, MVFR, IFR or LIFR conditions are expected (L in LIFR and M in MVFR indicate "low" and "marginal"). Terminal forecasts are written in the following form:

CEILING: Identified by the letter "C" (for lowest layer with cumulative sky cover greater than 5/1Ø)
CLOUD HEIGHTS: In hundreds of feet above the station (ground)
SKY COVER AMOUNT (including any obscuration)
CLOUD LAYERS: Stated in ascending order of height
VISIBILITY: In statute miles (omitted if over 6 miles)
WEATHER AND OBSTRUCTION TO VISION: Standard weather and obstruction to vision symbols are used
SURFACE WIND: In tens of degrees and knots (omitted when less than 6 knots)

EXAMPLE OF TERMINAL FORECAST

DCA 221Ø1Ø: DCA Forecast 22nd day of month - valid time 1ØZ-1ØZ.
1ØSCT C18 BKN 5SW—3415G25 OCNL C8 X 1/2SW: Scattered clouds at 1ØØØ feet, ceiling 18ØØ feet broken, visibility 5 miles, light snow showers, surface wind from 34Ø degrees at 15 knots, Gusts to 25 knots, occasional ceiling 8 hundred feet sky totally obscured, visibility 1/2 mile in moderate snow showers.
12Z C5Ø BKN 3312G22: By 12Z becoming ceiling 5ØØØ feet broken, surface wind 33Ø degrees at 12 knots, Gusts to 22.
Ø4Z MVFR CIG: Last 6 hours of FT after Ø4Z marginal VFR due to ceiling.

AREA FORECASTS are 12-hour aviation forecasts plus a 6-hour categorical outlook prepared 3 times/day, with each section amended as needed, giving general descriptions of potential hazards, airmass and frontal conditions, icing and freezing level turbulence and low-level windshear and significant clouds and weather for an area the size of several states. Heights of cloud bases and tops, turbulence and icing are referenced ABOVE MEAN SEA LEVEL (MSL); unless indicated by Ceiling (CIG) or ABOVE GROUND LEVEL (AGL). Each SIGMET OR AIRMET affecting an FA area will also serve to amend the Area Forecast.

SIGMET, AIRMET and CWA messages (In-flight advisories) broadcast by FAA on NAVAID voice channels warn pilots of potentially hazardous weather. SIGMET's concern severe and extreme conditions of importance to all aircraft (i.e., icing, turbulence, and dust storms/sandstorms or volcanic ash). Convective SIGMET's are issued for thunderstorms if they are sufficiently strong, wide spread or embedded. AIRMET's concern less severe conditions which may be hazardous to aircraft particularly smaller aircraft and less experienced or VFR only pilots. CWA's (Center Weather Advisories) concern both SIGMET and AIRMET type conditions described in greater detail and relating to a specific ARTCC area.

WINDS AND TEMPERATURES ALOFT (FD) FORECASTS are 6, 12, and 24-hour forecasts of wind direction (nearest 1Ø° true N) and speed (knots) for selected flight levels. Forecast Temperatures Aloft (°C) are included for all but the 3ØØØ-foot level.

EXAMPLES OF WINDS AND TEMPERATURES ALOFT (FD) FORECASTS:
FD WBC 121645
BASED ON 1212ØØZ DATA
VALID 13ØØØØZ FOR USE 21ØØ-Ø6ØØZ. TEMPS NEG ABV 24ØØØ FT

	3ØØØ	6ØØØ	9ØØØ	12ØØØ	18ØØØ	24ØØØ	3ØØØØ	34ØØØ	39ØØØ
BOS	3127	3425-Ø7	342Ø-11	3421-16	3516-27	3512-38	311649	292451	283451
JFK	3Ø26	3327-Ø8	3324-12	3322-16	312Ø-27	2923-38	284248	28515Ø	285749

At 6ØØØ feet MSL over JFK wind from 330° at 27 knots and temperature minus 8°C.

TWEB (CONTINUOUS TRANSCRIBED WEATHER BROADCAST) - Individual route forecasts covering a 25-nautical-mile zone either side of the route. By requesting a specific route number, detailed en route weather for a 12- or 18 hour period (depending on forecast issuance) plus a synopsis can be obtained.

PILOTS. . . . report in-flight weather to nearest FSS. The latest surface weather reports are available by phone at the nearest pilot weather briefing office by calling at H+10.

U.S. DEPARTMENT OF COMMERCE—NATIONAL OCEANIC AND ATMOSPHERIC ADMINISTRATION—NATIONAL WEATHER SERVICE—REVISED JANUARY 1984
NOAA PA 73029

APPENDIX IV: Weather Extremes

		UNITED STATES	CANADA	WORLD
Temperature	MAXIMUM	**57°C (134°F) Death Valley, CA 10 July 1913**	**45°C (113°F) Midale and Yellow Grass, Sask. 5 July 1937**	**58°C (136°F) El Azizia, Libya 13 September 1922**
	MINIMUM	**−62.1°C (−79.8°F) Prospect Creek, AK 23 January 1971**	**−63°C (−81.4°F) Snag, Yukon Terr. 3 February 1947**	**−88.3°C (−126.9°F) Vostok, Antarctica 24 August 1960**
Precipitation	24-HOUR MAXIMUM	**1092 mm (43 in.) Alvin, TX 25–26 July 1979**	**489 mm (19.3 in.) Ucluelet, Vancouver Island, B.C. 6 October 1967**	**1870 mm (73.62 in.) Cilaos, La Reunion, Indian Ocean 15–16 March 1952**
	ONE MONTH MAXIMUM	**1817 mm (71.54 in.) Helin Mine, CA January 1909 2718 mm (107.00 in.) Kuki, Maui, HI March 1942**	**2235.5 mm (88.01 in.) Swanson Bay, B.C. November 1917**	**9230 mm (366.14 in.) Cherrapunji, Assam, India July 1861**
	ONE YEAR MAXIMUM	**14681 mm (578.00 in.) Kuki, Maui, HI 1950 4688 mm (184.656 in.) Wynochee Oxbow, WA 1931**	**8122.0 mm (319.8 in.) Henderson Lake, B.C. 1931**	**26461 mm (1041.78 in.) Cherrapunji, Assam, India August 1860–July 1861 22990 mm (905.12 in.) Cherrapunji, Assam, India 1861**
	ONE YEAR MINIMUM	**0.0 mm (0.0 in.) Death Valley, CA 1919 0.0 mm (0.0 in.) Bagdad, San Bernardino County, CA 18 August 1909–6 May 1912**	**81.2 mm (1.23 in.) Eureka, NWT 1956**	**0.0 mm (0.0 in.) Arica, Chile October 1903–December 1917**

		UNITED STATES	CANADA	WORLD
Snowfall	24-HOUR MAXIMUM	**193 cm (75.8 in.) Silver Lake, Boulder County, CO 14–15 April 1921**	**112.3 cm (44.2 in.) Kitimat, B.C. 18 February 1972**	
	SINGLE STORM MAXIMUM	**480 cm (189.0 in.) Mt. Shasta Ski Bowl, CA 13–19 February 1959**		
	ONE MONTH MAXIMUM	**991 cm (390.0 in.) Tamarack, CA January 1911**		
	SEASON MAXIMUM	**2850 cm (1122.0 in.) Paradise Ranger Station, Mt. Rainier, WA 1971–72**		
Atmospheric Pressure	MAXIMUM	**1068 mb (31.43 in.) Barrow, AK 3 January 1970**	**1068 mb (31.43 in.) Mayo, Yukon Terr. 1 January 1971**	**1084 mb (32.005 in.) Agata, Siberia 31 December 1968**
		1063 mb (31.40 in.) Helena, MT 9 January 1962		
	MINIMUM	**892 mb (26.35 in.) Long Key, FL 2 September 1935**	**940.2 mb (27.77 in.) St. Anthony, Nfld. 20 January 1977**	**870 mb (25.69 in.) Typhoon Tip, Pacific Ocean 12 October 1979**

APPENDIX V:

Selected Climatic Data for United States and Canada

T = daily average temperature (°C) P = monthly precipitation (millimeters)

		J	F	M	A	M	J	J	A	S	O	N	D	ANNUAL
Montgomery, Alabama	T	8.3	9.9	13.9	18.3	22.2	26.1	27.8	27.2	24.9	18.3	12.8	9.4	18.3
	P	107	114	150	112	102	89	122	81	119	58	76	122	1250
Fairbanks, Alaska	T	−25	−20	−13.3	−1.1	8.9	15.0	16.7	13.9	7.2	−3.9	−15.6	−23.3	−3.3
	P	13	10	10	8	15	33	46	48	28	18	18	18	264
Phoenix, Arizona	T	11.1	13.3	16.1	20.0	25.0	30.0	33.3	32.2	29.4	22.8	16.1	11.7	21.7
	P	18	15	20	8	3	5	18	25	15	15	13	20	180
Little Rock, Arkansas	T	4.4	6.7	11.1	16.7	21.1	25.6	27.8	27.2	23.3	17.2	10.6	6.1	16.7
	P	99	97	119	137	135	94	91	79	109	71	112	107	1250
Los Angeles, California	T	13.9	15.0	15.6	16.7	18.3	20.6	23.3	23.9	22.8	20.6	17.2	14.4	18.3
	P	94	76	61	30	5	1.0	0.0	3	8	5	46	51	376
San Francisco, California	T	8.9	11.1	11.7	12.8	14.4	16.1	16.7	17.2	17.8	16.1	12.2	9.4	13.9
	P	117	81	66	38	8	3	1.0	1.0	5	28	61	91	500
Denver, Colorado	T	−1.1	1.1	3.3	8.3	13.9	19.4	22.9	21.7	17.2	11.1	3.9	0.6	10.0
	P	13	18	30	46	64	41	48	38	30	25	20	15	389
Hartford, Connecticut	T	−3.9	−2.2	2.8	9.4	15.0	20.6	22.8	21.7	17.2	11.1	5.6	−1.7	10.0
	P	89	81	107	102	86	86	79	102	99	89	102	107	1128
Wilmington, Delaware	T	−0.6	0.6	5.6	11.1	16.7	21.7	24.4	23.9	20.0	13.3	7.8	2.2	12.2
	P	79	76	99	86	81	89	99	102	91	74	84	89	1052
Washington, D.C.	T	1.7	2.8	7.8	13.3	18.3	22.8	25.6	25.0	21.1	14.4	8.9	3.3	13.9
	P	81	71	97	86	99	102	107	130	91	81	79	86	1105
Miami, Florida	T	19.4	20.0	22.2	23.9	25.6	27.2	27.8	28.3	27.8	25.6	22.8	20.0	24.4
	P	53	51	48	79	165	234	152	178	206	180	69	48	1463
Atlanta, Georgia	T	5.6	7.2	11.1	16.7	20.6	24.4	26.1	25.6	22.8	16.7	11.1	6.7	16.1
	P	124	112	150	112	102	86	119	86	81	64	86	107	1234
Honolulu, Hawaii	T	22.8	22.8	23.3	24.4	25.6	26.1	26.7	27.2	27.2	26.7	25.0	23.3	25.0
	P	97	69	89	38	30	13	13	15	15	48	81	86	597
Boise, Idaho	T	−1.1	2.2	5.0	9.4	13.9	18.9	23.9	22.2	17.2	11.1	4.4	0.0	10.6
	P	41	28	25	30	30	25	8	10	15	20	33	33	297
Chicago, Illinois	T	−6.1	−3.3	2.2	9.4	15.0	20.6	22.8	22.2	18.3	12.2	4.4	−2.2	9.4
	P	41	33	66	94	81	104	91	89	86	58	53	53	848

Indianapolis, Indiana	T	−3.3	−1.1	4.4	11.1	16.7	22.2	23.9	22.8	19.4	12.8	5.6	0.0	11.1
	P	66	64	91	94	94	102	109	89	69	64	76	76	996
Des Moines, Iowa	T	−7.2	−4.4	1.7	9.9	16.7	22.2	24.4	23.3	18.3	12.2	3.9	−3.3	10.0
	P	25	28	56	81	102	108	81	104	79	56	38	25	782
Topeka, Kansas	T	−3.3	0.0	5.5	12.8	18.3	23.3	26.1	25.0	20.0	13.9	6.1	0.0	12.2
	P	23	25	56	79	102	130	102	94	86	71	46	33	848
Louisville, Kentucky	T	0.0	2.2	7.2	13.9	18.3	23.3	25.6	24.4	21.1	14.4	7.8	2.8	13.3
	P	86	81	119	104	107	91	104	84	86	66	89	89	1107
Baton Rouge, Louisiana	T	11.1	12.2	15.6	20.0	23.9	26.7	27.8	27.2	25.6	20.0	15.0	11.7	20.0
	P	117	127	117	142	122	79	180	127	112	66	102	127	1417
Portland, Maine	T	−5.6	−5.0	0.0	6.1	11.7	16.7	20.0	19.4	15.0	8.9	3.3	−3.3	7.2
	P	97	91	102	99	84	79	71	71	84	97	119	114	1105
Baltimore, Maryland	T	0.6	1.7	6.1	12.2	17.2	22.2	25.0	24.4	20.6	13.9	7.8	2.2	12.8
	P	74	74	94	86	86	99	99	117	89	79	79	86	1062
Boston, Massachusetts	T	−1.1	−0.6	3.3	9.4	15.0	20.0	23.3	22.2	18.3	12.8	7.2	1.1	11.1
	P	99	94	104	94	89	74	69	94	86	86	107	114	1113
Detroit, Michigan	T	−5.0	−3.3	1.7	8.9	14.4	20.0	22.2	21.1	17.2	11.1	4.4	−2.2	9.4
	P	48	43	64	81	71	86	79	81	56	53	58	64	1016
Minneapolis, Minnesota	T	−11.7	−7.8	−1.7	7.8	14.4	20.0	22.8	21.7	16.1	10.0	0.6	−7.2	7.2
	P	20	20	43	51	81	104	89	91	64	46	33	23	671
Jackson, Mississippi	T	7.8	9.4	13.3	18.3	22.8	26.1	27.8	27.2	24.4	18.3	12.8	9.4	18.3
	P	127	114	150	147	122	74	112	94	91	66	107	137	1341
Springfield, Missouri	T	0.0	2.2	7.2	13.3	18.3	22.8	25.6	25.0	21.1	14.4	7.2	2.2	13.3
	P	41	53	86	102	109	119	91	71	107	81	74	66	1003
Helena, Montana	T	−7.8	−3.3	0.0	5.6	11.1	15.6	20.0	18.9	13.3	7.2	−0.6	−4.4	6.1
	P	15	10	18	28	43	51	25	30	46	15	13	15	290
Omaha, Nebraska	T	−6.1	−2.7	2.8	11.1	17.2	22.8	25.6	23.9	18.9	13.3	4.4	−2.8	10.6
	P	20	48	48	74	109	102	91	104	89	53	33	20	770
Las Vegas, Nevada	T	7.2	10.0	12.8	17.8	22.8	28.9	32.2	31.1	26.6	20.0	12.2	7.2	18.9
	P	13	13	10	5	5	3	10	13	8	5	10	8	107
Concord, New Hampshire	T	−6.7	−5.6	0.0	6.7	12.8	18.3	21.1	19.4	15.0	8.9	2.7	−4.4	7.2
	P	71	64	74	76	74	74	74	84	79	79	94	86	927
Newark, New Jersey	T	−0.6	0.6	5.0	11.1	16.7	22.2	25.0	24.4	20.0	13.9	7.8	2.2	12.2
	P	79	76	107	91	91	74	97	109	94	79	91	86	1074
Albuquerque, New Mexico	T	1.7	3.9	7.2	12.8	17.8	22.8	25.0	23.9	20.0	14.4	6.7	3.3	13.3
	P	10	10	13	10	13	13	33	38	20	23	10	13	206
New York, New York	T	0.0	0.6	5.0	11.1	16.7	21.7	25.0	23.9	20.0	14.4	8.3	2.2	12.2
	P	81	79	107	97	97	81	97	102	94	86	104	97	1120
Raleigh, North Carolina	T	4.4	5.6	9.4	15.6	19.4	23.3	25.6	25.0	21.7	15.6	10.0	5.6	15.0
	P	91	86	94	74	94	94	112	112	84	69	74	79	1062
Bismarck, North Dakota	T	−13.9	−10.0	−3.3	5.6	12.8	17.8	21.1	20.6	13.9	7.8	−1.7	−9.4	5.0
	P	13	10	18	38	56	76	51	51	43	20	13	13	391

		J	F	M	A	M	J	J	A	S	O	N	D	ANNUAL
Cleveland, Ohio	T	−3.3	−2.8	2.8	8.9	14.4	20.0	22.2	21.1	17.8	11.7	5.6	−0.6	10.0
	P	64	56	76	84	84	89	86	86	74	61	71	71	894
Oklahoma City, Oklahoma	T	2.2	5.0	9.4	15.6	20.0	25.0	27.8	27.2	22.8	16.7	9.4	4.4	15.6
	P	25	33	53	74	140	99	76	61	86	69	38	30	785
Portland, Oregon	T	3.9	6.1	7.8	10.0	13.9	16.7	20.0	19.4	17.2	12.2	7.8	5.0	11.7
	P	157	99	91	58	53	38	13	33	41	76	132	163	950
Pittsburgh, Pennsylvania	T	−2.8	−1.7	3.3	10.0	15.6	20.0	22.2	21.7	17.8	11.1	5.6	−0.6	10.0
	P	74	61	91	84	89	84	97	84	71	64	58	66	922
Providence, Rhode Island	T	−2.2	−1.7	2.8	8.9	14.4	19.4	22.2	21.7	17.8	11.7	6.1	0.0	10.0
	P	104	94	109	102	89	71	76	102	89	97	107	114	1151
Columbia, South Carolina	T	7.2	8.3	12.2	17.8	22.2	25.6	27.2	26.7	23.9	17.2	12.2	8.3	17.2
	P	109	102	132	91	97	112	137	142	107	66	64	89	1247
Rapid City, South Dakota	T	−6.1	−3.3	0.6	7.2	13.3	18.3	22.8	21.7	16.1	10.0	1.7	−3.3	8.3
	P	10	15	25	51	66	84	53	36	25	20	13	10	414
Nashville, Tennessee	T	2.8	4.4	9.4	15.6	20.0	24.4	26.1	25.6	22.2	15.6	9.4	5.0	15.0
	P	114	102	142	114	117	94	97	86	94	66	89	117	1232
Dallas–Fort Worth, Texas	T	6.7	8.9	13.3	18.9	23.3	27.8	30.0	30.0	26.1	20.0	13.3	8.9	18.9
	P	41	48	61	91	109	66	51	46	84	64	46	43	747
Houston, Texas	T	11.1	12.8	16.1	20.6	23.9	27.2	28.3	28.3	25.6	21.1	15.6	12.2	20.0
	P	81	81	69	107	119	102	84	94	124	94	86	94	1138
Salt Lake City, Utah	T	−1.7	1.1	5.0	9.4	15.0	14.4	25.0	23.9	18.3	11.7	4.4	−1.1	11.1
	P	36	33	43	56	38	25	18	23	23	28	30	36	388
Burlington, Vermont	T	−8.3	−7.8	−1.7	6.1	12.8	18.3	21.1	19.4	15.0	8.9	2.8	−5.0	6.7
	P	46	43	56	69	76	91	86	99	81	71	71	61	856
Richmond, Virginia	T	2.8	3.9	8.3	14.4	18.9	23.3	25.6	25.0	21.1	15.0	9.4	4.4	14.4
	P	81	79	91	74	91	91	130	127	89	94	84	86	1120
Seattle, Washington	T	5.0	6.7	7.8	10.0	13.3	16.1	18.3	18.3	16.1	12.2	7.8	6.1	11.7
	P	150	107	94	64	43	38	23	36	51	86	137	160	1013
Charleston, West Virginia	T	0.6	2.2	7.2	12.8	17.8	21.7	23.9	23.3	20.0	13.3	7.2	2.7	12.8
	P	89	79	102	89	94	84	137	107	76	66	74	84	1077
Madison, Wisconsin	T	−8.9	−6.7	−0.6	7.8	13.9	18.9	21.1	20.0	15.6	10.0	1.7	−5.5	7.2
	P	28	25	56	79	84	109	99	97	79	56	46	38	782
Cheyenne, Wyoming	T	−3.3	−1.7	0.0	5.6	11.1	16.7	20.6	19.4	14.4	8.9	1.7	−1.7	7.8
	P	10	10	25	30	61	51	48	36	25	18	13	10	338
Calgary, Alberta	T	−9.9	−8.8	−4.4	3.6	9.8	13.0	16.7	15.1	10.9	5.4	−2.2	−6.6	3.6
	P	17	20	26	35	52	88	58	59	35	23	16	15	444
Prince George, British Columbia	T	−11.3	−7.5	−2.3	4.3	9.7	12.9	14.9	13.7	10.1	4.8	−2.5	−6.6	3.3
	P	56	44	36	28	43	62	64	65	56	59	57	56	626
Winnipeg, Manitoba	T	−17.7	−15.5	−7.9	3.3	11.3	16.5	20.2	18.9	12.8	6.2	−4.8	−12.9	2.5
	P	26	21	27	30	50	81	69	70	55	37	29	22	517

Saint John, New Brunswick	T	−6.9	−6.4	−2.0	3.8	9.4	13.9	17.2	17.1	13.4	8.3	2.7	−4.3	5.4
	P	144	122	106	105	98	94	87	108	104	108	153	133	1362
Goose Bay, Newfoundland	T	−16.6	−14.9	−8.4	−1.6	5.1	11.9	16.3	14.7	10.1	3.2	−4.4	−12.9	0.2
	P	72	63	68	62	56	72	84	91	76	63	67	63	837
Frobisher Bay, Northwest Territories	T	−26.5	−25.5	−21.5	−13.7	−3.1	3.6	7.9	6.9	2.2	−4.7	−12.3	−20.5	−8.9
	P	22	26	19	21	19	33	53	53	43	34	33	24	380
Halifax, Nova Scotia	T	−3.3	−3.6	−0.1	4.8	9.9	14.4	18.5	18.8	15.5	10.3	5.3	−0.9	7.4
	P	141	119	113	112	109	94	94	96	117	120	143	126	1384
Toronto, Ontario	T	−3.9	−3.8	0.2	7.0	13.2	19.0	21.9	21.1	16.6	10.6	4.3	−1.8	8.7
	P	67	59	67	66	70	63	74	61	65	60	63	61	776
Montreal, Quebec	T	−8.7	−7.8	−2.1	6.2	13.6	18.9	21.6	20.5	15.6	9.4	2.3	−5.9	6.9
	P	87	76	86	83	81	91	102	87	95	83	88	89	1048
Regina, Saskatchewan	T	−16.9	−14.8	−8.1	3.4	11.2	15.3	19.3	17.8	11.9	5.1	−5.4	−12.3	2.2
	P	19	17	21	21	40	83	55	49	34	18	20	17	394
Whitehorse, Yukon	T	−18.1	−14.1	−7.6	−0.2	7.5	12.6	14.2	12.4	7.9	0.7	−8.2	−15.1	−0.7
	P	18	14	15	11	13	27	35	37	25	19	23	20	257

Data sources: National Climatic Center, Asheville, North Carolina and Atmospheric Environment Service, Canada

Glossary

absolute instability Ambient air layer that is unstable for both saturated air parcels and unsaturated air parcels.

absolute stability Ambient air layer that is stable for both saturated air parcels and unsaturated air parcels.

absolute temperature Temperature scale based on absolute zero and given in kelvins; 0 °C = 273 K, for example.

absolute zero The theoretical temperature at which all molecular activity ceases; 0 K.

absorptivity The efficiency of radiation absorption.

acclimatization The gradual adjustment of the body to new climatic or other environmental conditions, for example, the adjustment to low levels of oxygen at high altitudes.

acid rain Rain having a pH lower than 5.6, often due to the presence of sulfuric and nitric acids.

adiabatic processes (or assumption) Expansional cooling or compressional warming of air parcels in which there is no net heat exchange between the air parcels and the surrounding (ambient) air.

advection fog Ground-level clouds generated by advective cooling of a mild, humid air mass as it travels over a relatively cool surface.

aerosols Tiny liquid or solid particles of various compositions that occur suspended in the atmosphere.

AFOS Automation of Field Operations and Service, a computerized National Weather Service communications system that speeds transmission of weather data.

agroclimatic compensation Poor growing weather in one area is offset to some extent by better growing weather in other areas.

air density Mass per unit volume of air; about 1.275 km per cubic meter at 0 °C and 1000 millibars.

air mass A huge volume of air covering thousands of square kilometers that is relatively uniform horizontally in temperature and water vapor concentration.

air mass advection Horizontal movement of air or air masses from one place to another.

air mass climatology Description of climate in terms of frequency of occurrence of various types of air masses.

air mass thunderstorms Thunderstorms that develop almost randomly within a mass of maritime tropical air.

air parcels Unit masses of air—a single gram, for example.

air pollutants Gases and aerosols in air in concentrations that threaten the well-being of living organisms or that disrupt the orderly functioning of the environment.

air pollution episode Times when atmospheric conditions do not favor dilution of air pollutants, and the pollutants are thus a hazard to human health.

air pressure The cumulative force exerted on any surface by the molecules composing air.

air pressure tendency Change in air pressure with time; on a weather map, the air pressure change over the prior 3 hours.

albedo The fraction of radiation striking a surface that is reflected.

Aleutian low A semipermanent subpolar cyclone situated over the North Pacific in winter.

altimeter An aneroid barometer calibrated to read altitude or elevation.

altithermal episode Of or related to a period 5000 to 7000 years ago when global temperatures were warmer than at present.

altocumulus clouds Middle clouds consisting of patches or puffs of roll-like clouds forming a wavy pattern.

altocumulus lenticularis cloud A lens-shaped altocumulus cloud; a mountain-wave cloud generated by the disturbance of horizontal airflow caused by a prominent mountain range.

altostratus clouds Middle layer clouds that are uniformly gray or white.

ambient air The air surrounding a cloud, or the air surrounding rising or sinking air parcels.

ambient temperature Temperature of the surrounding (ambient) air.

aneroid barometer A portable instrument that utilizes a flexible metal chamber and spring to measure air pressure.

anomalies Departures of temperature, precipitation, or other weather elements from long-term averages.

Antarctic Circle Latitude 66 degrees 33 minutes S. Poleward of this latitude, there are 24 hours of sunlight at the summer solstice and 24 hours of darkness at the winter solstice.

anticyclone A dome of air that exerts a relatively high pressure compared with the surrounding air; same as a "high." In the Northern Hemisphere, surface winds in an anticyclone blow clockwise and outward.

aphelion The time of the year when the earth is farthest from the sun (about July 4).

apparent temperature index A measure of the combined effect of temperature and humidity on human comfort.

arctic (A) air A very cold and dry air mass that forms primarily in winter over the Arctic basin, Greenland, and the northern interior of North America.

Arctic Circle Latitude 66 degrees 33 minutes N. Poleward of this latitude, there are 24 hours of sunlight at the summer solstice and 24 hours of darkness at the winter solstice.

arctic high Anticyclone originating in the source regions for cold, dry arctic air.

Arctic sea smoke Fog that develops when extremely cold, dry air flows over a large body of open water; a type of steam fog.

atmosphere A thin envelope of gases (also containing suspended solid and liquid particles and clouds) that encircles the globe.

atmospheric effect Preferred name for so-called "greenhouse effect."

atmospheric stability Property of an air layer that imparts a buoyant force to air parcels moving vertically within the air layer; depends on the temperature profile of the air layer.

atmospheric windows Infrared wavelength bands for which there is little or no absorption by constituent gases of the atmosphere.

aurora australis Southern Hemisphere equivalent of the aurora borealis.

aurora borealis Lights visible at night in the Northern Hemisphere, and produced by electrical activity in the ionosphere; northern lights.

auroral zone Geographical area where the aurora is visible; in the Northern Hemisphere, centered on northwestern Greenland.

back (or backing) Counterclockwise shift in wind direction with time; for example, a wind shift from northeast to north.

barograph A recording instrument that provides a continuous trace of air pressure variation with time.

barometer An instrument used to monitor variations in air pressure. See also aneroid barometer; mercurial barometer.

Basic Weather Network (BWN) Stations that provide weather data primarily for aircraft operations and to supplement weather forecasting.

Beaufort scale A scale of wind strength based on visual assessment of the effects of wind on seas and vegetation.

Bergeron process Process of precipitation formation in cold clouds whereby ice crystals grow at the expense of supercooled water droplets.

black body A perfect radiator; a material that absorbs 100 percent of the radiation striking it.

blizzard warning Falling or blowing snow accompanied by winds of over 55 km (35 mi) per hour.

blocking system A cutoff cyclone or anticyclone that blocks the usual west-to-east progression of weather systems.

bora A cold katabatic wind that originates in Yugoslavia and flows onto the coastal plain of the Adriatic Sea.

boundary layer A thin zone of still air adjacent to the skin that insulates the body from heat loss.

Bowen ratio The ratio of energy available for sensible heating to energy available for latent heating.

Boyle's law When the temperature is held constant, the pressure and density of an ideal gas are directly proportional.

British thermal unit (Btu) The quantity of heat needed to raise the temperature of one pound of water one degree Fahrenheit.

calorie The amount of heat needed to raise the temperature of one gram of water one Celsius degree (from 14.5 to 15.5 °C).

centrifugal force A force directed outward, away from the center of a rotating object; equal in magnitude to the centripetal force but in the opposite direction.

centripetal force An inward-directed force that confines an object to a circular path; equal in magnitude to the centrifugal force but in the opposite direction.

Charles's law With constant pressure, the temperature of an ideal gas is inversely proportional to the density of the gas.

chinook winds Air that is adiabatically compressed as it is drawn down the leeward slope of a mountain range. As a consequence, the air is mild and dry.

chromosphere Portion of the sun above the photosphere; consists of ionized hydrogen and helium at 4000 to 40,000 K.

cirrocumulus A high cloud exhibiting a wavelike pattern of small white puffs; composed of ice crystals.

cirrostratus A high layered cloud, composed of ice crystals, that forms a thin white veil over the sky.

cirrus High clouds occurring as silky strands and composed of ice crystals.

climate Weather of some locality averaged over some time period plus extremes in weather behavior observed during the same period.

climatic norm (or normal) Average plus extreme weather at some locality for some period, usually 30 years.

climatology The study of climate.

cloud condensation nuclei (CCN) Tiny solid and liquid particles on which water vapor condenses.

cloud seeding An attempt to stimulate natural precipitation processes by injecting nucleating agents, such as silver iodide, into clouds.

clouds The visible products (ice crystals and water droplets) of condensation and deposition of water vapor in the atmosphere.

cold-air advection Flow of air from relatively cool localities to relatively warm localities.

cold clouds Clouds at temperatures below 0 °C (32 °F), composed of ice crystals or supercooled water droplets or both.

cold front A narrow zone of transition between relatively cold, dense air that is advancing and relatively warm, less dense air that is retreating.

cold-core anticyclones (or highs) Shallow high pressure systems that coincide with domes of relatively cold, dry air.

cold-core cyclones (or lows) Cyclones that occupy relatively cold columns of air; migrating midlatitude low pressure systems that intensify with altitude.

collision-coalescence process Precipitation growth in clouds whereby drops join together upon impact.

compressional warming A temperature rise that accompanies a pressure increase on a volume of air, as when air subsides within the atmosphere.

condensation Process by which water changes phase from a vapor to a liquid.

conditional stability Ambient air layer that is stable for unsaturated air parcels and unstable for saturated air parcels.

conduction Flow of heat in response to a temperature gradient within an object or between objects that are in physical contact.

confluence zone Zone where air streams having different origins come together.

continental drift Continents carried along on top of gigantic crustal plates that drift very slowly over the globe.

continental polar (cP) air Relatively dry air mass that develops over the northern interior of North America; very cold in winter and mild in summer.

continental tropical (cT) air Warm, dry air mass that forms over the subtropical deserts of the southwestern United States.

convection Air circulation in the vertical plane in which warm air parcels rise and cool air parcels sink.

convergence A wind pattern whereby there is a net inflow of air toward some point.

cooling degree-day units An index that measures the need for air conditioning when average daily air temperatures rise above 65 °F (18 °C); computed by subtracting 65 °F from the average daily temperature in °F.

Coriolis effect A deflective force arising from the rotation of the earth on its axis; affects principally synoptic-scale and global-scale winds. Winds are deflected to the right of the initial direction in the Northern Hemisphere, and to the left in the Southern Hemisphere.

corona Outermost region of the sun; consists of highly rarefied gases at 1 to 2 million K.

cumulonimbus clouds Thunderstorm clouds that form as a consequence of deep convection in the atmosphere.

cumulus clouds Clouds that develop as a consequence of convective air currents; resemble puffs of cotton floating in the sky.

cumulus congestus An upward building convective cloud with vertical development between that of a cumulus cloud and a cumulonimbus.

cumulus stage Initial stage in the life cycle of a thunderstorm cell; consists of towering cumulus clouds with an updraft throughout.

cup anemometer An instrument used to monitor wind speed. Wind rotation of cups generates an electric current calibrated in wind speed.

cutoff high Anticyclonic circulation system that separates from the prevailing westerly airflow and therefore remains stationary.

cutoff low Cyclonic circulation system that separates from the prevailing westerly airflow and therefore remains stationary.

cyclogenesis Birth and development of a cyclone, a low pressure system.

cyclone A weather system characterized by relatively low air pressure compared with the surrounding air; same as a low. Surface winds blow counterclockwise and inward in the Northern Hemisphere.

dart leaders Surges of negative electrical charge that follow the conductive path formed by the initial stepped leaders and return stroke of a lightning flash.

degree-day units *See* cooling degree-day units; heating degree-day units.

deposition Process by which water changes phase directly from a vapor into a solid. A good example is frost formation.

deposition nuclei Particles on which water vapor is deposited directly as ice.

dew Water droplets formed by condensation of water vapor on a surface.

dew point or dew-point temperature Temperature to which air must be cooled at constant pressure to achieve saturation (if above 0 °C or 32 °F).

diffuse insolation Solar radiation that is scattered or reflected by atmospheric components (clouds, for example) to the earth's surface.

dilution A reduction in concentration through mixing, as when polluted air mixes with cleaner air.

direct insolation Solar radiation that is transmitted directly through the atmosphere to the earth's surface without interacting with atmospheric components.

divergence A wind pattern whereby there is a net outflow of air from some point.

Doppler radar Conventional weather radar that has the added capability of determining the detailed motion of target precipitation based on the frequency shift between the outgoing and returning radar beam.

downburst A strong and potentially destructive thunderstorm downdraft; also called a microburst.

downdraft Downward moving air, usually within a thunderstorm cell.

drainage basin A fixed geographical region from which a river and its tributaries drain water.

drizzle A form of liquid precipitation consisting of water droplets less than 0.5 mm (0.02 in.) in diameter; falls from low stratus clouds.

dry adiabatic cooling Expansional cooling of rising unsaturated air parcels. No net heat exchange occurs between the air parcels and the ambient air.

dry adiabatic lapse rate Rising unsaturated air parcels cool at the rate of about 10 °C per 1000 m of uplift (or 5.5 °F per 1000 ft.).

dry adiabats Sloping lines (labeled in K) on a Stüve thermodynamic diagram that represent the temperature change of unsaturated air parcels subjected to adiabatic expansion or compression.

dry deposition Removal of suspended particulates from the air through impaction and gravitational settling; a natural means whereby air is cleansed of pollutants.

dust devil Swirling mass of dust caused by intense heating of dry surface areas that triggers air currents.

dust dome An accumulation of visibility-restricting aerosols in the air over an urban-industrial area.

dust plume A dust dome that elongates downwind.

earth-atmosphere system The earth's surface and atmosphere considered together.

effective emissivity A correction factor, dependent on the radiational characteristics of the earth-atmosphere system, that permits application of black body radiation laws to the earth-atmosphere system.

Ekman spiral Strengthening and shifting of direction of the horizontal wind with altitude through the friction layer.

El Niño An anomalous warming of surface ocean waters off the coasts of Ecuador and Peru and extending westward over the eastern tropical Pacific.

electromagnetic radiation Energy transfer in the form of waves that have both electrical and magnetic properties. Occurs even in a vacuum.

electromagnetic spectrum Range of radiation types arranged by wavelength or by frequency or both.

emissivity The efficiency of radiation emission by some object as compared with a black body.

emittance The rate at which a black body radiates energy across all wavelengths.

Eppley pyranometer The standard instrument for measuring solar radiation that strikes a horizontal surface. Calibrates the temperature response of a special sensor in terms of radiation.

equation of state *See* gas law.

equatorial trough A broad east-west belt of low pressure near the equator.

equinoxes The first days of spring and autumn when day and night are of equal length at all latitudes. The noon sun is directly over the equator.

evaporation Water changes phase from a liquid to a vapor at a temperature below the boiling point of water.

evapotranspiration Vaporization of water through direct evaporation from wet surfaces and the release of water vapor by vegetation.

expansional cooling A temperature drop that accompanies a pressure reduction on a volume of air.

eye of a hurricane An area of almost cloudless skies, light winds, and gently subsiding air at a hurricane center.

eye wall A circle of cumulonimbus clouds surrounding the eye of a mature hurricane.

F-scale Tornado intensity scale.

faculae Small, bright, relatively hot spots on the sun.

fair-weather bias Fair-weather days outnumber stormy days almost everywhere.

flash flooding A sudden rise in river or stream levels causing flooding.

foehn wind European term for chinook wind; warm, dry wind that flows into the Alpine valleys of Austria and Germany.

fog A cloud in contact with the earth's surface that reduces visibility to less than 1.0 km (0.62 mi).

fog dispersal Clearing of fog either by increasing the air temperature (thereby lowering the relative humidity) or by cloud seeding.

forced convection Convection aided by topographic uplift.

free convection Convection triggered by intense solar heating of the earth's surface.

freezing nuclei Particulates that cause liquid cloud droplets to freeze.

freezing rain Supercooled raindrops that freeze on impact with cold surfaces.

frequency Number of crests or troughs of a wave that pass a given point in a given period of time, usually 1 second.

friction The resistance an object encounters as it comes into contact with other objects.

friction layer Zone of the atmosphere, from the earth's surface to an altitude of about 1 km, where frictional resistance is primarily confined.

front A narrow zone of transition between air masses of contrasting density, that is, air masses of differing temperature or differing mixing ratio or both.

frontal thunderstorms Thunderstorms associated with lifting of air along frontal surfaces.

frontal uplift Uplift of air along either a cold or warm frontal surface.

frost Ice crystals formed by deposition of water vapor on a surface.

frost point Temperature to which air must be cooled (if at 0 °C or below) at constant pressure to achieve saturation.

frostbite Freezing of skin tissue.

fumigation Atmospheric stability condition that favors trapping of air pollutants within an air layer at the earth's surface.

funnel cloud A tornadic circulation that does not reach the ground.

gamma radiation Electromagnetic radiation with very short wavelength and great penetrating power.

gas law Relationship among the variables of state; pressure is proportional to the product of density and temperature.

geologic time A span of millions or billions of years in the past.

geostrophic wind Unaccelerated horizontal wind that flows in straight paths above the friction layer. Geostrophic wind is the consequence of a balance between the horizontal pressure gradient force and the force due to the Coriolis effect.

geosynchronous satellite A satellite that orbits the earth at the same rate as the earth's rotation, so it always scans the same region of the earth.

glacial climate Conditions favorable to the initiation and growth of glacial ice.

global-scale circulation The largest spatial scale of weather phenomena; includes global wind belts and semipermanent pressure systems.

glory Concentric rings of color about the shadow of an observer's head that appear on the top of a cloud situated below the observer. The glory is caused by the same refraction and internal reflection as a rainbow.

gradient The change in some factor (such as temperature or pressure) with distance.

gradient wind Large-scale, horizontal and frictionless wind that blows parallel to curved isobars.

graupel Granules of ice or compact snow; a form of precipitation.

gravitation A force of attraction between all objects that depends on the mass of the object and the distance between objects.

gravitational settling Aerosols drifting toward the earth's surface under the influence of gravity.

gravity The force that holds all objects on the earth's surface; the net effect of gravitation and centrifugal force due to the earth's rotation.

"greenhouse effect" Although nearly transparent to solar radiation, the atmosphere is much less transparent to infrared radiation. Terrestrial infrared radiation is absorbed and reradiated primarily by water vapor and carbon dioxide, thereby slowing the loss of heat from the earth-atmosphere system.

Greenwich Mean Time (GMT) A worldwide time reference used for synchronizing weather observations; the time at 0 degrees longitude, the prime meridian, which passes through Greenwich, England.

gust front The leading edge of a mass of relatively cool air that flows out of the base of a thunderstorm cloud and spreads along the ground well in advance of the parent thunderstorm cell; a mesoscale cold front.

haboob A dust storm formed by the downdraft of a desert thunderstorm.

Hadley cell Air circulation in tropical and subtropical latitudes of both hemispheres resembling a huge convective cell with rising air over the equator and sinking air in the subtropical anticyclones.

hail or hailstones Precipitation in the form of rounded or jagged chunks of ice, often characterized by internal concentric layering. Hail is associated with thunderstorms that have strong updrafts and relatively great moisture content.

hair hygrometer An instrument used to monitor relative humidity by measuring the changes in the length of human hair that accompany humidity variations.

halo Ring of light about the sun (or moon) caused by refraction of sunlight by tiny ice crystals suspended in the upper troposphere.

heat The total molecular energy of a given amount of a substance.

heat lightning Light reflected by clouds from thunderstorms occurring beyond the horizon.

heat of fusion Heat released when water changes phase from liquid to solid; 80 calories per gram.

heat of melting Heat required to change the phase of water from solid to liquid; 80 calories per gram.

heat of vaporization Heat required to change the phase of water from liquid to vapor; 540 to 600 calories per gram, depending on the temperature of the water.

heating degree-day units An index of space heating needs on days when average air temperature falls below 65 °F (18 °C); computed by subtracting the day's average temperature from 65 °F.

heliostats Computer-controlled mirrors that track the sun and focus rays on a single heat-collection point on a tower; component of a power tower system.

hertz Wave frequency of one cycle or wave per second.

heterosphere The atmosphere above 80 km (50 mi) where gases are stratified, with concentrations of the heavier gases decreasing more rapidly with altitude than concentrations of the lighter gases.

high *See* anticyclone.

hoarfrost Fernlike crystals of ice that form by deposition of water vapor on twigs, tree branches, and other vegetation.

homeothermic Organisms, such as humans, that can regulate their core temperature to some extent (±2 °C in the case of humans) regardless of variations in the temperature of the surrounding air.

homosphere The atmosphere up to 80 km (50 mi) in which the proportionality of principal gaseous constituents, such as oxygen and nitrogen, is constant.

hook echo A distinctive radar pattern that often indicates the presence of a severe thunderstorm and perhaps tornadic circulation.

horizontal pressure gradient Variations in air pressure from one location to another along a constant altitude (such as sea level).

horse latitudes Areas of calm winds associated with subtropical anticyclones; near 30 degrees latitude.

hot-wire anemometer An instrument that measures wind speed based on the heat loss of air flowing by a sensor.

hurricane Intense warm-core, oceanic cyclones that originate in tropical latitudes; also called typhoons in the Pacific Ocean. Winds are in excess of 120 km (74 mi) per hour.

hydrologic cycle Ceaseless flow of water among terrestrial, oceanic, and atmospheric reservoirs.

hydrostatic equation An expression of hydrostatic equilibrium that relates a change in pressure to a change in altitude.

hydrostatic equilibrium Balance between the vertical air pressure gradient force (directed upward) and the force of gravity (directed downward).

hygrograph An instrument that provides a continuous trace of relative humidity with time.

hygroscopic nuclei Tiny particles of matter that have a special chemical affinity for water molecules, so condensation may take place on these nuclei at relative humidities under 100 percent.

hyperthermia Physiological and behavioral responses (such as muscle cramps and unconsciousness) that occur when the human core temperature rises more than 2 °C (3.6 °F) above the normal body temperature of 37 °C (98.6 °F).

hypothermia Physiological responses (such as shivering and muscular rigidity) that occur when the human core temperature drops more than 2 °C (3.6 °F) below the normal body temperature of 37 °C (98.6 °F).

ice-forming nuclei (IN) Particulates that become active at temperatures well below freezing. The ice-forming nuclei include freezing nuclei and deposition nuclei.

ice pellets Frozen raindrops that bounce on impact with the ground; also called sleet.

Icelandic low A subpolar cyclone situated over the North Atlantic and best developed in winter.

ideal gas A gas that follows the kinetic-molecular theory precisely, that is, a gas that follows the behavior predicted by Boyle's law and Charles's law exactly.

ideal gas law Same as the gas law.

impaction Removal of particulates from the air through their impact on buildings and other objects at the earth's surface.

index of continentality Degree of maritime influence on continental air temperatures.

Indian summer A period of mild, sunny weather that occurs in autumn over eastern North America after the first freeze.

infrared radiation Radiant heat emitted by most objects at wavelengths longer than visible red light.

insolation *In*coming *sol*ar radi*ation*. *See also* diffuse insolation; direct insolation.

interglacial climate Conditions that favor the melting of glacial ice.

internal energy Molecular activity of the mixture of gases comprising an air parcel.

International Meteorological Organization (IMO) Founded in 1879, this is the predecessor of the World Meteorological Organization.

intertropical convergence zone (ITCZ) Discontinuous belt of thundershowers paralleling the equator and marking the convergence of the Northern and Southern Hemisphere surface trade winds.

ion An electrically charged atom or molecule.

ionosphere Region of the upper atmosphere from 80 to 900 km (50 to 600 mi) that contains a relatively high concentration of ions (charged particles).

isobars Lines on a map joining localities reporting the same air pressure.

isothermal Constant temperature.

ITCZ *See* intertropical convergence zone.

January thaw A period of relatively mild weather around January 20 to 23 that occurs primarily in New England; an example of a singularity in the climatic record.

JAWS Joint Airport Weather Studies; an investigation of microbursts in the vicinity of Denver's Stapleton International Airport conducted during the summer of 1982.

jet maximum An area of accelerated air flow along a jet stream; same as a jet core.

jet streams Relatively narrow ribbons of very strong winds embedded in the planetary winds aloft.

katabatic wind Downslope flow of cold, dense air under the influence of gravity.

kinetic energy Energy possessed by any object in motion.

Kirchhoff's law A perfect absorber is also a perfect emitter of radiation at the same wavelength.

lake breeze A relatively cool mesoscale wind directed from a lake toward land in response to differential heating between land and lake; develops during the day.

lake effect snows Snowfall induced along the lee shore of a lake by convergent flow of very cold air advected across the relatively mild open waters of a large lake.

land breeze A relatively cool mesoscale wind directed from land to sea or from land to lake in response to differential heating between land and water body; develops at night.

latent heat transfer Movement of heat from one place (the ocean, for example) to another (the atmosphere, for example) as a consequence of the phase changes of water. Heat is required for evaporation and sublimation at the earth's surface, and heat is released in condensation and deposition within the atmosphere.

latitude Distance measured in degrees north or south of the equator. The latitude of the equator is 0 degrees and that of the poles is 90 degrees N and 90 degrees S.

law of energy conservation Energy is neither created nor destroyed but can change from one form to another.

lifting condensation level (LCL) The altitude to which air must be lifted so that expansional cooling leads to condensation (or deposition) and cloud development; corresponds to the base of cumulus-type clouds.

lightning A flash of light produced by an electrical discharge in response to the buildup of an electrical potential between cloud and ground, between clouds, or within a single cloud.

Little Ice Age A period of relatively cold conditions in many regions of the globe from about 1450 to 1850.

LLWSAS Low-Level Wind Shear Alert System; an array of ground-level anemometers intended to detect microbursts at airports.

long waves Same as Rossby waves in the westerlies.

longitude The distance measured in degrees east or west of the prime meridian (0 degrees), which passes through Greenwich, England. The maximum longitude is 180 degrees W or 180 degrees E.

low *See* cyclone.

low-level jet stream A surge of maritime tropical air in the lower troposphere northward out of the Gulf of Mexico.

mammatus clouds Clouds exhibiting pouchlike, downward protuberances; may indicate turbulent air.

maritime polar (mP) air Cool, humid air mass that forms over the cold ocean waters of the North Pacific and North Atlantic.

maritime tropical (mT) air Warm, humid air mass that forms over tropical and subtropical oceans.

Maunder minimum A 70-year period from 1645 to 1715 when sunspots were rare.

mercurial barometer A mercury-filled tube used to measure air pressure; the standard barometric instrument, which features a high level of precision.

meridional component North-south direction parallel to a line of longitude.

meridional flow pattern Flow of westerlies in a series of deep troughs and sharp ridges; westerlies exhibiting considerable amplitude.

mesocyclone A stage in the development of a tornado; consists of a spinning cylinder of air 10 to 20 km (6.2 to 12.4 mi) in diameter within the updraft of a severe thunderstorm.

mesopause Transition zone between the mesosphere below and the thermosphere above.

mesoscale convective complex A nearly circular cluster of many thunderstorms covering an area of many thousands of square kilometers.

mesoscale systems Weather phenomena operating at the local scale; include thunderstorms and sea breezes, for example.

mesosphere Thermal subdivision of the upper atmosphere in which temperatures decline with altitude; situated between the stratosphere below and the thermosphere above.

meteorology Scientific study of the atmosphere and atmospheric processes.

METROMEX Metropolitan Meteorological Experiment; a study that indicated an enhancement of precipitation downwind of St. Louis.

microclimate Means and extremes in weather operating on a very small spatial scale.

microburst A strong and potentially destructive thunderstorm downdraft that strikes the ground and spreads laterally; produces wind shear that may pose a hazard to low-flying aircraft.

micrometer (μm) A millionth of a meter.

microscale weather The smallest spatial subdivision of atmospheric circulation.

microwave Electromagnetic energy having wavelengths in the 0.1 to 300 mm range.

millibar (mb) The conventional meteorological unit of air pressure. The average sea-level air pressure is 1013.25 millibars.

mirage An image formed when the atmosphere acts as a huge lens to refract (bend) light rays. The refraction is usually the consequence of a steep vertical temperature gradient.

mist Very thin fog in which visibility is greater than 1.0 km (0.62 mi).

mistral A katabatic wind that flows from the Alps down the Rhone River Valley of France to the Mediterranean coast.

mixing depth Vertical distance between the ground and the altitude to which convection extends.

mixing layer Surface layer of the atmosphere in which air is thoroughly mixed by convection. Mixing depth is the thickness of the mixing layer.

mixing ratio Mass of water vapor per mass of dry air; expressed as grams per kilogram.

mock suns Same as parhelia.

moist adiabatic lapse rate A variable rate of cooling applicable to rising saturated air parcels. This rate is less than the dry adiabatic lapse rate because some of the cooling is compensated for by the release of latent heat that accompanies the phase change of water vapor.

moist adiabats Sloping, curved lines on a Stüve thermodynamic diagram that represents the temperature change of saturated air parcels subjected to expansional cooling.

molecules The smallest subdivision of a substance that exhibits all of the characteristics of that substance. Molecules are composed of atomic and subatomic particles.

monsoon active phases Cloudy periods with frequent deluges of rain.

monsoon circulation Characterizes regions where seasonal reversals of winds cause wet summers and dry winters.

monsoon dormant phases Sunny and hot periods that interrupt the rainy monsoon episodes.

mother-of-pearl clouds Same as nacreous clouds.

mountain breeze A shallow, gusty downslope flow of cool air that develops at night in some mountain valleys.

mountain-wave clouds Stationary clouds situated downwind of a prominent mountain range and formed as a consequence of the disturbance of the wind by the mountain range.

nacreous clouds Rarely seen clouds that form in the upper stratosphere; may be composed of ice crystals or super-cooled water droplets.

National Substation Program A network of 11,590 cooperative weather stations that record weather data for hydrologic, agricultural, and climatic purposes.

National Weather Service (NWS) The agency of NOAA responsible for weather data acquisition, data analysis, forecast dissemination, and storm watches and warnings.

Newton's first law of motion An object at rest or in straight-line unaccelerated motion remains that way unless acted upon by a net external force.

Newton's second law of motion A net force is required to cause a unit mass of a substance to accelerate (or decelerate), that is, force = mass × acceleration.

Newton's third law of motion For every action (or force), there is an equal and opposite reaction (or force).

nimbostratus Low, gray, layered clouds that resemble stratus clouds but are thicker and yield more substantial precipitation.

NOAA National Oceanic and Atmospheric Administration, the U.S. federal administrative unit in the Department of Commerce that includes the National Weather Service.

noctilucent clouds Clouds that occur in the upper mesosphere, are rarely seen, and are probably composed of meteoric dust.

Norwegian cyclone model The original description of the structure and life cycle of a midlatitude low pressure system, first proposed during World War I by researchers at the Norwegian School of Meteorology at Bergen.

nuclei Tiny solid or liquid particles of matter on which condensation or deposition of water vapor can take place.

occluded front A front formed when a cold front catches up to a warm front; represents the final stage in the life cycle of a midlatitude cyclone.

occlusion The final stage in the life cycle of a midlatitude cyclone. Occlusion occurs when the cold front catches up with the warm front.

orographic lifting Rising motion of air induced by topography.

orographic precipitation Rainfall or snowfall from clouds, induced by topographic uplift.

ozone shield Within the stratosphere, ozone filters out potentially lethal intensities of ultraviolet radiation from the sun.

parhelia Two bright spots of light appearing on either side of the sun; each is separated from the sun by an angle of 22 degrees. Parhelia are caused by refraction of sunlight by ice crystals; same as mock suns and sun dogs.

penumbra Outer, lighter area of a sunspot.

perfect radiator A hypothetical object that absorbs all of the radiation that strikes it; a black body.

perihelion The time of year when the earth's orbital path brings it closest to the sun (about January 3).

persistence Tendency for weather episodes to continue for some period of time.

pH scale A system used to specify the range of acidity and alkalinity. A pH of 7 is neutral; acids have pH values less than 7, and alkaline substances have pH values greater than 7.

photochemical smog A noxious, hazy mixture of aerosols and gases produced when sunlight acts on the oxides of nitrogen and hydrocarbons in automobile exhaust.

photosphere The visible surface of the sun.

photosynthesis The process whereby plants use sunlight, water, and carbon dioxide to manufacture their food.

Planck's law The rate at which radiation is emitted by a black body depends on the absolute temperature of the body and on the specific wavelength of the radiation.

planetary albedo The fraction of solar radiation that is scattered and reflected back into space by the earth-atmosphere system; equivalently, "earthshine."

polar front Transition zone between cold polar easterlies and mild midlatitude westerlies.

polar front jet stream A jet stream situated in the upper troposphere between the midlatitude tropopause and the polar tropopause and directly over the polar front.

polar highs Anticyclones originating in the source regions for continental polar air.

poleward heat transport Flow of heat from tropical to middle and high latitudes as a consequence of latitudinal imbalances in radiant heating and cooling. Poleward heat transport is accomplished by air mass exchange, storms, and surface ocean currents.

power tower system Computer-controlled mirrors track the sun and focus its energy on a heat collection point situated on a tower.

precipitation Water in solid or liquid form that falls to the earth's surface from clouds.

pressure gradient Change in air pressure with distance.

pressure gradient force Force that causes air parcels to move as the consequence of an air pressure gradient from regions of high pressure to regions of low pressure.

primary air pollutants Substances that are pollutants immediately upon entry into the atmosphere.

Project Whitetop A cloud seeding project, carried out in Missouri during the 1950s, which may have reduced precipitation because of overseeding.

psychrometer An instrument used to determine relative humidity. The psychrometer depends on the difference in readings between a dry-bulb thermometer and a wet-bulb thermometer.

radar *Ra*dio *d*etection *a*nd *r*anging. An instrument that sends and receives microwaves for the purpose of determining the location and movement of areas of precipitation in the atmosphere.

radar echo Microwaves scattered by distant rain or snow back to a receiver where they are displayed as a bright spot on a cathode ray tube.

radiation Energy transport via electromagnetic waves traveling at the speed of light and capable of traveling through a vacuum.

radiation fog A ground-level cloud formed by the nocturnal radiational cooling of a humid air layer so that its relative humidity approaches 100 percent.

radiation frost Freezing temperatures at ground level that are induced by nocturnal radiational cooling.

radiation temperature inversion Cooling of a surface air layer by loss of infrared radiation, so the coldest air is at the earth's surface and the air temperature increases with altitude.

radiational cooling A temperature drop, most pronounced at night, that is induced by a loss of infrared radiation.

radio waves Long-wavelength, low-frequency electromagnetic waves.

radiosonde A small balloon-borne instrument package equipped with a radio transmitter that measures vertical profiles of temperature, pressure, and relative humidity in the atmosphere.

rain A form of precipitation consisting of liquid water droplets having diameters between 0.5 mm (0.02 in.) and 5.0 mm (0.2 in.).

rainbow An arch of colors formed by refraction and internal reflection of sunlight by raindrops in the direction opposite the sun. The raindrops refract sunlight into its constituent colors.

rain gage A device—usually a cylindrical container—for measuring rainfall.

rain shadow Region of reduced precipitation on the lee side (the side facing away from the wind) of a mountain range.

rawinsonde A radiosonde tracked from the ground by radar to measure variations in wind direction and wind speed with altitude.

reduction to sea level An adjustment applied to surface air pressure readings in order to eliminate the influence of station elevation.

refraction The bending of a light ray as it passes from one medium to another (from air to water, for example). The bending is due to the differing speeds of light in the two different media.

relative humidity How close an air sample is to saturation at a specific temperature; expressed as a percentage.

return stroke An electrical discharge that flows upward along the lightning path as soon as a downward leader stroke reaches the ground.

rime An opaque, granular layer of ice formed by the rapid freezing of supercooled water.

roll cloud A low, cylindrically shaped and elongated cloud occurring behind a gust front; associated with but detached from a cumulonimbus cloud.

Rossby waves Series of long-wavelength troughs and ridges formed by the westerlies as they encircle the globe.

Saffir/Simpson Hurricane Scale Hurricane intensity scale; 1 is minimal, 5 is extreme.

Santa Ana wind A hot, dry chinook-type wind that blows from the desert plateaus of Utah and Nevada toward coastal southern California.

saturation mixing ratio Maximum concentration of water vapor in a given volume of air at a specific temperature.

saturation vapor pressure The maximum possible vapor pressure in a sample of air at a specific temperature.

scavenging of pollutants Removal of pollutants from air by rainfall and snowfall.

scientific method A systematic form of inquiry that involves observation, speculation, and reasoning.

sea breeze A relatively cool mesoscale wind directed from sea toward land in response to differential heating between land and sea; develops during the day.

secondary air pollutants Pollutants generated by chemical reactions occurring within the atmosphere.

semipermanent pressure systems Areas of relatively high average air pressure and areas of relatively low average air pressure that persist at the global scale. These pressure cells exhibit some seasonal changes in location and in surface pressures.

sensible heat transfer Movement of heat from one place to another as a consequence of conduction or convection or both.

severe weather systems Storms that are potentially destructive and disruptive.

severe thunderstorms Thunderstorms accompanied by locally damaging winds, frequent lightning, or large hail.

sheet lightning Bright flashes across the sky due to cloud-to-cloud electrical discharges.

shelf cloud A low, wedge-shaped and elongated cloud that occurs along a gust front; associated with and attached to a cumulonimbus cloud.

short waves Relatively small ripples (troughs and ridges) superimposed on Rossby waves.

single-station forecasts Weather forecasts based on observations at one location.

singularity A weather event that occurs on or near a certain date with unusual regularity; the January thaw is an example.

sleet Same as ice pellets.

snow A type of precipitation consisting of an assemblage of ice crystals in the form of flakes.

smog See photochemical fog

solar altitude The angle of the sun above the horizon.

solar cells Photovoltaic cells that convert sunlight directly into electricity.

solar collectors Panels that collect and concentrate insolation for subsequent conversion to heat or electricity.

solar constant The flux of solar radiational energy falling on a surface positioned at the top of the atmosphere and oriented perpendicular to the solar beam when the earth is at an average distance from the sun; about 2.00 calories per square centimeter per minute.

Solar Disk Sextant A space instrument designed to measure the radius of the sun with great precision; changes in the solar radius are linked to changes in the sun's energy output.

solar flare Gigantic disturbance on the sun that sends out into space high-velocity streams of electrically charged subatomic particles.

solar irradiance The total radiative energy output of the sun.

solar wind A stream of charged subatomic particles (mainly protons and electrons) flowing out into space from the sun.

solstice When the sun is at its maximum poleward location relative to the earth (latitudes 23 degrees 30 minutes North and South); first days of summer and winter.

soundings Continuous altitude measurements that provide profiles of such variables as temperature, humidity, and wind speed.

southern oscillation Opposing swings of surface air pressure between the eastern and western tropical Pacific Ocean.

specific heat Amount of heat required to raise the temperature of 1 gram of a substance by 1 °C at sea-level atmospheric pressure.

split flow pattern Westerlies to the north have a wave configuration that differs from that of westerlies to the south.

Spörer minimum A period of reduced sunspot activity from 1410 to 1540.

squall line A line of intense thunderstorms occurring parallel to and ahead of a fast-moving, well-defined cold front.

stable air layer Air layer characterized by a vertical temperature profile such that air parcels return to their original altitudes following upward or downward displacements.

station model A conventional representation on a weather map, using standard symbols, of weather conditions at some locality.

steam fog The general name for fog produced when cold air comes in contact with relatively warm water; has the appearance of rising streamers.

Stefan-Boltzmann law Total emittance of a black body is directly proportional to the fourth power of the absolute temperature of the radiating body.

stepped leaders The initial electrical discharge in a lightning flash; consists of a negative electrical charge that travels from a cloud base to within 50 m (150 ft.) of the ground.

storm surge A hurricane-induced rise in sea level, which precedes the storm as it approaches the shoreline.

stratocumulus clouds Low clouds consisting of large, irregular puffs or rolls arranged in a layer.

stratopause Transition zone between the stratosphere and the mesosphere.

stratosphere The atmosphere's thermal subdivision situated between the troposphere and mesosphere; primary site of ozone formation. Within the stratosphere, air temperature in the lower part is constant with altitude, and then temperature increases with altitude.

stratus clouds Low clouds that occur as a uniform gray layer stretching from horizon to horizon. They may produce drizzle, and where they touch the ground, they are classified as fog.

streamlines The flow pattern of air moving horizontally; consists of lines that are drawn parallel to wind direction.

Stüve thermodynamic diagram A graphical display of the relationships among several atmospheric variables including pressure, temperature, and saturation mixing ratio.

sublimation Process by which water changes from a solid into a vapor without passing through the liquid phase.

sublimation nuclei Same as deposition nuclei.

subpolar lows High latitude semipermanent cyclones marking the convergence of global-scale surface southwesterlies of midlatitudes with surface northeasterlies of polar latitudes; Icelandic low and Aleutian low.

subsidence temperature inversion An elevated temperature inversion formed by air sinking gradually over a wide area and warmed by adiabatic compression; occurs on the eastern flanks of the subtropical anticyclones.

subtropical anticyclones Semipermanent warm-core high pressure systems centered over subtropical latitudes of the Atlantic, Pacific, and Indian oceans.

subtropical jet stream A zone of unusually strong winds situated between the tropical tropopause and the midlatitude tropopause.

sun dogs Same as parhelia.

sunspots Relatively large, dark blotches that appear on the face of the sun.

supercells Severe thunderstorm cells.

supercooled droplets Cloud droplets that remain liquid even though the air temperature is below the freezing point.

supersaturation An air sample having a relative humidity greater than 100 percent.

synoptic weather network (SWN) Weather observing stations that provide data primarily for weather map preparation and weather forecasting.

synoptic-scale weather Weather phenomena operating at the continental or oceanic spatial scale; includes migrating high pressure and low pressure systems, air masses, and fronts.

teleconnection A linkage between weather changes occurring in widely separate regions of the globe.

temperature A measure of the relative molecular activity of a substance.

temperature gradient Temperature change with distance.

temperature inversion An extremely stable air layer in which temperature increases with altitude, the inverse of the usual temperature profile in the troposphere.

thermal conductivity Property of a material to conduct heat.

thermal equator Latitude of highest mean annual temperature; about 10 degrees N.

thermal inertia, or thermal stability Resistance to change in temperature.

thermal low Same as a warm-core low; develops as a consequence of intense solar heating of the ground, which in turn heats the air, thereby lowering its density.

thermograph A recording instrument that gives a continuous trace of temperature with time.

thermometer An instrument used to measure temperature.

thermoregulation Processes in some organisms that maintain a nearly constant core temperature; in humans of 37 °C.

thermosphere Outermost thermal subdivision of the atmosphere in which temperatures increase with altitude.

thunder The sound accompanying lightning. Thunder is produced by violent expansion of the air due to the intense heating caused by a lightning discharge.

thunderstorm A mesoscale weather system produced by strong convective air currents that extend deep into the troposphere. Thunderstorms are always accompanied by lightning and thunder, and feature locally heavy rainfall (or snowfall) and gusty surface winds. The most intense phase of the life cycle of a thunderstorm cell, which features both updrafts and downdrafts, lightning, thunder, and heavy rainfall, is called the *mature stage*. In the final phase, or *dissipating stage*, downdrafts spread through the entire cell and clouds vaporize.

tilted updraft An updraft that is inclined so that precipitation is deflected away from the main updraft. The tilted updraft therefore persists, and the thunderstorm builds to great altitudes. (Precipitation falling against an updraft will weaken it.)

tipping bucket rain gage A device that accumulates rainfall in increments of 0.01 in. by containers that alternately fill and empty (tip).

tornado A small mass of air that whirls rapidly about an almost vertical axis. The tornado is made visible by clouds, dust, and debris sucked into the system.

tornado alley Region of maximum tornado frequency in North America; a corridor stretching from central Texas northeastward into Missouri.

trade winds Prevailing global-scale surface winds in tropical latitudes. The trade winds blow from the northeast in the Northern Hemisphere and from the southeast in the Southern Hemisphere.

transpiration Process by which water vapor escapes from plants through leaf pores.

Tropic of Cancer Latitude 23 degrees 27 minutes N; a solstice position of the sun.

Tropic of Capricorn Latitude 23 degrees 27 minutes S; a solstice position of the sun.

tropical depression The initial stage in the development of a hurricane; winds are less than 60 km (37 mi) per hour.

tropical storm A tropical cyclone having wind speeds of 60 to 120 km (37 to 74 mi) per hour; prehurricane stage.

tropopause Zone of transition between the troposphere below and the stratosphere above; top of the troposphere.

troposphere Lowest thermal subdivision of the atmosphere in which air temperature normally drops with altitude; site of most weather.

turbidity Dustiness of the atmosphere.

typhoons Hurricanes that form in the western tropical Pacific.

ultraviolet radiation Short-wave, high energy solar radiation, much of which is absorbed in the upper atmosphere. Ultraviolet radiation is involved in the formation of stratospheric ozone.

umbra Central dark area of a sunspot.

unstable air layer An air layer characterized by a vertical temperature profile such that air parcels accelerate away from their original altitudes to follow upward or downward displacements.

updraft Upward moving air, usually in a thunderstorm cell.

upslope fog Ground-level cloud formed as a consequence of the expansional cooling of humid air that is forced to ascend a mountain slope.

urban heat island The relative warmth of a city compared with surrounding areas.

valley breeze A shallow, upslope flow of air that develops during the day within mountain valleys facing the sun.

vapor pressure That portion of the total air pressure exerted by the water vapor in a sample of air.

variables of state Temperature, pressure, and density of air.

veer (or veering) Clockwise shift in wind direction with time. For example, a wind shift from south to southwest.

vertical pressure gradient Decrease of air pressure with altitude.

virga A shaft of rain or snow falling from a distant cloud that vaporizes before reaching the ground.

virtual temperature An adjustment applied to the real air temperature to account for a reduction in air density due to the presence of water vapor.

visible light Electromagnetic radiation having wavelengths in the range of about 0.40 (blue) to 0.70 (red) micrometers.

vorticity Measurement taken at some point in a fluid of the rotational tendency of that fluid.

warm air advection Flow of air from a relatively warm locality to a relatively cool locality.

warm clouds Clouds at temperatures above 0 °C (32 °F); composed of liquid water droplets only.

warm front A narrow zone of transition between relatively warm air that is advancing and relatively cool air that is retreating.

warm-core anticyclone A high pressure system occupying an extensive column of subsiding warm, dry air.

warm-core low, or thermal low A surface, synoptic-scale stationary cyclone that develops as a consequence of intense solar heating of a large, relatively dry geographical area.

water budget Balance sheet for the inputs and outputs of water to and from the various global water reservoirs.

watershed Same as drainage basin.

waterspout A tornadolike disturbance that travels or forms over a large body of water; usually much weaker than a tornado but associated with a cumulonimbus cloud.

wave cyclone A low pressure system that develops along a polar front; cyclonic circulation causes a wave to form along the front.

wavelength Distance from successive crest to crest or from successive trough to trough of a wave.

weather The state of the atmosphere in terms of such variables as temperature, cloudiness, precipitation, and radiation.

weather modification Any change in weather that is induced by human activity, either intentionally or nonintentionally. Cloud seeding and fog dispersal are examples of intentional weather modification, while enhancement of rainfall by air pollution is an example of nonintentional weather modification.

weather warnings Issued when hazardous weather is observed.

weather watches Issued when hazardous weather is considered possible based on current or anticipated atmospheric conditions.

weighing bucket rain gage A device that is calibrated so that the weight of rainfall is recorded directly in terms of rainfall in millimeters or in inches.

westerlies Global-scale prevailing west-to-east winds in the troposphere of midlatitudes from about 30 degrees to 60 degrees.

wet-bulb depression On a psycrometer, the difference in readings between the wet-bulb thermometer and the dry-bulb thermometer; used to obtain relative humidity.

Wien's displacement law The higher the temperature of a radiating object, the shorter are the wavelengths of its radiated energy.

wind chill equivalent temperature A theoretical air temperature at which the heat loss from exposed skin under calm conditions is equivalent to the heat loss at the actual air temperature and under the actual wind speeds.

wind shear An abrupt change in wind speed or direction with distance.

wind vane An instrument used to monitor wind direction by always pointing into the wind.

windsock A large, conical, open bag designed to indicate wind direction and relative speed; usually used at small airports.

World Meteorological Organization (WMO) Coordinates the weather data collection and analysis by more than 145 member nations; based in Geneva, Switzerland.

World Weather Watch (WWW) International weather monitoring network coordinated by the World Meteorological Organization.

X rays Highly energetic, short-wavelength electromagnetic radiation.

zonal component East-west motion of air.

zonal flow pattern Flow of the westerlies almost directly from west to east; westerlies exhibit little amplitude.

Index